T0342268

**Integration of Renewable
Sources of Energy**

Integration of Renewable Sources of Energy

Felix A. Farret
M. Godoy Simões

Second Edition

This edition first published 2018
© 2018 John Wiley & Sons, Inc.

The right of Felix A. Farret and M. Godoy Simões to be identified as the author(s) of this work has been asserted in accordance with law.

Registered Office
John Wiley & Sons, Inc., 111 River Street, Hoboken, NJ 07030, USA

Editorial Office
111 River Street, Hoboken, NJ 07030, USA

For details of our global editorial offices, customer services, and more information about Wiley products visit us at www.wiley.com.

Wiley also publishes its books in a variety of electronic formats and by print-on-demand. Some content that appears in standard print versions of this book may not be available in other formats.

Library of Congress Cataloguing-in-Publication Data

Names: Farret, Felix A., author. | Simões, M. Godoy, author.
Title: Integration of renewable sources of energy / Felix A Farret, M. Godoy Simões.
Other titles: Integration of alternative sources of energy
Description: 2nd edition. | Hoboken, NJ : John Wiley & Sons, Inc., 2018. |
Revised edition of: Integration of alternative sources of energy / Felix A. Farret, M. Godoy Simões. | Includes bibliographical references and index.
Identifiers: LCCN 2017007716 (print) | LCCN 2017008975 (ebook) | ISBN 9781119137368 (cloth) | ISBN 9781119137375 (pdf) | ISBN 9781119137399 (epub)
Subjects: LCSH: Power resources. | Renewable energy sources.
Classification: LCC TJ163.2 .F37 2017 (print) | LCC TJ163.2 (ebook) | DDC 621.042–dc23
LC record available at https://lccn.loc.gov/2017007716

Cover image: © Floriana/Gettyimages
Cover design by Wiley

Set in 10/12pt Warnock by SPi Global, Pondicherry, India

10 9 8 7 6 5 4 3 2 1

To my children Matheus, Angelica, Patrick, and Samara, parts of my life and my knowledge - FAF

To Ahriel, Lira, Rafael, Luiz, and Deborah, all with love and deep happiness to be my family. - MGS

Contents

Foreword for the First Edition

Integration of Renewable Sources of Energy is an important work about technology that has the potential to advance environmental goals and eventually support a sustainable future for society. Several countries such as Denmark, Australia, Spain, Germany, the United Kingdom, and others have begun the transition away from fossil fuels and nuclear energy as a response to increasing concerns about fuel supplies, global security, and climate change.

Renewable energy includes all sources and technologies that minimize environmental impacts relative to conventional hydrocarbon resources and economical issues related to fossil fuel resources. Therefore, fuel cells, natural gas, and diesel might be alternatives in respect to coal or nuclear power. Renewable energy sources are those that are derived from the sun or other natural and replenishing processes including solar (light and heat), wind, hydro (fall and flow), sustainable biomass, wave, tides, and geothermal energy. Throughout this book, the fundamentals of the technologies related to the integration of such alternative and renewable energy sources are reviewed and described with authority and skill and from the critical engineering point of view for the end user of energy.

In Chapters 1–15, the authors cover the principles of hydroelectric, wind, solar, thermal, and photovoltaic power plants. Induction generators—important electric machines for wind and hydropower generation—are described in detail. The chapters on fuel cells and biomass are of paramount importance in the current age of the hydrogen economy. Literature on microturbines is scarce; therefore, the authors make important contributions to the technical description of such important devices that have contributed, in the last few years, toward shifting traditional generation to distributed generation. A comprehensive evaluation of storage technology is complemented by the description of integrating control and association of sources into microgrids. There are two chapters authored by outside contributors that describe the standards and interconnection issues of alternative energy sources to the grid and principles of economic optimization. The book is complemented by three appendices covering some important sources of power and heat directly related to the rational use of energy, namely, diesel, geothermal, and Stirling engines.

I applaud the initiative taken by the authors to more closely cover, in this timely book, the electrical rather than the mechanical aspects of energy sources. I am sure this work will contribute in understanding how to integrate renewable energy sources for home, commercial, rural, and industrial applications.

I strongly recommend this textbook to a wide audience, including engineering educators and students of electrical and mechanical engineering.

Marian P. Kazmierkowski
Professor in Power Electronics
Warsaw University of Technology, Warsaw, Poland

Foreword for the Second Edition

The world is undergoing a dramatic evolution in its energy system structure as more and more power is now coming from renewable and sustainable energy sources, and this development is expected to be even more progressive in the next decades—it is driven by a goal of not only reducing carbon emission but also obtaining more security of supply for every nation and continent. Despite the fluctuating prices for carbon-based energy—although currently the prices are rather low—a lot of national programs are implemented in order to use more renewables. Furthermore, final energy prices for some of the renewables are more competitive than those of conventional fueled systems and such technologies will accelerate the implementation even more. Especially energy cost of the wind power technology and photovoltaics is very low, and it is a matter of scale to push that further down. In 2016 it was expected that more than 100 GW installation would be done for these technologies and that was more than half of the total installed new electrical power capacity generation for this year. Most of the renewable energy sources need power to operate properly, and we have also seen continuous improvement in this technology, which has become smaller, cheaper, more efficient, more reliable, and smarter, so energy sources facilitate a smooth and intelligent connection to the electrical grid.

The second edition of this book is very timely and covers many important aspects of integrating renewable energy sources. It starts with an overview of all renewable sources that exist today, and then the basics about thermodynamics are covered to get a full understanding of energy. A wide range of different relevant technologies are covered, including hydropower, solar power, wind power, fuel cell-based power, photovoltaics, geothermal power, microturbines, and ocean power systems as well as biomass-based systems. Different energy storage systems are also covered, as well as conventional power generators like induction and synchronous generators (also with permanent magnets) in some of the presented systems, energy sources that have to be integrated, and methods of proper interconnection of the electrical power-based generator systems. The book ends with examples of a micro-power system.

The book gives a solid introduction to the many aspects of present and future energy technology for the society and an inspiration for future research. It also demonstrates that the energy system requires not a single solution but a number of them with different energy carriers in order for the society to fully rely on renewable energy sources.

Enjoy reading!

Frede Blaabjerg
Professor
Aalborg University, Denmark

Preface for the First Edition

Our goal in writing this book is to discuss the "electrical side" of renewable energy sources. From the beginning, we felt that the approach would be a challenge that is very difficult to fulfill. Most of the current technical works explore just one or two types of alternative energy sources, but the "integration of sources" is our main objective. We also noticed that most of the works on this subject were exclusively concerned about those parts of the primary energy directly related to extraction and conversion of power and not really about processing energy and giving the user a final product. This product—energy—should be ready to feed a new reality and the dreams powered by renewable or alternative sources of energy. Those sources of energy have a lot in common. However, when discussing hydropower, wind, solar, and other sources of energy, the complexity of their aspects is soon realized.

Not long ago, huge power generation plants dominated the whole field of energy production. It seemed there was no way to develop and deploy small and disperse alternatives. For several decades, small plants nearly vanished. This was a worldwide trend because of the argument that electrical efficiency and concentrated sites would be incontestably the best economical and rational factors to generate electricity. Throughout those decades, small rural communities and remote areas were simply outside the scope of the centralized model. Population growth and the development of nations soon made our society realize that more and more energy production would be necessary for continuous industrial growth.

On the other hand, the availability of energy would not be enough or would be so distant from the perspective of consumption that central power plants would have devastating effects on ecology, scenery, and quality of life. For a future with sustainable energy, massive fossil fuel-powered plants are not economical; most of them waste more than 50% of the primary energy due to irrecoverable thermal losses. In addition, they demand the use of massive coolers and heat sinkers to guarantee operational conditions, quality, and stability of the final product. The energy must be transported throughout long and congested transmission lines: waste that is no longer reasonable.

Small, dispersed generating units do have the possibility of adding representative amounts of energy to the network without a noticeable long-term impact on the environment and economic investments. Current computer and power electronics technology supports the integration and distributed generation of energy to sites that have been so far neglected. The Earth has sunlit deserts, windy remote locations, offshore sea sites, glaciers, and streams, with each environment contributing to the world energy frame. They are not wastelands anymore.

Who can guarantee for how long we will have plentiful and available petroleum, at the right quantity corresponding to demand, for the years to come? How can mankind believe in a sustainable future if the only admissible alternative is nuclear energy? How can we quantify only economic benefits to mankind from fossil fuel and nuclear power and neglect other important issues such as avoided healthcare costs, air quality, space, sea and river poisoning, sound and visual pollution, or an unpredictably safe future? Nobody wants to go to war to sustain our future.

In this book, we review and organize all pertinent subjects in an orderly way. After a general introduction on the subject in Chapter 1, we review some general principles of thermodynamics in Chapter 2. Initially, we choose to explore the most common alternative energy sources such as hydro, wind, solar, fossil fuel, solar thermal, and fuel cell power plants (fuel cell plants are expected to soon become widespread). After we cover the basis of the primary sources, our approach is to show how to integrate them for electrical power production and integration into the main grid; these subjects are covered in Chapters 3–8. The means of interfacing the primary energy through microturbines, induction generators, and power electronics for electrical power are dealt with in Chapters 9 and 10. Chapter 11 is a special chapter on energy storage systems. Our main concern is the means of integrating, transforming, and conditioning the energy sources into more useful electrical applications, rather than coping with the distribution of sources throughout the electrical network. Power transmission is left for the reader to study in other related books. Problems related to system operation, maintenance, and management are briefly tackled in Chapters 13 and 14. These chapters also refer the reader to other more specialized texts and literature on the subject. Chapter 14 discusses the standards for interconnection. Finally, the last chapter gives the reader the opportunity to learn more about the HOMER™ Micropower Optimization Model, which is a computer model developed by the US National Renewable Energy Laboratory (NREL) to assist in the design of micropower systems and to facilitate the comparison of power generation technologies across a wide range of applications. HOMER™ software can be downloaded from the NREL website.

This book is especially dedicated to those people who believe that there is a way to work for a clean, long-lasting, and beautiful world and to those students, engineers, and professionals who believe that engineering is decisive in its contribution to this journey and who dedicate their professional lives to this mission. Many of these readers will be found in senior years of study or first graduate-level programs in advanced courses on energy, electrical, environmental, civil, chemical, and mechanical engineering and agronomic sciences.

Felix A. Farret, Santa Maria, Rio Grande do Sul, Brazil
M. Godoy Simões, Golden, CO, USA
Spring 2005

Preface for the Second Edition

It has been 10 years since we released the first edition of our book *Integration of Alternative Sources of Energy*. The friendship, partnership, and academic interaction of the authors remain solid when in 2002 we decided to make renewable and alternative sources of electrical power an easy and understandable subject for our students both at Colorado School of Mines, USA, and at the Federal University of Santa Maria, Brazil. For the past 15 years or so, Drs. Farret and Simões have been working in several joint books and papers, sharing research and educational materials, exchanging ideas, and cooperating with students; the very book that you now hold in your hands is our joint piece of art that we developed along these years. We are really happy that we can share with you what we learned and what we made available in this textbook, because the future that we believe has to be a sustainable one.

The second edition is our masterwork. We hope that students and engineers all over the world can appreciate and get in a deeper learning on this area of renewable and alternative energy systems. In the second edition, we polished our knowledge about alternative sources of energy with many comments received from our readers. We included new applications and a discussion of new sources, and we expanded details that were not so obvious in the first edition but made us to get a better comprehension for this book. There were many suggestions of so many topics and ideas to explore in this second edition that we could keep working on. But we think that this version has a good coverage and stature to be useful in teaching senior undergraduate and first-year graduate courses. We corrected, enriched, enlarged, and detailed many topics that at the beginning were difficult to understand, since this area is very multidisciplinary, and anyone to master these topics must have a very curious mind. Our work is such as the one performed by a curator, since "curare" from Latin means to "take care," that is, a curator interprets a heritage material for a certain art collection. As in a traditional curator's concern, we decided that in this book we would make available the cultural heritage necessary to educate students in all aspects of renewable and alternative energy systems. Several topics are fruits of research conducted in our academic career, while others were curated in order to make such an educational bridge that can be useful for the current generation of students and young engineers.

In the second edition, we added three new chapters about renewable sources of energy to be better explored in the next few years. One is geothermal energy, particularly the available thermal heat in shallow underground, which is not yet recognized and not fully explored but has a potential paradigm transformation, since it is available all over the world. The other one is electrical generators that involve permanent magnet synchronous machines, particularly with designs aiming to convert energy directly without inefficient gearboxes and operating at very low speed, compatible with the sources of nature, such as wind, water streams, ocean and sea water, or fluids in the industrial, commercial, and residential settlings. In addition, we included a chapter about the other promising types of electrical power generation such as semiconductor thermocouples, sea wave energy conversion systems, tidal power, gas microturbines, and piezo-electric harvesting and magneto-hydrodynamic energy systems. These possibilities are fascinating, and we expect that people all over the world will become excited to contribute with these energy sources and place everything at their right place and that we will live in a sustainable society, free of fossil fuel.

We envision that sources of energy should be classified into three main categories: those coming from the sun, those coming from the Earth, and those considered ambiguous (see our Table 1.5 on Chapter 1). Those coming from the sun, like hydro, wind, photovoltaic, and solar thermal power and photosynthesis, can be welcomed since they are "clean" for our planet, whereas those coming from the earth like coal, uranium, big plantations, and petroleum are not welcome, since they can contribute to pollution. In some ambiguous ones, like geothermal, fuel cell, and biomass, the surface geothermal can be very welcome to mankind because the sun has evenly warmed up the Earth's surface since billions of years ago, which will continue for more billions of years further. On the other hand, deep geothermal is not really a good source since obtaining boiling water from the earth core to feed steam turbines and exchange it with cold water to fill back the source where the steam was originally trapped may not be sustainable and not even possible for a continuous and worldwide use. Fuel cells fed by hydrogen from petrol hydrocarbons are obviously not welcome but still are marketed by giant multinational companies. Fuel cells are good solutions when the hydrogen is acquired by splitting water molecules through heat from sun irradiation concentrated by mirrors, but some economics studies are further needed to decide the direct use of irradiation heat for hydrogen and other applications. Biomass used in bio-digesters and fed from rubbish in order to produce electrical power is welcome. However, using biomass from wood burning is not. Wood is such an important gift of our nature, and human beings must use them with care and respect. It would be better to have a tree becoming a piece of furniture than just being burned. Plantations can exhaust the soil after several cycles but if a rational cycle is selected, maybe a good compromise is achieved. These matters are discussed in deep details in several chapters of this book.

We would like to thank many students, colleagues, and related professionals for their support and cooperation. We are particularly grateful to Frank Gonzatti, Lucas Feksa, Vinicius Kuhn, Fredi Ferrigolo, Luciano de Lima, Maro Jinbo, Carlos De Nardin, Felipe Fernandes, Adriano Longo, Luis Manga, Antonio Ricciotti, Emanuel Antunes, Ciro Egoavil, and Danilo Iglesias Brandão. Their contributions, in one or another way, were really important; they have been memorable in our hundreds or thousands of hours of discussions about renewable sources of energy, along with coffee or some other nice drink and a nice chat, or technical meeting. Thank you to all our friends and colleagues who made possible our second edition to become such an established literature reference in this important field of renewable and alternative energy systems. Thanks to all our students. A special thank-you to our families, who provided us time and space to devote uncountable hours in writing and preparing this textbook.

Felix Alberto Farret, Santa Maria, Rio Grande do Sul, Brazil
M. Godoy Simões, Denver, CO, USA
Spring of 2017

Acknowledgements

A book like this is not just two people's work. There are so many contributions in larger or smaller scale that it would be impossible to list all the people who contributed and inspired us. The subjects dealt in this second edition are so widespread that it is hard work name them all. Nevertheless, a few of our colleagues and friends were decisive in the quality of this text. We would like to offer our utmost recognition and respect to: Adriano Longo, Bimal K. Bose, Farnaz Harirchi, Sudipta Chakraborty, Antonio Ricciotti, Ben Kroposki, Carlos De Nardin, Emanuel Antunes, Felipe Teixeira, Frank Gonzatti, Fredi Ferrigolo, Lucas Ramos, Luciano de Lima, Maro Jinbo, Peter Lilienthal, Vinicius Khun, and Viviane da Silva. We exchanged many ideas and thoughts with our colleagues, professionals, and some of our very smart students from whom we got an oppurtunity to learn and we are very grateful for their contributions. We thank the staff at John Wiley & Sons in particular Brett Kurzman and Anumita Gupta for their full dedication to this project; they have encouraged and filled us with so a lot of enthusiasm. Thanks to our families, who were deprived of our company and attention for many months, thousands of hours; they never complained but rather gave us their support and understanding. Finally, we express our gratitude to our schools: The institutional support of the Federal University of Santa Maria and the Colorado School of Mines, who were always present, encouraging us and making our activities much fulfilling. Thanks to you all.

1

Alternative Sources of Energy

1.1 Introduction

The basic human needs to survive are air, water, food, space to live, and energy, as well as the ability to reproduce, and humans have been constantly searching for means to harvest and convert energy to hence survive. But the interrelation of energy with other needs has not been so evident as in the recent years. When the industrial revolution in Europe caused an evolution of societies and large areas of increasing population density, people realized that factors such as comfortable housing and energy would be relevant to the development of a country. Fossil fuels have become essential in modern societies, and new strategies have been developed to guarantee their uninterrupted supply. In the last 250 years, our population, and correspondingly the demands for industrial and commercial goods, has increased. We have to consider that we live on a planet of constant size and constrained resources, and increased population and their demands may have consequences: economic constraints, new frontiers, wars, international agreements, and heavy pollution [1–3]. Engineers and scientists are working toward the optimized use of resources. Humans are excavating the lands for charcoal, petroleum, gas, uranium, and other minerals, polluting the atmosphere, rivers, oceans, and food sources. Burning fossil fuels and thermal energy conversion just increase entropy and contribute to exhaustion of our planet's energy resources.

In the past the approach to generate large amounts of electrical energy was realized by means of constructing large power plants, which were considered more efficient than smaller ones on an economic scale, such as the Three Gorges Dam in China (18 GW with structure for 22.5 GW), Itaipu Binacional in Brazil (14.0 GW), Sayano–Shushenskaya Dam in Russia (6.4 MW), Churchill Falls Generating Station in Canada (5.43 GW), and Guri Dam in Venezuela (2.0 GW). However, such large power plants caused immense floods, massive power transmission lines and towers, air pollution, modified waterways,

Integration of Renewable Sources of Energy, Second Edition. Felix A. Farret and M. Godoy Simões.
© 2018 John Wiley & Sons, Inc. Published 2018 by John Wiley & Sons, Inc.

devastated forests, large population densities in cities, and wars for the dominion of energy resources. Because of these trends in development, distances to energy sources are increasing, material capacities are reaching their limits, fossil reserves are being exhausted, and pollution is becoming widespread. New alternatives must be devised if humanity is to survive today and for the centuries to come.

1.2 Renewable Sources of Energy

The Earth receives solar energy as radiation from the sun in quantity that far exceeds the needs of the entire humankind. The sun generates wind, rain, rivers, and waves by heating the plane. Along with rain and snow, sunlight is necessary for plants to grow. Biomass, the organic matter that makes up plants, in general can be used to produce electricity, transportation fuel, and chemicals. Plant photosynthesis (essentially, the chemical storage of solar energy) creates a range of biomass products, from wood fuel to rapeseed, which can be used for heat, electricity, and liquid fuels.

Hydrogen can also be extracted from many organic compounds, as can water. Hydrogen is the most abundant element on Earth, but it does not occur naturally in gas form. It is always combined with other elements, such as oxygen to form water. Once separated from another element, hydrogen can be burned as a fuel or converted into electricity.

The sun also powers the evapotranspiration cycle, which allows water to generate power in hydro schemes—the largest source of renewable electricity today. Interactions with the moon produce tidal flows, which can produce electricity.

Although humans have been tapping into renewable energy sources (such as solar, wind, biomass, geothermal, and water) for thousands of years, only a fraction of their technical and economic potential has been captured and exploited. Yet renewable energy offers safe, reliable, clean, local, and increasingly cost-effective alternatives for all our energy needs. It can dismantle the power promoted by petroleum, coal, and radioactive materials [2–6].

Research has made renewable energy more affordable today than it was 30 years ago. Wind energy has declined from 40 cents per kilowatt hour (¢/kWh) to less than 5¢. Electricity from the sun through photovoltaics (which literally means "light electricity") has dropped from more than $1/kWh in 1980 to nearly 15¢/kWh today. Ethanol fuel costs have plummeted from $4/gal in the early 1980s to $1.20 today. As a result, renewable energy resource development will result in new jobs, local power plants, and less dependence on oil and radioactive materials from foreign countries [5–7].

There are some drawbacks in developing renewable energy solutions. An example is when solar thermal energy is used, because solar rays are captured through collectors (often huge mirrors) and solar thermal generation requires large tracts of land, and affects natural environment. The environment is also affected when buildings, roads, transmission lines, and transformers are built. In addition, the fluid often used for solar thermal generation is toxic, and spills can occur. Solar or photovoltaic cells are produced using the same technologies as those used to create silicon chips for computers, and this manufacturing process also uses toxic chemicals. In addition, toxic chemicals are used in batteries that store solar electricity through nights and on cloudy days. Manufacturing this equipment also has environmental effects. Therefore, even though the renewable power plant does not release air pollution or use fossil fuels, it still has an effect on the environment.

Wind power has also some drawbacks, involving primarily land use. For example, the average wind farm requires 17 acres to produce 1 MW of electricity (about enough electricity for 750–1000 homes). However, farmers and ranchers can use the land beneath wind turbines. Wind farms can cause erosion in desert areas, and they affect natural views because they tend to be located on or just below ridgelines. Bird deaths also result from collisions with wind turbines and wires. This is the subject of ongoing research. Ultimately, combined with energy efficiency, renewable energy can provide everything fossil fuels offer in terms of energy services: from heating and cooling to electricity, transportation, chemicals, illumination, and food drying.

Energy has always existed and has been used and transformed in one form or another. For example, the energy in a flashlight's battery becomes light energy when the flashlight is turned on. Food, the most natural stored chemical energy, resides in fat tissues and cells as potential energy. When the body uses that stored energy to do work, it becomes kinetic energy. Telephones transform a voice into electrical energy variations, which flow over wires or are transmitted through air. Other telephones change this electrical energy into sound energy through speakers. Cars use stored chemical energy in fuel to move, and they change chemical energy into heat and kinetic energy. Toasters change electrical energy into heat. Computers, television sets, and DVD players change electrical energy into coordinated types of mechanical movement and image and sound energy to reproduce the ambient of life. That means that as soon human beings are awake in the morning, they begin to use more energy than that keeping them alive to switch lights on, for a bath, morning cooking, heaters on, car on, going to work, and so on.

In all such transformations of energy, intermediary transformations are involved. For example, consider the case for a home computer. Electricity allows self-organization of the main processor, according to a preestablished program, to convert ventilator movement to the cooling process for the main processing unit and the motherboard. The alternating current (ac) source power after being distributed to all houses is converted into integrated direct current (dc) power to feed peripheral plates. After many electric processes, the monitor produces a luminous energy on screen. Many processes and intermediary sources are integrated into a simple computer. They produce heat, light, movement, and circulation of electrical current to make it an impressively organized machine. This diversity of energy forms is an example of the changes happening in power systems since the primary source conversion.

1.3 Renewable Energy versus Alternative Energy

All forms of energy are renewable after a lapse of time. For example, coal can be renewed in nature after millions, perhaps billions, of years. Sugar cane would take no more than one year to be replanted. Therefore, a source of energy is considered renewable if the time it takes to be recovered is referred to human life duration. Furthermore, a renewable energy source cannot run out and causes so little damage to the environment that its use does not need to be restricted. On the other hand, no energy system based on mineral resources is renewable because, one day, the mineral deposits will be used up. This is true for fossil fuels and uranium. The debate about when a particular mineral resource will run out becomes irrelevant in this context. A renewable energy source is replenished continuously.

Renewable energy sources—solar, wind, biomass (under specific conditions), and tides—are based directly or indirectly on solar energy. Hydroelectric power is not necessarily a renewable energy source because large-scale projects can cause ecological damage and irreversible consequences. Geothermal heat is renewable but must be used cautiously to guard against irreversible ecological effects.

There is no shortage of renewable energy because it can be taken from the sun, wind, water, plants, and garbage to produce electricity and fuels. For example, the sunlight that falls on the United States in one day contains more than twice the energy the country normally consumes in a year. California has enough wind gusts to produce 11% of the world's wind electricity.

Clean energy sources can be harnessed to produce electricity and process heat, fuel, and valuable chemicals with less effect on the environment than fossil fuel would cause. Emissions from gasoline-fueled cars and factories and other facilities that burn oil affect the atmosphere through the greenhouse effect. About 81% of all US greenhouse gases (GHGs) are carbon dioxide emissions from energy-related sources.

At the International Climate Convention in Kyoto (1997), it was agreed that the developed nations of the world must reduce their GHG emissions. The European Union (EU) committed to reducing emissions of carbon dioxide (CO_2) by 8% from levels in 1990 by the year 2010. The United States was to reduce emissions by 6% and Japan by 7%. These agreements were laid down in the Kyoto Protocol and aimed for a society that uses renewable energies, not fossil fuels [8–10].

Table 1.1 lists the relative gaps between GHG emissions in the non-emissions trading system (ETS) sectors for the first commitment period and the original 2008–2012 Kyoto targets, including land-use change and forestry (LUCF) with and without the use of Kyoto mechanisms, as well as the gap between GHG emissions (along with LUCF) and the Chalmers publication library (CPL)

Table 1.1 Comparison between the original Kyoto credits and the expected actual results (2014–2020).

Country	With Kyoto credits	Without Kyoto credits
Austria	−0.6	+20
Belgium	−2.3	+2
Bulgaria	−39.9	−42
Croatia	−5.8	−6
Czech Republic	−3.8	−17
Denmark	−0.2	+3
Estonia	−10.6	−46
EU15	−5.4	−4
Finland	−4.3	−2
France	−6.3	−7
Germany	−4.3	−5
Greece	−7.9	−8
Hungary	−30	−35
Ireland	−6.3	−3
Italy	+0.7	+1
Latvia	−14.3	−16
Lithuania	−15.7	−44
Luxembourg	+1.7	+23
The Netherlands	−1.5	+3
Poland	−24.3	−25
Portugal	−15.5	−13
Romania	−39.2	−40
Slovakia	−3.8	−15
Slovenia	5.7	−1
Spain	−2.8	+10
Sweden	−18.1	−18
United Kingdom	−11.7	−12

Table 1.2 Emission of CO_2/kWh by renewable sources of power.

Renewable source	Emissions of CO_2/kWh (g)
Waste incineration	600
Biogasifier	−3,800
Biomass	−4,000
Photovoltaic cells	120
Wind turbine	10
Hydraulic power station	25
Nuclear power station	55
Gas-fired power station	400
Coal-fired power station	1,160

Source: Ref. [11–14]. © European Union, 1995–2013.

target with (and issued ERUs) and without Kyoto credits (program for research and innovation in 2014–2020). Gaps are expressed in percentage of base year emissions (including ETS and non-ETS). Negative and positive values, respectively, indicate over-delivery or shortfall as basis for the new EU program for research and innovation in 2014–2020.

It is understandable that the world worries about emissions because our environment is unable to absorb them all. Table 1.2 lists some renewable sources of energy and their approximate production, or absorption, of CO_2/kWh.

Because every source is more or less intensive in what it produces, special measures have to be considered when considering global energy solutions. These include availability, capability, extraction costs, emissions, and durability. Table 1.3 shows indicators of renewable energy technologies, and Table 1.4 illustrates the intensity and frequency characteristics of some renewable sources.

The atomic energy industry seems to be profiting from concerns about GHGs and global climate change. The reason is that most people believe that nuclear energy does not emit GHGs. However, the waste of nuclear energy is either stored in long-lasting containers and thrown into the sea or kept in underground caves. Nevertheless, in the developed northern hemisphere, nuclear energy has little political or social support. The United States has not built a single reactor since the accident in Harrisburg, Pennsylvania, in 1979. In addition, there is no expansion of nuclear power generation in any EU member state. On the contrary, there is support for reduction in and closure of their atomic programs. Eight Western European countries (Denmark, Iceland, Norway, Luxembourg, Ireland, Austria, Portugal, and Greece) have never had a nuclear energy program and have instead favored the alternative programs of

Table 1.3 Indicators of renewable energy technologies.

Renewable energy technology	Volatility (approximate time variation)	Resource availability	Range of generation cost (EU cents/kWh)	Preferred voltage level of grid connection (kV)
Biogas	Year	High	5.18–26.34	1.30
Biomass	Year	High	2.87–9.46	1.30, except co-firing
Geothermal electricity	Year	Low: country specific	3.34–6.49	10.110
Large hydro power: run-of-river power plants	Months	Low	2.53–16.37	220.380
Storage power plants	Months	Low	Not considered	220.380
Small hydro power	Months	High	2.69–24.93	10.30
Landfill gas	Year	Low	2.50–3.91	1.30
Sewage gas	Year	Medium	2.85–6.24	1.30
Photovoltaics	Days, hours, seconds	High	47.56–165.32	<1
Solar thermal electricity	Days, hours, seconds	Low: country specific	12.48–66.97	1.30
Tidal	12 hours	High	Not considered	10.380
Wave	Weeks	High	9.38–45.16	10.380
Wind onshore	Hours, minutes	Low: country specific	4.63–10.80	30.380
Offshore	Hours, minutes	Low: country specific	6.09–13.39	110.380

Source: Ref. [11, 12]. © European Union, 1995–2013.

renewable energy (see Figure 1.1). Outside Europe, only China, South Korea, Japan, Taiwan, and South Africa aspire to expand the share of nuclear power generated in their countries [11–13, 16, 17].

Today, the atomic energy industry is targeting developing countries, and the Kyoto Protocol is paving the way. The protocol provides for the use of "flexible instruments," which were introduced so that wealthy nations could achieve their emission reductions in other countries by paying royalties to compensate for pollution levels. One instrument is the Clean Development Mechanism (CDM), which facilitates the financing of clean technologies (through investment in solar energy, wind turbines, hydroelectric power stations, and energy-saving

Table 1.4 Intensity and frequency characteristics of renewable sources.

System	Major periods	Major variables	Power relationship	Comment	Approximate time variation
Direct sunshine	24 hours, 1 year	Solar beam radiance G_b^* (W/m^2), beam angle from vertical q_z	$P \propto G_b \cos\theta_z$; $P_{max} \cong 1\,\mathrm{kW/m}^2$	Daytime only, highly fluctuating	Hours to seconds
Diffuse sunshine	24 hours, 1 year	Cloud cover, perhaps air pollution	$P < \sim 300\,\mathrm{W/m}^2$	Significant energy, however	Day
Biofuels	1 year	Soil condition, solar radiation, water, plant species, wastes	stored energy 10 MJ/kg	Many variations, linked to forestry and agriculture	Year
Wind	1 year	Wind speed v_0 nacelle height above ground z, height anemometer mast h	$P \propto v_0^3$ $\dfrac{v_z}{v_h} = \left(\dfrac{z}{h}\right)^b$	Highly fluctuating b ≈ 0.15	Minutes to hours for wind farms
Wave	1 year	Reservoir height H_s, wave period T	$P \propto H_s^2 T$	High power density $\approx 50\,\mathrm{kW/m}$ across wave front	Week
Hydro	1 year	Reservoir height H, water volume flow rate	$P \propto HQ$	Established resource	Months
Tidal	12 hours, 25 minutes	Tidal range R; contained area A; estuary length L; depth h	$P \propto R^2 A$	Enhanced tidal range if $L/\sqrt{h} \approx 36{,}000m^{1/2}$	12 hours
Deep geothermal	None	Temperature of aquifer or rock formation, hence temperature difference from ambient	$P \propto (\Delta T)^2$	Very few suitable locations for electricity generation	None
Surface geothermal	Seasons	Temperatures of low depth ground ($h > 1\,\mathrm{m}$, typically $h > 15\,\mathrm{m}$)	$P \propto (\overline{\Delta T} - ke^{-h})$	Almost everywhere at an average temperature $\overline{\Delta T}$	Half a year

Source: Twidel, 2003—the symbols are standard in the technologies [15]. © European Union, 1995–2013.

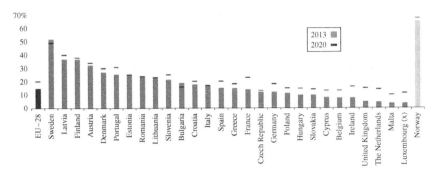

Figure 1.1 Share of renewable powers in gross final energy demand, 2013 and 2020 (dash mark) (%). *Source*: From Refs. [11, 12]. © European Union, 1995–2013.

technologies) and the transfer of these technologies from the northern to the southern hemisphere. Wealthy nations can use emission reductions achieved via the CDM to meet their Kyoto commitments, but the same cannot be said with respect to developing nations. Developing countries gain access to clean, endemic sources and compromise their future in much the same way, as did the northern hemisphere. Is this the ideal win–win solution? The atomic energy industry claims that nuclear energy can be used as an effective solution in the struggle to prevent climate change.

From socioeconomic and environmental points of view, renewable energy increases supply security, has the lowest environmental effect of all energy sources, allows for local solutions, and offers sustainable energy development worldwide. Renewable energy also offers wider opportunities for investment, avoided fuel costs, CO_2 emissions savings, and new jobs. In general, renewable energy technologies are important because of the income that results from manufacturing, project development, servicing, and, in the case of biomass, rural jobs and income diversification for farmers.

From what has been discussed in this section, we can divide the sources of energy into three categories: the ones coming from the sun, the ones coming from the Earth, and the ambiguous (see Table 1.5). Those coming from the sun, like hydropower, wind power, photovoltaics, solar thermal power, and photosynthesis, are welcome and clean. Those coming from the Earth, like coal, uranium, plantations, and petroleum, are not welcome and polluting. Ambiguous ones include geothermal, fuel cells, and biomass. Surface geothermal is welcome because sunshine has warmed the Earth's surface for billions of years. In compensation, the deep geothermal is not welcome since taking boiling water from the Earth's core to feed steam turbines and exchanged it for cold water to fill back where the steam was may not be a good idea in a continuous and worldwide spread. Fuel cells fed by hydrogen from petrol hydrocarbons are not welcome unless if hydrogen is obtained by splitting water molecules using

Table 1.5 Origin of the energy sources.

Origin	Primary sources
Sun (renewable)	Hydro, wind, photovoltaics, solar thermal, photosynthesis, tidal, sea wave, MHD
Earth (not renewable)	Coal, uranium, and petroleum
Ambiguous (can be either types)	Geothermal, plantations, fuel cells, biomass
Harvesting types	Piezoelectric, thermocouple
Space	Cosmic rays

heat from sun irradiation concentrated by mirrors. Biomass used in biodigestors and derived from rubbish to produce power is welcome, but if it is from burnt wood, it is not. Plantations can exhaust the soil after several cycles, but if a rational cycle is selected, it may not. These matters are discussed in greater detail in the next chapters.

1.4 Planning and Development of Integrated Energy

Many studies show that the global wind resource technically recoverable is more than twice the projection for the world's electricity demand in 2020. Similarly, theoretical solar energy potential (see Chapters 5 and 6 for solar thermal and solar or photovoltaics) corresponds to almost 90,000,000 metric tons of electricity per year, which is almost 10,000 times the world total primary energy supply [2–4]. The rapid deployment of renewable energy technologies, and their wider deployment in the near future, raises challenges and opportunities regarding their integration into energy supply systems, as deeply discussed in the Paris Climate Change Conference (COP 21 in November 2015).

The planning and development of integrated energy must consider the environment itself, existence of energy sources, system needs, and local needs where it is desirable to install a renewable energy source. The capability of the grid supply, the electrical and mechanical behavior of the load, the distributed generation (DG) sources, and the effects on the regional economy defines how successful the investment may be.

1.4.1 Grid-Supplied Electricity

The thickness, and hence cost, of conducting cable is inversely proportional to the voltage of power; therefore, high voltages are preferred for electricity supply along transmission and sub-transmission lines. The practical limits relate to safety issues, especially sparking and insulation at high voltage. In practice, the voltage of long distance transmission is 50–750 kV. Local area distribution is 6–50 kV, and supply to consumers is 100–500 V. Internally, in equipment, it is 3–48 V.

The grid electricity is converted from the primary source typically by

Moving wires in magnetic fields (Faraday effect)
Photovoltaic generation with sunlight (photovoltaic effects)
Chemical transformations as in fuel cells and batteries (electrochemical effects)

Transformation of transmission voltages is possible with ac to ac using transformers. However, between dc and dc or between ac and dc, it is necessary to use electronic interfaces, which have become increasingly reliable and cheaper because of solid-state power electronics. Transmission with ac power has more loss per unit of distance than with dc power due to stray capacitances and inductances along lines, which increase current losses. Nevertheless, the ease of transformation means that the majority of power transmission is accomplished with high-voltage ac up to 350 km. The economic facilities of high-voltage dc transmission systems favor distances greater than 500 km.

To regulate power, voltage, speed, and frequency, each method depends on instantaneous matching of load to generation. Generation is distinguished by its economic and physical ability to vary the voltage to match load levels. Examples are

1) Base load generation (difficult or expensive to vary; e.g., nuclear power, large coal, and large biomass)
2) Peak generation (easier to vary quickly but may be expensive; e.g., gas turbines and fuel cells)
3) Standby generation (easy to rapidly increase generation from off or idling modes; e.g., diesel, fuel cells, and gas turbines)
4) Intermittent generation (e.g., run-of-the-river hydroelectricity, photovoltaics, wind, and most renewables, except biomass and geothermal)

Note that fuel cells can be used as peak or standby generation, depending on the availability of hydrogen or fuel gas and the maximum excess power that can be withdrawn from the cell without overly compromising its useful life. The same note is applied to reservoir hydropower, which may be either base (plenty of water) or peaking (limited water) load. Note also that intermittent does not mean unpredictable availability but guarantees that load always equals generation.

1.4.2 Load

It is important to realize that electricity users do not want electricity alone. They want service, such as transportation (vertical or horizontal), lighting, water, welding, motor movement, communication, or ambient conditioning.

The success of the electricity supply must be judged by the availability, quality, and cost of the service. The quality of a service (e.g., heating or water supply) tends to be measured by an intensive parameter (e.g., temperature or pressure) and the availability of that parameter. The cost of the service is measured by an extensive parameter (e.g., energy or kilowatt hours) linked to its availability. The desire for the service presents a demand on the grid system, which power engineers see as load.

Ordinary consumers use the name of the service (e.g., television) for the function instead of the word "load," which is most efficient to use for the consequent electricity consumption and "demand" for the desire of consumers for service. This subtle distinction is maintained in this book.

Satisfactory service can be maintained without the continuous consumption of electricity. For instance, water of a satisfactory temperature can be supplied from a previously heated tank. If the value of the intensive parameter is maintained (i.e., shower temperature), the consumer is satisfied even if the electrical supply is interrupted. A demand that is satisfied by intermittent power is an interruptible load, also called a switchable load. If load and tariff management are used to optimize a power system (e.g., to increase the penetration of renewable energy), it is quite acceptable to use it to induce new trends in energy use.

1.4.3 Distributed Generation

DG is the application of small generators, typically 1–10 MW, scattered throughout a system to provide electrical energy closer to consumers. Current DG power sources include hydropower, wind, photovoltaics, diesel, fuel cells, and gas turbines. Renewable and other generators located downstream in a distribution network and involving small, modular electricity generation units close to the point of consumption are defined in this book as DG.

In this section, we provide a qualitative analysis of the issues that drive the effects of DG on a transmission system. More technical details and deeper insights are dealt with in Chapter 13, where we also define the services that DG can provide to distribution systems. In this section, the transmission services that DG is technically capable of providing are identified, and guidelines are developed to enable DG to participate in markets for these services.

Studies, reports, and experts in the field of DG [1–4] refer broadly to the benefits that DG can provide to transmission and distribution systems. The amount of generation relative to system total load, or penetration, is the most important factor in determining the influence of DG on transmission operation. A single 2-MW generator may have a considerable effect on the operation of a distribution system but goes entirely unnoticed on a transmission system. On the other end of the spectrum, if a fully mature DG market supplies 30% or more of total customer load, the effect and importance to transmission operation will be undeniable. Tougher questions are "What are the effects at penetration levels between the two extremes" and "how should they be treated in respect to system control and economic valuation?" This is addressed by focusing on (i) localized transmission benefits that a relatively small penetration of well-sited DG can provide and (ii) the benefits to the larger transmission system that can feasibly be achieved by growing DG penetrations.

DG can provide services to the distribution system such as capacity support, contingency capacity support, loss reduction, voltage support, voltage regulation, power factor control, phase balancing, and equipment life extension. DG can be defined as generation located at or near a load. Combined heat and power (CHP) is associated with prime movers that provide shaft power to generators and encompasses two broad categories: reciprocating engines and turbines. (Fuel cells may soon make a significant entrance on the CHP stage as well, but they are not yet ready for prime time.) CHP systems (also known as cogeneration) are generally developed by a user to avoid the purchase of power from the grid or by the energy service provider that retails the power to the site.

CHP is considered a subset of DG and can be used when there is a potential for profitable use of thermal energy. CHP is an energy cascade that captures energy normally rejected as part of a process. In the traditional case, steam is raised with a boiler on site, and power is purchased from the local utility. The thermal energy in the steam is then employed for another use.

1.5 Renewable Energy Economics

To meet the demand for a broad range of services (e.g., household, commerce, industry, and transportation needs), energy systems are needed. An energy supply sector and the end-use technology to provide these energy services are also necessary. In the United States, EU, Russia, and Japan, the electricity supply system is composed of large power units—mostly fossil fueled and centrally controlled—with average capacities of hundreds of megawatts. Conversely, renewable energy sources are geographically distributed and, if embedded in distribution networks, are often closer to customers and therefore subject to smaller losses.

In the power sector, most utilities have limited experience interconnecting numerous small-scale generation units with their distribution networks. Complicating matters, the possible level of renewable power penetration depends on the existing electrical infrastructure. For example, transporting to land the power produced by a large offshore wind farm is (economically) possible only where sufficient electricity grid capacity is available. In some locations, new electricity infrastructures have been set up to provide high penetration levels of up to 100% electricity from renewable powers.

Distributed electricity generation, close to the end customer, differs fundamentally from the traditional model of a large power station that generates centrally controlled power. The DG approach is new and replaces the concept of economy of scale (using large units) with the economy of numbers (using many small units), although it has yet to prove itself.

Far from being a threat, DG-based renewable energy can reduce transmission and distribution losses as well as transmission and distribution costs, provide consumers with continuity and reliability of supply stimulate competition in supply, adjust prices via market forces, and be implemented in a short time and with gradated resources because of its modular nature. The International Monetary Fund (IMF) predicts appreciable savings for the transmission and distribution grids because of the increased use of DG. This is a significant and driving argument when recent blackouts in the United States, Brazil, and Italy are taken into account.

1.5.1 Calculation of Electricity Generation Costs

When calculating generation costs, a distinction must be made between existing and potential plants. For existing plants, the running costs (short-term marginal costs) are relevant only for the economic decision as to whether to use the plant for electricity generation. Conversely, for new capacities, long-term marginal costs are important.

1.5.1.1 Existing Plants

Annual running costs are split into fuel costs and operation and maintenance (O&M) costs. Fuel costs are a function of the fuel price of the primary energy carrier and efficiency. O&M costs refer to electricity output, hence must be coupled with full-load hours. In general, one average operation time (full-load hour) is taken for each technology band. Analytically, generation costs for existing plants are given by

$$C = C_{var} = C_{fuel} + \tilde{C}_{O\&M} - R_{heat} = \frac{p_{fuel}}{\eta_{el}} + \frac{C_{O\&M}}{H} \cdot 1000 - p_{heat} \frac{\eta_{heat}}{\eta_{el}} \cdot \frac{H_{heat}}{H_{el}}$$

(1.1)

where:

C = Generation costs per kWh in EU/MWh
C_{var} = Running costs per energy unit in EU/MWh
C_{fuel} = Fuel costs per energy unit in EU/MWh
$\tilde{C}_{O\&M}$ = O&M costs per energy unit in EU/MWh
R_{heat} = Revenues gained from purchase of heat in EU/MWh
p_{fuel} = Fuel price primary energy carrier in EU/MWh$_{primary}$
p_{heat} = Heat price in EU/MWh$_{heat}$
η_{el} = Efficiency—electric generation
η_{heat} = Efficiency—heat generation
H_{el} = Full-load hours—electricity generation per annum in h/year
H_{heat} = Full-load hours—heat generation per annum in h/year

The full-load hours represent the equivalent time of full operation for a year. This is calculated for a power plant by dividing the amount of electricity generated per year by the plant's nominal power capacity. For theoretical cost–resource curves, this reflects an important aspect: the suitability of sites. In the case of wind energy, the full-load hours are determined by the wind speed distribution and the rated wind speed of the machines. Knowing the expected full-load hours, the quantity of electricity to be generated can be calculated. Hence, costs per unit are determined. The number of full-load hours divided by the number of hours in a year (8765 hours, on average) equals the system capacity (dimensionless).

1.5.1.2 New Plants

Electricity generation costs consist of variable costs and fixed costs. Generation costs are given by

$$C = C_{var} + \frac{C_{fix}}{q_{el}} = \left(C_{fuel} + \frac{C_{O\&M}}{H_{el}} \cdot 1000 - R_{heat} \right) + \frac{1000 \cdot I \cdot CRF}{H_{el}} \tag{1.2}$$

where:

C = Generation costs per kWh in EU/MWh
C_{var} = Running costs per energy unit in EU/MWh
C_{fix} = Fixed costs in EU
q_{el} = Amount of electricity generation in MWh/year
C_{fix}/q_{el} = Fixed costs per energy unit in EU/MWh
C_{fuel} = Fuel costs per energy unit in EU/MWh
$\tilde{C}_{O\&M}$ = O&M costs per energy unit in EU/MWh
R_{heat} = Revenues gained from purchase of heat in EU/MWh
I = Investment costs per kW in EU/kW
CRF = Capital recovery factor ($CRF = (z \cdot (1+z)^{PT})/((1+z)^{PT} - 1)$)
z = Interest rate
PT = Payback time (PT) of the plant in years
H_{el} = Full-load hours—electricity generation per annum in h/year

Fixed costs occur whether or not a plant generates electricity. These costs are determined by investment costs (I) and the capital recovery factor.

1.5.1.3 Investment Costs

Investment costs differ according to technology and energy source. In general, investment costs per unit of capacity for renewable energy systems are higher than for conventional technologies based on fossil fuels. In addition, differences exist among renewable energy technologies (e.g., investment costs per unit of capacity for small hydropower plants are generally at least twice those for wind turbines).

Investment costs decrease over time and are usually derived annually. It is usual to consider renewable powers as having zero fuel costs, apart from biomass (biogas, solid biomass, and sewage and landfill gas), so running costs are determined by O&M costs only. Therefore, the running costs for renewable energy systems are normally low compared with those of fossil fuel systems.

1.5.1.4 Capital Recovery Factor

The capital recovery factor allows investment costs incurred in the construction phase of a plant to be discounted. The amount depends on the interest rate and the PT of the plant. For the standard calculation of generation costs, these factors may be set as follows for all technologies:

PT of all power plants: 15 years
Interests rate (z): 6.5%

Different interest rates may be applied in any economic study. The interest rate depends on stakeholder behavior and is a function of a guaranteed political planning horizon of the promotion scheme of technology of the investor category.

Generation costs are calculated per unit of energy output, so fixed costs must be related to generation. Hence, fixed costs per unit of output are lower if the operation time of the plant—characterized by full-load hours—is high. Deriving generation costs for CHP plants is similar to calculating them for plants that produce electricity only. Both short-term marginal costs (i.e., variable costs) and fixed costs must be considered for new plants. Of course, variable costs differ between CHP and conventional electricity plants because the revenue from heat power must be considered in the former. In general, no taxes are included in the various cost components.

1.6 European Targets for Renewable Powers

Worldwide, several scenarios share the goal of sustainability in general or in the energy field. Thus, groundbreaking targets toward this goal are important for renewable energy and end-use energy efficiency. Such targets can guide policy-makers during decision-making and send important signals to investors, entre-preneurs, and the public. Case studies have demonstrated how concrete targets can lead to increased impact in various fields. In the case of renewable energies, policymakers formulate concrete policies and support measures to foster their development. Investors develop related strategies and renewable businesses as targets convince them that their investment will yield the projected returns.

Renewable energy is available in many environmental energy flows, harnessed by a range of technologies. The parameters used to quantify and analyze these forms are listed in Table 1.4.

A study by C. Kjaer [5] emphasizes 10 requirements for any community-wide mechanism to create a sound investment climate for renewable powers:

1) Compatibility with the "polluter pays" principle
2) High investor confidence
3) Simplicity and transparency in design and implementation
4) High effectiveness in deployment of renewable powers
5) Encouragement of technological diversity
6) Encouragement of innovation, technological development, and lower costs
7) Compatibility with the power market and with other policy instruments
8) Facilitation of a smooth transition ("grandfathering")
9) Encouragement of local and regional benefits, public acceptance, and site dispersion
10) Transparency and integrity by protecting consumers and avoiding fraud and free riding

1.6.1 Demand-Side Management Options

In the transport sector, biofuels are just beginning to be developed in Europe. However, in some countries, such as Brazil, sugarcane and oily plants already play important roles in the energy matrix. Moreover, the integration of renewable powers requires the adaptation of an infrastructure that has grown over a century of development based exclusively on fossil fuels. Besides the gradual substitution of vehicles in circulation, it is necessary to develop a new supply chain for the production and distribution of biofuels, hence requiring a substantial investment. However, development of the fossil fuel-based transport system also required investment, which was subsidized by the public sector in many countries.

In the heating sector, the full integration of renewable energy requires an adaptation of historical infrastructures. In many parts of Europe, it is already possible to construct buildings completely independent of fossil fuels or electricity for heating needs. This is achieved using state-of-the-art renewable heating and cooling applications linked with energy efficiency measures and demand-side management (see Appendices A and B).

A substantial economic restriction to the integration of renewable heating (i.e., solar thermal, biomass, and geothermal) results in the long lifetime of buildings. The installation of renewable heating systems is more cost-effective during the construction of a building or when the overall heating system is being refurbished. This means that there is a small window of opportunity for cost-effective integration of renewable heating. If this opportunity is lost, a building will remain dependent on fossil fuels or electricity to cover its heating demand for decades. For this reason, it is essential that all possible measures be taken to ensure that renewable heating sources are installed in all new buildings. It is also necessary to promote the use of renewable heating systems whenever a conventional heating system is being modernized.

Renewable heating sources can also be used for cooling. An increasing number of successful systems are being installed, based mainly on solar thermal and geothermal energy. The growing demand for cooling is affecting electricity systems in Europe, and several countries are now reaching peak electricity demand in summer instead of winter. These problems can be mitigated by the development and commercialization of renewable cooling technologies.

The existing infrastructure and market dominance of conventional heating and cooling technologies create a substantial barrier to the growth of renewable heating. Biomass heating and cooling can be competitive in areas where the fuel supply chain is well developed, but this is not yet the case in many parts of the world. Solar thermal systems can be good economic investments, but in many areas, users are not aware of this. In addition, most heating installers are trained only on conventional heating systems and therefore encourage customers to stick to conventional heating. The integration of distinct energy intensity profiles can be considered in these cases.

The choices that millions of citizens make for their homes and offices are crucial to the future integration of renewable energies in the heating and cooling sectors. Raising awareness among the public and training the professionals involved (e.g., conditioning and acclimatization installers, building engineers, architects, and managers of heat-intensive buildings or devices) are therefore very important.

Increasing use of renewable energies must be accompanied by energy efficiency and demand-side management measures at the customer end. Renewable energy development and energy efficiency are interdependent. The EU has always stressed the need to renew commitment at the community and member-state levels to promote energy efficiency more actively. In light of the Kyoto agreement to reduce carbon dioxide emissions, improved energy efficiency, together with increased use of renewable powers, will play a key role in meeting the EU's Kyoto target economically (see Table 1.1). In addition to a significant positive environmental effect, improved energy efficiency will lead to a more sustainable development and enhanced security of supply as well as many other benefits. An estimated economic potential for energy efficiency improvement of more than 18% of present energy consumption still exists today in the EU because of market barriers that prevent the satisfactory diffusion of energy-efficient technology and the efficient use of energy. This potential is equivalent to more than 1900 TWh, roughly the total final energy demand of Austria, Belgium, Denmark, Finland, Greece, and the Netherlands combined.

Special emphasis should be placed on urban areas, where a high proportion of energy is consumed. Urban areas are characterized by highly developed infrastructures, which do not always easily allow a rapid increase in renewable energy generation. The fact that electrical network infrastructures are generally over-dimensioned in urban areas can, in some cases, allow a high penetration of photovoltaic generators and wind energy without changing the existing cabling, transformer stations, and so on. However, in general, the future energy infrastructure will need to be designed from the beginning to accommodate renewable energy effectively at a high level. The small contributions that every home make by using energy derived directly from nature (such as wind, heat, coolant, light, photovoltaic electricity, and clean air) will make the biggest difference in the end.

1.6.2 Supply-Side Management Options

The European renewable energy industry has already reached an annual turnover of $10 billion and employs 200,000 people. Europe is the global leader of renewable energy technologies, and the use of renewable powers has a considerable effect on the investments made in the energy sector. Renewable energy replaces imported fuels, with beneficial effects on the balance of payments. Although, per unit of installed capacity, renewable energy technology is more capital intensive, when the external costs that have been avoided are taken into account, investing in renewable powers turns out to be cheaper for society than business-as-usual investments in conventional energy. Renewable energy technologies are often on a smaller scale than fossil fuel and nuclear projects, and they can be brought online more quickly and with lower risks. Finally, deployment of renewable powers creates more employment than do other energy technologies.

The development of smarter, more efficient energy technology over the past decades has been spectacular. Technologies have improved, and costs have fallen dramatically (see Figure 1.2) as estimated by the leveled cost that an equal-valued fixed revenue delivered over the life of the asset's generating profile would cause the project to break even [6, 7]. This can be roughly calculated as the net present value of all costs over the lifetime of the asset divided by the total electrical energy output of the asset.

The levelized cost of electricity (*LCOE*) is given by

$$LCOE = \frac{\text{sum of costs over lifetime}}{\text{sum of electrical energy produced over lifetime}} = \frac{\sum_{n=1}^{y} \frac{I_n + M_n + F_n}{(1+r)^n}}{\sum_{1}^{n} \frac{E_n}{(1+r)^n}}$$

(1.3)

where:

I_n = Investment expenditures in the year n
M = O&M expenditures in the year n
F_n = Fuel expenditures in the year n
E_n = Electrical energy generated in the year n
r = Discount rate
y = Expected lifetime of system or power station in years

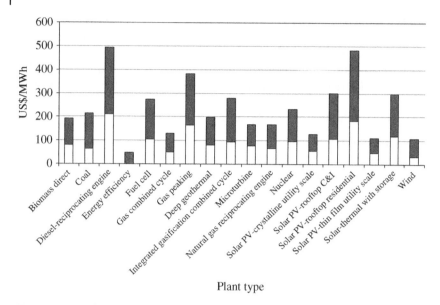

Figure 1.2 List of the minimum and maximum NREL-LCOEs. *Source*: © Investment bank Lazard, 2015.

The examples of wind and solar or photovoltaic cells are striking. Investment costs for wind energy declined by around 3% per annum over the past 20 years. For solar or photovoltaic cells, unit costs have fallen by a factor of 10 in the past 20 years (stimulated initially by the space program).

In EU, renewable powers already make up a significant share of total energy production. Germany, for example, has doubled its renewable output in the past five years to 8% of the total electricity production. Denmark now has got 18% of its electricity from wind power alone and created an industry with more jobs than in the electricity sector. Spain has leapt from using virtually no renewable powers a few years ago to become the second biggest wind power country in Europe, with 6000 MW of capacity. Countries such as Finland, Sweden, and Austria have supported the development of very successful biomass power and heating industries through fiscal policies, sustained R&D support, and synergistic forestry and industrial policies. In addition to saving significant carbon dioxide emissions, equipment from all three countries is exported worldwide. Table 1.6 presents some typical technical and economical characteristics of selected renewable energy technologies, and Table 1.7 presents the US Electricity Production Mix in 2015 and Table 1.8 presents some selected global indicators of renewable energy.

Table 1.6 Technical and economical characteristics of selected renewable energy technologies.

Type of source	Unit capacity kW	Electrical efficiency %	Thermal efficiency %	Lifetime Years	Full-load operating Hours
Gas diesel engines	3–10,000	30–45	45–50	15	5,000
Microturbines	25–250	15–35	50–60	15	5,000
Stirling engines	10–150	15–35	60–80	15	5,000
Steam engines	0.5–10,000	15–35	40–70	15	5,000
Wind power	0.1–5,000	40–50	—	20	2,500
Fuel cells	0.5–2,000	38–55	40–70	15	5,000

Table 1.7 US electricity production mix, 2015.

Primary source	Production (%)
Coal	33
Natural gas	33
Nuclear	20
Hydropower	6
Other renewables	7
Biomass	1.6
Geothermal	0.4
Solar	0.6

Source: Ref. [8–10].

1.7 Integrating Renewable Energy Sources

Integration of renewable energy sources involves integrating in a system any energy resource that naturally regenerates over a short period. This time scale is derived directly from the sun (such as thermal, photochemical, and photoelectric energy), indirectly from the sun (such as wind, hydropower, and photosynthetic energy stored in biomass), or from other natural movements and mechanisms of the environment (such as geothermal and tidal energy). In the long term, renewable energies will necessarily dominate the world's energy supply system for the simple reason that there is no alternative. Mankind

Table 1.8 Selected global indicators of renewable energy.

Selected global indicators	2014	2015	Units
Investment in new renewable capacity (annual)	270	285	Billion USD
Existing renewables power capacity, including large-scale hydro	1,712	1,849	GWe
Existing renewables power capacity, excluding large hydro	657	785	GWe
Hydropower capacity (existing)	1,055	1,064	GWe
Wind power capacity (existing)	370	433	GWe
Solar PV capacity (grid-connected)	177	227	GWe
Solar hot water capacity (existing)	406	435	GWth
Ethanol production (annual)	94	98	Billion liters
Biodiesel production (annual)	29.7	30	Billion liters
Countries with policy targets for renewable energy	164	173	

cannot survive indefinitely off the consumption of finite energy resources, concentrate supplies on some points on Earth, or carelessly spread its population over the world.

Today, the world's energy supply is based largely on fossil fuels and nuclear power. These sources of energy will not last forever and have proved to be a major cause of environmental problems. Environmental effects of energy use are not new, but it is increasingly well known that they range from deforestation to local and global pollution. In less than three centuries since the industrial revolution, mankind has burned away roughly half of the fossil fuels accumulated under the Earth's surface for hundreds of millions of years. Nuclear power is also based on limited resources such as uranium, and the use of nuclear power creates such incalculable risks that nuclear power plants cannot be ensured.

Renewable sources of energy are in line with an overall strategy of sustainable development. They help reduce and not create dependence on energy imports, thereby ensuring a sustainable security of supply. Furthermore, renewable energy sources can improve the competitiveness of industries, at least in the long run, and have a positive effect on regional development and employment. Renewable energy technologies are suitable for off-grid services; they can serve remote areas of the world without expensive and complicated grid infrastructure.

The ability to integrate electricity generated from renewable powers into grid supplies is governed by several factors, including

The variation with time of power generated
The extent of the variation (availability)

The predictability of the variation
The capacity of each generator
The dispersal of individual generators
The reliability of plants
The experience of operators
The technology for integration
The regulations and customs for embedded generation

Despite these difficulties, the experience for the past 25 years has shown that ever-increasing amounts of electricity from renewable powers can be integrated into grid supplies without significant financial penalty. The standard response of grid operators that are accustomed to large-scale centralized generation is that intermittent and dispersed renewable energy generation cannot be so integrated. However, given the requirement to accept specific renewable energy generation, the technology and methods have followed successfully. Examples include

Electrical safety equipment and grid–fault disconnectors
Grid-linked inverters for photovoltaic or solar cells and power from buildings
Doubly fed induction generators for variable speed wind turbines
Voltage reinforcement on rural power lines
Co-firing of steam boilers with biomass
Gas turbines for the output of gasifiers

The outstanding example of ever-increasing integration of renewable energy generation into the grid is Jutland, western Denmark, due to their willing application of new technologies and practices. In the early 1980s, the limit for wind power exported to the grid was considered to be 20% of the total supply. However, by 2003, about 40% of the annual electricity supply was from wind, and at times, significant areas were supplied totally by wind power.

Nevertheless, there are fundamental limitations for any renewable energy generation technology and plant; for instance, the sun never shines at night. In addition, in the middle of large towns and cities, the surface roughness for wind move is not acceptable for small towers. Therefore, it is essential to integrate renewable energy generation options with control and storage such that they complement each other.

1.7.1 Integration of Renewable Energy in the United States

The United States currently relies heavily on coal, oil, and natural gas for its energy. Fossil fuels are nonrenewable: they draw on finite resources that will eventually dwindle and become too expensive or too environmentally damaging to retrieve. In contrast, renewable energy resources such as hydropower, wind energy, and solar energy are replenished constantly and will never run out.

As in any other place, most renewable energy in the United States comes directly or indirectly from the sun. Sunlight, or solar energy, can be used directly for heating and lighting, to generate electricity, and for cooling as well as for a variety of commercial and industrial uses. The sun's heat also drives winds (whose energy is captured with wind turbines) and evaporates waters, turning them into rain or snow, which then flows into rivers or streams and whose energy may be captured in water dams.

Other renewable sources include geothermal energy, which is tapped from the Earth's internal heat for electric power production and heating and cooling buildings, and the oceans' tides, which come from the gravitational pull of the moon and sun. In fact, ocean energy comes from a number of sources. In addition to tidal energy, there is the energy of the oceans' waves, which are driven by tides and winds. The sun also warms the surface of oceans more than it warms ocean depths, which creates a temperature difference that can be used as an energy source. All these forms of ocean energy can be used to produce electricity.

In contrast to fossil energy, renewable energy is an attractive source for several reasons: clean environment, long-lasting life, increased jobs, increased comfort, and industry and energy self-sufficiency through decreased dependence on other nations. An economy that uses less energy also produces less pollution, and an energy-efficient economy can grow without using more energy. Energy efficiency means using less energy to accomplish the same task resulting in spending less money on energy by homeowners, schools, government agencies, businesses, and industries. The money that would have been spent on energy can instead be spent on consumer goods, education, services, and products. From 1970 to 2000, US energy consumption grew only 45% although the US gross domestic product increased 160%. In other words, the energy used per dollar of gross domestic product decreased 44% from 1970 to 2000. By 1999, GHG emissions from energy use had risen 13% above the levels in 1990. During that period, energy use increased 14.9%.

1.7.2 Energy Recovery Time

The cost of electricity depends entirely, or largely, on the size of power stations. Between 1960 and 1980, the ideal size of a station rose from 400 to 1000 MW. These days, 5 MW is regarded as ideal because small-scale power generation permits a flexible response to energy demand and return of capital. Small-scale units such as wind turbines, photovoltaic cells, fuel cells, and bio-gasification plants represent the future.

Regardless of the type of primary source, it takes energy to convert energy from one type into another. The lower the specific energy content, the more energy intensive the conversion process is. When the specific energy content is low, the energy process chain uses more energy than it generates in electricity.

Most of the primary energy extracted today has a profitable content that makes conversion cost-effective.

However, if any energy were to gain momentum, a point would come when the specific energy conversion would no longer be cost-effective. The amount of time a power plant needs to operate before all the energy consumed in the chain has been earned back (and the power plant begins to produce net energy), or the energy recovery time, is highly dependent on the specific energy content of the primary source. It is difficult to compare this figure with the energy recovery time for fossil fuel-powered power stations. A fossil fuel power station has to recover only the electricity used for construction and other constituent processes in the chain. In such a case, the recovery time for power stations fired by gas and oil is 0.09 of a full-load year (approximately 0.13 of a calendar year); for coal-fired power stations, it is 0.15 of a full-load year (approximately 0.21 of a calendar year) [4, 16, 17]. Nevertheless, unlike modern gas-fired power stations that generate and supply commercial heat, alternative sources of energy such as nuclear power plants, wind turbines, and photovoltaic systems can generate only electricity. All the energy used in the chain is recovered in the form of electricity, which increases recovery time considerably. As a frame of reference, assume that fossil fuel-fired power stations must recover the energy used in their construction only in the form of electricity. This results in a recovery time of 0.7 of a full-load year for gas- or oil-fired power stations, which is approximately one calendar year. Coal-fired power stations have a longer recovery time. Table 1.9 is a comparative list of the recovery time of selected sources of energy.

Improvements in conversion yields and production methods will help reduce the recovery time for photovoltaic systems in the future. Photovoltaic technology is at a peak of development and now is in the sharply rising section of the learning curve, which means that prices will fall significantly, as more capacity is commercialized. It is conceivable that the recovery time for photovoltaics will drop to less than one year as technical progress continues. Nuclear energy, on the other hand, is a mature technology; the price of nuclear power will not decrease as more nuclear power stations are built. In the past, there were even

Table 1.9 Recovery time of selected sources of energy.

Alternative source	Recovery time (years)
Wind	0.62–0.90
Gas and oil	1
Photovoltaic system	1.5–3
Nuclear power station	10–18

Source: Ref. [4]. © European Union, 1995–2013.

cost hikes of approximately 14% a year until the mid-1980s. Since 1979, no new nuclear power stations were ordered in the Organization of Economic Cooperation and Development (OECD) countries, which ended the competitive time in which further price rises could occur. Clearly, the recovery time for nuclear power stations is much longer than that of other power stations and will never decrease. In contrast, the recovery time for photovoltaic systems, in particular, is certain to decrease if new technologies and materials are used.

Environmental issues such as the greenhouse effect have focused attention on fossil fuel combustion and electricity generation around the world. In Australia, 47% of the annual emissions of greenhouse carbon dioxide come from fossil fuel-fired power plants [16, 17]. As coal-based plants are retired, due to age and greenhouse concerns, there is an opportunity for renewable energy generation sources to grab a larger share of the global electric energy market.

Wind systems, solar systems, storage components, and complete energy systems are now commercially available from many suppliers [5] to fill niche markets.

Fundamental research (especially in the production of thin-film solar or photovoltaic devices, hydrogen from sun-mirrored heat, and new forms of batteries) is occurring in many countries, and these activities are steadily reducing the cost of renewable energy systems. However, there are still issues to resolve before such systems gain a bigger portion of the electric energy marketplace. Such systems must lower the overall cost of delivered energy, gain acceptance by a conservative industry, and convince the industry's customers that renewable energy systems are safe, reliable alternatives to conventional grid-supplied power.

Another issue is that although the energy supply may be free, the cost of using wind and solar energy is not because structures and energy collectors must be built and energy storage must be provided. Any initiative that increases the energy collected or stored will lead to a reduction in the price of energy delivered from a complete system [5, 8]. Balanced and optimized are terms frequently used to indicate that a system is designed to size the renewable, storage, and fuel-based components to deliver minimal all-of-life costs in a specific site and for a specific customer-loading pattern. Such a system operates to maximize renewable energy capture and to minimize all-of-life costs of components.

1.7.3 Sustainability

Humans require only a few basic needs in order to survive and be sustainable: air, water, food, space, reproduction, and energy. Everything else is exceeding human demands. The Fifth Environmental Action Programme of 2000 established a EU legislation and defined sustainable development as "that which meets the needs of the present without compromising the ability of future generations to meet their own needs" [12, 13]. The policy objectives underlying this definition were to ensure compatibility between economic growth and efficient and secure energy supplies together with a clean environment.

Environmentally polluting by-products are produced by conventional energy generation, which also depends on finite energy sources that are gradually being depleted. However, energy is essential for socioeconomic progress in developing and industrialized countries, and the demand for energy will increase with global population. For example, waste to energy (WTE) or energy from waste (EFW) is the process of generating energy in the form of electricity and/or heat from the primary treatment of waste. Incineration or combustion of organic materials such as waste with energy recovery is the most common WTE implementation. All new WTE plants in OECD countries incinerating waste (residual MSW, commercial, industrial, or RDF) must meet strict emission standards, including those on nitrogen oxides (NO_x), sulfur dioxide (SO_2), heavy metals, and dioxins.

Targets established in the EC white paper of 1997 foresees a 12% share of renewable powers in Europe's total energy consumption by 2010 (double the 1997 share). Individual targets for each renewable energy technology are also set. Annual growth rates between 1995 and 2001 show that one sector (wind) is far beyond the target and that others (i.e., hydro, geothermal, and photovoltaics) are in line with expectations. To reach the overall and sector targets (which is feasible), specific support actions have to be taken soon for technologies that lag behind, such as biomass and solar thermal. Therefore, the deployment status of energy consumption by energy source in the United States did not improve much since 1995 and worsen in some cases as illustrated in Table 1.10.

Given the present state of market progress and political support, the expectation is that if strong additional support measures are adopted, the overall contribution of renewable energy consumption in 2020 will be 20%. These estimates are based on a conservative annual growth scenario for the technologies. To reach the target, strong energy efficiency measures have to be taken to stabilize energy consumption between 2010 and 2020. Further prediction would include the Hubbert curves and the energy depletion curves (see Figure 1.3). These novelties in the energy market have opened discussions about what would be necessary socially, politically, and economically for a country to adapt to new environmental surroundings.

In particular, when the director of the National Aeronautics and Space Administration's Goddard Institute for Space Studies enlightened the US Congress to the fact that human-induced global warming was detectable in the climate record, some skeptical members of the Congress argued that the data were unclear and inconclusive. According to the Goddard Institute, the consequences predicted for global warming include worldwide floods, droughts,

Table 1.10 US energy use and consumption by energy source.

	Production	Use
Source	1995 (%)	2014 (%)
Coal	21.69	18.00
Natural gas	24.54	26.60
Petroleum	27.22	35.10
Nuclear	7.95	8.27
Hydroelectric	3.83	2.56
Biomass	3.25	4.49
Solar	0.08	0.32
Wind	0.03	1.60
Deep geothermal	0.36	—

Source: © Renewable Energy Manual, 2014.

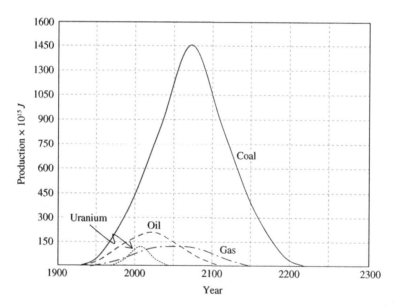

Figure 1.3 Energy depletion curves.

rising sea levels, category 5 hurricanes, and typhoons. These effects, though, were widely debated. Agreement was barely reached when deep reductions in carbon emissions cannot be made economically without the use of energy-efficient and renewable energy technologies.

Since then, standards have become necessary to regulate new power interconnections with distribution systems. The Institute of Electrical and Electronics Engineers (IEEE) Standard for Interconnecting Distributed Resources with Electric Power Systems is the first in the 1547 series of planned interconnection standards [7–9]. There are major obstacles to an orderly transition to the use and integration of distributed power resources with electric power systems, as discussed in Chapter 15. Examples include the lack of uniform national standards and tests for interconnection operation and certification, uniform national building, and electrical and safety codes. Resolving this requires time to develop and promulgate consensus. The 1547 standard is a milestone for the IEEE standard-setting process and demonstrates a model for ongoing success for further national standards and for moving forward in modernizing the national electric power system.

1.8 Modern Electronic Controls for Power Systems

Renewable and alternative energy sources must eventually be integrated with existing electric systems. Power electronics are a crucial enabling technology toward this end. Power electronics are part of electronic application systems that encompass the entire field of power engineering, from generation to transmission and distribution to transportation, storage systems, and domestic services. The progress of power electronics has generally followed microelectronic device evolution and influenced the current technological status of renewable energy conversion.

Figure 1.4 depicts the 2015 US energy flow in the net primary consumption given in quads and exajoules. A quad is 1 quadrillion (1015) Btu, and an exajoule is 1018 J.

The power produced by renewable energy devices such as photovoltaic cells and wind turbines varies on hourly, daily, and seasonal bases because of the variation in the availability of the sun, wind, and other renewable resources. This variation means that power is sometimes not available when it is required and that on other occasions there is excess power. The variable output from renewable energy devices also means that power conditioning and control equipment is required to transform this output into a form (i.e., voltage, current, and frequency) that can be used by electrical appliances. Therefore, energy must be stored and power electronics used to convert this energy.

Power-processing technology can be classified according to the energy, time, and transient response required for its operation. As the cost of power electronics falls, system performance improves. Applications are proliferating, and it is expected that this trend will continue with high momentum in this century. Modern industrial processes, transportation, and energy systems benefit tremendously in productivity and quality enhancement with the help of

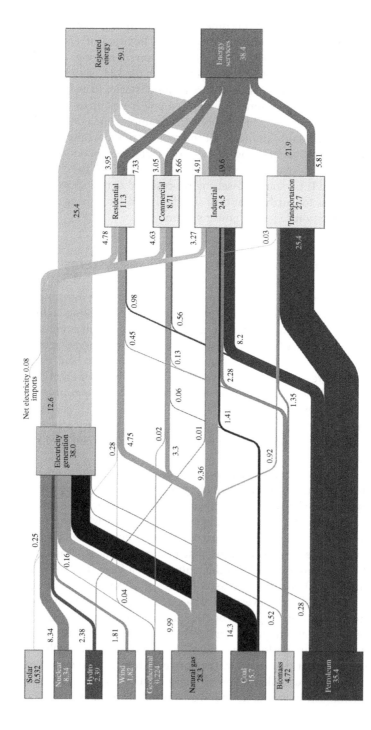

Figure 1.4 Estimated US consumption in 2015: 97.5 quads. *Source:* Data is based on DOE-EIA0035(2016/6) [9]. © Lawrence Livermore National Laboratory, 2016.

power electronics, on which efficient energy conversion from renewable sources depends leading to the future emphasis of the environmentally clean sources of power—such as wind, photovoltaics, and fuel cells.

In this book, we are concerned with how alternative and renewable energy can be integrated electrically. Power electronic technology plays a major role in the injection of electrical power to the utility grid, as discussed in Chapter 12. If only photovoltaic and fuel cell systems are used, a dc-link bus could be used to aggregate them, and ac power could be integrated through dc-to-ac conversion systems (inverters). If only hydro or wind power is used, variable-frequency ac voltage control can be aggregated into an ac link through ac-to-ac conversion systems that can be created through several approaches discussed in this book.

Of course, alternative energy sources such as diesel and gas can also be integrated with renewable powers. They have a consistent and constant fuel supply, and the decision to operate them is based more on straightforward economics. Gas microturbines and diesel generators are commercially available with synchronous generators that supply 60 Hz, and a direct interconnection with the grid is typically easier to implement. When integrating and mixing these sources, a microgrid can be based on a dc- or ac-link structure. The design of such a microgrid must incorporate energy storage with seamless control integration of source, storage, and demand.

1.9 Issues Related to Alternative Sources of Energy

Figure 1.5 reunites all modern issues related to implantation of an alternative source of energy developed in this book. Beginning with the selection of the primary source, this has to do with consumers, proximity, availability, or facilities for extraction of electrical power. From there it is necessary to select the type of energy conversion, for instance, from mechanical, radiation, or chemical to electricity. This electricity may be in dc or ac, whose magnitude and frequency must be controlled.

For alternative sources of energy, it is fundamental to maximize the extraction of power from the primary source so as to compensate the investment. It is quite possible to be able to manage the heat usually generated by any power supply to be converted into electricity. Use of heating source was not very usual in the conventional power plants by the distance between the source and the consumer. Particularly, fuel cells may almost double their efficiency if electricity and heat are consumed. Almost all alternative sources of energy can be complementary to each other as, for example, wind and photovoltaic or hydro and fuel cells due to their availability period in the nature. A decision has to be made if the small power plant has to be standing alone or interconnected to the other distribution system.

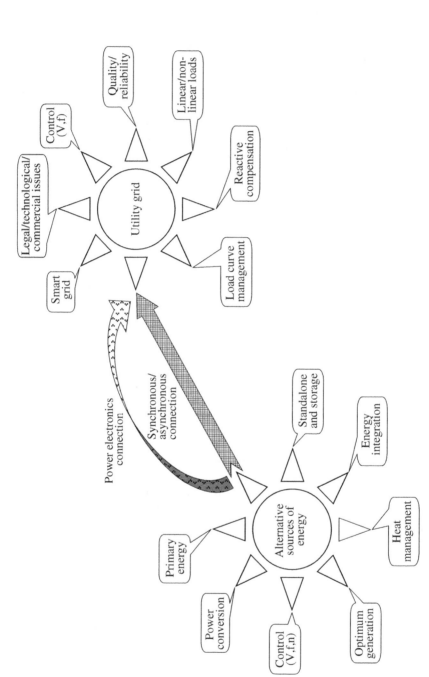

Figure 1.5 Issues related to interconnection of alternative sources of energy.

Another important decision is whether to connect or not the alternative power plant to the mains and how to do it (directly or through power electronics). Synchronous and asynchronous generators can be both connected directly to the mains grid without many problems. Dc generators such as photovoltaic, dc rotating machines, and fuel cells definitely have to use power electronics to control the frequency, rotation, and voltage levels (see Table 1.11).

Current applications of smart grid in distribution systems are taking into consideration the existence or absence of alternative sources of energy. These sources can be used to cope with transient problems or as backup for critical loads, like hospitals, alarms, and security installations (Table 1.12).

Table 1.11 Applications of primary sources.

Primary source	Conversion	Usual application
Biomass	Mechanical	Electrical power
Combined cycle	Mechanical	Electromechanical
Deep geothermal	Mechanical	Electrical power
Diesel	Mechanical	Electrical power
Fuel cells	Chemical	Electrical power
Fluidized bed	Mechanical	Electromechanical
Horizontal ocean thermal	Mechanical	Electrical power
Hydropower	Mechanical	Electrical power
Integrated gasification	Mechanical	Electromechanical
Magneto-hydrodynamics	Electrical	Electrical power
MHD	Electrical	Electrical power
Microturbine	Mechanical	Electrical power
Nuclear	Thermomechanical	Electrical power
Photovoltaics	Electrical	Electrical power
Piezoelectric	Electrical	Electrical power
Sea tidal power	Mechanical	Electrical power
Sea wave power	Mechanical	Electrical power
Solar thermal	Direct	Heating
Surface geothermal	Direct	Conditioning
Thermocouple	Electrical	Electrical power
Traditional boiler	Mechanical	Electrical power
Vertical ocean thermal	Mechanical	Electrical power
Wind power	Mechanical	Electrical power
WTE	Combustion	Electrical/heat power

Table 1.12 US average estimated levelized cost (2013 $/MWh) of electricity (LCOE) entering service in 2020.

Plant type	Capacity factor (%)	Levelized capital cost	Fixed O&M	Variable O&M (including fuel)	Transmission investment	Total system LCOE	Subsidy	Total LCOE including subsidy
Dispatchable technologies								
Conventional coal	85	60.4	4.2	29.4	1.2	95.1		
Advanced coal	85	76.9	6.9	30.7	1.2	115.7		
Advanced coal with CCS	85	97.3	9.8	36.1	1.2	144.4		
Natural gas-fired								
Conventional combined cycle	87	14.4	1.7	57.8	1.2	75.2		
Advanced combined cycle	87	15.9	2.0	53.6	1.2	72.6		
Advanced CC with CCS	87	30.1	4.2	64.7	1.2	100.2		
Conventional combustion turbine	30	40.7	2.8	94.6	3.5	141.5		
Advanced combustion turbine	30	27.8	2.7	79.6	3.5	113.5		
Advanced nuclear	90	70.1	11.8	12.2	1.1	95.2		
Geothermal	92	34.1	12.3	0.0	1.4	47.8	-3.4	44.4
Biomass	83	47.1	14.5	37.6	1.2	100.5		
Non-dispatchable technologies								
Wind	36	57.7	12.8	0.0	3.1	73.6		
Wind offshore	38	168.6	22.5	0.0	5.8	196.9		
Solar PV	25	109.8	11.4	0.0	4.1	125.3	-11.0	114.3
Solar thermal	20	191.6	42.1	0.0	6.0	239.7	-19.2	220.6
Hydroelectric	54	70.7	3.9	7.0	2.0	83.5		

Legal, technological, and commercial issues are related to the decision of either building up or not a power plant to inject power into the grid.

Regulations try to make a safer, standardized, and reliable injection of power into the grid. With respect to the technological side, one wants to make sure whether the harmonic, sags, tilts, droops, and other common events in small power plants are not going to affect substantially the other consumers. Commercial units must satisfy all regulations, including quality and reliability, and have available spare parts everywhere and be long-lasting units because blackouts are very disturbing when affecting many people.

The small power plant control in power systems is mostly related to load management, reactive compensation, and coping with nonlinear loads. The power supply depends essentially on a careful control design taking into account cost, reliability, and power supply quality.

Increasing efficiency of a facility will not resolve the humanly energetic needs if we keep increasing the number of facilities. Durability and usage sharing are thinkable strategies. As discussed in the COP21, the forms of energy originated from the sun will be perhaps the only way out if we want to survive in our planet.

References

1 M. Davies and T. Oreszczyn, The unintended consequences of decarbonising the built environment: a UK case study, Energy and Buildings, Vol. 46,· pp. 80–85, 2010.

2 J. Depledge, Tracing the Origins of the Kyoto Protocol: An Article-by-Article Textual History, UNFCCC/TP/2000/2, Framework Convention on Climate Change, United Nations, Bonn, Germany, 1996.

3 Wiertzstraat, Wise, Coming Clean: How Clean Is Nuclear Energy? Stitching GroenLinks in the European Union/The Group of the Greens/European Free Alliance, Brussels, Belgium, October 2000.

4 Renewable energy statistics, http://ec.europa.eu/eurostat/statistics-explained/, Eurostat (t2020_31), accessed March 4, 2017.

5 EREC (Integration of Renewable Energy Sources), Targets and benefits of large-scale deployment of renewable energy sources, presented at the Workshop on Renewable Energy Market Development Status and Prospects, Workshops in the New Member States, European Commission, April–May 2004.

6 Energy Quest, http://www.energyquest.ca.gov/story/chapter19.html, accessed March 4, 2017.

7 Estimation suggested by Bank Lazard's Levelized cost of energy analysis—version 9.0, November 2015.

8 T.S. Basso, IEEE 1547 Standard for Interconnecting Distributed Resources with Electric Power System, IEEE Press, Piscataway, NJ, 2014.

9 Estimated US Energy use in 2013, Lawrence Livermore National Laboratory, Energy and Environment Directorate, DOE-EIA0035-5(2014–3), California, U.S., March 2014.

10 Final Report from the Commission to the European Parliament and the Council Progress Towards Achieving the Kyoto and Eu 2020 Objectives, Decision No. 280/2004/EC, Com(2013) 698, European Commission, Brussels, October 2013, pp. 1–21.

11 C. Kjaer, Position Paper on the Future of EU Support Systems for the Promotion of Electricity from Renewable Energy Sources, news release, European Wind Energy Association, Brussels, Belgium, November 2004.

12 Eurostat statistics explained, http://ec.europa.eu/eurostat/statistics-explained/index.php/, accessed March 4, 2017.

13 Independent Statistics and Analysis, US Energy Information Administration (EIA), Washington, DC, 2015.

14 U.S. Energy Information Administration, Annual Energy Outlook 2015, DOE/EIA-0383(2015), U.S. Energy Information Administration, Washington, DC, April 2015.

15 H. Auer, C. Obersteiner, L. Weissensteiner, and G. Resch, Guiding a least cost grid integration of RES-electricity in an extended europe, EIE/04/049/S07.38561, system operation cost and grid reinforcement/extension cost allocated to large-scale RES-E integration, GreenNet-EU27, Intelligent Energy—Europe (EIE), Key action: VKA5.3—Grid System Issues, Deliverable D5a, Vienna, Austria, 2005.

16 Communication from the Commission to the Council and the European Parliament—The Share of Renewable Energy in the EU—report in accordance with Article 3 of Directive 2001/77/EC, Evaluation of the Effect of Legislative Instruments and Other Community Policies on the Development of the Contribution of Renewable Energy Sources in the EU and Proposals for Concrete Actions, SEC 366, European Commission, Brussels, May 26, 2004.

17 Commission of Restructuring of British Energy, A quantitative assessment of direct support schemes for renewables, IP/04/1125, September 22, 2004; Working Group on Renewables and Distributed Generation, Ref. 2003-030-0741, Table 4, p. 22, January 2004.

2

Principles of Thermodynamics

2.1 Introduction

It is very important to understand the conversion of energy from one form to another, the availability of energy to do work, and the flow of heat for engineering applications. The laws and applications of thermodynamics are the foundation to make analysis and design involving transformations of work, heat, and internal energy, by boilers, steam turbines, internal combustion engines, refrigerators, air conditioners, and other thermal (such as geothermal) or mechanical devices.

Human activity requires heat supply for several sectors: (i) *residential*, for appliances and lighting, space heating, water heating, and air conditioning; (ii) *commercial*, for lighting, space heating, office equipment, water heating, air conditioning, ventilation, and refrigeration; (iii) *industrial*, for water and steam boilers, direct process energy, and machine drives; and (iv) *transportation*, for personal automobiles, light and heavy trucks, air transport, heat water recovery transport, pipe transport, rail transport, and many others. In all these sectors, in addition to primary energy sources, the use of electrical energy is very extensive.

Fluids play an important role in modern storage and energy conversion systems. Fluids are the medium where the energy conversion takes place, such as those in wind or hydro turbines, as well as tidal or steam systems in large power production, or in gas turbines and microturbines. Furthermore, thermosolar applications used solar energy to heat a chamber with a fluid in the focus of a large mirror; geothermal production of energy based on the underground temperature or ocean temperature differential depends on fluids. Fluids make systems to interact and exchange energy and sometimes mass (matter).

A system is considered open if it exchanges matter and energy with its surroundings, and it is considered isolated if there is no interaction with its surroundings. Closed systems can still exchange energy, but not matter, with their surroundings. A "working fluid" is the matter connecting and performing

Integration of Renewable Sources of Energy, Second Edition. Felix A. Farret and M. Godoy Simões.
© 2018 John Wiley & Sons, Inc. Published 2018 by John Wiley & Sons, Inc.

the boundaries of such systems. Such matter can be solid, liquid, vapor (or gaseous), or in plasmatic phases. Either a perfect gas or a substance that works as a combination of liquid and vapor can be considered as the "working fluid" in engineering thermodynamic applications [1–3].

A very important assumption can be made as related to the surroundings of two subsystems, particularly when one is much larger than the other: the heat capacity of the surroundings is so large that any transfer of heat from the system to the surroundings does not change in practice the temperature of the larger system. When that happens, any interaction of the original system with the surroundings does not change practically the intensive parameters of the surroundings. Work performed on the surroundings by changing its volume makes the pressure in the surroundings to not change in practice because of a great difference in the system volume and their surroundings.

A system whose heat capacity and volume might be considered infinite when compared with smaller subsystems is called a *heat and/or volume reservoir*. Our atmosphere is a good example of such a reservoir. Geothermal energy conversion is underground and can also be assumed as a heat reservoir.

The understanding of thermodynamics can make the designer recognize what is possible or not regarding thermal and mechanical processes. The primary application of thermodynamics is in predicting the input energy required and output energy expected. To understand limitations and constraints concerning fluid flows, establishing limit for how hot and/or how temperatures are in the system. It is very important to the comprehension of thermodynamics principles, as they apply to understand the sustainability for our society.

2.2 State of a Thermodynamic System

The state of a system is a set of thermodynamic variables known as properties. There are extensive properties that depend on the amount of material (mass and volume), and there are intensive properties that do not depend on the amount of material. Numerical computation of those variables requires their units be given in a homogeneous system, that is, consistent with their metrics. The metric units adopted by the International Organization for Standardization (ISO) are called Système international d'unités (SI). However, thermodynamic problems and industrial processes still use the US customary system (USCS) units. For example, air-conditioning systems, furnaces, kitchen appliances, and water heaters are mostly measured by British thermal unit (Btu) ratings. In USCS, both force and mass are called by the same name (pounds). Therefore, it is necessary to distinguish them by calling them either pound-force (lbf) or pound-mass (lbm). The pound-force is the force that accelerates 1 lbm at $32.174 \, \text{ft/s}^2$, that is, the standard acceleration of free fall, or standard gravity, expressed in feet and seconds ($9.806 \, \text{m/s}^2$). Thus, the second law of Newton using these units (force = mass · acceleration) makes $1\,\text{lbf} = 32.174\,\text{lbm-ft/s}^2$. The expression $32.174\,\text{lbm-ft/(lbf-s}^2)$, designated as g_c, is used to resolve expressions involving both mass and force expressed as pounds. For instance, the second law of Newton could be expressed as $F = m(a/g_c)$, where F is given in lbf, m is in lbm, and a is in ft/s^2. Similar expressions for other quantities are:

Kinetic energy: $KE = m(v^2/2g_c)$ with KE in ft-lbf.
Potential energy (PE): $PE = mgh/g_c$ with PE in ft-lbf.
Fluid pressure: $p = \rho gh/g_c$ with p in lbf/ft^2.
The specific weight: $SW = \rho g/g_c$ in lbf/ft^3.

In these examples, g_c should be regarded as a unit conversion factor. It is frequently not written explicitly in engineering equations, but it is required to produce a consistent set of units. Note that the conversion factor g_c [lbm-ft/(lbf-s^2)] should not be confused with the local acceleration of gravity, g, which has different units (m/s^2) and may be either its standard value (9.807 m/s^2) or some other local value. If any problem is presented in USCS units, it may be necessary to use the constant g_c in the relevant equation in order to have a consistent set of units.

Energy is an extensive property, and thermodynamic principles provide ways to calculate the change in energy for a closed system. The dimensions of energy are the same for work. Engineers may use the SI system where the unit is the joule (J), where $1\,\text{J} = 1\,\text{Nm} = 1\,\text{W/s}$. However, in USCS there are two main units: (i) the Btu, used for heat processes, and the foot-pound-force (ft-lbf), used for the studies of mechanical work. One Btu equals approximately the amount of energy required to raise the temperature of 1 lbm of liquid water by 1 °F from the room temperature.

Temperature is a measure of the average kinetic energy. Pressure is defined as a force acting on a given unit area. The USCS units are pounds per square inch gauge (psig). Normally the letter "g" is assumed and omitted and the unit is only called psi. The SI unit is pascal (Pa) or $1\,\text{N/m}^2$, and the conversion factor is $1\,\text{psi} = 6894.8\,\text{Pa}$. Absolute pressure, the pressure above the lowest possible pressure, is the sum of the gauge pressure and the atmospheric pressure. Pressure-measuring devices generally measure gauge pressure rather than absolute pressure (i.e., gauge pressure is the difference between a referenced pressure and atmospheric pressure). By convention, measurements in psi-a always include the "a." A constant atmospheric pressure of 14.7 pounds per square inch absolute is usually assumed, corresponding to a gauge pressure of 0 psig. When pressures are very close to atmospheric pressure, as in most of the heating and air-conditioning applications, one may follow the usual practice of expressing pressure in inches or meters of water. A pressure of 1 in. of water is the pressure at the bottom of a column of water 1 in. high. One psi is equivalent to 27.7 in. of water, and 1 atmosphere (atm) is equal to 1.01325×10^5 Pa.

Energy of a system is taken in practice as the relative hotness or coldness of a substance. Temperature is one of the intensive parameters very common in the everyday life; measured with thermometers, the oldest ones are made of mercury or alcohol. Two empirical temperature scales are in common use, namely, the *Celsius scale used worldwide and* basically impressed by European standards and the *Fahrenheit scale* only used in the United States. Celsius chose two reproducible phenomena, the freezing and boiling of water, and assigned them 0 and 100 °C, respectively. Fahrenheit did a similar experiment, but he chose the temperature 0 °F to the freezing point of a mixture of water and salt ammoniac, and the value of 100 °F was assigned to the temperature of his (sick) wife. Fahrenheit also divided his scale into 100 equal degrees. It is considered that the range from 0 to 100 °F is where most of human activity can be performed outdoors, and outside that range it is very difficult for human beings to survive without protection. It follows from the comparison of the two scales that 0 °C corresponds to 32 °F and 100 °C corresponds to 212 °F. The units of temperature can be changed from degrees Fahrenheit (°F) converted to Celsius as $T_C = 5(T_F - 32)/9$.

The unit of thermodynamic temperature or absolute temperature is the kelvin; it is the fraction 1/273.15 of the thermodynamic temperature of the triple point of water. The conversion of degrees Celsius to kelvin is $T_K = T_C + 273.15$. The thermodynamic temperature is used in the SI system because it is related to the energy possessed by matter. However, temperature itself does not express energy. Temperature should also not be confused with the heat output of a device. A solar collector at near-zero flow conditions can generate very high temperatures. The energy collected for this condition is quite low when compared to a high-flow moderate-temperature condition.

Heat is transferred from an object at higher temperature to an object at a lower temperature. The prerequisite of the transfer of heat is a temperature difference. The final temperature of those two bodies will be between the two original temperatures. Next important observation is that if we wish to restore the temperatures of those two bodies to their original ones, a refrigeration device will be required, which will need input of energy or consumption of work. It is similar to the physical observation that when an apple falls from a tree, some work will have to be performed to lift the apple again to the original height. Some other variables are important in the definition of thermodynamic problems, for example, specific weight, specific volume, entropy, and enthalpy.

Specific weight is defined as the weight per unit volume. The USCS units are pounds per cubic foot (lb/ft^3), and the SI unit of specific weight is kN/m^3. Because weight is a force, the specific weight is related to the density by Newton's second law. A common value is the specific weight of water on Earth at 5 °C, which is 9.807 kN/m^3 or 62.43 lb/ft^3. The term specific gravity is used for relative density, that is, the ratio of the density (mass of a unit volume) of a substance to the density of a given reference material.

It is important to observe that the density of substances varies with temperature and pressure. Specific volume is defined as the reciprocal of the density. The specific volume v of a system is the volume occupied by a unit mass of the system, and

$$v = \frac{1}{\rho} = \frac{V}{m} = \frac{volume}{mass}$$

The SI unit of specific volume is m^3/kg; other units are:

$1\,m^3/t = 1\,m^3/ton = 0.001\,m^3/kg$
$1\,L/kg = 1\,lit/kg = 1\,dm^3/kg = 0.001\,m^3/kg$
$1\,cm^3/g = 0.001\,m^3/kg$
$1\,in.^3/lbm = 3.6175 \times 10^{-5}\,m^3/kg$
$1\,ft^3/lbm = 0.0625\,m^3/kg$

The average density of human blood is $1060\,kg/m^3$, and the specific volume for that density is $0.00094\,m^3/kg$. Notice that the average specific volume of blood is almost the same as to the water: $0.0010\,m^3/kg$.

Weight rate of flow or mass flow rate is the mass of a substance that passes per unit of time; in SI system, its unit is in kg/s; it can be defined in USCS units as pounds per hour (lb/h), slug per second, or pound per second. Several problems may use force per time instead of mass flow rate, and there are mixed uses; for example, the flow of water in an open channel can be expressed in units of volume per time, but industrial liquid flows are usually specified in gallons per minute (gal/min). Airflow is generally measured in cubic feet per minute (ft^3/min).

The volumetric flow rate, sometimes called volume velocity, is the volume rate of flow F per unit area A, and the units vary with the application. In heat transfer, the units may be feet per hour (ft/h) for both liquids and gases, whereas in fluid mechanics, the velocity of liquids is generally in feet per second (ft/s), and the velocity of gases is in feet per minute (ft/min) or meters per second (m/s). The general relationship, with units of flow and area selected to give the desired units for the velocity, is $(V) = (volume\ flow\ rate)/(pipe\ or\ duct\ are)$. When the mass flow rate is known and the density can be assumed constant, there is an easy way to get the volumetric flow rate $Q = \dot{m}/\rho =$ mass flow rate in kg/s divided by density in kg/m^3.

Entropy is a quantity that measures how much system energy is available for conversion to work. The second law of thermodynamics implies that all natural processes of adiabatic systems (any thermodynamic system is part of a greater adiabatic system) proceed to a direction where the entropy of the adiabatic system increases, that is, the *entropy of the Universe tends to a maximum.*

Suppose a system with zero heat transfer to their surroundings (completely insulated). Then the entropy can be defined as an infinitesimal reversible change taken from a quantity of heat dQ at absolute temperature T. Its entropy is increased by $dS = dQ^0/T$, where the superscript "0" in the differential of heat denotes that the integral should be calculated during a *reversible* process. The area under the absolute temperature–entropy graph for a reversible process represents the heat transferred in the process, that is,

$$S_2 - S_1 = \int_1^2 \frac{dQ^0}{T}$$

For an adiabatic process, there is no heat transfer, and the temperature–entropy graph is a straight line; the entropy remains constant through the process. Any process in which there is *no change in entropy* is called as *isentropic*. S denotes entropy, and s denotes both the specific and molar entropies. The unit kJ/K is used for entropy, kJ/(kg·K) for specific entropy, and kJ/(kmol·K) for molar entropy. Therefore, specific entropy of a system is the entropy of the unit mass of the system and has the dimension energy/mass/temperature. Other possible units for specific entropy are

$1\,kJ/kg \cdot K = 1000\,J/(kg \cdot K)$
$1\,erg/g \cdot K = 10^{-4}\,J/(kg \cdot K)$
$1\,Btu/lb \cdot °F = 4186.8\,J/(kg \cdot K)$
$1\,cal/g \cdot °C = 4186.8\,J/(kg \cdot K)$

Analysis of engineering systems requires conducting the study of energy balance in order to obtain the operating values for temperature, pressures, and mass flow rate of flowing streams entering and leaving the engineering system. That is why it is required to use enthalpy, that is, for systems with flow. For systems without flow, the analysis can be done only by the internal energy. As long as there is flow, there is flow work, similar to the "force × velocity" of linear mechanical systems or "torque × angular speed" of rotating mechanical systems. When there are flowing streams, the fluid in the stream will exert about their neighbor packets the energy of that fluid, which is the internal energy plus the flow work. The name given to (internal energy + flow work) is enthalpy. Therefore, in engineering analysis when enthalpy is used instead of internal energy, everything is included, and the entropy should be treated only as thermodynamic property of a fluid. The enthalpy of a body is the sum of its internal energy and the product of its volume and the pressure exerted upon it.

In order to arrive to enthalpy as internal energy, one can consider a quasi-static process, which means the pressure in the system is equal to the external pressure during the process, then it is possible to calculate the flow of heat that causes the volume of this system to change, that is, work (W) in the isobaric process:

$$W = \int_{V_i}^{V_f} dW = -p \int_{V_i}^{V_f} dV = -p\left(V_f - V_i\right)$$

where:

V_i and V_f = The initial volume and final volume of the system

The specific internal energy is the internal energy of a system per unit mass of the system and thus has the same dimensions as energy/mass or enthalpy. To understand the concept of internal energy, we need to look at macroscopic and microscopic forms of energy.

PE and kinetic energy are macroscopic forms of energy. They can be visualized in terms of position and velocity of objects. In addition to these macroscopic forms of energy, a substance possesses several microscopic forms of energy. Microscopic forms of energy include those due to the rotation, vibration, translation, and interactions among the molecules of a substance. None of these forms of energy can be measured or evaluated directly, but techniques have been developed to evaluate the change in the total sum of all these microscopic forms of energy. These microscopic forms of energy are collectively called internal energy, customarily represented by the symbol U.

In engineering applications, the unit of internal energy is the Btu, which is also the unit of heat. The specific internal energy measures the energy content of a system due to its thermodynamic properties, such as pressure and temperature. The change of internal energy of a system depends only on the initial and final states of the system and not in any way on the path or manner of the change. This concept is used to define the first law of thermodynamics.

Continuing with the idea of calculating the flow of heat that causes the volume of a system to change, that is, calculating work (W) in an isobaric process, we can consider pressure constant, so

$$dQ = dU + pdV = d\left(U + pV\right) \tag{2.1}$$

Therefore, the quantity in equation (2.1) is called enthalpy. It takes the internal energy as a function of state, and as an extensive quantity, whose unit is joule (J). In this way, the infinitesimal amount of heat in an isochoric process is equal to the differential of internal energy, whereas in the isobaric process, it is equal to the enthalpy differential, such as equation (2.2):

$$h = u + Pv \tag{2.2}$$

$$dQ = dH \tag{2.3}$$

Enthalpy (from the Greek meaning "to heat") can be defined for a homogenous solid, with constant mass, as the mass of the system (m) multiplied by the specific enthalpy of the system, $H = mh$. It is measured in joules and is defined in terms of thermodynamic state functions (i.e., a state function, so enthalpy changes are path independent). The specific enthalpy of a working fluid is defined in terms of its intensive or unit mass properties, given by $h = u + Pv$, where u is the specific internal energy (the amount of energy per unit of mass of a substance), P is the pressure, and v is the specific volume (m^3/kg). The SI unit of specific enthalpy is J/kg. In the past, it was common to use the term *heat content* to be the thermodynamic quantity equal to the internal energy of a system plus the product of its volume and pressure. The term heat content is no longer used in modern thermodynamics, although it is still common—the preferred term is enthalpy.

Heat capacity is defined by the ratio of heat Q required to raise the temperature of an object in a small amount ΔT (i.e., $C = \Delta Q/\Delta T$). The specific heat c is the heat capacity divided by the mass m of the object; thus, $c = \Delta Q/m\Delta T$. Since an infinitesimal variation in entropy dS may be expressed as the ratio of an infinitesimal amount of the absorbed heat divided by the system temperature (i.e., $dS = dQ/T$), the following expression (2.3) for entropy in terms of specific heat capacity holds:

$$dS = \frac{dQ}{T} = \frac{m \cdot c \cdot dT}{T} \tag{2.4}$$

A very common situation occurs for several fluids (an example is boiling water or melting ice): when two phases (solid–liquid or liquid–gas) are present during heat transfer at constant pressure, a temperature change will not occur during the transition. The change in heat content of the substance must be obtained from tables of properties of the substance. The heat required to change the phase of an object with mass m is defined as equation (2.4):

$$Q = \pm mL \tag{2.5}$$

The heat required for phase change (L) is a quantity called heat of fusion for a phase change from solid to liquid, heat of vaporization or condensation for a phase change from liquid to gas, or the heat of sublimation for a phase change from gas to a solid. They are in general called latent heat. The sign of heat depends on the direction of the phase change. When ice is heated from 0 to 32 °F, the heat required in Btu/lb (as shown in Figure 2.1) is

$$\frac{Q}{m} = c\Delta t = 0.49(32 - 0)\text{Btu/lb} \tag{2.6}$$

Figure 2.1 Temperature variation with phase change.

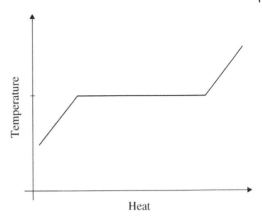

The specific heat of water is (in gram/degree Celsius)

State of water	Specific heat in cal/g·°C
Ice	0.50
Water	1.00
Steam	0.48

while the latent heat for water is

Change of state of water	Latent heat in cal/g
Melting/freezing	80
Boiling/condensation	540

Therefore, equation (2.5) can be calculated in SI as (2.6):

$$\frac{Q}{m} = c\Delta t = 0.5\big[0 - (-17.778)\big] = 8.89 \text{J/g} \tag{2.7}$$

As more heat is added, the temperature remains at 32 °F or at 0 °C until all of the ice has been changed into water. The energy required to affect the change, termed the latent heat of fusion or melting, is 80 cal/g or 144 Btu/lb. Suppose that more heat is added, atmospheric pressure is constant, and the temperature rises to 212 °F or 100 °C. The heat required to raise the water temperature is

$$\frac{Q}{m} = c\Delta t = 1.0(212 - 32) = 180 \text{Btu/lb} = 1.0(100 - 0) = 100 \text{J/g} \tag{2.8}$$

Table 2.1 Specific weight and specific heat (thermodynamic properties).

Substance	Specific weight w (lb/ft³) $1\,kg/m^3 = 0.062428\,lb/ft^3$ $1\,lb/ft^3 = 16.018463\,kg/m^3$	Specific heat C_p (Btu/lb·°F)—same value in (cal/g·°C)
Air	0.075 (0 psig, 70 °F)	0.24
Aluminum	165	0.22
Brick	125	0.20
Cast iron	450	0.12
Concrete	144	0.22
Flue gas	0.041 (0 psig, 500 °F)	0.26
Glass	155	0.19
Ice	57.5	0.49
Steam	0.037 (0 psig, 212 °F)	0.48
Steel	490	0.12
Stone	165	0.21
Water	62.5	1.00
White pine	27	0.67

For further addition of heat, the temperature remains at 212 °F until all the water has been converted into steam. The energy required for this change of phase, called the latent heat of vaporization or boiling, is 970 Btu/lb at an atmospheric pressure of 14.7 psi or is 540 cal/g at sea level. Air contains various amounts of moisture as water vapor at very low pressures. The presence of water vapor has little effect on the specific heat of the air, but if water is added to the air (clothes drier) or condensed from the air (air conditioner, dehumidifier), an energy transfer of about 1060 Btu is required per pound of water. A brief list of values of specific heat is given in Table 2.1. The specific heat of gases (and to a much lesser extent, of liquids and solids) depends on the process by which heat is added or removed.

2.2.1 Heating Value

The heating value represents the chemical energy stored in a fuel. It can be measured by burning a small sample of the fuel in a controlled oxygen environment. The heat transferred to a water jacket surrounding the sample is termed the higher heating value (HHV) of the fuel. When a fuel is burned, the moisture present is converted into water vapor. The latent heat of vaporization must be subtracted from the HHV, and the result is defined as the lower

Table 2.2 Higher heating value for selected fuels.

Fuel	HHV
Anthracite coal	13,000 Btu/lb
Bituminous coal	13,500 Btu/lb
Lignite	6,500 Btu/lb
Peat	1,200 Btu/lb
Fuel oil	137,000 Btu/gal
Natural gas	1,050 Btu/ft^3
Propane	92,000 Btu/gal
Butane	102,050 Btu/gal
Wood, mixed varieties	7,200 Btu/lb
Hardwoods	25 $\times$ 10^6 Btu/cord
Softwoods	16 $\times$ 10^6 Btu/cord
Carbon (C)	14,600 Btu/lb
Hydrogen (H$_2$)	62,000 Btu/lb
Sulfur (S)	4,050 Btu/lb

heating value (LHV). For example, peat (decayed vegetable matter) is the first step in the transformation of plants and trees into coal. The HHV is very low due to the high moisture content. Therefore, peat must be dried before being used as a fuel. Lignite is a better fuel because geological conditions have reduced the moisture content, increased the carbon content, and raised the heating value to a very usable level.

Table 2.2 shows the chemical energy in some fuels, where the HHV is usually given in USCS units as Btu/lb for solids, Btu/gal for liquids, and Btu/ft^3 for gases. In the SI system, the name of heating value is calorific value and has units of J/kg or J/m^3, but most of the fuels are defined in industry by their Btu content.

Table 2.2 shows the heating value of several fuels. Bituminous and anthracite coals are products of final stages in geologic formation and have heating values above 12,000 Btu/lb with moisture below 5%. The anthracites contain about 90% carbon and are found only in mountainous regions where thrust pressures were very large. Petroleum contains about 85% carbon and 12% hydrogen; therefore, natural gas can be found in conjunction with crude oil (it can also be found independently).

The primary component for natural gas is methane, and the heating value varies from 1000 Btu/ft^3 (volume measured at 14.7 psig and 60 °F) to 1100 Btu/ft^3. Because of such variation, it is often sold in *therms* (1 therm = 100,000 Btu) rather than on a volume basis. Propane and butane are liquefied petroleum gases (LPGs); they are stored under pressure as liquids but vaporize by heat transferred from their surroundings and will burn as a gas. Both propane and butane are very common for daily use. The pressure of propane storage is 189 psig at 100 °F and 38 psig at 0 °F. Butane is stored at lower pressures: −52 psig at 100 °F and −7 psig at 0 °F. Very pure forms of butane, especially isobutane, can be used as refrigerants and have largely replaced the ozone-layer-depleting halomethanes, for instance, in household refrigerators and freezers.

The heating value of wood averages around 7200 Btu/lb at 20% moisture. Wood is usually sold by *cord*s (one cord is a stack 8 ft × 4 ft × 4 ft). Of course, a cord has variation in their heating value, depending on the density of the wood.

2.2.2 First and Second Laws of Thermodynamics and Thermal Efficiency

The first law of thermodynamics is expressed as the net heat added to a system minus the net produced work by the system must be equal to the increase in stored energy within the system. The second law of thermodynamics, on the subject of energy conversion, is that work may not be produced spontaneously by a cyclic engine only with a heat reservoir. Therefore, most power plants, gas turbines, and jet and car engines (all cyclic engines) must be in contact with another heat reservoir where it rejects heat. For a cyclic engine to operate, typically a substance may start as a solid. It is heated up at constant pressure until it all becomes gas, depending on the prevailing pressure. The matter will pass through various phase transformations, that is, (i) solid; (ii) mixed phase of liquid and solid; (iii) subcooled or compressed liquid (means it is not about to vaporize); (iv) wet vapor or saturated liquid–vapor mixture, the temperature will stop rising until the liquid is completely vaporized; and (v) superheated vapor (a vapor that is not about to condense). In order to characterize all these possible variations, a plot of intensive properties can be used, that is, in simple and homogeneous thermodynamic systems, intrinsic properties can be represented in a Cartesian three-dimensional space, depicted as a surface as in Figure 2.2.

The thermodynamic properties of a pure substance can be related by the general relationship $f(P,V,T) = 0$, which represents a surface in (P, V, T) space. It is more useful for qualitative insights than actual quantitative evaluations. The state of a working fluid is defined completely by knowing two independent properties of the fluid. This makes it possible to plot the state changes on two-dimensional diagrams such as:

- A pressure–volume (P-V) diagram
- A temperature–entropy (T-s) diagram
- An enthalpy–entropy (h-s) diagram

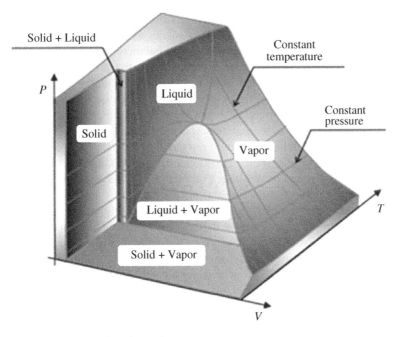

Figure 2.2 P-V-T surface for a substance.

When considering a transition between two points of a thermodynamic state, the following processes are often used:

- Adiabatic $dq = 0$ no heat transfer
- Isentropic $ds = 0$ constant entropy
- Isothermal $dT = 0$ constant temperature
- Isochoric $dv = 0$ constant volume
- Isobaric $dP = 0$ constant pressure

2.3 Fundamental Laws and Principles

When energy is transferred into or out of a system, principles of thermodynamics describe the nature of such changes, the external effects in terms of work and heat, and the properties of matter in the system, such as fluid pressure and temperature. The law of conservation of energy is useful for this analysis. Energy can be neither created nor destroyed (i.e., the total energy within the universe remains constant). Energy can be stored in a number of ways and can be transferred from one system to another by various processes.

The first law can be used in a general manner, considering that in addition to work and heat, electrical energy and chemical energy, as distinct energy forms, can move in or out of systems. They should be balanced by corresponding changes in other energy forms. The second law of thermodynamics is a straightforward application of a principle of physics: in a closed system, it is not possible to finish any real physical process with as much useful energy as it had originally, that is, something is always wasted and a perpetual motion machine is impossible. The second law was formulated after nineteenth-century engineers noticed that heat could not pass by itself from a colder body to a warmer body. According to the philosopher of science Thomas Kuhn, two scientists, Rudolf Clausius and William Thomson (Lord Kelvin), first put the second law into words.

The Clausius statement is that "heat cannot itself pass from a cold to a hot body." The quantum physicist Richard P. Feynman mentioned in his lectures that French physicist Sadi Carnot discovered the second law 25 years earlier than Clausius and Thomson. Therefore, it seems that the second law was proposed before the first law. The first law does not prohibit energy conversion at 100% efficiency. On the other hand, the second law of thermodynamics prohibits a conversion with 100% efficiency. We can easily observe these considerations with our current technology. For example, an air conditioner transfers energy from a cold body (the interior) to a hot body (the surroundings). The first law would permit the process to operate without input work as long as the heat rejected to the surroundings balanced the heat gained by the interior. The second law prohibits such a device and imposes minimum limits on the input work required.

There is a statement defined as Kelvin–Planck statement of the second law, which is equivalent in consequences to the Clausius statement: *It is impossible for any device to operate in a cycle and produce work while exchanging heat only with bodies at a single fixed temperature.*

A consequence of the second law is the Carnot principle, in which it is shown that with heat supplied at a temperature T_H and rejected at a temperature T_L, the maximum efficiency η at which the heat supplied can be converted into work is

$$\eta_{max} = 1 - \frac{T_L}{T_H} < 1 \tag{2.9}$$

where:

T_L and T_H = Thermodynamic temperatures (in kelvin)

The Carnot cycle uses only two thermal reservoirs—one at high temperature T_H and the other two at temperature T_L. If the process goes through, such as the working fluid during the cycle is reversible, the heat transfer must take place with no temperature difference. The Carnot cycle consists of a reversible isothermal expansion, a reversible adiabatic expansion, a reversible isothermal compression, and a reversible adiabatic compression. Then, the efficiency can be calculated as (net work done)/(energy absorbed as heat).

Equation (2.10) is used to calculate the efficiency, where the temperature is expressed in the Kelvin scale. Since a degree on the Fahrenheit scale is 1/180 of the interval between the freezing point and the boiling point of water, the absolute zero is defined as −459.67 °F. Equation (2.11) shows a modification of Carnot efficiency calculation using the Fahrenheit scale (do not use Celsius for this calculation):

$$\eta_{mas} = 1 - \frac{T_L}{T_H}, \text{ for temperature in the Kelvin scale} \tag{2.10}$$

$$\eta_{mas} = \frac{T_H - T_L}{T_H + 460}, \text{ for temperature in the Fahrenheit scale} \tag{2.11}$$

In a steam boiler, the efficiency of energy conversion from the fuel to steam is limited to 100% by the first law limits (practical considerations set the limit at around 90%). If the steam is used for space or process heat, the limitation is again 100%, a value that can be approached as closely as possible. If the steam is used to produce work (and indirectly, electrical energy) with a heat engine, the maximum efficiency is limited by the second law.

2.3.1 Example of Efficiency in a Power Plant

In a steam electric power station, heat is supplied at 1000 °F (steam temperature) and rejected to the condenser cooling water at a temperature of 60 °F (Figure 2.3). What is the maximum possible efficiency of conversion?

$$\eta_{max} = \frac{1000 - 60}{1000 + 460} = 64\% \tag{2.12}$$

Thus, 36% of the energy supplied is wasted to the cooling water. This loss, by the second law, cannot be reduced except by changes in operating temperatures. The actual efficiency is considerably less, and when combined with the boiler conversion efficiency and losses in the conversion from shaft work to electrical energy, the overall efficiency of the modern steam electric station is around 35%. At the time electrical energy is delivered to the point of use, less than one-third of the energy in the fuel remains.

A heat engine is a device that converts heat energy into mechanical energy. More exactly, it is a system that operates continuously in which only heat and work may pass across its boundaries. The operation of a heat engine can be represented by a thermodynamic cycle such as Otto, Diesel, Brayton, Stirling, or Rankine cycle. For electrical power conversion, the most relevant cycles are

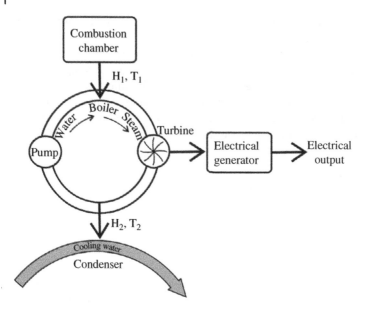

Figure 2.3 Steam electric power station maximum efficiency.

the Rankine (vapor–liquid system, typical of steam power plants) and the Brayton (gas turbine-based power plants).

The Carnot cycle is an idealized representation of the operation of a steam engine. To understand it, an ideal gas could be imagined to be confined in a cylinder by a piston and allowed to absorb heat from a reservoir held at T_H and reject heat to a reservoir held at T_L. The gas goes through a cyclical path (the Carnot cycle) depicted in Figure 2.4, which performs work on the surroundings. The overall effect is to convert heat into useful work. Assume absolute temperature or thermodynamic temperature (kelvin); the following transformations occur along the cycle:

- Reversible isothermal expansion at T_H, segment 2 to 3
- Reversible adiabatic expansion (from T_H to T_L), segment 3 to 4
- Reversible isothermal compression at T_L, segment 4 to 1
- Reversible adiabatic compression (from T_L to T_H), segment 1 to 2

Figure 2.5 shows the scheme of a simple power plant; it consists of a boiler, turbine, condenser, and pump. Fuel is burned in the boiler and heats the water to generate steam. The steam is used to rotate the turbine, which powers the generator. Electrical energy is generated when the generator windings rotate within a magnetic field.

After the steam leaves the turbine, it is cooled down to their liquid state in the condenser. The pump pressurizes the liquid before going back to the boiler.

Figure 2.4 Ideal Carnot cycle.

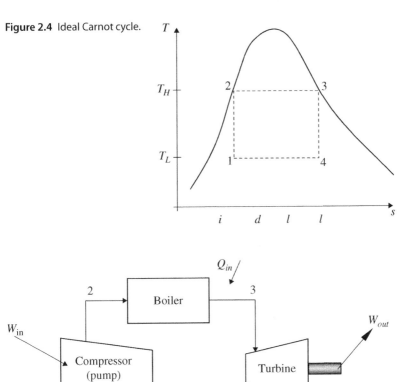

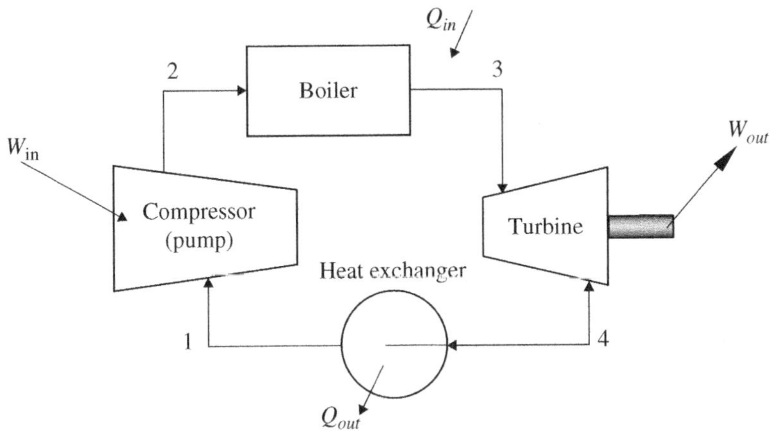

Figure 2.5 Principles of a simple power plant.

The clockwise path on Figure 2.4 can evaluate the power cycle balancing as follows:

- The heat absorbed by the working fluid during isothermal process 2–3 (high-temperature reservoir) is

$$q_{23} = T_H \left(s_3 - s_2 \right) \Rightarrow Q_{in}$$

- During isothermal process 4–1 (low-temperature reservoir), the heat given up by the working fluid is

$$q_{41} = T_L \left(s_1 - s_4 \right) = T_L \left(s_2 - s_3 \right) = -T_L \left(s_3 - s_2 \right)$$

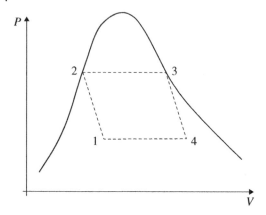

Figure 2.6 Pressure versus volume curve for an energy conversion cycle.

- The work produced by the cycle is

$$W_{cycle} = q_{23} + q_{41} = T_H\left(s_3 - s_2\right) - T_L\left(s_3 - s_2\right)$$

The thermal efficiency can be calculated by

$$\eta = \frac{work\ produced\ by\ the\ cycle}{heat\ supplied\ to\ the\ cycle} = \frac{q_{23} + q_{41}}{q_{23}} = \frac{T_H - T_L}{T_H} \qquad (2.13)$$

The product of pressure multiplied by volume represents a quantity of work. This can be represented by the area under a P-V curve as plotted in Figure 2.6, where the area enclosed by the four curves represents the net work done by the engine during one cycle. By using the second law of thermodynamics, it is possible to show that no heat engine can be more efficient than a reversible heat engine working between two fixed-temperature limits. Due to mechanical friction and other irreversibilities, no cycle can actually achieve this level of efficiency.

2.3.2 Practical Problems Associated with Carnot Cycle Plant

The Carnot cycle is not a practical engineering design because of some real-life problems:

- The isentropic expansion in a turbine from segment to 3 to 4—a major concern is related to the quality of the steam inside the turbine because a high moisture content results in blade erosion.
- The isentropic compression process in a pump from 1 to 2—it is very difficult to design a condenser and transmission line system that precisely control the quality of the vapor to achieve an isentropic compression.
- The two-phase pump compressor that is required is very difficult to implement.

Figure 2.7 Modification of cycle with high-pressure steam.

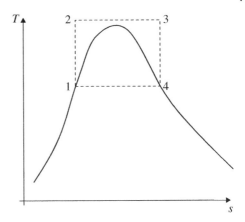

A possible solution for these problems is depicted in Figure 2.7. Such a cycle requires compression (1–2) of the liquid at a very high pressure (exceeding 22 MPa for steam), which is not practical. In addition, maintaining a constant temperature above the critical temperature is also very difficult since the pressure will change continuously. A major problem is the maximum temperature limitation for the cycle and metallurgical limitations imposed by materials used within the boiler and turbine.

2.3.3 Rankine Cycle for Power Plants

W.J.M. Rankine, a Scottish engineer, made possible a thermodynamic cycle, which has his name, and it is the cornerstone of modern steam power plants. The main components of a Rankine cycle steam plant are:

- A boiler that generates steam, usually at a high pressure and temperature
- A turbine that expands the steam to a low pressure and temperature, thereby producing work
- A generator driven by the steam turbine
- A condenser that cools the steam to a liquid so that it can be pumped back into the boiler
- A feed pump
- Feed heaters, which preheat the water before it enters the boiler
- A reheater, which is part of the boiler, which reheats the steam after it has partially expanded

Figure 2.8, a temperature–entropy diagram, illustrates the state changes for the Rankine cycle. Such a cycle avoids transporting and compressing two-phase fluid by trying to condense all fluid exiting from the turbine into saturated liquid before it is compressed by a pump. An actual power plant introduces regenerative and reheat modifications in the basic Rankine cycle.

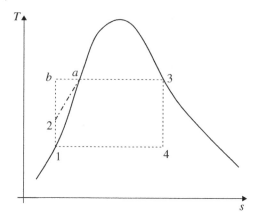

Figure 2.8 Ideal Rankine cycle.

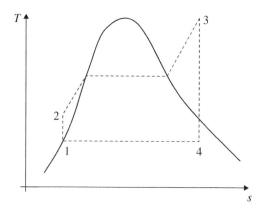

Figure 2.9 Rankine cycle with superheating.

It is evident in the T-s diagram that the ideal Rankine cycle is less efficient than a Carnot cycle for the same maximum and minimum temperatures. The Rankine cycle work is represented by the area 2-a-3-4-1-2, which is less than the Carnot cycle work represented by the area 2-b-3-4-1-2. Figure 2.9 illustrates a typical superheated Rankine cycle, the following path describing the cycle:

- 1–2: Pump ($q = 0$); isentropic compression (pump)

$$W_{pump} = h_2 - h_1 = v\left(p_2 - p_1\right)$$

- 2–3: Boiler ($W = 0$); isobaric heat supply (boiler)

$$Q_{in} = h_3 - h_2$$

- 3–4: Turbine ($q = 0$); isentropic expansion (steam turbine)

$$W_{out} = h_3 - h_4$$

- 4–1: Condenser ($W = 0$); isobaric heat rejection (condenser)

$$Q_{out} = h_4 - h_1$$

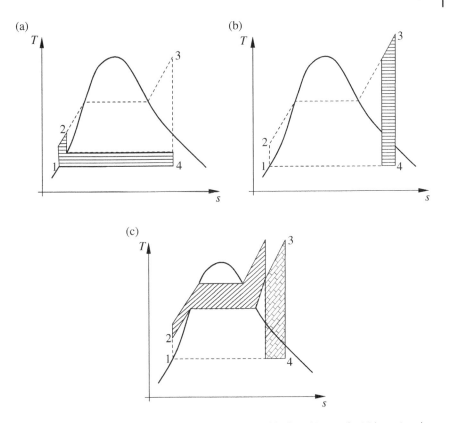

Figure 2.10 Thermal efficiency improvement with modified Rankine cycle: (a) lowering the condensing pressure (lower condensing temperature, lower T_L), (b) superheating the steam to higher temperature, and (c) increasing the boiler pressure (increasing boiler temperature, increases T_H).

Figure 2.10 shows three ways to improve the thermal efficiency: (i) lowering the condensing pressure (lower condensing temperature, lower T_L), (ii) superheating the steam to higher temperature, and (iii) increasing the boiler pressure (increasing boiler temperature, increases T_H). By increasing the steam temperature, the point 3 is shifted up and results in dry saturated steam from the boiler, which goes through a second bank of smaller-bore tubes within the boiler, until the steam reaches the required temperature. Increasing the steam temperature increases the cycle efficiency, and reduces the moisture content, for the turbine exhaust. Sometimes two stages of reheating are connected in tandem.

The best way to increase the boiler pressure without increasing the moisture content in the exiting vapor is to reheat the vapor after it exits from a first-stage turbine; then this vapor should be redirected into a second turbine. Regeneration helps to improve the Rankine cycle efficiency by preheating the feedwater into the boiler. Open or closed feedwater heaters can achieve regeneration. In open feedwater heaters, a fraction of the steam exiting a high-pressure turbine is mixed with the feedwater at the same pressure. In a closed system, the steam bled from the turbine is not mixed directly with the feedwater. Therefore, the two streams can be at different pressures. Such industrial details are outside the scope of this book.

Fluids other than water and steam may be used in the Rankine cycle if they are considered more appropriate for a particular process. For example, ocean thermal conversion plants depend for their operation on the temperature difference between the hotter surface water (up to 27 °C) and the colder deep water (down to 4 °C). The working fluid for these plants is ammonia because of its low boiling point, but the cycle used is called as *Kalina cycle*.

2.3.4 Brayton Cycle for Power Plants

The basic gas turbine cycle was initially proposed by George Brayton around 1870 and is currently used for gas turbine engine-based aircraft propulsion and electric power generation. Gas turbines may be used as stationary power plants to generate electricity or as stand-alone units, or may be in conjunction with steam power plants taking their high-temperature exhaust. In these plants the exhaust gases serve as a heat source for the steam. Steam power plants are considered external combustion engines, in which the combustion takes place outside the engine. Figure 2.11 shows a gas turbine where air is drawn into a compressor, which raises the temperature and pressure. The high-pressure air proceeds into the combustion chamber, where the fuel is

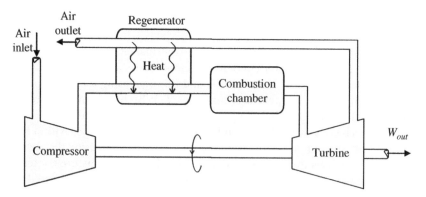

Figure 2.11 Brayton cycle-based gas turbine.

burned at constant pressure. The changes in properties of the working fluid brought about by the addition of fuel or the presence of combustion products are ignored.

The resulting high-temperature gases enter the turbine and expand to atmospheric pressure through a row of nozzle vanes. This expansion causes the turbine blade to spin, which then turns a shaft inside a magnetic coil. When the shaft is rotating inside the magnetic coil, electrical current is produced. The following points are considered in an analysis of the Brayton cycle:

- The working substance is air, which is treated as an ideal gas throughout the cycle.
- The combustion process is modeled as a constant-pressure heat addition.
- The exhaust is modeled as a constant-pressure heat rejection process.
- The gas turbine is a rotatory device working at a nominal steady state.
- Spark ignition is used for startup, since air compressor output temperature is not high enough to ignite the fuel.

In the ideal air-standard Brayton cycle, air is assumed to follow the four processes depicted in Figure 2.12. The cycle path is composed of isentropic compression, constant-pressure heat input from the hot source, isentropic expansion, and constant-pressure heat rejection to the environment. In the combustion chamber, heat is added to the gas at constant pressure, the density decreases, and the specific volume and temperature increase. Entropy is also increased, since combustion is not a reversible process. In the turbine the situation is the opposite of that in the compressor.

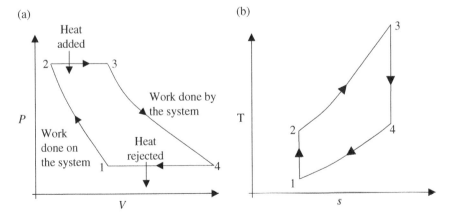

Figure 2.12 Ideal Brayton cycle.

The pressure decreases and specific volume increases. The temperature decreases and in an ideal expansion the entropy is constant. The encircled area in the P-V diagram represents the net produced work by the gas turbine. The efficiency of the gas turbine is the ratio of the net produced work and heat power added:

$$\eta = \frac{W_{net}}{Q_{in}} = 1 - \frac{1}{r_p^{(k-1)/k}} \tag{2.14}$$

where:

r_p = The pressure ratio in the turbine
k = The ratio of specific heat

At a higher pressure ratio, the gas turbine has higher efficiency. The equation is valid only for ideal gas turbines with no friction and reversible processes. A gas microturbine used for stationary power generation is a combination of a small gas turbine and a directly driven high-speed generator placed on the same shaft, without a gearbox and optimized mechanical design to minimize friction losses, usually applied in combined heat and power applications. Most designs have heat recuperation with excellent thermal efficiencies, good power output efficiencies, and high reliability. Chapter 8 covers microturbines in detail.

2.3.5 Geothermal Energy

There are two types of geothermal energy: the surface geothermal and the deep geothermal [4–7]. A deep geothermal power plant is an open thermodynamic system that receives a mass flow rate of the geothermal fluid. That can be (i) superheated steam (such as geysers in California or Larderello in Italy), (ii) steam and water mixture (such as in Wairakei, New Zealand), or (iii) geopressured liquid water at high temperature (such as Port Aransas, Texas). Those three can be classified as (i) high enthalpy geothermal energy, with the temperature of the liquids at 150 °C usually dry steam; (ii) middle enthalpy geothermal energy, with temperature between 80 and 150 °C; and (iii) low enthalpy geothermal energy, with temperatures below 80 °C.

One important issue for geothermal power plants is to make them deliver maximum work and power. The modeling considerations are usually very abstract and difficult to formulate. We can introduce a term called rate of entropy generation, $\dot{\Theta}$, which is always positive and should be related to the mass flow rate of geothermal fluid $\dot{m}$; reservoir entropy s; specific enthalpy h, with atmospheric discharge temperature T_e; specific entropy s_e; and specify enthalpy h_e [4–6].

Applying the laws of thermodynamics, we can write equation (2.15), resulting in equation (2.16), which gives the rate of entropy $\dot{\Theta}$:

$$\dot{Q} - \dot{W} = \dot{m}\left(h_e - h\right) \tag{2.15}$$

and

$$\dot{m}\left(s_e - s\right) - \frac{\dot{Q}}{T_e} = \dot{\Theta} > 0 \tag{2.16}$$

In order to maximize the power, the following two conditions are required:

1) The rate of entropy generation $\dot{\Theta}$ is minimized to a value close to zero.
2) The state of effluents is at the same state of the environment.

It must be noted that for vapor–liquid systems, such as steam water, the numerical value for the exergy at dead state is very close to zero, which implies on equation (2.14):

$$W_{max} \approx m\left(h - T_e s\right) \tag{2.17}$$

Another very promising low enthalpy solution is the surface geothermal energy resource. The surface geothermal energy is a renewable source that comes from solar radiation. This energy keeps the continents and oceans warmed up, and it is stored under the earth's surface as heat at low depths. At depths equal to or greater than 15 m, the temperature of the subsoil is insensitive to any temperature variation at the earth surface. At depths greater than 15 m, this temperature does not suffer as much the effects with seasonal variations and becomes equivalent to the average annual temperature on the ground surface. For example, in the United States, the underground temperature at these depths varies between 10 and 21 °C, depending on the geographical position and altitude. This thermal energy reservoir at constant temperature can be used with or without heat pumps reducing electrical consumption [5–8]. A small installation around a house can provide enough energy to acclimate the internal environment of an area room with 10 m². For this kind of application, a peak power tracking can be implemented using hill climbing control (HCC) in order to optimize the pump pressure of the water traveling in the pipes underground.

2.3.6 Kalina Cycle

Low-temperature (100–150 °C) geothermal heat sources have very large potential as renewable energy source. The amount of energy available is huge, but due to the low temperature, the direct electricity conversion efficiency is low. For this energy conversion, two types of cycles are discussed in the literature. The first one is the Organic Rankine cycle (ORC), which is similar to a steam Rankine cycle but uses an organic working fluid instead of water. Organic fluids have lower critical temperatures than water and are better suited for power production from low-temperature heat sources. A second type is the Kalina cycle, where a mixture of ammonia and water is used and the concentration of ammonia changes in the cycle. A Kalina cycle always requires a separator, where two-phase fluid is separated into its liquid a vapor part.

The Kalina thermodynamic cycle can be used with a working fluid composed of 70% ammonia and 30% water mixture through a cyclic engine, for electrical power generation. The ammonia/water mixture in the condenser is at 3 bar and 10 and 21 °C. It enters a pump, its pressure rises up to 13 bar, and its temperature rises. The pump can be powered by solar energy or by another local renewable energy source (wind or hydro). The ammonia/water mixture needs to be heated up to 89 °C (superheated vapor) in order to expand to a 13-bar turbine of 90% isentropic efficiency. This is accomplished by passing this mixture through heat exchangers. The ammonia/water mixture flow is taken to the turbine at a prescribed kJ/kg rate in order to produce a given shaft power, and the cycle is closed.

Figure 2.13 shows the scheme for the Kalina cycle. Initially the working fluid, with an intermediate concentration of ammonia, comes from the end of the thermodynamic cycle, and then it is pumped to high pressure, preheated in a recuperator that connects the out stream of the turbine. In sequence this fluid is heated by the brine (geothermal external fluid) with a heat exchanger; vapor starts at this point, but the working fluid still has a low concentration of ammonia. After the heating, a controlled mix with a system that previously separated the ammonia from the output of the turbine (separator) is engaged for allowing a two-phase region and optimal mix of water and ammonia (which are already in vapor phase). This working fluid is expanded in the turbine, cooled in the recuperator, and cooled in the condenser. During the expansion the mix changes, because the ammonia has a lower temperature of condensation, and the changing makes the turbine operate in a better energy conversion performance at low temperature. Because such a mixture of fluids is used, the evaporation at constant pressure does not take place at constant temperature, so less irreversibilities are created in the Kalina cycle than in a subcritical ORC. The working fluid is heated by the cooling brine, and the mix of ammonia is controlled by a separator.

2.3.7 Energy, Power, and System Balance

Energy E may be defined as the "capacity to do work," such as the lifting of a weight. To lift a 10-lb weight, a distance of 5 ft requires $5 \times 10 = 50$ ft-lb of energy. In the metric systems we have to consider the gravitational field strength (9.81 N/kg on Earth), also called as the acceleration of gravity,

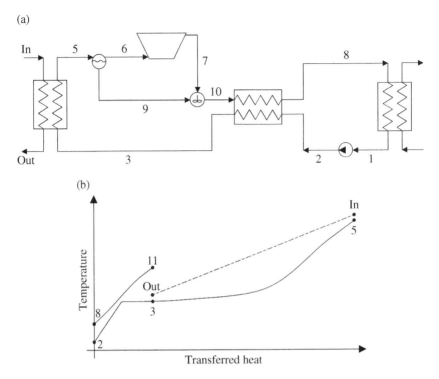

Figure 2.13 (a) Kalina cycle scheme; (b) Kalina cycle temperature–heat diagram.

and such *PE* is calculated as equation (2.15), with mass (kg) × gravitational field strength (N/kg or m/s^2) × height (meters), and the energy in SI is given in Joules:

$$PE = m \cdot g \cdot h \tag{2.18}$$

Energy may be converted from one form to another, stored, and transferred from one body to another. Energy may take a variety of forms: chemical, electrical, mechanical (work, kinetic, potential, flow), and thermal (heat, internal). Most problems in energy conversion require detailed knowledge of the subject and careful attention to limitations of the fundamental relationships. It is worthy to note that for residential and commercial applications, the range of working fluids is limited (air, water, and flue gas). Flue gas is a mixture of air and gas products resulting from combustion flowing through a chimney. For most residential and commercial applications, the pressures are generally moderate and the processes involved are relatively simple. Consequently, it is possible to simplify the subject considerably while retaining the ability to solve most energy problems in this field.

Residential and industrial applications are concerned with only three forms of energy: chemical, electrical, and heating [9, 10]. Other forms of energy may be of interest, but they are not usually involved in industrial problems:

- Chemical energy (*CE*) is energy such as that stored in fuels, usually identified by the term HHV and measured in Btu. 1 Btu is the energy required to increase the temperature of 1 lb of water 1 °F; 1 Btu = 1055.06 J, and the relation to power in Watts is the time rate of Btu or the integral of power, that is, 1 Btu = 0.293071 Wh.
- Electrical energy (*EE*) is a form of energy entering a system as electricity, measured in kilowatt-hours (1 kWh = 3413 Btu).
- Heat (*Q'*) is a form of thermal energy that is transferred across system boundaries by a temperature difference, measured in Btu.

If a hot body is placed close to a cold body, heat flows from the higher- to the lower-temperature body, and the energy content of that body must increase. For most problems (solid bodies or contained liquid and constant-pressure gas), the energy change within the body can be treated as a change in the heat content of that body, that is, the temperature of the substance increases in proportion to the quantity of energy transferred. For problems where the working fluid changes their state, the enthalpy must be calculated, instead of the heat content.

Power is the rate of energy transfer or conversion. Horsepower (hp), kilowatts (kW), and Btu/h are units of power, not energy. Application of the first law involves energy balances. Usually, change of energy (energy rate) is used and is generally more convenient. Table 2.3 gives some useful energy and power conversion factors.

The statement of the first law is a powerful tool for solving energy problems and helps toward a better understanding of energy transformations. For example, internal energy stored in molecular bonds can be converted into kinetic energy; PE can be converted to a kinetic or an internal energy; and so on. Energy can also be transferred from one point to another or from one body to a second body. Energy transfer can occur by flow of heat, by transport of mass

Table 2.3 Energy and power conversion factors.

Energy	Power
1 kWh = 3,413 Btu	1 hp = 2,545 Btu/h
1 Btu = 778 ft-lb	1 kW = 3,413 Btu/h
1 hp-hr = 2,545 Btu	1 hp = 0.746 kW
1 kWh = 1.34 hp-hr	1 hp = 33,000 ft-lb/min

(transport of mass is known as convection), or by performance of work. Consider a stream with a mass flow rate $\dot{m}$. The material in the stream carries their kinetic, potential, and internal energy with it. Therefore, rates of energy transport (units: energy/time, e.g., J/s, Btu/h) that accompany the material flow in a process stream can be calculated as follows:

$$\dot{E}_k = \frac{1}{2}\dot{m} \cdot u^2 \quad \dot{E}_p = \dot{m} \cdot g \cdot z \quad \dot{E}_i = \dot{m} \cdot \hat{U}$$

So, the rate of energy for an input or output stream is expressed by equation (2.17); the units of energy/time are equivalent to power:

$$\dot{E}_k = \frac{1}{2}\dot{m} \cdot u^2$$

$$\dot{E}_{k+p+i} = \dot{m} \cdot \left[\hat{U} + \frac{1}{2}u^2 + g \cdot z \right] \tag{2.19}$$

The general energy balance for a process can be expressed as

Accumulation of energy in system = input of energy into system − output of energy from system.

Rate of Energy Accumulation in a System =

$$\sum_{\text{input streams}} \dot{m}_i \left(\hat{U} + \frac{u^2}{2} + g \cdot z \right) - \sum_{\text{output streams}} \dot{m}_o \left(\hat{U} + \frac{u^2}{2} + g \cdot z \right) + \dot{Q} - \dot{W} \tag{2.20}$$

The energies can be transferred into or out of the system by flow of mass through process streams that bring the various forms of energy with them. In addition, they can be transferred by performance of work or by flow of heat. In the first term on the right of equation (2.18), it runs over all incoming streams, and in the second the index or it runs over all outgoing streams; rate of heat is positive, and rate of work is negative in this equation.

The first law of thermodynamics as applied to residential and commercial energy problems, without involving any work (i.e., turbines or mechanical shaft power), is simply stated as equation (2.21), that is, the energy added to a system equals the energy leaving the system plus the increase in stored energy within the system:

$$E_1 = E_2 + \Delta E_S \tag{2.21}$$

where the subscripts 1, 2, and s refer to energy entering, leaving, and stored, respectively. Expanding that equation for the forms of energy to be most concerned with and using energy rates gives equation (2.22):

$$Q_1 + CE_1 + EE_1 + H_1 = Q_2 + CE_2 + EE_2 + H_2 + \Delta H_S \qquad (2.22)$$

where:

Q = Heat transferred by temperature difference across system boundaries, Btu/h
CE = Chemical energy, Btu/h = (W) (HHV of fuel in Btu/lb)
EE = Electrical energy in Btu/h = 3413 (kW)
H = Heat content of fluids crossing system boundaries, Btu/h = $W(C_p)(T)$
ΔH = Change in stored heat content, Btu/h = $M(C_p)(T'/t)$
M = Weight of the storage material in pounds
T'/t = Change in temperature T (°F) in a time period t (hours)

2.4 Examples of Energy Balance

This section covers some examples of energy balance, most of them based on the use of the first law of thermodynamics, which states that energy cannot be created or destroyed but it may change from one form to another. Several engineering problems can be resolved by using simple modeling based on this first law. For example, when materials are used for cooling purposes, it is important to remove energy quickly without the coolant changing their temperature range too much. Water is, of course, ideal, but other materials are used. On the other hand, sometimes it is important to insulate the loss of heat energy, for example, storage heaters rely on the large mass and the high specific thermal capacity of the hot materials. The heater needs to retain energy for a significant time for releasing it into a room later on the day.

The first law of thermodynamics deals with heat, work, and internal energy. The second law of thermodynamics deals with the direction and efficiencies of thermodynamic processes, because in several situations the conversion is not reversible. A very simple daily use of the second law is to understand how brakes operate to stop a car, that is, applying the brakes will convert mechanical energy into heat, but that is not reversible in the case of braking a car with friction pads. It is possible to convert mechanical energy into heat with 100% efficiency. However, it is not possible to convert heat into mechanical energy with 100% efficiency. This section deals with simple examples to illustrate energy balance.

2.4.1 Simple Residential Energy Balance

The following examples were considered from Ref. [1]. Suppose that the residence shown in Figure 2.14 is kept at a constant temperature with no

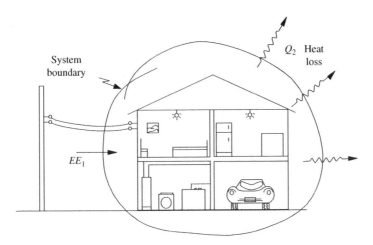

Figure 2.14 Residence at constant temperature.

change in energy stored and only two energy flows crossing the boundaries. Assume an average electrical power use of 1.5 kW. What is the heat loss from the house?

Use equation (2.22) with $EE = Q_2$ (all other terms are zero since there is no other energy flow outside the boundary): $Q_2 = 1.5 \times 3413$ so the heat loss is 5120 Btu/h. Even though nothing was stated about how the energy was being used in the house, the energy balance tells us that 100% of all electrical energy entering the house is being converted into heat, no matter what the primary use. It may be for operating a refrigerator, stove, fan, radio, TV, lights, or a power tool. All of the electrical energy supplied to these devices must appear as heat after the overall energy conversion. The electrical demand of lights and appliances can easily exceed an average of 1 kW (3413 Btu/h), perhaps one-third of the energy required to heat a small, well-insulated house at 30 °F. This loss of indirect electrical heat is usually quite noticeable during a winter power outage. If an air conditioner or heat pump was being used, a third energy flow would be shown crossing the boundaries, and the results would be different.

2.4.2 Refrigerator Energy Balance

In using energy balances the boundaries can be drawn around any device, group of devices, or components of a device. The refrigerator shown in Figure 2.15 could be the system of interest. Electrical energy is flowing in, as is the heat gain Q_1 through the insulated walls. The heat flow Q_2 from the condenser is out of the system. Suppose that the electrical input averages 150 W (0.15 kW). Does the refrigerator add heat or subtract heat from the house? How many Btu/h?

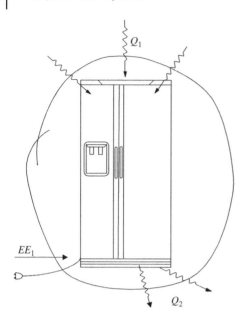

Figure 2.15 Refrigerator energy balance.

From equation (2.22) we have

$$EE_1 + Q_1 = Q_2$$

$$Q_2 - Q_1 = EE_1 = 3413 \times 0.15 = 512 \, \text{Btu/h}$$

The net effect is to heat the house, since the heat given off by the condenser (Q_2) exceeds the cooling (Q_1) by 512 Btu/h. The heat rejected by the condenser was taken as $Q_2 - Q_1$, which is the energy flow from the condenser to the surroundings due to a temperature difference. It is more likely that the condenser was cooled by a forced-air system (two airflows crossing the boundaries) so that the Q_2 term should have been omitted and the terms H_1 and H_2 added. The results would have been the same, since a heat balance of the air inside the condenser would show that $Q_2 = H_2 - H_1$. Where fluids flow into and out of a system and information is not available or required on flow rates and temperature, the simplifying substitution of Q for ΔH is desirable. The system boundaries could have been drawn to include only the compressor, condenser, and evaporator. The results would have been the same, since the heat gain by the evaporator equals the heat gain through the walls. Other boundaries could have been drawn but would not have yielded the desired information.

2.4.3 Energy Balance for a Water Heater

An electric water heater offers a good example of the use of stored energy and heat content in the energy balance (Figure 2.16). The 4.5-kW water heater shown is supplied with cold water at 60 °F and delivers hot water at a

Figure 2.16 Water heater energy balance.

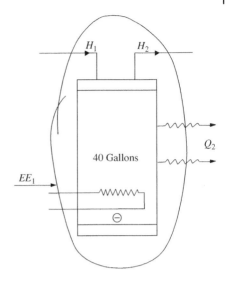

thermostat setting of 130 °F. Loss through the insulation (Q_2) is 300 Btu/h. Assume 8.3 lb of water per gallon:

1) Can the unit supply the 130 °F water continuously at a rate of 20 gal/h (166 lb/h)?
2) If the cost of electrical energy is 4 cents/kWh, what does it cost to heat 1 gal of water at these conditions?
3) If the heater is initially at 70 °F, how long will it take the temperature to reach 130 °F? (Neglect water use during the warm-up period.)

From equation (2.8),

$$H_1 + EE_1 = H_2 + Q_2$$

$$H_1 = 166 \times 1.0 \times 60 = 9960 \, \text{Btu/h}$$

$$H_2 = 166 \times 1.0 \times 130 = 21{,}600 \, \text{Btu/h}$$

so that $9960 + 3413 \times P_{kW} = 21{,}600 + 300$, and solving for the power required gives $P_{kW} = 3.5 \, \text{kW}$. Since this is less than the 4.5 kW available, the 20-gal/h rate is no problem. Since 20 gal/h requires 3.5 kW, 1 gal requires $3.5/20 = 0.18 \, \text{kWh}$ and the cost is $0.18 \times 0.04 = \$0.72$. There is no flow under these conditions, but there is an increase in stored energy. From equation (2.22) once more,

$$EE_1 = Q_2 + \Delta H_S$$

$$3413 \times 4.5 = 300 + \frac{40 \times 8.3 \times 1.0 \times (130 - 70)}{t}$$

Solving for t gives $t = 1.32 \, \text{h}$.

2.4.4 Rock Bed Energy Balance

The 30,000-lb well-insulated rock bed shown in Figure 2.17 is used for energy storage in a solar air system. If the bed is initially at 150 °F and the average airflow required to heat the building is 100 ft^3/min, calculate the average temperature of the rock bed after 24 h of operation. Assume an ideal bed such that the outlet temperature remains at 150 °F until the average temperature drops to the return air temperature of 70 °F.

From equation (2.22),

$$H_1 + EE_1 = H_2 + Q_2$$

$$H_1 = 166 \times 1.0 \times 60 = 9960 \, \text{Btu/h}$$

$$H_2 = 166 \times 1.0 \times 130 = 21,600 \, \text{Btu/h}$$

Solving for the temperature change gives $T' = 33$ °F. The negative sign indicates a temperature decrease. Since T' is the change in temperature, the final temperature of the bed is $150 - 33 = 117$ °F. This is the average temperature of the bed—the top is still at 150 °F and the bottom at 70 °F. We could calculate the location of the line separating the two temperatures by equating the energy at the average temperature to the sum of the energy stored at the high and low temperatures. The results would indicate that the line was located at 41% of the distance from the bottom to the top.

2.4.5 Array of Solar Collectors

In a 300-ft^2 array of solar collectors, the fluid absorbs energy at a maximum rate of 160 Btu/ft^2-h. What airflow rate should be used to increase the air temperature by 50 °F? For liquid-cooled collectors, what water flow would give a 10 °F rise?

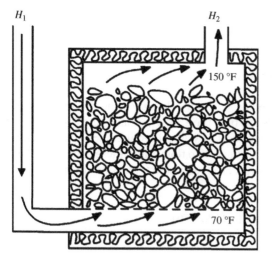

H_1 H_2

150 °F

70 °F

Figure 2.17 Rock bed energy balance.

From the first law, the increase in heat content of the flowing fluid must equal the heat added:

$$H_2 - H_1 = Q_1$$

$$WC_p(T_2 - T_1) = Q_1$$

$$W = \frac{Q_1}{C_p \Delta T}$$

For air:

$$W = \frac{160 \times 300}{0.24 \times 50} = 4000 \text{ lb/h}$$

$$\frac{4000}{4.5} = 889 \text{ cfm}$$

For water:

$$W = \frac{160 \times 300}{1.0 \times 10} = 4800 \text{ lb/h}$$

$$\frac{4800}{500} = 9.6 \text{ gpm}$$

2.4.6 Heat Pump

The heat pump can be understood as an air conditioner. During the summer, the air on the room side is cooled, while air outside is heated; during winter a cycle reversing valve is used to switch the functions of the evaporator and condenser, and the refrigeration flow is reversed through the device. The capacity of air conditioners and heat pumps is measured in tons of refrigeration effect; 1 ton of capacity is equivalent to 12,000 Btu/h. Figure 2.18 illustrates the heat pump concept. In the cooling mode, the indoor heat exchanger functions as an evaporator, with heat flow from the room to the cold evaporating refrigerant. In the outdoor heat exchanger (condenser), heat is removed from the hot compressed vapor, and liquid refrigerant is returned through a restriction to the indoor unit. In the heating mode, the roles of the heat exchangers are reversed, with the outdoor unit functioning as the evaporator and the indoor unit functioning as the condenser.

In heating mode use:

$$Q_H = Q_C + EE_1$$

The useful heat delivered to the building is the sum of the electrical energy required for operation plus the heat transferred to the outdoor unit

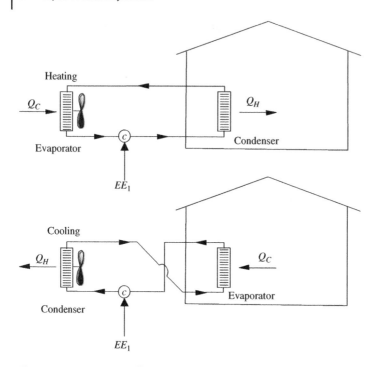

Figure 2.18 Heat pump efficiency evaluation.

from the surroundings. The ratio of the useful energy output to the electrical energy input is determined by the coefficient of performance:

$$(COP)_{heat} = \frac{Q_H}{EE_1}$$

In the cooling mode use:

$$Q_C = Q_H - EE_1$$

and

$$(COP)_{cool} = \frac{Q_C}{EE_1}$$

2.4.7 Heat Transfer Analysis

Although thermodynamics is a powerful tool, there are practical problems that lack sufficient information to permit use of the first law. For example, industrial systems may recover waste heat in the flue gas from heaters and boilers as in Figure 2.19. For this analysis, heat transfer techniques are required to study

Figure 2.19 Flue gas heat recovery system.

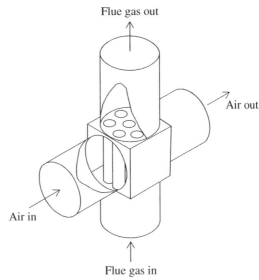

Flue gas out

Air out

Air in

Flue gas in

conduction, radiation, and convection in order to estimate steady-state fluid temperatures. The energy engineer needs to understand a very interdisciplinary area for analysis, modeling, and design of heat and mass transfer processes, with application to common practical problems. A holistic perspective and integration of alternative energy sources with architectural projects considering climate aspects and natural solutions is also required for a successful sustainable energy system design.

2.4.8 Simple Steam Power Turbine Analysis

A simple steam turbine used for electric power generation is depicted with the diagram of Figure 2.20. Saturated water entering the pump is compressed to a high pressure, and the compressed water is fed to the steam generator, also called boiler. The compressed water is heated to superheated steam state in the steam generator, which is a large heat exchanger where heat is transferred from the hot combustion gases to the water. The superheated steam, leaving the steam generator at a high pressure and temperature, enters the turbine, expands, and rotates the turbine shaft. The shaft work output of the rotating turbine is used to produce electricity by rotating an electrical machine (synchronous or induction).

The wet steam leaving the turbine is condensed to saturated water state in a condenser, which is a large heat exchanger where the heat is transferred from the steam to cooling water. The saturated water leaving the condenser enters the pump, thus making a cyclic flow through the pump, steam generator, turbine, and condenser. The combustion gases leaving the steam generator

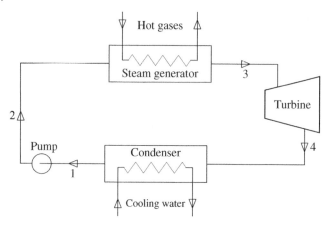

Figure 2.20 Diagram for a simple steam power plant.

have a lot of pollutants, that is, the greenhouse gas carbon dioxide, going to the atmosphere. The cooling water leaves the condenser at an elevated temperature; it is then cooled in cooling towers by transferring considerable amount of heat to the atmosphere and increasing thermal pollution and entropy in our planet.

In the steam turbine power plant, shown in Figure 2.19, assume that saturated water at 0.08 bar flowing at a rate of 50 kg/s is compressed by the pump to 70 bar. The compressed water is fed to the steam generator where it is converted to superheated steam at 450 °C at the same pressure. The superheated steam expands through the turbine and leaves as wet steam at 0.08 bar. The wet steam is condensed to saturated water state in the condenser, completing the cycle:

a) Determine the enthalpies of the flow at the inlets and the outlets of the turbine and the pump, assuming reversible adiabatic flows through the turbine and the pump.

Assume the labels 1, 2, 3, and 4 to mark the states of water/steam entering or leaving the components of the steam turbine power plant. At state 3, which is the turbine inlet, we have superheated steam at 70 bar and 450 °C. The specific enthalpy found from a superheated steam table is $h_3 = 3287$ kJ/kg. At state 4, which is the turbine outlet, we have wet steam at 0.08 bar. Since the flow through the turbine is assumed to be reversible adiabatic, $s_4 = s_3 = s$ at 70 bar and 450 °C $= 6.632$ kJ/kg·K. For state 4 at 0.08 bar and $s_4 = 6.632$ kJ/kg·K, the dryness fraction can be calculated using

$$x_4 = \frac{s_4 - s_f}{s_{fg}} = \frac{6.632 - 0.593}{7.634} = 0.7911$$

The specific enthalpy is then

$$h_4 = h_f + x_4 \times h_{fg} = 174 + 0.7911 \times 2402 = 2074 \, \text{kJ/kg}$$

At state 1, which is the pump inlet, we have saturated water at 0.08 bar. The specific enthalpy is $h_1 = 174 \text{kJ/kg}$. At the pump outlet (State 2), the compressed water is at 70 bar, assuming the flow through the pump is a reversible adiabatic process. That is, $s_2 = s_1 = s_f$ at 0.08 bar, which is 0.593 kJ/kg·K. State 2 at 70 bar and $s_2 = 0.593 \, \text{kJ/kg·K}$ is a compressed water state since s_f at 70 bar is 3.122 kJ/kg·K. Therefore, in order to find h_2 with a table for compressed water data, it is also possible to use the convenient method discussed next. Combining the work input to a pump with adiabatic flow with the work input to a pump with reversible flow, the expression for h_2, as equation (2.23), and h_2 is calculated by equation (2.24):

$$\dot{m}\left(h_2 - h_1\right) \approx \dot{m} v_1 \left(P_2 - P_1\right) \tag{2.23}$$

$$h_2 = h_1 + v_1 \left(P_2 - P_1\right) \tag{2.24}$$

The results are summarized below.

State	1	2	3	4
Condition	Saturated water	Compressed water	Superheated steam	Wet steam
P (in bar)	0.08	70	70	0.08
T (in °C)	41.5	—	450	41.5
h (in kJ/kg)	174	181	3287	2074
x	0	—	—	79.11%

b) Determine the power production at the turbine, power requirement of the pump, and the back work ratio, defined as the ratio of the pump work to the turbine work.

The power production can be calculated, again assuming flow through an adiabatic turbine as

$$\dot{W}_{output} = \dot{m}\left(h_3 - h_4\right) = 50 \times \left(3287 - 2074\right) \text{kJ/s} = 60.65 \, \text{MW}$$

The power requirement for the pump (assuming flow through and adiabatic pump) is

$$\dot{W}_{input} = \dot{m}\left(h_2 - h_2\right) = 50 \times \left(182 - 174\right) \text{kJ/s} = 0.35 \, \text{MW}$$

The back work ratio can be calculated as

$$bwr = \frac{pump\ work\ input}{turbine\ work\ output} = \frac{0.35\,MW}{60.65\,MW} = 0.6\%$$

The power consumption of the pump to compress water from 0.08 to 70 bar is only about 0.6% of the power production of the turbine. That is almost all the work produced by the turbine is still available for electricity generation. This is an advantage that a steam turbine power plant has over a gas turbine power plant.

c) Determine the heat input at the steam generator and the thermal efficiency of the steam turbine plant, defined as the ratio of net work out to heat in.

The heat input to the steam generator is calculated by assuming the fluid flowing through the steam generator as

$$\dot{Q}_{steam\ generator} = \dot{m}(h_3 - h_2) = 50 \times (3287 - 181)\,kJ/s = 155.30\,MW$$

The thermal efficiency of the steam turbine plant is calculated as

$$\eta_{thermal} = \frac{net\ work\ out}{heat\ in} = \frac{60.65\,MW - 0.35\,MW}{155.30\,MW} = \frac{60.30}{155.30} = 38.8\%$$

Therefore, only about 38.8% of the 155.30 MW heat input to the steam generator is available for electricity generation.

d) The heat rejected at the condenser is calculated by using the characteristics through the condenser as

$$\dot{Q}_{out\ condenser} = \dot{m}(h_4 - h_1) = 50 \times (2074 - 174)\,kJ/s = 95.0\,MW$$

The percentage of the heat input in the steam generator that is rejected by the condenser is

$$\frac{95.0\,MW}{155.30\,MW} = 61.2\%$$

meaning that for each MW of electric power generation, there is at least about 1.6 (= 95.0/60.30) MW of power wasted. All this wasted energy dispersing through the environment is **thermal pollution**.

e) Calculate the cooling water mass flow rate if the cooling water temperature is raised from 32 °C to about 2.5 °C less than the temperature of the wet steam condensing in the condenser. Take cooling water C_p as 4.2 kJ/kg·K.

The steam condenses in the condenser at the saturated temperature at 0.08 bar, which is 41.5 °C. The cooling water temperature is raised from 32 to 39 °C. Assuming the cooling water absorbs all the 95.0 MW of heat rejected at the condenser, it is possible to calculate the mass flow rate of the cooling water as

$$\dot{m}_{cooling\ water} = \frac{95.0\,\text{MW}}{4.2(39-32)} = 3231\,\text{kg/s} = 11.63 \times 11.63 \times 10^5\,\text{kg/h}$$

2.5 Planet Earth: A Closed But Not Isolated System

Energy cannot be destroyed, so how has the public developed the perception that we have a shortage of energy? The scarcity is not in total energy but in energy forms or levels that are suitable for transfer to useful work or heat. When fuel is burned in a boiler, the products of the process (primarily steam and hot flue gases) have the same energy content as the fuel supplied. The flue gases mix with the atmosphere and are no longer available for useful work or heat. A portion of the energy in the steam can be utilized in the production of work or heat, but the end result of these processes will eventually appear as energy rejected to the environment, so that all of the original energy is, in effect, lost.

During the process of conversion of some form of energy into another intermediate form, which serves a useful purpose, a question arises as to the maximum efficiency of the conversion process. Is energy waste to the environment always a by-product of energy conversion? If so, is there a predictable or limiting portion of the energy converted that can never be recovered regardless of the skill or ingenuity of the designer?

The two laws of thermodynamics can be stated together. The total energy content of the universe is constant, and the total entropy is continually increasing. Entropy is thus a measure of the amount of energy that is no longer capable of performing work. An increase in entropy involves a reduction of the energy available to do work in the future. Increasing entropy can also be considered as a change from an ordered state to a disordered state. One example of increasing entropy is water falling over a dam. When the water is above the dam, it has some PE due to gravity, which can be used to generate electricity or turn a wheel to perform some useful task. Once the water has fallen to the level below the dam, its total energy is the same (as the fall warms the water, increasing its thermal energy), but it no longer has the same capacity to do work. The water has moved from what is referred to as an available or free energy state (high-grade energy) to an unavailable or bound energy state (low-grade energy). This change in the energy state of the water as it falls over the dam is an increase in entropy. Unavailable energy is most often in the form of thermal energy.

The second law of thermodynamics tells us that the entropy of an isolated system tends to increase with time. An isolated system is one in which matter and energy cannot either enter or exit. The Earth is certainly not an isolated system because it receives a constant flow of energy from the sun and from outer space; however, with the exception of a few meteorites, artificial space, satellites, and some cosmic dust, Earth is a closed system when it comes to matter. The order (entropy) in a closed but not isolated system can be maintained over time, but only if there is an external source of energy. Thus, ultimately, the sun energy allows highly ordered and complex organisms and civilizations to develop on Earth.

From the beginning of life on Earth until the last several hundred years, all organisms, including human beings, utilized the available energy of the sun to provide the necessities of life. During this period, energy from the sun was used to maintain the entropy in our planetary system. Plants use sunlight directly, the process of photosynthesis converting the energy from the sun into plant matter, creating order. Animals and humans use sunlight indirectly by eating plants and other animals to survive. Humans in the past also used indirect solar energy in the form of wood for fuel and building materials and flowing wind and water as well as animals for performing work. As long as life on Earth used sunlight in an evenly distributed manner and at a rate less than that of incoming solar radiation, order was maintained all over the planet.

As populations increased in certain parts of the world in the past several thousand years, humans started to cut down more trees, build up their houses and roads, and harvest more crops in a year than could be regenerated through the natural process of photosynthesis. Such practices led to deforestation and desertification of previously fertile land. Human population increases, and the use of fossil fuels since the industrial revolution has greatly increased the rate at which humans are adding to the entropy of the planet.

Fossil fuels such as coal, oil, and hydrocarbons are complex and highly ordered chains of hydrocarbons, formed originally by photosynthesis from sunlight millions of years ago. When humans burn fossil fuels to provide energy to drive cars or generate electricity, the second law of thermodynamics states that a penalty must be extracted, and this penalty is an increase in entropy. In most populated parts of the world, energy from fossil fuels is consumed at an enormous rate. Fossil fuel consumption therefore results in a large increase in entropy, and this entropy shows up in the form of pollution.

The substances that human beings classify as pollutants have always been present on the planet, because Earth is a closed system. The reason that these materials cause a negative impact on the environment, therefore, is not that they exist but that they have been dispersed or concentrated so much throughout the world's ecosystems in a very disordered fashion. To illustrate this, two pollutants of major concern will be considered, carbon dioxide (CO_2) and sulfur dioxide (SO_2). CO_2 and SO_2 are both given off when coal is burned to provide heat or to generate electricity. CO_2 is the most important greenhouse gas contributing to global warming, and SO_2 is the main cause of acid rain.

Carbon and sulfur have no impact on the environment when they are locked up in the highly ordered form of coal. It is when fossil fuels are burned, and the combustion products are released in the atmosphere, that these compounds become pollutants.

Solar energy can be used to counterbalance increases in entropy caused by the use of fossil fuels. As long as the rate of entropy increase due to fossil fuels is less than the rate of entropy decrease from solar radiation, the net entropy on the planet will not increase. Since the amount of solar radiation reaching Earth is quite large, why cannot humans continue to use vast quantities of fossil fuels?

It is virtually impossible to channel solar radiation such that it can counteract the negative impact of pollution caused by fossil fuels. For example, it is diffi- cult to imagine using the sun's energy to reduce the acidity of lakes and forest soils damaged by acid rain. A better solution would be to use solar radiation to power our societies directly. This would be more efficient than using fossil fuels to create our highly ordered civilization and then trying to eliminate the pollution and entropy by using sunlight. Renewable energy technologies such as solar energy (photovoltaic cells, active and passive solar heating, etc.), bio- mass (plant matter), wind power, and hydropower use sunlight either directly or indirectly to produce the high-quality energy demanded by industrialized societies, as discussed in subsequent chapters.

References

1 C.B. Schuder, Energy Engineering Fundamentals: With Residential and Commercial Applications, Van Nostrand Reinhold, New York, 1983.
2 A.W. Culp, Jr., Principles of Energy Conversion, McGraw-Hill, New York, 1979.
3 S.W. Angrist, Direct Energy Conversion, Allyn & Bacon, Boston, MA, 1982.
4 D. Walraven, B. Laenen, and W. D'haeseleer, Comparison of thermodynamic cycles for power production from low-temperature geothermal heat sources, Energy Conversion and Management, Vol. 66, pp. 220–233, 2013.
5 A.J. Longo, F.A. Farret, F.T. Fernandes, and C.R. De Nardin, Instrumentation for surface geothermal data acquisition aiming at sustainable heat exchangers, IEEE International Conference of the Industrial Electronics, IECON2014, Dallas, TX, EUA, November 2014.
6 C.R. De Nardin, F.T. Fernandes, A.J. Longo, F.A. Farret, L.P. Lima, and D.R. Cruz, Underground geothermal energy to reduce the residential electricity consumption, 3rd IET Renewable Power Generation Conference (RPG™), Ramada Naples, via Galileo Ferraris, Naples, Italy, September 2014.
7 F.T. Fernandes, F.A. Farret, C.R. De Nardin, A.J. Longo, and J.G. Trapp, PV efficiency improvement by underground heat exchanging and heat storage, 3rd Renewable Power Generation Conference (RPG 2014), Naples, Vol. 1, 2014, p. 1–6.

8 C. De Nardin, F.T. Fernandes, A. Longo, S. Cunha, L.P. Lima, F.A. Farret, and E.M. Ferranti, Reduction of electrical load for air conditioning by electronically controlled geothermal energy, IECON 2013—39th Annual Conference of the IEEE Industrial Electronics Society, Vienna, Vol. 1, 2013, pp. 1850–1851.

9 F.P. Incropera and D.P. DeWitt, Fundamentals of Heat and Mass Transfer, 5th ed., John Wiley & Sons, Inc., New York, 2001.

10 K.F.V. Wong, Thermodynamics for Engineers, CRC Press, Boca Raton, FL, 2000.

3

Hydroelectric Power Plants

3.1 Introduction

Hydropower technology enables a large production of electric power. In addition to large hydropower projects implemented by governments, there are many private investments. The majority of rural properties have water creeks and small water heads that can be used as primary energy water-based resource. Such energy is commonly used for machine operation, generation of electricity, and water storage in elevated reservoirs. Small electric power plants can also be used with waterwheels or, depending on the flow or the available head, with turbines. The first known turbine generator set, with approximately 8 hp, driven from a gross water head of 10 m was constructed in 1882 in order to supply carbon-filament light bulbs. The first known small hydroelectric power plant was built in Northern Ireland in 1883, consisting of two turbines of 52 hp (1 horsepower = 746 watts) to supply energy for an electric train.

In the same year in Diamantina, Brazil, a hydroelectric energy system of 12 kW (two generators driven by waterwheels) for an elevation of 5 m was established. Ever since, the diversified evolution of the energy sector has been influencing all sectors of everyday life. From that time, a variety of hydroelectric power plants became a reality [1].

Hydroelectric power plants can be driven from a water stream or from an accumulation reservoir. Run-of-river hydroelectric plants (those without accumulation reservoirs) are built along a river or a stream, without lake formation, for the intake of water. Therefore, the river course is not altered, and its minimum flow will be the same or higher than that of the turbine output power. The excess water must be diverted, and in this case, the water volume is not totally used. Costs are lower than those for reservoirs, with less environmental impact. However, the plant is unable to store energy. On the other hand, there is no need for extensive hydrological studies, just simple measurements for estimating the availability of power [2–4].

Integration of Renewable Sources of Energy, Second Edition. Felix A. Farret and M. Godoy Simões.

The development of a hydroelectric power plant based on accumulation reservoir demands complex hydrological and topographical studies for determining the site elevation. Such data is of great importance, because it is directly used to establish the watercourse flow in order to find the power calculation to be produced by the power plant [1–4]. The use of water is total, which means the whole reservoir volume can convert energy, and the civil works have a big effect on the environment proportional to the size of the plant. A reservoir system also has a higher generation potential than that of a water stream, compensating for the larger capital investment. The lake formed by water accumulation can be used for other purposes, such as recreation, creation of fisheries, irrigation, and urbanization. Useful data for hydroelectric power plant projects might include:

- Local availability of materials necessary for the construction of a dam (when applicable)
- Labor availability and qualification for cost reduction
- Reasonable distance of the power plant from the public network to allow future interconnection through a single-phase ground return or a single-phase, three-phase, or low-voltage dc link

3.2   Determination of the Available Power

Turbines and waterwheels use the energy that can be evaluated as the sum of the three forms of energy, derived from thermodynamics energy balance and arriving at Bernoulli's principle. Bernoulli's principle is based on the sum of all forms of energy in a fluid along a streamline as being the same at all points on that streamline. This requires that the sum of kinetic energy, potential energy, and internal energy remains constant. Thus an increase in the speed of the fluid occurs with a simultaneous decrease in the sum of the static pressure, potential energy, and internal energy. If the fluid is flowing out of a reservoir, the sum of all forms of energy is the same on all streamlines because in a reservoir the energy per unit volume (the sum of pressure and gravitational potential ρgh) is the same everywhere [3–6].

The results of expression (3.1) remain constant for every cross section and water location in a channel:

$$\frac{v^2}{2g} + h + \frac{p}{\rho g} = \frac{P}{\rho g Q} \tag{3.1}$$

where:

$v =$ Water flow speed in meters/second
$g =$ Gravity constant ($= 9.81\,\text{m/s}^2$) in meters/square second
$h =$ Height of the water in meters

p = Pressure of the water in Newton/square meters
ρ = Density of the water in kilograms per cubic meters ($\cong 1000.00\,\text{kg/m}^3$)
P = Pwer in kilogram-meter per second ($1\,\text{hp} = 75\,\text{kgm/s} = 746\,\text{W}$)
Q = Flow of the watercourse (in m^3/s)

For ordinary modern turbines, the effective power at their input may be obtained from equation (3.1) (neglecting the terms in v and p) for the potential energy in the watercourse as

$$P_t = \eta_t \rho g Q H_m \tag{3.2}$$

where:

η_t = Turbine simplified efficiency (for standard turbines it is taken as 0.80)
H_m = Water head

The available flow of a watercourse, in m^3/s, is expressed by

$$Q = Av \tag{3.3}$$

where:

A = The watercourse normal area

Energy capture from the flow of large rivers is a very attractive idea. However, the designer must take into account the density of energy that can be obtained from the running water. For example, the power density of a 1.0-m/s stream (by equation (3.1)) is approximately $500\,\text{W/m}^2$ at the river cross section. But only 60% of that energy can theoretically be recovered for most small hydroelectric power plants, and usually, machines using this type of energy are not optimized and their efficiency rarely reaches 50%. That is, only $250\,\text{W/m}^2$ of the energy through the intercepted area across the water stream might be available. As predicted by the combination of equations (3.1) and (3.3), the power available from a stream is proportional to the cube of the speed. If the stream speed is 2 m/s, the density would rise to $2000\,\text{W/m}^2$ and so on. However, such speeds are not usually very common in streams from which energy can be extracted.

In practice, the maximum power a site can offer depends on their topographical and hydrological characteristics of elevation and available flow. Notice that the useful power in the machine shaft should be the same as the maximum electric power that can be generated at that place. Depending on the geometry and material dimensions used in the machines and piping of the power plant, this electric power can be determined in kilowatts ($\eta = 0.60$, $\rho \approx 1000\,\text{kg/m}^3$, $g = 9.81\,\text{m/s}^2$), approximately, by equation (3.4).

$$P = 6.0QH \quad \text{in kW/m}^2 \tag{3.4}$$

where:

H = The gross head (m), that is, the difference between the levels of the crests of water in the reservoir (dam) and the river at the powerhouse site

The pump flow and pumping pressure height can be calculated, respectively, by

$$Q_b = 0.75Q$$
$$H_b = 0.55H$$

(3.5)

The most usual methods for topography and hydrology works are discussed in Section 3.3. There are references providing further detailed routines for these calculations [3, 4].

3.3 Expedient Topographical and Hydrological Measurements

3.3.1 Simple Measurement of Elevation

Two very simple methods are used to measure the natural water head of a hydroelectric station. For the first method, it is enough to use a carpenter level and two flat metric rulers, one 3–4 m long and another 2 m long. The inferior tip of the smaller ruler is placed vertically at the level of the water (Figure 3.1). Then the larger ruler is placed on the ground and the carpenter level controls its horizontal position. It measures the height h_i and marks the level of a point i where the tip of the larger ruler rests on the ground. Restarting from that position, the operation is repeated between several pairs of elevation points, making sure that they were verified in all level differences, h_k. The height of the total head is the straight summation of the level differences, h_k ($k = 1,2,\ldots,n$):

$$h = h_1 + h_2 + h_3 + \cdots + h_n$$

(3.6)

For the second method, two metric rulers (about 2 m long) are used, together with a plastic tube (flexible, transparent, 1 cm in internal diameter, and at least 6 m long). The rulers are placed at two points vertically, between which

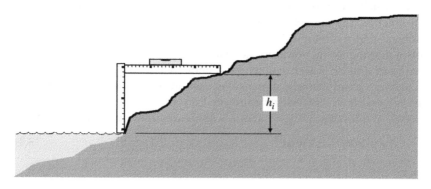

Figure 3.1 Using two rulers to measure an elevation.

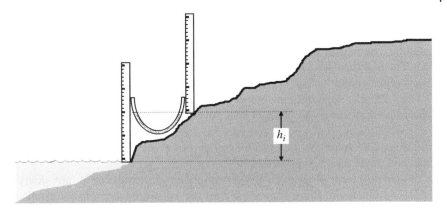

Figure 3.2 Using a ruler and a flexible tube to measure an elevation.

the elevation is measured (Figure 3.2). With the help of the plastic tube (full of water), points of equal level are determined on each ruler, creating a horizontal plane of reference. Summation of all differences between the ground heights of every two points gives the total difference in level, h, as in equation (3.6).

3.3.2 Global Positioning Systems for Elevation Measurement

Global positioning systems (GPSs) provide specially coded satellite signals that can be processed in a GPS receiver, enabling the receiver to compute area, elevation, velocity, and time. Four GPS satellite signals are used to compute positions in three dimensions and time offset in the receiver clock. Although there are many thousands of civilian GPS users worldwide, the original system was designed for and operated by the US military and funded and controlled by the US Department of Defense.

The control segment consists of a system of tracking stations located around the world. The space segment of the system consists of GPS satellites, space vehicles (SVs), which send radio signals from space. The nominal GPS operational constellation consists of 24 satellites that orbit the Earth in 12 hours. There are often more than 24 operational satellites as new ones are launched to replace older satellites. The satellite orbits repeat almost the same ground track, as the Earth turns beneath them, once each day. The orbit altitude is such that the satellites repeat the same track and configuration over any point approximately every 24 hours (4 minutes earlier each day).

There are six orbital planes (with nominally four SVs in each), equally spaced (60° apart) and inclined at about 55° with respect to the equatorial plane. This constellation provides the user with between five and eight SVs visible from any point on Earth.

The master control facility is located at Schriever Air Force Base (formerly Falcon AFS) in Colorado. These monitor stations measure signals from the

SVs, which are incorporated into orbital models for each satellite. The models compute precise orbital data (ephemeris) and SV clock corrections for each satellite. The master control station uploads ephemeris and clock data to the SVs. The SVs then send subsets of the orbital ephemeris data to GPS receivers over radio signals.

The GPS user segment consists of GPS receivers and the user community. The receivers convert SV signals into position, velocity, and time estimates. Four satellites are required to compute the four dimensions x, y, z (position), and time. GPS receivers are used for navigation, positioning, time dissemination, and other research purposes.

The current technology of GPS is very popular today, available for any smartphone or tablet, and there are several apps and softwares that can be installed to help with such measurement [5–8].

3.3.3 Pipe Losses

There are four main causes of losses in pipes conducting water or any viscous fluid from a higher potential point, say, a water dam, to a lower potential point, such as a hydraulic turbine: (i) the specific mass ρ of the fluid (kg/m), (ii) the average speed v (m/s), (iii) the pipe diameter D (meters), and (iv) the viscosity coefficient η (kg/ms). Those are all related quantitatively by the Reynolds number:

$$N_R = \frac{\rho v D}{\eta} \tag{3.7}$$

For the specific case of water, which is the basis for all hydraulic turbines, $\rho = 1000\,\text{kg/m}$ and $\eta = 10^{-3}\,\text{kg/ms}$ at $20\,^\circ\text{C}$. The Reynolds number N_R is a no-dimensional number; it does not depend on a system of units. The number is used to establish the speed limit at which a fluid can run in a pipe without turbulent flow. It is common practice to establish N_R within the following limits: for $N_R = 2000$ and below, the flow is laminar; for N_R from 2000 to 3000, the flow is unstable; and for N_R above 3000, the flow is certainly turbulent.

Part of the loss of energy in pipes is due to the roughness of their internal walls. However, it is important to note that turbulent flow is not caused by the roughness of the pipe but by the conditions at which the fluid runs in the pipe. That means that a smooth pipe can have turbulent flow and a rough pipe can have smooth flow, because when there is smooth pipe flow, the viscosity of the fluid forms a laminar sub-layer covering the roughness of the pipe. As the speed of the fluid increases in transitional pipe flow, the sub-layer is reduced in thickness, smoothing out some of the roughness. Finally, in rough pipe flow, the laminar sub-layer is very thin and the roughness of the pipe is well exposed to the water flow, causing heavy pipe losses of otherwise useful kinetic energy. Pipe friction is proportionally the inverse of the Reynolds number.

Pipe losses are extremely sensitive to the diameter of the pipe not only by affecting the Reynolds number inversely but also by contributing to the increase in velocity of the fluid and the associated losses. It is a complex task to quantify these contributions mathematically, so it is usual in hydrology studies to unite all losses in a head equivalent h_i. The h_i equivalent is defined as a term proportional to the pipe length l and friction and to the kinetic energy given by Bernoulli's expression (equation (3.1)) and inversely proportional to the pipe diameter. This expression is known as the Darcy–Weisbach formula:

$$h_i = \lambda \left(\frac{l}{D}\right)\frac{v^2}{2g} \tag{3.8}$$

The friction factor λ is provided by the pipe manufacturer.

If equation (3.3) is used to determine the square of the water velocity, it will be observed that the diameter affects equation (3.8) in its fifth power. The result from this equation has to be added linearly to the results from equation (3.6) to obtain the effective pipe head loss. The final value of practical interest is the effective hydraulic gradient, expressed as

$$\nabla h = \frac{h + h_i}{l} \tag{3.9}$$

3.3.4 Expedient Measurements of Stream Water Flow

Among the expedient methods used to measure the flow of a ditch, a stream, or a river, the floats are the most simple. It is also possible to make rectangular or triangular spillways, or we can measure the water resistivity between two points [3–5].

3.3.4.1 Measurement Using a Float

In measurements with a float, a straight passage of watercourse with a uniform bed is chosen where the water flows calmly. If possible, it is measured a length $l > 10\,\text{m}$ upstream of the river course, marking the beginning and the end. Points of interest are then marked by two strings tied to stakes nailed on the margins perpendicular to the ditch or stream axis. Next, the float is placed some meters upstream from the beginning of the length chosen and in the middle of the riverbed. The float can be a closed bottle ballasted with water at about one-third of its volume to keep it vertical. A chronometer measures the time, in seconds, that the float takes to travel the length chosen.

The cross-sectional areas, limited by the levels of water crest and the ditch or stream bottom, should be determined for at least the initial and final points of the measurements. If the length of that passage is too long, it is advisable to determine the areas of one or more intermediate cross sections and use the average value. The flow Q is calculated by the formula

$$Q = \frac{0.8L\bar{A}}{t} \tag{3.10}$$

where:

0.8 = Coefficient of the surface speed correction for an average speed
L = Length of the passage to measure the flow in m
$\bar{A}$ = Average area of the cross sections in m^2
t = Time of the float course in seconds across the measured section

The speed correction coefficient for the superficial speed across the section is necessary because of the speed differences at the various points of the river cross section, such as between waters closer to the margins and the riverbed as opposed to those in midstream, due mostly to viscosity and friction.

3.3.4.2 Measurement Using a Rectangular Spillway

Measurement using a rectangular spillway leads to more precise results than with a float. It demands more work, and it is limited to cases where the conditions of the watercourse allow the implementation. For a spillway, the watercourse is blocked by a vertical board with a rectangular central aperture of known area, sufficient for the passage of all the water. The spillway width should be from one-half to two-thirds of the watercourse width. The aperture cuts should be chamfered in the water flow direction. After all the fencing panel rifts and the spillway are secured, an upstream stake is nailed to the waterbed 1 or 2 m distant, with its top at the crest of the spillway (the lower side of the aperture) (see Figure 3.3a,b). Once water is draining normally through the spillway, the height of the water level, H, is measured above the top of the stake. In this case, the water discharge can be calculated from

$$Q = 1.84(L - 0.2H)H^{3/2} \tag{3.11}$$

where:

L = Aperture width of the spillway (m) (it should be larger than $3H$)
H = Head in meters of the upstream water level above crest of the spillway at the location of the stake

(a)   (b)

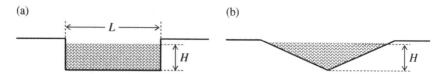

Figure 3.3 Dimensions of: (a) rectangular spillway; (b) triangular spillway.

3.3.4.3 Measurement Using a Triangular Spillway

A spillway with a triangular "V-shaped" aperture is used when the discharges are very small (lower than 200 L/s) or when the stream has a smaller width than depth, which precludes the installation of a spillway of rectangular section. The installation measurement procedures for this type of spillway are the same as those noted for a rectangular spillway; however, the flow is calculated by

$$Q = 1.4H^{5/2} \qquad (3.12)$$

where:

H = The head of the upstream water level (m) above the lower vertex of the spillway at the location of the stake (see Figure 3.3b)

3.3.4.4 Measurement Based on the Dilution of Salt in the Water

Water flow can also be determined using the resistivity alteration caused by salt dilution along a river course [2]. In one method, salt is poured in the water all at once. The variation in salt concentration in the course is measured downstream during a certain period. In a second method, a salt solution is poured in the water at a certain constant rate, and the concentration is measured downstream.

To implement the first method, use approximately 3 kg of salt for every estimated m³/s of water flow determined by

$$Q = \frac{\bar{V}}{T}, \quad \text{in m}^3/\text{s} \qquad (3.13)$$

where:

$\bar{V}$ = The volume of water at an average cross section of the river $\bar{A}$ and the downstream length of the river used in the measurement
T = The time period during which the readings are made

A mass M (kg) of salt is initially dissolved in a bucket of water and poured in the river all at once. About 50 m or more downstream, the water resistivity (and thus the conductivity) is measured periodically for 5 or 6 minutes such that it can be graphed (Figure 3.4).

The mass of salt in a bucket, which has volume V_0, has an initial concentration given by

$$C_0 = \frac{M}{V_0}, \quad \text{in kg/m}^3 \qquad (3.14)$$

where:

C_0 = The initial concentration of salt in the bucket and
V_0 = The bucket volume occupied by the salt solution

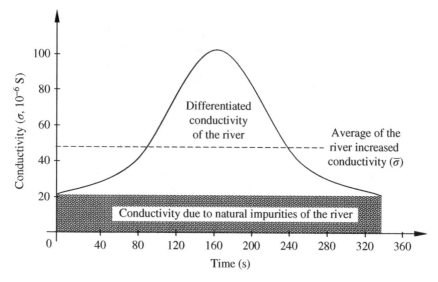

Figure 3.4 Typical curve of conductivity alterations at a point on a river.

All the salt (M) in the water stream would provide in the stream an average salt concentration of

$$\bar{C} = \frac{M}{\bar{V}} \quad \text{in kg/m}^3 \tag{3.15}$$

where:

$\bar{V}$ = The volume of water given by the average cross section of the river and the downstream length of the river used for the measurement

However, due to the proportionality k between the concentration of salt $\bar{C}$ and the water conductivity $\bar{\sigma}$ (Ω^{-1}), it is possible to establish

$$k = \frac{\bar{C}}{\bar{\sigma}} = \frac{C_0}{\sigma_0} \tag{3.16}$$

Based on the classical equation for conductivity,

$$\sigma_0 = \frac{1}{R_0} \frac{\ell}{S}$$

and on equation (3.16), it is easily shown that

$$\frac{R_0}{R} = \frac{\bar{\sigma}}{\sigma_0} = \frac{\bar{C}}{C_0} \tag{3.17}$$

The average conductivity of the water, $\bar{\mu}$, is proportional to the average salt concentration in the water, $\bar{C}$. This means that

$$\bar{\mu} = \frac{1}{T}\int_0^T \mu(t)\,dt = \frac{1}{T}A \tag{3.18}$$

where:

$A = \int_0^T \mu(t)\,dt =$ The area under the curve given in Figure 3.4 discounted the natural impurities of the river

After combining equations (3.13), (3.15), and (3.18), the following equation is defined:

$$Q = \frac{\bar{V}}{T} = \frac{M/\bar{C}}{A/\bar{\sigma}} = \frac{M}{kA} \tag{3.19}$$

where the area A under the curve in Figure 3.4 is calculated in $(\Omega^{-1} \cdot s)$ and the scale factor k in $kg/m^2 \cdot \Omega^{-1}$ is obtained through measurement of C_0, the initial concentration of salt occupying volume V_0 in the bucket.

The second method, used to measure the flow of a typical turbulent watercourse, is conducted by gradually pouring a bucket's initial salt concentration C_0 upstream (given by equation (3.14)) at a constant rate $q = V_0/T$ (m³/s). The average salt concentration in the river, $\bar{C}$, may be determined using equation (3.15). For these measurements, a relatively long, straight passage of river (some 50 m long) is chosen, into which the salt is poured. With an ordinary multimeter or an ohmmeter (capable of measuring up to $0.01\,\Omega$), the resistivity of the downstream water is measured at the end of the length chosen for testing. It is helpful to remember that the resistivity (or resistance) is the inverse of conductivity (or conductance) and is proportional to the concentration of salt. This method is based on the idea that the concentration of salt in the stream is reduced from C_0 to $\bar{C}$ and that Q is much larger than q. If those two assumptions are not met, a more precise form of the equation should be sought [2].

As the salt is being diluted in the water at a constant rate, its concentration in the water, and therefore the conductivity, will change as shown in Figure 3.5. This means that the mixture will be complete at a certain point downstream after some period of time (say, 200 s). At that selected point, the concentration of salt will have changed from C_0 to $\bar{C}$ so measuring indirectly the water flow through the conductivity of the water [2]; that is,

As a result, during the steady-state concentration shown in Figure 3.5, the rate of high concentration of salt being poured into the water will be proportional to the volume of water V just passing downstream at a rate of $Q = \bar{V}/t$, that is,

$$Q = q\left(\frac{C_0}{\bar{C}}\right) \quad \text{in m}^3/\text{s} \tag{3.20}$$

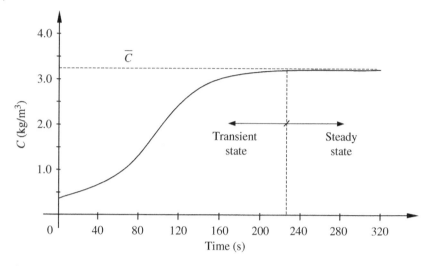

Figure 3.5 Variation of salt concentration down the river flow.

This method is fast, simple, and precise to about 7%.

$$\frac{\bar{V}}{V_0} = \frac{C_0}{\bar{C}}$$

3.3.5 Civil Works

Civil works include construction of a dam (or water storage facility), a water intake (to divert the river from its normal course), power system of pipes (piping), a reservoir, forced piping, a balance chimney, a machine house, or a spillway (to return water to its normal course). Such facilities cost a lot of money and require a civil engineer who is expert in the mechanics of soils, geology, and structures to set up the project and to follow through the building stages [2–4]. To avoid unnecessary facilities when setting up a small power plant could necessitate employment of a specialist to minimize building expenses, the time until the plant can be operated, and possibly, maintenance problems.

Whenever sufficient height and water flow are available for a small power plant, a dam may not be needed. The water will be used in essentially the same proportion, as it is available in the river. Such a system is known as a run-of-river plant. A portion of the stream flow is simply diverted to the powerhouse and, once used, is returned to the river. The large majority of small power plants are of the run-of-river type. This concept is sometimes misunderstood as being that of a turbine put directly into the river as a lower bucket type of waterwheel, which would be the case for a flow turbine and not for a run-of-river turbine.

3.4 Hydropower Generator Set

A hydro generating set is formed by a primary hydraulic machine and a generator together with such auxiliary equipment as a regulator (electromechanical, electro-electronic, or electronic), a water admission valve, an electrical command board, and an inertial flywheel. In Chapters 12 and 13, we discuss in general terms the equipment and electrical characteristics of small power plants. In this section, we simply present some peculiarities of hydroelectric power plants.

3.4.1 Regulation Systems

Regulation systems maintain voltage and rotation at constant levels, thereby maintaining the frequency of the generating unit within the variation limits of the electric network demands. A centrifugal pendulum of the mechanical type (i.e., a servomechanism working under pressurized oil) commands conventional automatic speed regulators of the type used in small hydroelectric power plants, the simplest of which are the inertial flywheels (see Section 11.5).

When acquiring a regulator, its cost, the distance of the powerhouse to the load, and the load type should be taken into account. When considering cost, it is advisable to choose a regulator for turbine powers above 20 kW. In smaller turbines, regulation can be achieved by load adjustment, by manual control of water flow, or through a simplified control for small integrated generating units [5–8]. Because voltage regulation is controlled by only one electric parameter, it is covered in more detail in Chapter 15.

3.4.2 Butterfly Valves

Butterfly valves are used in small hydroelectric power plants with metallic forced piping. They block water flow to the turbine during maintenance work and provide an additional resource for blocking the turbine in case of system failure. They are installed at the powerhouse by means of flanged connections, between the forced piping and the snail-shaped box housing the turbine.

3.5 Waterwheels

Waterwheels are primitive and simple machines, usually built of wood or steel, with shovels of steel blades fixed regularly around their circumference [5, 6]. The water pushes the shovels tangentially around the wheel. The water does not exert exactly a thrust action or shock on the shovels as is the case with turbines. The water action on the shovels develops torque on the shaft, and the wheel rotates. As one can infer, such machines are relatively massive, work

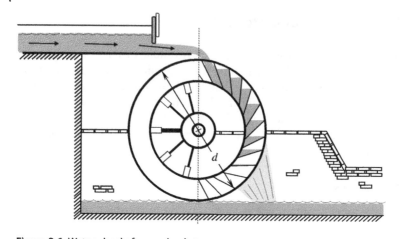

Figure 3.6 Waterwheel of upper buckets.

with low angular speeds, and are of low efficiency, due to losses by friction, turbidity, incomplete filling of the buckets (or cubes), and other causes.

In most cases, waterwheels are handcrafted and a great variety of models exist [3, 5, 6]. The most common is the wheel with upper buckets, so called because it consists of fixed shovels around a wheel, filled with water coming in at the top of the machine (Figure 3.6). This type of wheel is also hydraulic motor driven by potential or gravity, because the water accumulates in the small compartments of the half-wheel out of the water flow. The weight of the water provokes a motor torque in the sense of its downstream drainage.

A second type of wheel is one with coulisse buckets, which uses potential and kinetic energy. Water drains through a mouthpiece tangent to the wheel, and the water thrust on the shovels develops a torque on the wheel's shaft, causing it to rotate.

In another type, known as a lower pushed wheel, water causes a thrust on the lower part of the wheel (see Figure 3.7). Such wheels can be mounted floating on a body of water or inserted into the water flow. This assembly is advantageous because it follows the potential variation of the stream and maintains almost constant torque.

In the case of bucket wheels, considering input and output at about the same atmospheric pressure ($\Delta p \approx 0.0$), the effective shaft power (equation (3.1)) becomes

$$P_p = \eta \rho g Q H_m \tag{3.21}$$

where:

η = The efficiency (practical value = 0.60)
H_m = The head difference between crests of the water stream at the input and output of the channel

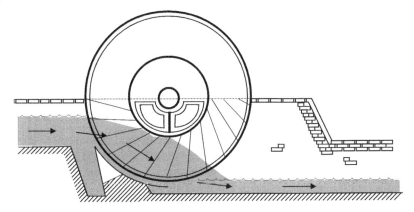

Figure 3.7 Waterwheel of lower buckets.

Notice that the upper bucket wheel diameter is the minimum head of the water required to move it, whereas the lower bucket wheel is adapted for smaller heights, on the order of the wheel shovel length or less. In this case, the resulting kinetic energy power is given by

$$P_k = \frac{\eta \rho A v^3}{2} \tag{3.22}$$

where:

η = The efficiency of the wheel
A = The normal cross-sectional area of the body of water intercepted by the waterwheel shovels

It should be noted that waterwheels have a very low rate of rotation, so to produce electricity, they require speed multipliers (around 1:20), introducing mechanical losses.

The tangent speed of the water wheel is, approximately, that of the water in movement at the input mouthpiece of the wheel without load on the shaft. In other words, $v \cong \sqrt{2gh_0} = 4.43\sqrt{h_0}$ (from 3 to 3.5 m/s), h_0 being the average height of the stream at the water admission channel (from 0.50 to 0.70 m). The potential energy term is negligible.

The use of waterwheels for electric power generation presents some important advantages: (i) production and conservation goals are easily met; (ii) operation is not affected by dirty water or by solids in suspension; and (iii) their motor torque value is aided in a certain way with an increase in load because during the decreased rotation under load, the buckets have more time to fill with water and thus increase the moment applied on the machine shaft.

The greatest disadvantage of waterwheels to generate electricity is their low angular speed of operation, which causes the generator to work with speed multipliers at high efficiency reduction rates, thus resulting in appreciable losses of energy. Another disadvantage is the usual need for some speed regulation, which could complicate a lot such a simple energy-generating system.

3.6 Turbines

Turbines can be considered as wheels driven by some fluid in movement that makes them rotate by action of the energy contained in them (in potential or kinetic form) on slats that can have several formats. The starting operation of the various types in use at present differs according to the form of energy used to drive them [4, 9–11].

Several types of turbines are manufactured in a number of sites worldwide with projects and construction somehow perfected. They are more compact, made from melted metal, and usually operate at high rates of rotation and with high mechanical efficiency. Use of the following types of turbines is recommended:

- Pelton turbine (PT)
- Francis turbine (FT)
- Michell–Banki turbine (MBT)
- Kaplan or propeller hydraulic turbine
- Deriaz turbine (DT)

For general guidance, Table 3.1 lists various types of runners and heads associated with each type referred to by the specific speed, n_s, based on the optimum point, defined as

$$n_s = n\sqrt{\frac{Q}{\sqrt{H^3}}}$$

Table 3.1 Summary of head and specific speed ranges.

Turbine runner	Specific speed (n_s)	Head (m)
Pelton	0–30	400–2000
Francis	20–120	50–500
Michell–Banki	30–210	1.5–150
Mixed-flow	120–180	20–80
Kaplan (vertical)	180–260	8–50
Bulb pit (horizontal)	260–360	0–10

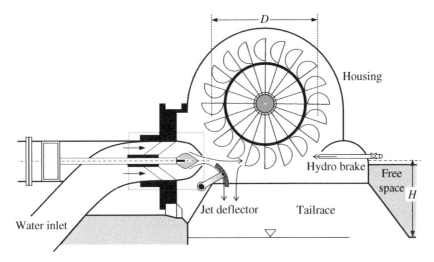

Figure 3.8 Cross section of a Pelton turbine.

3.6.1 Pelton Turbine

A PT is a turbine of free flow (action). The potential energy of the water becomes kinetic energy through injectors and control of the needles that direct and adjust the water jet on the shovels of the motive wheel. They work approximately at the atmospheric pressure. Thus, the lower limit of the net height, H, is affected by the impact of the jet on the shovels of the motive wheel. In PTs equipped with multiple injectors, needles, and motive wheels of large diameter, the situation is more complex, needing a better understanding of the manometric pressure.

In small hydroelectric power plants, operation of a PT can result in reasonable economy when it operates with flows above 30 L/s and heads from 20 m. The necessary combination of head with design flow to obtain a requisite PT power is shown in Figure 3.8. The wheel diameter of the Pelton crown of shovels, D (see Figure 3.9), as a function of the net head, H, the flow, Q, and the specific rotation, n_s, can be calculated by

$$D = \frac{97.5Q}{n_s^{0.9} H^{1/4}} \tag{3.23}$$

where:

$n_s = 240(a/D)z^{1/2}$ (rpm)
a = Diameter of the water jet (Figure 3.9) (meters)
D = Diameter of the rotor circle at the point of jet incidence, that is, the diameter of the crown of shovels at the impact points (meters)
z = Number of jets

(a)

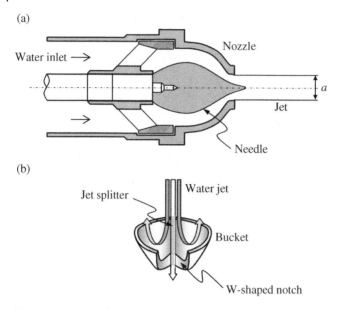

(b)

Figure 3.9 Details of PT jet control action: (a) injector nozzle, needle, and jet diameter; (b) bucket and jet splitter.

The diameter of the jet at the injector output can be calculated by

$$a = 0.55Q^{1/2}H^{1/4} \tag{3.24}$$

where:

H = Water fall head

Figure 3.8 indicates that in many situations there is a normal need to deflect some amount of incoming water to resolve transient load situations. Such situations are related to the control of water flow during sudden appreciable load changes, to ease synchronization of the generator with other machines of the network, or, in extreme cases, during heavy load rejection. The action of the deflector is to minimize the shock intensity of a sudden reduction in the speed of the water in the pipelines, called water hammer. These phenomena are caused by abnormal generator load rejection. With water hammer, a very high increase in upstream pressure may damage the pipelines and the turbine itself.

Most modern Pelton designs include a flow splitter (Figure 3.9). The flow splitter reduces the axial thrust that would be caused by water escaping the shovel. By splitting the shovel, the water jet exerts equal force on both sides of the shovel, thus reversing the water at almost 160°, to avoid striking back on the incoming water. A counter jet of water, known as a hydro brake, may also be used to help reduce the rotor speed in such operations, as illustrated in

Figure 3.9. Maintenance and repair of PTs are expeditious due to the relatively simple construction of the rotor and its easy access.

The injector aperture has a smaller diameter according to manufacturer specifications. To optimize turbine efficiency and ensure that the water exits freely after the incidence of the jet on the shovels, the Mosonyi condition should be observed:

$$\frac{D}{a} > 10 \tag{3.25}$$

The optimal diameter of the rotor circle is obtained as

$$D_{opt} = 5.88Q^{1/2}z^{-9/2}H^{-1/4}$$

3.6.2 Francis Turbine

The field of application of FTs in small hydroelectric power plants is from 3 to 150 m of height and from 100 L/s of design flow of the motive water (Figure 3.10). According to local conditions, the turbine shaft can be horizontal or vertical.

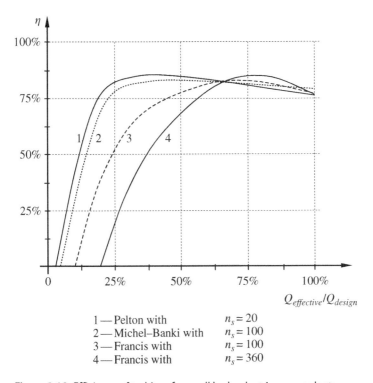

η
100%
75%
50%
25%
0
25%
50%
75%
100%
Q_effective/Q_design
1
2
3
4
1—Pelton with n_s = 20
2—Michel–Banki with n_s = 100
3—Francis with n_s = 100
4—Francis with n_s = 360

Figure 3.10 Efficiency of turbines for small hydroelectric power plants.

The horizontal position is the most suitable, because it eases direct connection of generators in conventional manufacturing (synchronous or asynchronous). Vertical shafts are difficult to repair and maintain, and more space is needed above the machine to connect the generators, resulting in a greater weight and housing, which increases costs and, in addition, necessitates some means of erection.

FTs are mostly used because of their wide range of suitable heads, characteristically ranging from 3 to 600 m. At the high head range, the flow rate and output must be large; otherwise, the runner becomes too small for reasonable fabrication. At the low-head end, propeller turbines are usually more efficient unless the power output is kept within certain limits.

PTs have the advantage of a better efficiency curve than that of FTs (Figure 3.10). That curve is more horizontal for a $Q_{effective}/Q_{design} = 25\%$, which is explained by the small variation in flow speed as a function of the flows themselves and wear caused by sensitivity to materials in suspension in the moving water.

Wearing caused by fine grains in suspension is less for FTs and MBTs than for PTs, as shown in the following sections. The following disadvantages of FTs are observed with respect to PTs:

1) Disassembly and assembly of an FT for maintenance and repair are very difficult.
2) The turbine is very sensitive to cavitation.
3) The efficiency curve of the turbine is not optimum, particularly at flows much smaller than the design flows (Figure 3.10).
4) The turbine is more sensitive to materials in suspension dragged by the water, which causes possible wear and consequent efficiency reduction.
5) The fast closure times of this turbine cause more intense water hammer, to cope with which the tube specifications must be over dimensioned.
6) The FT is not stable for operation at power 40% lower than its maximum power.
7) Complex mechanisms of control demand expert work during maintenance and recovery.

The advantage of an FT over a PT is due to the suction tube, which makes usable the totality of the elevation between upstream and downstream used for generation of hydroelectric energy. The diameter of the rotor can be approximately determined by the relationship

$$D = \frac{(0.16n_s + 35.1)H^{1/2}}{n} \tag{3.26}$$

where:

n = Number of revolutions (effective speed of rotation) (rpm)
n_s = Rated specific rotation (rpm)

The specific rotation is given by

$$n_s = \frac{n\left(1.36P_t\right)^{1/2}}{H^{5/4}}$$
(3.27)

where:

P_t = Turbine power (kW)
H = Height of the water head (meters)

Empirically, for heads between 10 and 50 m, the rotation of an FT is established between 900 and 1200 rpm, and for heads higher than 50 m, the range of values is 1200–1500 rpm.

Having thus verified the estimated diameter of the wheel, one can determine the approximate dimensions of the turbine set (the snail-shaped box), with details shown in Figures 3.11 and 3.12. Table 3.2 presents the main proportions for construction of an FT. In the table notice that a distinction is made between the length of vertical and horizontal aspiration tubes, L_v and L_h, respectively.

In small hydroelectric power plants, turbines are generally already equipped with a tube of vertical aspiration; in such cases, there is no need for a concrete aspirator tube.

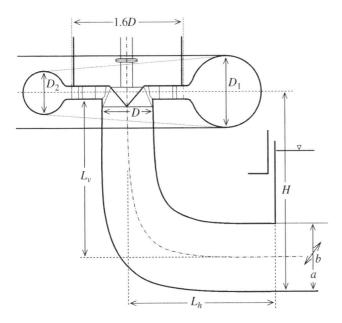

Figure 3.11 Dimensions of a spiral Francis turbine.

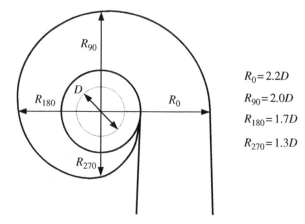

$R_0 = 2.2D$

$R_{90} = 2.0D$

$R_{180} = 1.7D$

$R_{270} = 1.3D$

Figure 3.12 Estimate of dimensions of a Francis turbine.

Table 3.2 Approximate dimensions for a Francis turbine.

Description	Variable	Dimension (m)
Smaller diameter	D_1	$0.6D$
Larger diameter	D_2	$1.0D$
Height of the aspirator tube	a	$1.2D$
Width of the aspirator tube	b	$3.0D$
Area of the aspirator tube	ab	$3.6D^2$
Length of the vertical aspiration	L_v	$(5 - n_s/200)D$
Length of the horizontal aspiration	L_h	$2.2D$
Maximum height	H	$(3.4 - n_s/400)D$

3.6.3 Michell–Banki Turbine

MBTs (also called radial thrust) have a capacity of up to 800 kW. Their flows vary from 25 to 700 L/s (according to the machine dimensions), with head heights in the range 1–200 m. The number of slats installed around the rotor varies from 26 to 30, according to the wheel circumference, whose diameter is from 200 to 600 mm. An MBT can be installed with an output free from the water or with a suction tube, in which the entire water elevation is used. The MBT set is shown in Figure 3.13, where the path followed by the water flow around the turbine rotor can also be observed.

Compared with other types of turbines, the MBT with its rotor division in cells (longitudinal segments of the rotor) in the proportion 1 : 2 presents a big advantage. In other words, this multicellular turbine can be operated at one- or

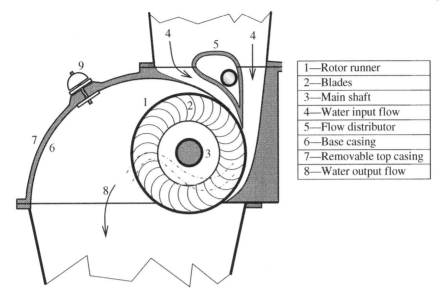

1—Rotor runner	
2—Blades	
3—Main shaft	
4—Water input flow	
5—Flow distributor	
6—Base casing	
7—Removable top casing	
8—Water output flow	

Figure 3.13 Michell–Banki turbine with vertical input.

two-thirds of its capacity (in the presence of low or average flows) or at full capacity (in the presence of design flows, i.e., three-thirds).

By that disposition of variable flow capacities, any water flow is useful with optimum efficiency. The efficiency curves of the MBT are almost horizontal (Figure 3.14). Due to this tuning possibility, the turbine can be operated at 20% of its full power. The tuning device for variable flows works with rotatory slats, which can also serve as elements of turbine closure when the head does not exceed 30 m, as illustrated in Figure 3.14 for the Michell–Banki–Ossberger type. This is an interesting feature in many areas of recent electrification, where there are only conditions of low perceptual use of the turbine and where the demand for energy increases gradually. In these cases, a small hydroelectric power plant will reach its full potential of energy production only after some time. The major advantages of the MBT are fast assembly of the machine set, less demand for civil structures, and easy access to all elements of the equipment during maintenance.

3.6.4 Kaplan or Hydraulic Propeller Turbine

A Kaplan turbine (KT) is related to a low head or water stream, limited power, and optimum flow variations during the year and is recommended for heads from 0.8 to 5 m, approximately. To start a KT, a vacuum pump fills a siphon with water and forms an elevation between upstream and downstream. To stop the turbine, it is enough to stop the water flow through the relief valve on the distributor's upper part.

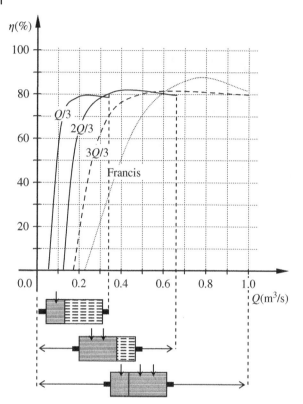

Figure 3.14 MBT efficiency curves for one or both cells.

In addition to being lower in cost than conventional types of low-head turbines (such as bulb and bulb-well types), the KT has the advantage of maintaining its electromechanical parts out of the water. This feature eases routine inspection and maintenance and adds safety in the case of floods. As its installation does not demand water reservoirs or prominent civil structures, the impact on the environment is negligible. According to the flow type (i.e., regulated or variable), the wheel shovels are fixed or adjustable, respectively (Figure 3.15). The efficiency of the turbine becomes higher if adjustable shovels are used because they adapt better to changes in the watercourse.

Either a pulley or a belt can drive generators with output powers below 100 kW. The turbine wheel

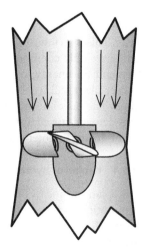

Figure 3.15 Kaplan turbine.

Figure 3.16 Flow-head curves for the Kaplan turbine.

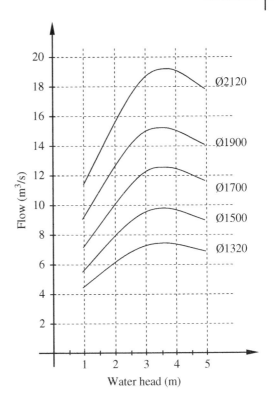

should be installed so that the lowest point of the wheel is above the upstream maximum level. That height is calculated for each project, as a function of cavitation, related directly to the downstream level of operation. The selection of a KT that adapts best to the flow and head of a watercourse can be based on a graph similar to the one in Figure 3.16.

A small power plant using a KT does not dispense some auxiliary services: the electric motor of the vacuum pump and the electric motor drivers of the wheel shovels, when applicable (approximately from 0.5 to 3.0 kW). A rotation multiplier should also be installed for high-speed generators. All of this contributes to additional losses.

3.6.5 Deriaz Turbines

When a turbine is neither radial nor axial, it is known as being of mixed flow, diagonal, or semi-axial. The Deriaz turbine, named after its inventor Paul Deriaz, is a type of water turbine similar to a KT but with inclined blades to adapt it for higher heads. It can be classified between KT and FT for the head range between 20 and 100 m with its runner blades adjustable to allow smooth and efficient operation over a wide range of heads and loads with uniform

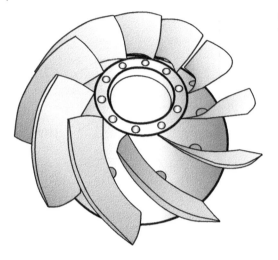

Figure 3.17 Rotor of a Deriaz turbine.

distribution of pressure and load across the blades. It is not subject to cavitations across the entire operating range, and its rotor looks like the one represented in Figure 3.17.

The DT is classified more precisely as a double-regulated, reaction, mixed-flow turbine. It has a hydraulic design and was the first kind of diagonal turbine to be developed. An early major DT system was installed at plants on both the Canadian and US sides of Niagara Falls.

The thrusting water follows an approximately conic surface around the runner. A DT, developed in the 1960s, can reach a capacity from 1 MW up to 200 MW. Its flow varies broadly from 1.5 to 250 m^3/s (according to the machine dimensions), with head heights in the range 5–1000 m. The runner diameter may be up to 7000 mm with six to eight runner blades. Diagonal turbines operate very economically as either turbines or pumps. They resemble a fast FT except for the size and move of the runner blades.

3.6.6 Water Pumps Working as Turbines

The reverse use of water pumps as turbines for small hydroelectric power plants became very popular because of their lower costs in several applications [3, 12]. These pumps, usually for small capacity, have been used for many years in industrial applications to recover energy that could be wasted. They have the following advantages:

1) They cost less because they are mass-produced for other purposes (i.e., as water pumps for buildings and residences).
2) Their acquisition time is minimal because they have a wide variety of commercial standards and are available in hardware stores and related shops.

However, there are some disadvantages: (i) they have slightly reduced efficiency compared with the same head height used for water pumping, and (ii) they are sensitive to the cavitation characteristics and operating range.

The main difference between the operations of those machines either as a pump or as a turbine is that fluid (water) flowing to the pumps is determined by the head height, of which there is no efficient control. On the other hand, a hydraulic turbine has efficient flow control. Such efficiency may pay back for their higher cost.

The use of pump–turbines in industries is justified by the need to dissipate some form of surplus rotating energy [12–14]. If the water flow has to be controlled for any reason, it can also be used to produce another form of energy when such constant flow is diverted for the turbine. Constant power is generated, which can be injected into the utility grid or used to convert energy when a load controller is available. If the water flow has a lot of variation, such conditions can be approached by the use of another pump–turbine with a different capacity (for instance, 1 : 2) in order to restrict the range of variation of the water flow in each or by electronic control by the load (see Chapter 12) [15–17].

3.6.7 Specification of Hydro Turbines

The choice of turbine type depends on the application range, primarily on height and water flow and other criteria, such as maintenance, turbine cost, and sensibility to materials in suspension. In Figure 3.18, the limits of operation recommended are represented for the aforementioned turbines, with their application fields a function of the head height, water flow, and power of each machine, as discussed in the following sections [12–14, 18].

In order to get a good purchase price for a turbine, several manufacturers should be consulted. Comparing values demands good specifications (as close as possible to that desired), so that small differences in the product characteristics do not unduly influence the final price. Super dimensioning should be avoided, as it dissuades manufacturers from offering equipment better adapted to each case. The following items should be available at the time of purchase:

1) Buyer's identification (name, profession, address, telephone, etc.)
2) Cost estimation or detailed specification
3) Sketch of the installation site for the power plant (if possible, with pictures), elements available in addition to natural features, and usable technical characteristics

For a hydro turbine, it is necessary to specify:

1) Available average, minimum, and maximum heights
2) Available flow for generation with its duration curve, site water storage capacity, information about historical variations, and so on

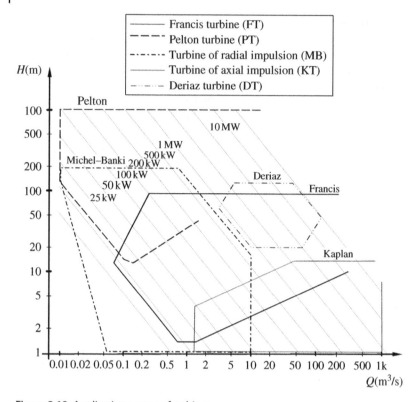

Figure 3.18 Application ranges of turbines.

3) Water quality (materials in suspension, abrasiveness, and other useful characteristics)
4) Present and future energy needs of the generator or turbine, according to the effective capacity of the site (otherwise, any specification becomes useless)
5) Possible need for an isolation valve for the turbine
6) Specifications for the type of regulator required
7) Specifications for the generator (see Chapter 10)
   a) Type (induction, synchronous, or direct current)
   b) Electric output (alternating or direct current, maximum power, voltage, number of phases, frequency)
   c) Climate (winds, temperature, and air humidity)
   d) Operating mode (manual, automatic, semiautomatic, or remote control, stand-alone or for injection in the network; in this last case, is there a sale, purchase, or exchange agreement with the electric power company?)
   e) Other indispensable equipment (control panels, protection and drivers, single- or three-phase transmission and distribution lines)

8) Location of the powerhouse, if there is one, with upstream head, distance from the water intake, dimensions, drainage height (minimum, maximum and average)
9) Specification of the diameter, length, configuration, and manufacturing material if a feeder of water is used under pressure for the turbine
10) Access to the site (boat, bike, highway, rails, walking path, or inhospitable access by land) for transport of equipment and maintenance planning
11) Existence of animals, fish, or any other form of life that should be preserved

References

1 Eletrobrás/DNAEE, Small Hydroelectric Power Plants Handbook (Manual de Pequenas Centrais Hidroelétricas), Eletrobras Co., Rio de Janeiro, Brazil, 1985.
2 A. Brown, Stream Flow Measurement by Salt Dilution Gauging, ITIS, Cambridge University, November 1983.
3 A.R. Inversin, Micro-hydropower Sourcebook, National Rural Electric Cooperative Association (NRECA), International Foundation, Washington, DC, ITDG Publishing, December 1986, 286 pp.
4 H. Lauterjung, Inventory of offer of hydropower plants, Taller de Microcentrales Hidroelectricas, 1990: Proceedings of Unesco-GTZ-PLL, Montevideo, Uruguay, November 1990, pp. 89–188.
5 W.G. Ovens, A Design Manual for Water Wheels, Vita, Zingem, Belgium, 1975.
6 Z. Souza, R.D. Fuchs, and A.H.M. Santos, Hydro and thermo electrical power plants, Electric Brazilian Power Plants, Federal School of Engineering of Itajubá, Minas Gerais, Brazil, and Edgard Blücher, São Paulo, Brazil, 1983.
7 G.P. Schreiber, Hydroelectric Power Stations, Vol. 3, Engevix Studies and Projects of Engineering, Edgard Blücher, São Paulo, Brazil, 1987.
8 Z. Souza, Hydroelectric Power Plants: Sizing of Components, Edgard Blücher, São Paulo, Brazil, 1992.
9 A.J. Macintyre, Hydraulic Motive Machines, Guanabara, Rio de Janeiro, Brazil, 1983.
10 The Small Notable, Catalog, MTV-1527, Betta Hydraulic Turbines, Franca, Brazil, 1991.
11 Types of hydropower turbines, https://energy.gov/eere/water/types-hydropower-turbines, energy.gov, Office of Energy Efficiency & Renewable Energy, accessed in March 2017.
12 F.A. Farret, Use of Small Sources of Electrical Energy (Aproveitamento de Pequenas Fontes de Energia Elétrica), Editora UFSM, Santa Maria-RS, Brazil, 2014.

13 W. Bolliger, J.A. Menin, and S. Pumps, Éxito de las bombas como turbinas, Sulzer Technical Review (Spanish ed.), February 1997, pp. 27–30.

14 E.L. Henn, Máquinas de fluído (Machines of Fluid), Universidade Federal de Santa Maria, Santa Maria, Brazil, 2001.

15 C.D. Mello, Jr. and F.A. Farret, Improvements in the structure and operation of an experimental micro power plant with electronic control by the load, Proceedings of the 12th CIGRÉ National Seminar on Production and Electric Power Transmission, Vol. 2, October 1993, pp. 46–51.

16 S. Chakraborty, M.G. Simões, and W.E. Kramer (eds.), Power Electronics for Renewable and Distributed Energy Systems, Springer-Verlag, London, UK, 2013.

17 M.G. Simões and F.A. Farret, Modeling and Analysis with Induction Generators, CRC Press, Boca Raton, FL, 2015.

18 F. A. Farret, M. Godoy Simões, and A. Michels, "Small Hydroelectric Systems," in Power Electronics for Renewable and Distributed Energy Systems: A Sourcebook of Topologies, Control and Integration, 1st ed., Springer-Verlag, London, 2013, Chapter 5, pp. 151–184.

4

Wind Power Plants

4.1 Introduction

The development of our modern society is contingent on the sustainability of energy. The vulnerability of the energy chain, only reliant on nonrenewable fossil fuel resources, may bring a collapse in our society if the exhaustion of natural reserves will really happen in the future.

Renewable sources are mostly welcome but dependent on where it is going to be used. In particular, wind energy is strongly advocated in several studies based on the following factors:

- Nonexistence of rivers or other energetic hydro resources in close proximity
- High costs of hydro- and thermoelectrical generation
- Areas with fairly high average wind speeds (>3 m/s)
- Need to feed remote loads, where taking a transmission network is not cost-effective
- Need for renewable, nonpolluting energy

The best locations for a wind power plant must have several attributes: high wind resource, close to the transmission line access, close to the load center, easy access to major highway for transportation, availability of inexpensive land, and environmentally suitable for wind power plant. Wind power energy is derived from solar energy. Uneven distribution of temperatures in different areas of the Earth allows the mass flow of air; such resulting movement of air mass is the source of mechanical energy to spin wind turbines and their respective generators. Across Europe, there are several leading wind energy markets: Germany, Spain, Denmark, Great Britain, Sweden, the Netherlands, Greece, Italy, and France. United States plays a major role in developing wind energy systems; Japan and Australia also give importance to the contribution of wind energy technology [1–5].

There are several European companies exporting products and technology. US wind resources are also appreciable and large enough to generate more

Integration of Renewable Sources of Energy, Second Edition. Felix A. Farret and M. Godoy Simões.
© 2018 John Wiley & Sons, Inc. Published 2018 by John Wiley & Sons, Inc.

than 4.4 trillion kWh of electricity annually. There is sufficient wind energy for powering generation on mountainous areas and deserts, as well as in the mid-lands, spanning the wind belt in the Great Plains states as shown in Figure 4.1. North Dakota alone is theoretically capable (if there were enough transmission capacity) of producing enough wind-generated power to meet more than one-third of US demand. According to Battelle Pacific Northwest Laboratory, wind energy can supply about 20% of US electricity, with California having the largest installed capacity. The US Department of Energy provides a,high-resolution wind resource map for 80-m height in the United States with links to state wind maps. Potentially windy sites can be located within a fairly large region and determining a potential site economic and technical viability analysis [6, 7].

In South America, Brazil has vast wind potential because of their extended Atlantic coast. The Brazilian northeast states have scarce hydroelectric resources. However, the integrated generation of wind and solar power is a powerful alternative [8–10]. The Brazilian wind power potential is evaluated at 63 trillion kWh/year. Table 4.1 shows a comparison of conventional power plants with wind power plants.

4.2 Appropriate Location

In order to select the ideal site for placement of wind power turbines, it is necessary to study and observe the existence of wind and make measurements to evaluate extraction of energy at a desired rate and cost. Although flat plains may have steady strong winds, for small-scale wind power, the best choice is usually along the dividing lines of waters (i.e., the crests of mountains and hills). In those geographical locations, there is good wind flow perpendicular to the crest direction [11–13]. Some basic characteristics to be observed for defining a site are as follows:

- Wind intensities in the area
- Distance of transmission and distribution networks
- Topography
- Purpose of the energy generated: residential, rural, industrial, public, commercial, and others
- Means of access

4.2.1 Evaluation of Wind Intensity

The energy captured by the rotor of a wind turbine is proportional to the cubic power of the wind speed. It is very important to evaluate the historical wind power intensity (W/m^2) to access the economic feasibility of a site. It requires taking into account seasonal as well as year-to-year variations in the local climate [14–16]. To estimate the wind power mechanical capacity, P, Bernoulli's

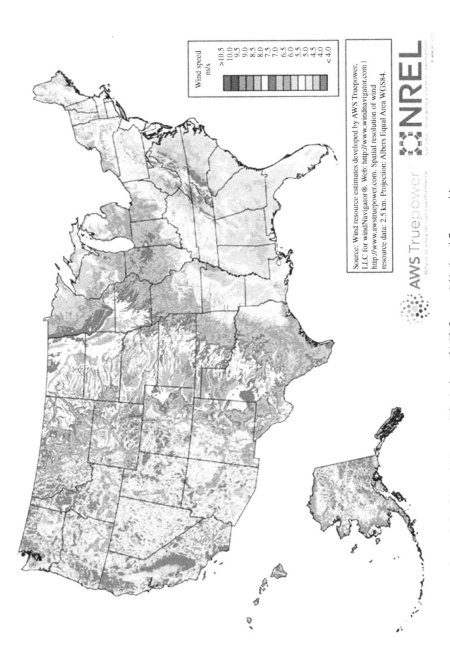

Wind speed
m/s

>10.5
10.0
9.5
9.0
8.5
8.0
7.5
7.0
6.5
6.0
5.5
5.0
4.5
4.0
<4.0

Source: Wind resource estimates developed by AWS Truepower,
LLC for windNavigator®. Web: http://www.windnavigator.com |
http://www.awstruepower.com. Spatial resolution of wind
resource data: 2.5 km. Projection: Albers Equal Area WGS84.

AWS Truepower
Where science delivers performance

NREL

Figure 4.1 Utility-scale land-based 80-meter US wind map [6, 7]. *Source:* © National Geographic.

Table 4.1 Comparison of conventional power plants with wind power plants.

Conventional power plant	Wind power plant
Single or several large (40 MW to 1000 MW+) generator	Many (hundreds) of wind turbines (1–5 MW each) covers a very large area
Prime mover: steam, combustion engine—nonrenewable fuel	Prime mover: wind turbine—wind
Controllability: adjustable up to max limit and down to min limit	Controllability: curtailment, ramp rate limit, output limit
Real power is scheduled with power droop capability	Real power follows the wind speed variation
Reactive power is controllable up to its maximum limit for a PQ bus, or it is used to control the voltage for a PV bus	Reactive power is controllable by the wind turbine control and/or plant level reactive compensation to control the reactive power (Q), the power factor (PF), or the voltage (V)
Located where convenient for fuel and transmission access	Located at wind resource, it may be far from the load center
Generator: synchronous generator	Generator: four different types (fixed speed, variable slip, variable speed, full converter)
Fixed speed—no slip: flux is controlled via exciter winding. Flux and rotor rotate synchronously	Types 3 and 4: variable speed with flux-oriented controller (FOC) via power converter. Rotor and flux do not rotate synchronously

equation (defined in Chapter 3) is used with respect to the mass flow derivative of their kinetic energy, K_e:

$$P = \frac{dK_e}{dt} = \frac{1}{2}v^2\frac{dm}{dt} \tag{4.1}$$

The mass flow rate per second is given by the derivative of the quantity of mass, dm/dt, of the moving air that is passing with velocity v through the circular area A swept by the rotor blades, as suggested in Figure 4.2. According to equation (3.3), for any average flow ($\bar{Q} = A\bar{v}$) of a fluid, the upstream flow of mass can be given in terms of the volume of air V as

$$\frac{dm}{dt} = \rho\frac{dV}{dt} = \rho A\bar{v} \tag{4.2}$$

where:

$\rho = m/V$ = The air density in kg/m^3 (= 1.2929 kg/m^3 at 0 °C and at sea level)
A = The surface swept by the rotor or blades (m^2)

The effective power extracted from wind is derived from the airflow speed just reaching the turbine, v_1, and the velocity just leaving it, v_2. The number of

Figure 4.2 Typical three-blade wind rotor. *Source:* © National Geographic.

blades must also be considered. Equation (4.2) considers the average speed $(v_1 + v_2)/2$ passing through area A of the rotor blades (i.e., effectively acting on the rotor blades). Therefore, equation (4.2) becomes

$$\frac{dm}{dt} = \frac{\rho A (v_1 + v_2)}{2} \tag{4.3}$$

The energy absorbed by the turbine is the difference in the kinetic energy in the wind speed (usually expressed in kgm/s = 9.81 W) just reaching and just leaving the turbine. A net wind mechanical power of the turbine is imposed by this kinetic energy difference, which may be estimated by equation (4.1) as

$$P_m = \frac{dK_e}{dt} = \frac{1}{2}\left(v_1^2 - v_2^2\right)\frac{dm}{dt} \quad \left(W/m^2\right) \tag{4.4}$$

Combining equations (4.3) and (4.4), the following power is calculated:

$$P_m = \frac{dK_e}{dt} = \frac{1}{2}\rho A\left(v_1^2 - v_2^2\right)\cdot\frac{1}{2}\left(v_1 + v_2\right)$$

whose upstream wind speed, v_1, is put in evidence to give

$$P_m = \frac{1}{4}\rho A v_1^3 \left(1 - \frac{v_2^2}{v_1^2}\right)\left(1 + \frac{v_2}{v_1}\right)$$

or

$$P_m = \frac{1}{2}\rho C_p A v_1^3 \tag{4.5}$$

where:

$C_p = \frac{1}{2}\left(1 - v_2^2/v_1^2\right)\left(1 + v_2/v_1\right) =$ The power coefficient or rotor efficiency η_{tip_end} as expressed by Betz.

If C_p is considered a function of v_2/v_1, the maximum of such a function can be obtained for $v_2/v_1 = 1/3$ as $C_p = 16/27 = 0.5926$. This value is known as the Betz limit, the point at which if the blades would be 100% efficient. However, at this condition, the wind turbine would no longer work because the air (having given up all its energy) would stop entirely after passing its blades. In practice, the collection efficiency of a rotor is not as high as 59%; a more typical efficiency is between 35 and 45%.

To achieve the best designs, several sources of data, such as meteorological maps, statistical functions, and visualization aids, should be used to support the analysis of the wind potential.

4.2.1.1 Meteorological Mapping

Meteorological maps show curves connecting the same intensity of average wind speed. Although such maps may yield valuable information, they are not sufficient for a complete analysis because such meteorological data acquisition stations do not look only for a determination of the wind potential for power energy use. Therefore, a final decision on the site can be made only after a convenient local selection procedure and thorough intensive data acquisition [17–20].

Local wind power is directly proportional to the distribution of speed, so that different places that have the same annual average speed can present very distinct values of wind power. Figure 4.3 displays a typical wind speed distribution curve for a given site. That distribution can be made monthly or annually. It is determined through bars of occurrence numbers, or percentile of occurrence, for each range of wind speed over a long period. It is usually observed as a variation of wind linked with climate changes in the area. It is typical in temperate climates to have summer characterized as a season of little wind and winter as a season of stronger winds.

When the wind speed is lower than 3 m/s (referred to as a calm period), the power becomes very limited for the extraction of energy, and the system should

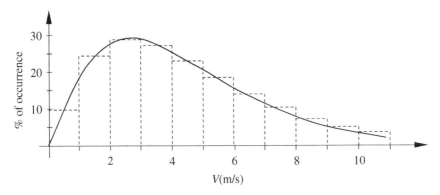

Figure 4.3 Wind speed distribution.

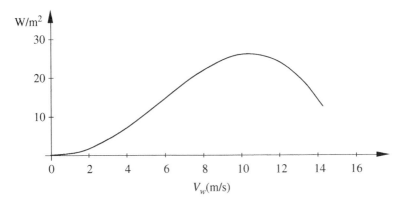

Figure 4.4 Wind power typical distribution.

be stopped. Therefore, for power plants, calm periods will determine the time required for energy storage. An easy way to define storage needs is to plot the wind data and mark where the wind velocity is less than 3 m/s and define what storage strategy should take over during those periods. As discussed next, power distribution varies according to the intensity of the wind and with the power coefficient of the turbine. A typical distribution curve of power assumes the form shown in Figure 4.4. Sites with high average wind speeds do not have calm periods, and there is not much need of storage. However, high wind speed may cause structural problems in a system or in a turbine [12, 21–23].

The worldwide wind pattern is mostly caused by three factors: (i) the low temperatures on the Earth poles with respect to equator temperatures, (ii) the low temperatures of the high altitudes of the atmosphere with respect to the Earth ground temperatures, and (iii) the Earth rotation with respect to the air mass involving the planet. The effect composition of these three factors cause

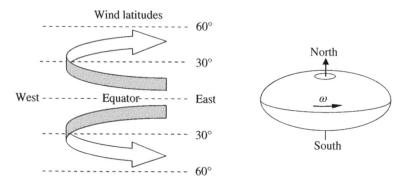

Figure 4.5 Global wind patterns.

a wind tendency as depicted in Figure 4.5. The temperature difference causes convection effect above the ground surface, which makes the airflow move upward on the equator and go downward about 30° north and south latitudes. This air convection can be distorted by the tangent speed differences of the Earth's spin such that the wind tends to move from east to west. Horizontal wind speed near the equator is very low because winds move more upward. The most pronounced intensities of wind go from west to east between latitudes 30° and 60° north and south of the equator, forming two natural wind tunnels by a motion caused by the circulation of east to west winds below the 30° latitudes (see Figure 4.5). Usually, west wind speeds are higher than east wind speeds. This general tendency is affected strongly by the ground formation (mostly by valleys and mountains). As a result, it is very useful to represent wind variation by a probabilistic vector of determined intensity and direction.

4.2.1.2 Weibull Probability Distribution

To establish a probability distribution, it is important to establish the duration curve of the wind speed for every hour of the day, every day of the year, a total of 8760 pieces of data. Figure 4.6 shows a curve formed by points marked in several speed ranges, in accordance with the number of hours accumulated for a particular wind intensity and above that range [13, 16, 24].

Wind flowing around the Earth is a random phenomenon. Therefore, a large sample of wind data taken over many years should be gathered, to increase confidence in the data available. This is not always possible, so shorter periods are often used. The data are often averaged over the calendar months and can be described by the Weibull probability function, given as equation (4.6) [11–13].

$$h(v) = \left(\frac{k}{c}\right)\left(\frac{v}{c}\right)^{k-1} e^{-(v/c)^k} \quad \text{for } 0 < v < \infty \tag{4.6}$$

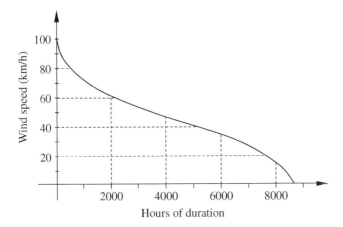

Figure 4.6 Accumulated duration of wind speed for one year period.

The Weibull function expresses the fraction of time; the wind speed is between v and $v + \Delta v$ for a given Δv. In practice, most sites around the world present a wind distribution for k (shape factor) within the range 1.5–2.5. For most of them, $k = 2$, a typical wind distribution found in most sites, is known as the Rayleigh distribution, given by

$$h(v) = \left(\frac{2}{c}\right)\left(\frac{v}{c}\right)e^{-(v/c)^2} \tag{4.7}$$

Factor c is defined as *scale factor*, related to the number of days with high wind speeds. The higher c is, the higher the number of windy days. This parameter is usually taken in order to represent the wind speed features for most practical cases.

Weather repeats in seasons from one year to the next. For the purpose of wind variation studies, this period is usually taken and divided into the total number of hours (8760 hours/year). The unit of h in equation (4.6) may be stated as a percentage of hours per year per meter per second.

Equation (4.7) can be plotted for various parameters c as shown in Figure 4.7. The values of h are the number of hours in a year that the wind speed is within the interval from v to $v + \Delta v$ divided by the speed interval Δv. This plot provides a useful and realistic view of the average speed.

There is a predominant wind speed throughout a year. The predominant speed can be considered in at least three different ways, with very distinct implications:

1) One could be defined as a single speed at which the wind blows most of the time. It is a simplistic method that does not take into account any secondary predominant speed, which may contain a lot of energy.

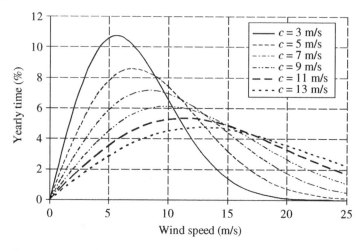

Figure 4.7 Rayleigh distribution of wind speed (parameter *c*).

2) A second way of defining wind speed is by the average speed experienced by a number of hours *h* throughout the year, given by equation (4.8):

$$V_{av} = \frac{1}{8760} \int_0^\infty hv\,dv \tag{4.8}$$

In this case, if $c = V_{av}$, the Rayleigh distribution given by equation (4.7) becomes equation (4.9):

$$h(v) = \frac{2v}{V_{av}^2} e^{-(v/V_{av})^2} \tag{4.9}$$

3) A third and more appropriate definition is to use the concept of root mean cube speed, V_{rmc}, analogous to the root mean square (effective value).
This definition is based on the idea that yearly average power varies with the cube of the wind speed experienced again by *h*; that is indicated by equation (4.10):

$$V_{rmc} = \sqrt[3]{\frac{1}{8760} \int_0^\infty hv^3\,dv} \tag{4.10}$$

Table 4.2 lists some popular formulas for the calculation of wind speed, root mean cubic power, and energy density over the number *n* of speed samples during the year, usually with $\rho_i = 1.225\,\text{kg/m}^3$ at $t = 15\,°\text{C}$.

Table 4.2 Common wind speed formulas.

Average speed	Root mean cubic speed	Cubic power density	Cubic energy density
$V_{av} = \dfrac{1}{n}\sum\limits_{i=1}^{n} v_i$	$V_{rmc} = \sqrt[3]{\dfrac{1}{n}\sum\limits_{i=1}^{n} v_i^3}$	$P_{rmc} = \dfrac{C_p}{2n}\sum\limits_{i=1}^{n}\rho_i v_i^3$	$E_{yw} = \dfrac{8{,}760 C_p}{2n}\sum\limits_{i=1}^{n}\rho_i v_i^3$
m/s	m/s	W/m^2	Wh/m^2/y

If it is assumed the power coefficient C_p as given by equation (4.5), it is possible to assume a value no higher than 0.5 for high wind speeds for two-blade turbines. Therefore, 0.5 can be conveniently used as the maximum practical rotor efficiency. Therefore, the maximum output power for the root mean cubic of the wind speed is estimated in accordance with equation (4.11).

$$P_{rmc} = \frac{\rho}{4} A V_{rmc}^3 \quad \text{in watts} \tag{4.11}$$

The total annual energy at the site can be obtained from equation (4.11) by multiplying it by 8760 hours. Therefore, the energy density per year and per swept area of the blades is given by

$$E_{yw} = 8760\left(\frac{P_{rmc}}{A}\right) = 8760\frac{\rho}{4}V_{rmc}^3 \quad \text{in Wh/m}^2 \tag{4.12}$$

The Rayleigh wind and energy density distribution is depicted in Figures 4.8a and b for $c = 7\,\text{m/s}$ and $c = 10\,\text{m/s}$, respectively. Note that for a minor change in the wind speed, there is a significant increase and shift in the power density along the speed axis. Such changes would not be noticed much if only the yearly maximum or average speeds were used for those estimations, because they retain almost the same value in the two distributions represented.

Based on these curves and on the load to be fed, it is possible to establish the turbine–generator power that would have the best utilization factor for a given wind condition. It is evident that the longer the wind behavior is observed in an area, the more appropriate will be the design for a small power plant. Variable-speed control is recommended for a good wind energy system, due to the cubic relation of the yearly energy to the wind speed.

4.2.1.3 Analysis of Wind Speed by Visualization

If a site being evaluated is not close to meteorological stations, a good practical suggestion is to observe the existing trees. Their deformation level will serve as a good indicator of the wind speeds in the area. The intensity of the wind

(a)

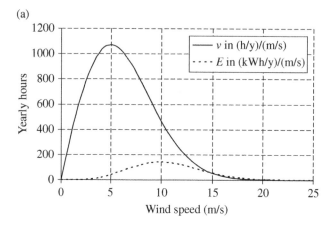

(b)

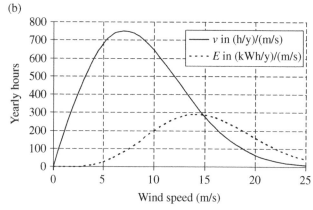

Figure 4.8 Rayleigh wind and energy density distribution ($k = 2$): (a) $c = 7$ m/s and (b) $c = 10$ m/s.

increases with height; therefore, trees of larger span are reached by more intense winds, which can harm their growth. The following levels of tree deformation are recognized:

- *Brush:* When branches are on the lee side (the side exposed to wind), especially in the absence of leaves, the winds are weak.
- *Flag:* Branches are the lee barriers (the static trunks in their original position, free to windward); in other words, the side from which the wind blows.
- *Lie down:* The wind is strong enough to produce permanent deformations in trunks and branches.
- *Shear:* The wind is always strong, to the point of breaking branches, giving the impression that they were cut uniformly.
- *Rug of trees:* The wind energy is so strong that it limits the growth of trees to some inches of the soil, giving the impression that the trees form a rug.

For small wind power installations, a visual assessment of the wind intensity is described in Table 4.3, which is an adaptation of the Beaufort table [9, 10]. The International Committee of Meteorology adopted such a table in 1874. It is important to note that wind measurement using anemometers did not begin until 1939.

4.2.1.4 Technique of the Balloon

Another very simple method to measure wind speed is by measurement of the speed of a balloon allowed to move freely in the air [10, 13]. Two points of a known reasonable distance on the ground are selected. The time required for the balloon to pass between the two points in a horizontal line is recorded, and the wind speed is inferred from that.

Table 4.3 Estimation of wind speed (Beaufort table).

Degree	Classification	Effects of the wind on nature	Speed (m/s)
0	Calm	Everything is still. Smoke goes up vertically	0.00–0.30
1	Almost calm	Smoke is dispersed. Weather vanes are still. Wind is felt on the face	0.30–1.40
2	Breeze	Wind is felt on the face. The noise of leaves agitated by the wind is heard. Weather vanes move	1.40–3.00
3	Fresh wind	Leaves and small branches of trees are agitated constantly. Flags are stretched out	3.00–5.50
4	Moderate wind	The wind lifts dust and paper from the ground. Small tree branches are agitated	5.50–8.00
5	Regular wind	Small trees with leaves begin to balance	8.00–11.00
6	Wind mildly strong	Large branches move, electrical lines whistle. It begins to be difficult to walk against the wind	11.00–14.00
7	Strong wind	Entire trees are agitated. It is definitely difficult to walk against the wind	14.00–17.00
8	Very strong wind	Branches of trees break. It requires a great effort to walk	17.00–21.00
9	Windstorm	Tiles are lifted	21.00–25.00
10	Gale	Trees are torn down. There is construction damage	25.00–28.00
11	Storm	The wind assumes characteristics of a hurricane (rarely happens far away from coasts)	28.00–33.00
12	Hurricane	The air is full of solid particles and drops of water. The sea is entirely whitish	33.00–36.00

4.2.2 Topography

The analysis of topography is very important when choosing a site. This is an important point when mountainous locations are under consideration. For those areas, the wind speed increases on the front side of the mountain and decreases on the opposite side. In plane areas, the position of tree barriers should be observed with respect to the wind, as their proximity to the place chosen for turbine installation is critical. There should be a maximum of free space between tree barriers and the power plant, to avoid any slowing down of the wind (see Section 4.3).

With increased altitude, the movement of a wind draft takes a more complex form, due to different land shapes. Thus, it is necessary to have a reasonable knowledge of the characteristics of the land to be occupied. The roughness of the ground surface causes wind shear, which changes significantly with altitude. Some data collected at Merida Airport in Mexico show that the wind speed can be four to five times higher at an altitude of about 450 m with respect to the ground surface, and then it starts to decrease again [12]. About 100 m from the surface, many places on Earth have a wind speed sufficient for wind energy, but the costs of such high installation severely limit such a decision in favor of using such heights. There are some recent projects using balloons with wind turbines, but the transmission of the collected energy to the ground is not still solved.

4.2.3 Purpose of the Energy Generated

Wind power turbines can be sited for electric power generation, pumping of water, grinding of grains, and other uses. For electricity generation, a turbine should be installed at the highest places or places of maximum wind. For pumping water, the installation is obviously constrained by the availability of water. Valleys can behave like wind tunnels, making them good places for the installation of pumping systems [12].

High-scale electric power generation demands installation of several turbines. The number of units will depend on the total capacity of each unit, economical sizing of the turbine (cost per capacity (kW)), and the effects of interference among the turbines.

4.2.4 Accessibility

A turbine site should have good access conditions, especially in cases of energy generation on larger scales, to ease installation, operation, and maintenance services. The site should be close to roads, highways, or even railways, since there is heavy machinery and equipment to be transported. There should be suitable safety conditions against vandalism and theft. There is a great impact on a project's total cost if highways, security systems, and other facilities must be constructed.

A favorable economics factor is that the high-speed characteristics of the wind for power plant installations lower the costs for the right-of-way. Generally, these lands are impractical for other uses, such as agricultural or animal farming.

4.3 Wind Power

As discussed in Section 4.2, only part of the full energy available from the wind can be extracted for energy generation, quantified by the power coefficient, C_p. The power coefficient is the relationship of the power extraction possible to the total amount of power contained in the wind [3, 13–16]. Figure 4.9 relates the power coefficient to the tip speed ratio, λ, defined as the relationship between the rotor blade-tip speed v and the free speed of the wind V_w for

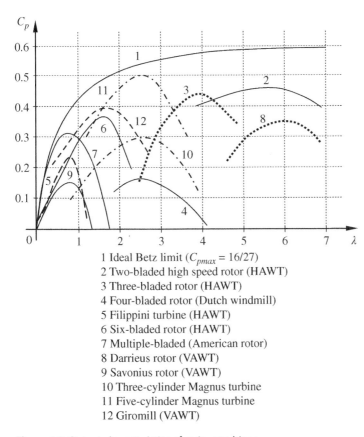

1 Ideal Betz limit (C_{pmax} = 16/27)
2 Two-bladed high speed rotor (HAWT)
3 Three-bladed rotor (HAWT)
4 Four-bladed rotor (Dutch windmill)
5 Filippini turbine (HAWT)
6 Six-bladed rotor (HAWT)
7 Multiple-bladed (American rotor)
8 Darrieus rotor (VAWT)
9 Savonius rotor (VAWT)
10 Three-cylinder Magnus turbine
11 Five-cylinder Magnus turbine
12 Giromill (VAWT)

Figure 4.9 Output characteristics of various turbines.

Table 4.4 Standard air constants.

Sea level standard atmospheric pressure	$P_0 = 101.325\,\text{kPa}$
Sea level standard temperature	$T_0 = 288.65\,\text{K}\ (15.5\,^\circ\text{C})$
Earth-surface gravitational acceleration	$g = 9.80665\ \text{m/s}^2$
Temperature lapse rate	$L = 0.0065\,\text{K/m}$
Ideal (universal) gas constant	$R = 8.31447\,\text{J/(mol}\bullet\text{K)}$
Molar mass of dry air	$M = 0.0289644\,\text{kg/mol}$
Air density	$\rho = 1.2929\,\text{kg/m}^3$

several wind power turbines. As stated earlier, in quantitative terms, the tip speed ratio (*tsr*) is defined as $\lambda = v/V_w = \omega R/V_w$, where ω is the angular speed of the turbine shaft and R is the length of each blade. This coefficient emphasizes the importance of knowing the purpose for which the energy will be used, to allow determination of the best selection for wind power extraction.

From equation (4.5), the turbine mechanical power can be given by

$$P_t = \frac{C_p \rho A V_w^3}{2} \quad \text{kg}\cdot\text{m/s} \tag{4.13}$$

4.3.1 Wind Power Corrections

The main parameters for wind speed corrections are related to the sea level as displayed in Table 4.4. The lapse rate or air adiabatic incremental rate (in kelvin/m) is the decrease of an atmospheric variable with height, the variable being temperature unless otherwise specified [13, 17, 25].

The air density ρ in Table 4.5 can be corrected by the gas law ($\rho = m/v = mP/nRT = MP/RT$, which, referred to standard atmospheric conditions 0 and 1, gives $\rho_1 = \rho_0(T_0/T_1)(P_1/P_0)$, say, $P_0 = 760\,mm, T_0 = 273\,K$) and expressed as

$$\rho = 1.2929\frac{273}{T}\frac{P}{760} \tag{4.14}$$

where:

$P = $ The atmospheric pressure (mmHg)
$R = $ Reynold's constant
$T = $ The kelvin absolute temperature

However, under normal conditions ($T = 296\,K$ and $P = 760\,\text{mmHg}$), the value of ρ is 1.192 kg/m^3, and at $T = 288\,K$ (15 °C) and $P = 760$ mmHg is 1.225 kg/m^3.

Table 4.5 Speed and power loss measured at the margin of trees.

Porosity	Loss of:	Distances in the direction against the wind in tree widths				
(%)	(%)	5	10	15	20	30
20	Speed	16	7	4	3	2
	Power	41	18	12	8	6
40	Speed	20	9	6	4	3
	Power	49	25	17	13	9
Height of the area of turbulent flow (in tree heights)		1.5	2.0	2.5	3.0	3.5
Width of the area of turbulent flow (in tree widths)		1.5	2.0	2.5	3.0	3.5

Source: Refs. [3, 17]. © European statistics.

For equation (4.14) we can assume a temperature derating factor of approximately $1\,°C$ for every $150\,m$. The influence of humidity might be neglected. If just the altitude h (in meters) is known ($10{,}000\,ft = 3048\,m$), the air density can be estimated by the two first terms of a series expansion of the exponential term like

$$\rho = \rho_0 e^{-\left(\frac{0.297}{3048}h\right)} \approx 1.225 - 1.194 \times 10^{-4}\,h \tag{4.15}$$

Similarly, the temperature and pressure above the sea level can be corrected with

$$T = T_0 - Lh$$

$$P = P_0\left(1 - \frac{Lh}{T_0}\right)^{gM/RL}$$

where:

$h =$ Altitude (m)
$L =$ Temperature lapse or air adiabatic incremental rate ($0.0065\,K/m$)

Therefore, from equation (4.13), the turbine torque is given by

$$T_t = \frac{P_t}{\omega} = \frac{\rho A R V_w^2 C_T}{2} \tag{4.16}$$

where the torque coefficient is defined as $C_T = C_p/\lambda$.

If $S = 1\,\text{m}^2$ and $\rho = 1.2929\,\text{kg/m}^3$, the maximum wind potential can be obtained from equation (4.13) (without taking into account the aerodynamic losses in the rotor, the wind speed variations in several points in the blade sweeping area, the rotor type, etc.) as being

$$\frac{P}{A} = 0.5926 \times 0.6464 \times V_w^3 = 0.3831 V_w^3 \tag{4.17}$$

where:

$P/A =$ The wind power per swept area (W/m^2)
$V =$ The wind speed (m/s)

The coefficient in equation (4.17) is usually quite smaller because of losses and uneven distribution of the wind on the blades, and it may be approximated by

$$\frac{P}{A} = 0.25 V_w^3 \tag{4.18}$$

4.3.2 Wind Distribution

The local wind power is proportional to the occurrence of the wind speed distribution, so that in different places with the same annual average speed, the wind power can have quite different values. A factor of speed loss, decreasing wind power, due to buildings and trees in the vicinity of small wind power plants, is known as site roughness. Obstructions and obstacles such as trees, wind barriers, man-made constructions, and forests decrease wind speed (i.e., they represent some porosity with respect to the wind). Figure 4.3 shows the typical form of the distribution of winds in a given location. Porosity is defined as the relationship between the open area and the total perpendicular area to the direction of wind flow. Porosity may be expressed as depicted in Table 4.3.

They can be expressed as in Table 4.3. For the porosity of the ground cover, the height of the turbine support tower can be selected. In practice, the tower height can be four times the turbine rotor diameter of a few kW to little more than a diameter of many turbines kW [15].

According to Hellmann's *exponential law* for the wind speeds, to improve the data accuracy, it should be taken into account that the anemometer measures the wind speed at a certain height, H_o. In order to estimate the wind velocity at a height H above crops and trees, it is possible to adjust the wind speed using the friction coefficients of Table 4.7 in the equations as follows:

$$\left(\frac{v}{v_0}\right) = \left(\frac{H}{H_0}\right)^{\alpha} \tag{4.19}$$

where:

α = Attrition coefficient
v_0 = Wind speed at a height H_o
v = Wind speed at a height H

If altitude z is considered, then

$$\left(\frac{v}{v_0}\right) = \frac{ln(H/z)}{ln(H_0/z)} \tag{4.20}$$

To complete the data for the final decision to locate a convenient wind site, the local capacity factor (*CF*) is defined, which is based on both the characteristics of the turbine and the site characteristics (typically 0.30 or above for a good site). Factor *CF* is the fraction of the year the turbine generator is operating at rated (peak) power, that is,

$$CF = \frac{P_{av}}{P_{peak}} \tag{4.21}$$

4.4 General Classification of Wind Turbines

Wind turbine engineers avoid building large machines with an even number of rotor blades, for reasons of extreme high torque that may cause instability and vibrations caused by blades passing in front on the tower (tower shade). Therefore, two-blade design is no longer manufactured as it has been in the past. A rotor with at least three blades can be considered approximately like a circular plate when calculating the dynamic properties of the machine [15–17]. Such a rotor with an even number of blades will cause stability problems in a machine with a stiff structure; at the very moment when the uppermost blade bends backward (because it gets maximum power from the wind), the lowermost blade passes into the wind shade in front of the tower. With an odd number of blades, this phenomenon is minimized [12–15].

Although one-blade wind turbines do exist, their commercial use is not widespread, because the problems noted for the two-blade design apply even more strongly to one-blade machines, regularly causing 100% notch power. In addition to higher rotational speed, noise, and visual intrusion problems, they require a counterweight to be placed on the other side of the hub from the rotor blade to balance the rotor. The main problem with the one-blade turbine is the electromechanical stress on the blades and support structures since every time a blade passes in front of the tower pole, it causes an electromechanical impact and a notch in the electrical generation for not capturing the wind. This is a disadvantage in regard to both noise and visual intrusion.

Compared with a two-blade design, this feature adds weight to the generating system without generating additional power.

Two-blade wind turbine designs save the cost of one rotor blade and its weight, they decrease the electromechanical stress when the blades pass in front of the support structures, but they increase the blade base traction by 3/2. However, they tend to have a difficulty in penetrating the market, partly because they require higher rotational speed to yield the same energy output. This makes its market acceptance difficult and causes more noise and visual intrusion. In recent years, several traditional manufacturers of two-blade machines have switched to three-blade designs.

The rotor of two- and one-blade machines must be sufficiently flexible to tilt, to avoid too-heavy shocks in the no-wind position of the turbine when the blades pass the tower. The rotor is therefore fitted on an axis perpendicular to the main shaft that rotates along with it. This arrangement may require additional shock absorbers to prevent the rotor blade from hitting the tower.

Most modern wind turbines are three-blade designs with the rotor position maintained on the wind side of the tower using electrical motors in their yaw mechanism. This design, usually called the classical Danish concept, tends to be a standard when other concepts are judged. The vast majority of turbines sold in world markets are of this design. It was introduced with the renowned 200 kW Gedser wind turbine remaining the largest in the world for many years.

Wind power plants can be defined also based on their electrical generator, that is, an induction machine (discussed in Chapter 12) or a permanent magnet synchronous generator (discussed in Chapter 13).

For very simple types of wind power battery charger [10–12], the wind vanes are just a set of blades coupled directly to a fixed-inertia flywheel and to a dynamo (or generator) shaft with permanent ceramic magnets on the rotor. On the other hand, high-speed wind power turbines are better for bulky generation of electricity and are relatively less costly. However, most turbines are quite a bit more complex.

In general, wind turbines can be divided into two groups: horizontal shaft and vertical shaft (with or without accessories). Horizontal-shaft turbines include:

- Blade type with one, two, or three blades (the most typical are three blades)
- Multiple-blade, farm, or spiked type, which can work with wind coming from the front or back (many variations exist, much as the multirotor type)
- Double opposite blade type, which can use sails in place of blades

Vertical-shaft turbines are subdivided based on the following working principle: some are based either on drags or on friction, and some use lifting (as in an airplane wing). Some turbines utilize both principles.

Drag turbines are machines whose surface executes movements in the wind direction; in other words, they work with the force of the wind drag acting on them. The following types are available:

- Savonius: single or multiple blades, with or without eccentricity (when the rotation shaft is or is not shifted with respect to the shaft that contains its gravity center)
- Blade, paddles, or oars
- Cup

Lifting turbines are machines whose rotor movement is perpendicular to the wind direction, and they are moved by the lifting action of the wind. Among the lifting turbines there are the triangular Darrieus, the Darrieus Giromill, and the Darrieus–Troposkien, as well as a Magnus effect turbine type.

4.4.1 Rotor Turbines

Blade turbines have high rotation with high efficiency. They have automatic regulation of turn speed through the attack angle as a function of the wind speed. The blades present variable sections and great strength against mechanical stresses. This type of turbine needs some orientation mechanism with direct action on turbines at low loads and with indirect action on turbines at higher loads [2, 3].

Blade turbines are those most used for electricity generation, feeding batteries, or injecting energy to the electric grid. They can be used to pump water as well, since they allow high-rotation driving. They possess the disadvantage of needing towers of great height for installation, on those towers there will be generators, control equipment, and the power transmission system.

Blade turbines can vary with respect to the conventional model, but such variations do not indicate notable differences in performance. One is the multirotor, with several turbines mounted on the same tower and interlinked by mechanical shafts. Another is the sail turbine, which follows the conventional model by sails rather than blades. These are made of special fabric (sails of embarkation), usually in four or six pieces. They are low in cost, but their efficiency is also lower. They operate at low rotation and high torque.

4.4.2 Multiple-Blade Turbines

Turbines with multiple blades compensate for their low operational rotation. They are simple to manufacture and have high torque. Such a turbine is usually mounted on a horizontal shaft for low power uses. They have lower efficiency than that of blade turbines, are generally used to pump water, and demand very high towers. Multiple-blade turbines for farm applications are manufactured with

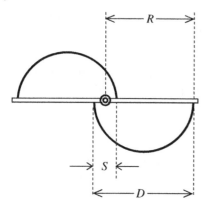

Figure 4.10 Cross section of a Savonius turbine.

metallic foils of uniform curved profile. In addition to the usual multiple-blade models, there is a cup type used in the construction of anemometers and toys.

4.4.3 Drag Turbines (Savonius)

Drag turbines operate on the principle of the friction caused by wind on the turbine blades. A Savonius turbine represents this design of vertical-axis type. This turbine model is of simple construction and consists of two parts like a barrel cut in the middle and fastened opposite each other by one of their opposed longitudinal edges (Figure 4.10). Savonius turbines are in wide use in rural areas for water pumping, attic ventilation, and agitation of water to prevent freeze-up during the winter. They are used little in electricity generation.

The best configuration for the Savonius half barrels is given by the relationships

$$R = D - 0.5S \tag{4.22}$$

$$S = 0.1D \tag{4.23a}$$

$$A = 2Rh \tag{4.23b}$$

where:

D = Diameter of each half-barrel
h = Height of the barrel
R = Radius projection exposed to the wind
A = barrel area exposed to the wind

The torque of a Savonius turbine is caused by the difference in pressure between the concave and convex surfaces of the blades and by the recirculation of wind coming from behind the convex surface. Its efficiency reaches 31%, but it presents disadvantages with respect to the weight per unit of power, because its constructional area is totally occupied by material. A Savonius rotor needs 30 times more material than is needed by a rotor of a conventional type.

Table 4.6 Output power (watts) of a Savonius turbine according to the wind speed.

Number of barrels (n_b)	1	2	3	4
	Total height of barrels, h (m)			
Wind speed V_w (m/s)	0.85	1.70	2.55	3.40
2	0.70	1.40	2.10	2.80
4	5.52	11.04	16.56	22.08
6	18.66	37.32	55.98	74.64
8	44.24	88.48	132.72	176.96

For the rudimentary installation of a Savonius turbine, metallic barrels of 200 L capacity with an H-shaped wood structure are used. The useful power can be determined for several wind speeds through the dimensions of the barrel, such as

$$D = 0.60\,\text{m}$$
$$h = 0.85\,\text{m}$$
$$R = 0.57\,\text{m}$$

To calculate the area exposed to the wind,

$$A = 2Rh = 2 \times 0.57 \times 0.85 = 0.96\,\text{m}^2$$

With this result and assuming a power coefficient C_p of 0.15 (as defined in Section 4.3), from equation (4.13) the power for n barrels is (see Table 4.6)

$$P = 0.6AC_p V_w^3 n = 0.6 \times 0.96 \times 0.15 \times V_w^3 n = 0.0864 V_w^3 n \quad (\text{W}) \tag{4.24}$$

4.4.4 Lifting Turbines

Lifting turbines operate through the lifting effect produced by wind. They may have horizontal or vertical shaft.

The most common models of vertical turbines are the triangular (delta) Darrieus turbine, the Giromill turbine, and the Darrieus–Troposkien turbine. The triangular Darrieus turbine has straight blades and variable geometry. The straight blades present disadvantages due to their natural bending. However, the variable geometry changes the power coefficient. Therefore, the rotation tends to be constant. All of the main components are close to the ground; also, the wind turbine itself is near the ground, unlike horizontal where everything is on a tower, facilitating maintenance. There are two types of action on the vertical-axis wind turbines: lift and drag. The lift-based designs are generally much more efficient than drag.

The Giromill turbine has straight, vertical blades and a rotating movement around its shaft. It is used for large-scale systems because it is considered a low-load system, operating with only a few hundreds of watts. It has an automatic mechanism that maintains the attack angle position that supplies the best working conditions [7].

Among lifting vertical-shaft turbines, the Darrieus–Troposkien turbine is best suited to wind power plants. The Darrieus–Troposkien turbine consists of blades, a shaft, a tower, and guide wires. The blades are curved with glide sections. The rotor is vertical and connects the top to the bottom of the blade. The tower is fixed on the ground by solid foundations that sustain the shaft. At the lower part of the shaft, rollers are fixed to guide wires to maintain the turbine in a vertical position. The other extremities are fixed on the ground. Studies of this turbine have proved its economic benefits and constructive simplicity. Among the vertical-axis types, the Darrieus turbines are the most used turbines for electric power generation.

4.4.4.1 Starting System

This type needs a special starting system, for which it may adapt an electric motor connected to the network, a dc motor fed by batteries recharged by a generator connected to the turbine itself or to an auxiliary turbine. For instance, a Savonius turbine can be coupled to the shaft of the Darrieus itself, primarily for small-load turbines.

4.4.4.2 Rotor

The rotor of the Darrieus–Troposkien turbine has curved blades with section glides of aerodynamic profile fixed on the shaft extremity of the rotor. The blades can be made of aluminum, fiberglass, steel, or wood. The rotor shaft can be tubular or latticed with an external cover to improve the aerodynamics.

4.4.4.3 Lifting

The rotor support is made of one tower and guide wires. The tower sustains the rotor through bearings and rollers. It can still shelter parts of the system, such as a gearbox, generators, or pumps. The guide wires are fixed through bearings with rollers at the upper part of the rotor shaft in one of the extremities; in the other, they are fastened to the ground. Guide wires are essential for turbines that handle small loads.

4.4.4.4 Speed Multipliers

The system of speed multiplication used in turbines to reach an electrical generating speed is carried out through gearboxes of parallel or perpendicular (smaller losses) shafts. A system of belts can also be a good solution. The cost of the multiplier depends on the multiplication rate. The cost of the generator increases with reduced rotation, as the number of poles or turns per coil must then be increased. Thus, the optimum value of the multiplication rate is a trade-off between the speed multiplication rate and the number of poles. Speed multiplication also causes a representative percentage of the total losses of a wind energy system. In extremely small systems, this loss may represent about 20% of the total loss. The most restrictive parameter for a very low speed wind turbine is the efficiency of speed multipliers (gearbox).

4.4.4.5 Braking System
Braking systems have both safety and maintenance purposes since a turbine must have some mechanical speed limitation. Sizing will determine the best system to be adopted (hydraulic, electromagnetic, or mechanic).

4.4.4.6 Generation System
The electricity generated can feed the grid directly or be stored in batteries. Due to the large variations in rotation, the use of induction generators is recommended only for stand-alone mode up to 15 kW (e.g., for places difficult to access). In these cases, self-excitation capacitors will be needed (see Chapter 12). Larger generators can be used to feed the grid directly as is the case of permanent-magnet synchronous generators (PMSG) (see Chapter 13).

4.4.4.7 Horizontal- and Vertical-Axis Turbines
In 2010, Pope *et al.* [21] evaluated two horizontal and two vertical wind power systems on the basis of energy and exergy, that is, using the laws of thermodynamics. It was concluded that by using exergy methods, a difference of 44–55% between energy and exergy efficiencies of vertical-axis wind turbines (VAWT). This indicates a scope for further improvement in the current VAWT designs. The following points summarize the advantages and disadvantages of horizontal-axis wind turbines (HAWT) and VAWT, based on the current technology. The following are advantages and disadvantages of either one.

Horizontal Axis Wind Turbine (HAWT)
Advantages:

- Blades are to the side of the turbine's center of gravity, helping on the weight and mechanical balance.
- Ability to wing warp, which gives the turbine blades the best angle of attack.
- Ability to fully pitch the rotor blades at high-speed winds to minimize damage.
- Tall tower allows it to be sited in forest above tree line, on uneven land, or in offshore locations and to have access to stronger wind in sites with wind shear.
- Most of them are self-starting.

Disadvantages:

- Dangerous and difficult to operate in near ground winds.
- Difficult to transport (20% of equipment costs).
- Difficult to install (require tall cranes and skilled operators).
- Affect radars in proximity.
- Local opposition to aesthetics.
- Because its height, it causes several bird deaths.
- Difficult maintenance.

Vertical Axis Wind Turbine (VAWT)
Advantages:

- Easy to maintain
- Lower construction and transportation costs
- Does not have any wind preferential direction, it is omnidirectional
- Most effective at mesas, hilltops, ridgelines, and passes

Disadvantages:

- Blades constantly spinning back into the wind causing drag.
- Low starting torque and may bidirectional schemes allowing the generator to start initially in motoring mode.
- Lower efficiency.
- Operate in lower and more turbulent wind.
- High mechanical stress, materials may go under fatigue and failure.

4.4.5 Magnus Turbines

Magnus turbine is very recently being considered. It is recommended when the wind speed is below 7 m/s or for very high wind speeds. Unlike conventional turbines, this turbine has rotating cylinders instead of blades that generate a torque on the generator shaft called Magnus effect. The Magnus effect can be observed when someone throws a ball with rotation, and the ball catches an uprising trajectory, very common in some sports. The Magnus effect is the result of the wind speed difference in one and the other side of the cylinder, respectively, opposing or favoring wind. The increase of the wind relative speed caused by this action causes a pressure increase on one side of the cylinder and a reduction on the other side, and therefore, a net torque. Commonly turbines are found with three or five rotating cylinders (see Figure 4.11) [16, 26]. As a term of comparison, the variables of power production control of a Magnus turbine are listed in Table 4.7.

4.4.6 System TARP–WARP

The current wind power technology trend is the increasing constructions of larger sweeping diameter of the blades on the rotor shaft as much as possible

Figure 4.11 Magnus turbine with non-smooth rotating cylinders (courtesy *CEESP-UFSM/IFSC*—Brazil). *Source*: Simões and Farret [14]. Reproduced with the permission of Taylor & Francis.

Table 4.7 Control variables of Magnus and Conventional turbines.

Variable definition	Magnus	Conventional
Cylinder angular speed	ω_c	
Rotor angular speed	ω_r	ω_r
Generator current	I_g	I_g
Angle between turbine shaft and wind	θ_w	θ_w
Pitch angle		β

in order to have a larger area for increased generated output power. On the other hand, new technologies of energy generation from the wind are being proposed, such as the toroidal accelerator rotor platform (TARP), which combines generation and transmission in one same tower. This concept of wind capture is based on the wind amplified rotor platform (WARP) [22]. Small wind turbines are grouped and installed in modules, to conform to the needs of distributed generation. This entire platform consists of a determined number of TARPs piled up one on top of another.

Some toroidal form of aerodynamic turbine shelter characterizes the working principle of the TARP. Wind accelerates around the shelter, amplifying the

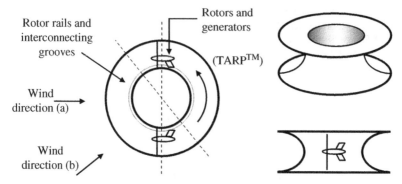

Figure 4.12 Effect of wind direction on TARP rotors.

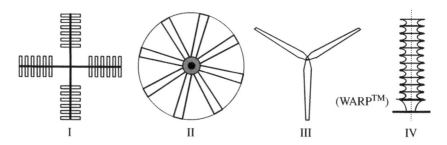

Figure 4.13 Technological generations of wind turbines.

density of wind power energy available. Each TARP structure supplies a field of increased outlying flow in all directions, impelling two wind power rotors of small diameter disposed about 180° from one another around the channel of toroidal flow so formed on the shelter (see Figures 4.12 and 4.13).

TARP rotors have a typical diameter of 3 m or less and are coupled directly to the generator through a system of brakes but without a speed multiplication gearbox. As shown in Figure 4.12, if the wind changes from direction (a) to direction (b), a torque will form on the rotating wheel shelter that moves until it is balanced in the new wind direction. The wind power structure also serves as a support, as protection, and as housing for the turbine controls and other internal subsystems. This configuration differs dramatically from the traditional single rotor with a mounted horizontal shaft in a tower, and it is claimed to be of high efficiency [6].

The design just described overcomes the traditional configuration of wind turbines through an odd combination of distribution/transmission, superior performance, easy operation, easy maintenance, high readiness, and reliability. It needs little land area, has a better appearance, has less interference and less electromagnetic noise in TV transmission, and reduces the mortality rate of birds. The estimated cost of a kilowatt-hour in developed countries is from 2 to 5 cents, depending on the wind power resources. Such systems still have high installation costs (due to the need for higher tower heights) and are not economically feasible for small power applications. Figure 4.13 shows the technological evolution of wind power turbines.

4.4.7 Accessories

Accessories are devices used to improve the efficiency of wind power turbines and are most common for blade turbines. They include solar wind generators, confined vortexes, diffusers, wind concentrators, plane guides, deflectors, venturi tubes, heating towers, and rotor accelerating shelters [11–17].

Solar wind generators are special accessories that are able to store the heat irradiated by the sun on black surfaces protected against convection effects. Warm-air circulation is guided by the chimney effect and ends up crossing a turbine when in its ascending movement. The confined vortex consists of a tower where, in its interior, the effects of a tornado are reproduced through the orientation of free wind heating. In a Spanish prototype, the tower sits in the center of a 7-km (4-mile)-radius circular glass building, as shown in Figure 4.14. Under the glass, the sun warms the air. As the warm air rises, it is drawn through turbines at the base of the tower, thus generating renewable electricity.

Table 4.8 Worldwide production of high-power turbines.

Manufacturer	Capacity (MW)	Blade length (m)	Swept area (1000 m^2)	Max wind speed (m/s)	Power density (m^2/kW)	Generator	Production
Fuhländer	3.00	120	11.310	25	0.038	PM	2012
Enercon	7.58	127	12.668	34	0.017	SYNC RT	2014
GE Energy	4.10	113	8.500	27	0.025	SYNC PM	2014
Nordex	6.00	150	17.672	25	0.030	PM	2013
Alston Power	6.00	150	17.672	25	0.03	PM	2015
REpower	5.00	126	12.469	—	0.025	ASYNC	2004
Vestas	8.00	164	21.125	—	0.027	PM	2014
WinWind	3.00	120	11.310	25	0.038	SYNC PM	
Gamesa	7.00	145	16.513	—	0.024		
Mitsubishi	7.00	165	21.383	—	0.031		
Wobben	3.00	82	5.281	34	0.018	SYNC RT	
CCWE	5.00	126	12.469	25	0.025		
Samsung	7.00	171	22.966	—	0.033		
Hitachi	5.00	126	12.469	25	0.025	SYNC PM	

Adapted from The Wind Power, M. Pierrot, 2016, http://www.thewindpower.net/turbine_en_7_nordex_2500.php, accessed March 6, 2017.

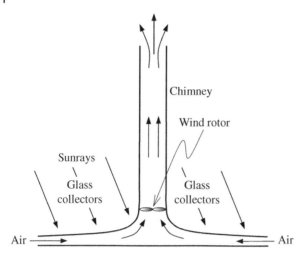

Figure 4.14 Principle of solar power tower. *Source*: From Refs. [3, 19–21]. © European statistics.

4.5 Generators and Speed Control Used in Wind Power Energy

As discussed in Section 4.3, the power of a wind generator varies with the cube of the wind speed. Therefore, if the wind speed doubles, the power will increase eight times. Wind generator efficiency is affected directly by the wind speed, whose rated value in turn depends on rotor design and rotation. Again, the rotation is expressed as a function of the blade-tip speed for a given wind speed. Therefore, to maintain a fixed generator speed on a high efficiency level is virtually impossible. This is the main reason why two-blade rotors (see Figure 4.9) can be used only for higher power turbines. Table 4.8 lists major manufacturers of large turbines in Europe, the United States, Japan, and China [14–18].

With respect to speed control, there are three different methods for safety control in wind turbines: stall regulation, pitch control, and active stall control. Around two-thirds of wind turbines are passive stall regulated, with the machine rotor blades bolted onto the hubs at a fixed attack angle. In these cases, the rotor blade is twisted when moving along its longitudinal axis to ensure that the blade stalls. By stall, it is meant that turbulence on the side of the rotor blade not facing the wind is created gradually rather than abruptly when the wind speed reaches its critical value. If stall control is applied, there are no moving parts in the rotor and a complex control system is needed, but stall control represents a very complex aerodynamic design problem and further design challenges in the structural dynamics of the entire wind turbine to avoid stall-induced vibrations [19–23].

On pitch control, an electronic controller checks the output power several times per second. When the output power becomes too high, it sends a

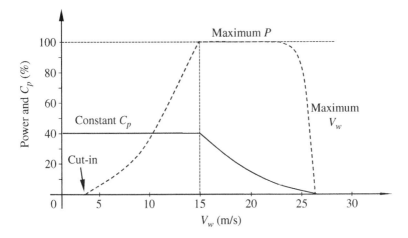

Figure 4.15 Speed control range for wind turbines.

command to the blade pitch mechanism, which immediately turns the hydraulic drive of the rotor blades slightly out of the wind direction, and vice versa when the wind speed drops again. The rotor blades are kept at an optimum angle by the pitch controller. In active pitch control, the blade pitch angle is adjusted continuously based on the measured parameters to generate the desired output power up to the maximum limit, as in conventional regulated pitch control. Electric stepper motors can be used for this purpose.

There are four speed bands to be considered in the operation of a wind turbine (see Figure 4.15). The first band goes from zero to the minimum speed of generation (cut-in). Below this speed, the power generated just supplies the friction losses.

The second band (optimized constant C_p) is the normal operation maintained by a system of blade position control with respect to the direction of wind attack (pitch control). In the third band, for high-speed winds, the speed is controlled such as to maintain a maximum constant output power (constant power) limited only by the generator capacity. Above this band (at wind speeds around 25 m/s), the rotor blades are aligned in the direction of the wind, to avoid mechanical damage to the electrical generator (speed limit).

The simplest way of driving a wind power turbine is (of course) by not using any special control form. In this case, the speed of the turbine is constant. Since the turbine is connected directly to the public grid, which operates at a fixed frequency, the electrical machine will have rotating flux synchronized with the grid. It would be best to be under a constant load. Power generated this way cannot be controlled since it is altered by the speed of the winds. So just a fraction of that total energy could be captured with extremely strong winds, and it would be not effective to size the power plant to the maximum power of the winds. There are benefits to incorporating some control in a wind power turbine, such as an increase in the energy capture and a reduction in the effect of dynamic loads [16, 23, 24, 27]. Furthermore, the speed actuators allow flexible adjustment of the operating point, and the project's safety margins can be decreased.

The more commonly used form of speed and power control for wind generators is the control of the attack angle of the turbine blades (when the generator is connected directly to the public grid). Another control form, which has received a lot of attention in the last few years, is the use of generators driven by turbines with fixed attack angles of the blades, connected to the public network through a power electronic interconnection. Control occurs on the load flow, which in turn, acts on the turbine rotation [2, 10, 15]. As the rotor speed must change according to the wind intensity, the speed control of the turbine has to command low speed at low winds and high speed at high winds, to follow the maximum power operating point (MPP) as indicated in Figure 4.16.

In addition to purchasing wind turbines, the following ancillary equipment (discussed in Chapters 11 and 12) might be necessary in the capital costs:

1) Yaw control, for maximum wind intensity
2) A support tower, to hold the wind capture system and the generator
3) A speed multiplier, necessary for the low operating rotations usually found in wind power turbines (up to approximately 500 rpm) when it is known that 60-Hz commercial generators (the pattern throughout the United States) with two poles need 3600 rpm and those with four poles need 1800 rpm

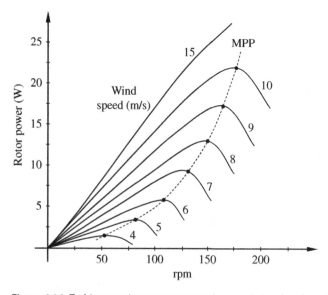

Figure 4.16 Turbine rotation versus power characteristic related to wind speed.

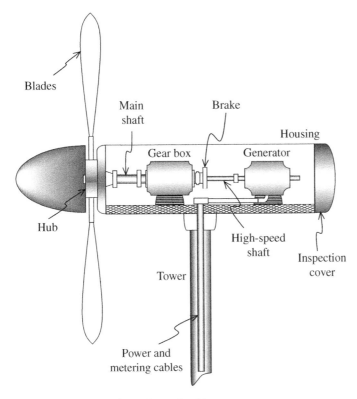

Figure 4.17 Nacelle of a blade wind turbine.

4) Voltage and frequency or rotation control
5) Voltage-raising and voltage-lowering transformers to match the generation, transmission, and distribution voltages at normal consumption levels (usually, 127/220 V or 220/380 V)
6) An electric distribution network for consumers
7) Protection systems for overcurrent, overspeed, overvoltage, atmospheric outbreaks, and other anomalous forms of operation

Figure 4.17 illustrates the nacelle of a wind turbine generator with the main devices used to drive the turbine.

4.6 Analysis of Small Generating Systems

Several types of generators can be coupled to wind power turbines: parallel and compound generators, dc and ac types, and especially, induction generators (see Chapter 12).

Once the installation site of a wind power plant has been selected, the next steps are to select the turbine rating, the generator, and the distribution system. In general, the distribution transformer is sized to the peak capacity of the generator according to the available distribution network capacity. As a practical rule, the output characteristics of a wind turbine power do not exactly follow those of the generator power, and they must be matched in the most reasonable way possible. Based on the maximum speed expected for a wind turbine, and taking into account the cubic relationship between wind speed and power, the designer must select the generator and gearbox to match these limits. The most sensitive point is the correct selection of the rated turbine speed for the power plant. If it is too low, generation for high-speed winds will not be possible. If it is too high, the power factor will be too low.

In an iterative design process to match the characteristics of commercially available wind turbines and generators to their cost, efficiency, and the maximum power generated, sometimes optimization and mathematical linear programming techniques are used. The maximum value of C_p should occur at approximately the same speed as that of the maximum power in the power distribution curve. So the tip speed ratio must be kept optimally constant at the maximum speed possible to capture the maximum wind power. This feature suggests that to optimize the annual energy capture at a given site, it is necessary that the turbine speed (the tip speed ratio) should vary to keep C_p maximized, as illustrated in Figure 4.18 for a hypothetical turbine. Obviously, the design stress must be kept within the limits of the turbine

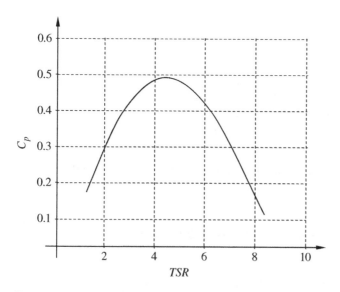

Figure 4.18 Example of maximum value of C_p.

manufacturer's data, since torque relates to the instantaneous power as $P = T\omega$ or as a relationship between the torque coefficient and the power coefficient $(C_T = C_p/\lambda)$ [28–31].

It should be pointed out that in the case of parallel generators connected to batteries, small rotation increments are linked to large increments in the output current. When the terminal voltage of the battery stays constant (increasing the nominal rotation by a determined percentage), the current suffers an increase of k times its initial value. As a result, the power also suffers an increase of k times its initial value [16, 24, 27].

If the voltage drop of the generator occurs at higher rates than the current increment, a dc compound generator of three brushes should be used rather than a parallel generator. For a compound generator, the design is executed so that the decrease in the characteristic is at the same rate as the number of demagnetization coil turns. For three-brush generators (due to the increase in coil current), the voltage of both the field and the magnetic field itself is reduced.

Due to the advantages of working as a motor or generator (and its low cost with respect to other generators), the induction machine offers enhanced conditions for wind power plants. This chapter considers an induction generator connected to an infinite bus. However, other configurations are fully discussed in Chapter 10 [14, 29].

4.6.1 Maximization of C_p

Once the turbine to be used in a wind power plant has been decided, the area A is swept by the blades. Similarly, the local air density ρ and wind speed V_w are not controllable. Therefore, the only way is to control the turbine speed ω, that is, controlling C_p. That means, for a given wind speed V_w, there is a direct relationship (equation 4.13) between P_t and C_p given by

$$P_t = \left(\frac{1}{2}\rho A V_w^3\right) \cdot C_p = k_1 \cdot C_p$$

Also, for a given wind speed, there is a direct relationship (Section 4.3) between the tip speed ratio (*tsr* or λ) and the turbine rotation n given by

$$tsr = \lambda = \frac{\omega R}{V_w} = \left(\frac{\pi p}{60} \cdot \frac{R}{V_w}\right) \cdot n = k_2 \cdot n$$

These two relationships above mean that the difference between Figures 4.16 and 4.18 is only a matter of scale factors k_1 and k_2 for a given wind speed V_w. Therefore, regulating the rotor speed n is a way of searching the maximum power as suggested in Figure 4.18.

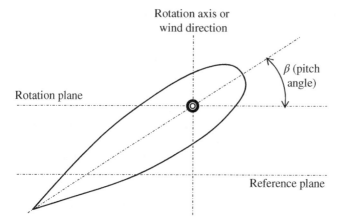

Figure 4.19 Pitch angle β.

A more detailed power coefficient may be approximated as [26]

$$C_p(\lambda,\beta) = c_1\left(\frac{c_2}{\lambda_i} - c_3\beta - c_4\right)e^{-c_5/\lambda_i} + c_6\lambda \qquad (4.25)$$

where:

$$\frac{1}{\lambda_i} = \frac{1}{\lambda + 0.08 \times \beta} - \frac{0.035}{\beta^3 + 1}$$

β = The pitch or attack angle (see Figure 4.19)

$\qquad c_1 = 0.5176$

$\qquad c_2 = 116$

$\qquad c_3 = 0.4$

$\qquad c_4 = 5$

$\qquad c_5 = 21$

$\qquad c_6 = 0.0068$

Other effects may be included in the determination of C_p, such as the whirlpool losses. In the idealized derivation of the Betz equation, the wind does not change its direction after it encounters the turbine rotor blades. This change in direction can be accounted for by a modified form of the power coefficient known as the Schmitz power coefficient $C_{pSchmitz}$, if the same airfoil design is used throughout the rotor blade. The theory developed by Schmitz and Glauert applies to wind turbines with four or less rotor blades because the air movement becomes too complex for a strict theoretical treatment and an empirical approach is adopted as [7, 16, 25, 26, 28–31]

$$C_{pS-G} \cong C_{pSchmitz}\eta_{profile}\eta_{tip\ end}\eta_{no.\ of\ blades} \tag{4.26}$$

There are still even more efficiencies factors involved:

1) Frictional losses in the bearings and gears
2) Magnetic drag and electrical resistance losses in the electrical generator

All these effects would result in,

$$C_p \cong C_{pSchmitz} \cdot \eta_{profile} \cdot \eta_{tip\ end} \cdot \eta_{no.\ of\ blades} \cdot \eta_{friction} \cdot \eta_{electrical} \tag{4.27}$$

The aerodynamic efficiency of a wind power machine of very small size varies, at most, between 40 and 45%, but in practice, we can consider the average efficiency of 35%. The efficiency of a dc generator is about 55–60%. Therefore, the total efficiency of a small system (a few hundred watts) is approximately 20%. For very low wind speeds, this will result in the loss of the excitation (magnetizing flux) demanded by the generator. In that case, it is advisable to choose permanent-magnet-based dc generators. It is possible to eliminate the switches in alternators of six or more poles by the use of permanent magnets. If the system is made for charging batteries, it is advisable to use simple rectifiers. The generator should present voltage and frequency proportional to the rotation to provide current regulation by the inductive reactance of the circuit.

In conclusion, the designer needs the following specifications before purchasing a wind power turbine [32–34]:

1) Wind intensities in the area and the duration curve
2) Topography
3) Purposes of the energy generated
4) Present and future energy needs for the generator and turbine according to the area's wind capacity (otherwise, any specification becomes useless)
5) Determined need for a turbine isolation valve
6) Specifications for the type of regulator
7) Specifications for the generator (see Chapters 11 and 12) according to:
   a) Type: induction, synchronous, or direct current
   b) Electric output: alternating or direct current, maximum power, voltage level, number of phases, frequency
   c) Climate: temperature and humidity
   d) Automation level: manual, automatic, semiautomatic, or remote control
   e) Operation: stand-alone or for injection in the net (in which case a sale, purchase, or exchange commercial agreement should exist with the electric power company)
   f) Number of phases: single- or three-phase transmission and distribution

8) Other equipment: control panels, protection, and driving
9) Location of the machine house (if there is one, distance and dimensions)
10) Access possibilities to the site: boat, bike, highway, rails, walking distance, suitability for land transportation of equipment, and maintenance planning

The cost of wind power plants has been changing in the last few years. By the year 2016, they cost around \$0.60/W and 2.5¢/kWh; in the United States wind power plants are competitive with fossil fuel generation.

References

1 P. Weis and T.S. Latorre (eds.), IEA Wind, 2013 Annual Report, Boulder, Colorado, United States, August 2014.

2 S. Gsänger, S. Sawyer, and C.S. Januario, Renewable Energy Technologies: Cost Analysis Series, Issue 5/5, Volume 1: Power Sector, International Renewable Energy Agency (IRENA), Bonn, Germany, June 2012.

3 I. Pineda, S. Azau, J. Moccia, and J. Wilkes, Wind in Power 2013: European Statistics, European Wind Energy Association (EWEA), Belgium, Brussels, February 2014.

4 Ad Hoc Group Report to the Executive Committee of the International Energy Agency, Long-Term Research and Development Needs for Wind Energy, for the Time Frame 2012 to 2030, IEA Wind, July 2013.

5 S. Gundtoft, Wind Turbines, Annual Report, Aarhus University School of Engineering, Denmark, June 2009, pp. 1–43.

6 National Geographic, http://news.nationalgeographic.com/news/ energy/2014/04/140416-solar-updraft-towers-convert-hot-air-to-energy/, accessed March 16, 2016.

7 W.A. Moran, Giromill Wind Tunnel Test and Analysis Volume 1—Executive Summary, Final report for the period June 1976–October 1977, Prepared for the USA Energy Research and Development Administration, Division of Solar Energy, under contract no. EY-76-C-02-2617.A001, McDonnell Aircraft Company, St. Louis, MO, October 1977, http://wind.nrel.gov/public/library/ gwt.pdf, accessed February 13, 2016.

8 R.R. Barcellos, Use of wind energy in RS using a turbine of vertical shaft, MSc dissertation, PPGEM-UFRGS, Porto Alegre, Brazil, 1981.

9 D.P. Sadhu, Studies about Wind Energy (Estudos sSbre Energia Eólica), Department of Mechanical Engineering, UFRGS, Porto Alegre, Brazil, 1983.

10 S. Adeodato and W. Oliveira, The richness of the best winds (A Riqueza dos Melhores Ventos), Globo Ciência, Vol. 63, pp. 20–25, 1996.

11 J.S. Rohatgi and V. Nelson, Wind Characteristics: An Analysis for Generation of Wind Power, Alternative Energy Institute, West Texas A&M University, Canyon, TX, 1994.

12 M.R. Patel, Wind and Solar Power Systems, CRC Press, Boca Raton, FL, 1999.

13 The Ruthland Windcharger: User's Handbook for the Series 910, Marlec Engineering Company, Northants, England, 1995.

14 M.G. Simões and F.A. Farret, Modeling and Analysis with Induction Generators, CRC Press (Taylor and Francis Group), Boca Raton, FL, 2015.

15 S. Heier, Grid Integration of Wind Energy Conversion Systems, 2nd ed., John Wiley & Sons, Inc., Hoboken, NJ, 2006.

16 M. Ragheb and A.M. Ragheb, "Wind Turbines Theory—The Betz Equation and Optimal Rotor Tip Speed Ratio," in R. Carriveau (ed.), Fundamental and Advanced Topics in Wind Power, InTech, Rijeka, Croatia, 2011, http://www.intechopen.com/books/fundamental-and-advanced-topicsin-wind-power/wind-turbines-theory-the-betz-equation-and-optimal-rotor-tip-speed-ratio, accessed March 6, 2017.

17 Wind Energy Applications Guide, American Wind Energy Association (AWEA), Washington DC, Jan 2001.

18 R.N. Meroney, Wind in the perturbed environment: its influence on WECS, presented at the American Wind Energy Association Conference, Boulder, CO, 11-14 May, 1977.

19 J. Ross, Wind in power 2013, European statistics, The European Wind Energy Association, February 2014.

20 R. Waddington, Grid Integration of Wind Energy Conversion Systems, translated by S. Heier, 2nd ed, 446 pp, John Wiley Co., June 2006, 446 pp.

21 K. Pope, I. Dincer, and G.F. Naterer, Energy and exergy efficiency comparison of horizontal and vertical axis wind turbines, Renewable Energy, Vol. 35, pp. 2102–2113, 2010.

22 A.L. Weisbrich, S.L. Ostrow, and J.P. Padalino, WARP: a modular wind power system for distributed electric utility application, IEEE Transactions on Industry Applications, Vol. 32, No. 4, pp. 778–787, 1996.

23 G.L. Johnson, Wind Energy System, Electronic ed., Prentice Hall, Manhattan, KS, 2001, http://www.rpc.com.au/pdf/contents.pdf.

24 M. Patel, Spacecraft Power Systems, CRC Press, Boca Raton, FL, 2004.

25 P. Novak, T. Ekelund, I. Jovik, and B. Schmidtbauer, Modeling and control of variable speed wind-turbine drive-system dynamics, Proceedings of the IEEE Control Systems Society Conference, pp. 28–38, 1995.

26 M. Jinbo, F.A. Farret, G. Cardoso Junior, D. Senter, and M.F. Lorensetti, MPPT of Magnus wind system with dc servo drive for the cylinders and boost converter, Journal of Wind Energy, Vol. 2015, pp. 1–10, 2015.

27 W.L. Schmidt, Wind power studies in Eastern Africa, Part I, Prepared for International Development Research Centre (IDRC), University of Waterloo, Ottawa, Canada, January 1977 (Filippini Turbine).

28 A. Barbero, J.A. García Matos, A. Cantizano, and A. Arenas, Numerical tool for the optimization of wind turbines based on Magnus effect, 9th World Wind Energy Conference and Exhibition (WWEC 2010), Istanbul, Turkey, June 15–17, 2010.

29 M. Godoy Simões, F.A. Farret, and F. Blaabjerg, "Small Wind Energy Systems," in F. Blaabjerg and D. Ionel (eds.), Renewable Energy Devices and Systems with Simulations in MATLAB and ANSIS, 1st ed., Taylor and Francis, Boca Raton, FL, 2016.

30 R.M. Hilloowala and A.M. Sharaf, A ruled-based fuzzy logic controller for a PWM inverter in a standalone wind energy conversion scheme, IEEE Transactions on Industry Applications, Vol. 32, No. 1, pp. 57–65, 1996.

31 C.A. Portolann, F.A. Farret, and R.Q. Machado, Load effects on dc-dc converters for simultaneous speed and voltage control by the load in asynchronous generation, Proceedings of the IEEE International Conference on Devices, Circuits and Systems, Caracas, Venezuela, 1995, pp. 122–127, 12-14 December 1995.

32 Wind Turbine Specification, http://www.hitachi.com/products/power/wind-turbine/specification/, accessed Mar 2017.

33 E. Hau, Wind Turbines, Springer, Heildelberg, Germany, 2013.

34 General Specification of Wind Turbines, Vestas Wind Systems A/S, R&D department, March 2004.

5

Thermosolar Power Plants

5.1 Introduction

The sun is a perennial, silent, free, and nonpolluting source of energy. It is responsible for all life forms on our planet. It has been the same for millions of years. It is the largest renewable source of energy for human beings with about 174 PW. The sunlight power reference potential is 1000 W/m^2 (1 sun). Its use for energy generation can be direct or indirect. Indirect solar energy is related to wind power, hydropower, photosynthesis, sea tidal energy, and the microbiological conversion of organic matter into fuels (topics for other chapters in this book).

Direct solar energy is used to heat water and houses (domestic, industrial, or commercial), cool ambient and air conditioning, and dry agricultural products and for distillation (mainly for the production of salt or brine by evaporation of seawater), electric power generation, and air conditioning with surface geothermal energy [1–5]. Thermal solar energy is most appropriate for areas in our planet that form the solar belt, that is, those areas up to 30° of latitude to either north or south of the equator. In those latitudes, the direct solar radiation is very high throughout the year.

The sun is not highly directional; in addition their intensity changes per time of day and per day of year. Therefore, a utilization factor coefficient is fundamental in describing feasibility studies for any solar power installation.

Two main types of solar technology exist for the conversion of solar energy into electricity: (i) large-scale applications using solar radiation directly to produce heat (subject of this chapter) and (ii) transformation of solar light directly into electricity that is possible using modules consisting of photovoltaic cells, the subject of Chapter 6. The former is the largest renewable source of energy, usable primarily in areas along the solar belt where solar energy is used on water heating and in indirect production of electric power in solar thermal power plants. Solar heating is an important and useful source of energy because it has minimum environmental impacts and complements electrically

Integration of Renewable Sources of Energy, Second Edition. Felix A. Farret and M. Godoy Simões.
© 2018 John Wiley & Sons, Inc. Published 2018 by John Wiley & Sons, Inc.

distributed generation. Direct production of heat through solar panels is quiet, flexible, secure, clean, cheap, usable as thermal storage, relatively durable (minimum 25 years), seasonal, and depends on site, for example, on roofs, building fronts, and desert places in the solar belt. Usual applications are related to low-temperature extraction of heat from the ground (surface geothermal power seen in Chapter 10) (~15–30 °C), water heating for home and business (~30–60 °C), high-temperature process-heating water for industry (~100–200 °C), and solar thermal power plants (~500 °C), as discussed in this chapter.

5.2 Water Heating by Solar Energy

Solar energy for heat capture can be divided into three phases: reception, transfer, and accumulation [2]. Heat capture is accomplished directly or through collecting plates. To have an idea of how much solar energy can be captured by a collecting plate surface, let's consider the following round calculation. The sun gives off a total of 3.90×10^{26} W. The Earth would intercept energy equal to a normal disk as big as the Earth's diameter. Earth's average radius is roundly 6,371,000 m. Therefore, the Earth's solar interception normal area is $(3.14)(3,185,500)^2 = 3.188 \times 1013 \text{m}^2$. As the amount of power crossing the Earth's orbit is 1361 W/m^2, the Earth intercepts 5.02×10^{16} W of the sun's energy. That is, the Earth intercepts 50 quadrillion watts daily that can be used as solar power. Similar calculation can be applied as basis for PV panels.

The solar collector intercepts solar radiation and transfers the collected energy to a thermal accumulation unit, which is an essential element because it provides for the solar-generated heat to be available during periods of low solar radiation and at night.

A typical example of a collecting plate consists of a blackbody with a large radiation index of absorption. A very conventional example of the use of solar energy is a flat collector for water heating as portrayed in Figure 5.1. These collectors are basically formed by a box of insulating material, usually made of fiberglass and resins of polyester, isolated internally with phenolic glass wool and a blackbody that covers a large copper exchanger [3, 4]. In order to increase thermal resistance and minimize losses, a crystal glass about 4 mm thick, perfectly isolated with glass wool or silicon, covers the device.

A practical system of water heating with solar collectors is illustrated in Figure 5.2. Cold water from the cold reservoir reaches the base of the solar collector pipes, which absorbs heat. Through thermal expansion and natural or forced convection, it returns to the reservoir. The water flow continues in this cycle, and the temperature increases gradually after each time it passes through the collector pipes [6, 7].

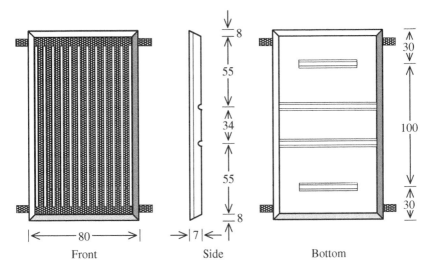

Figure 5.1 Dimensions of a typical solar collector (centimeters).

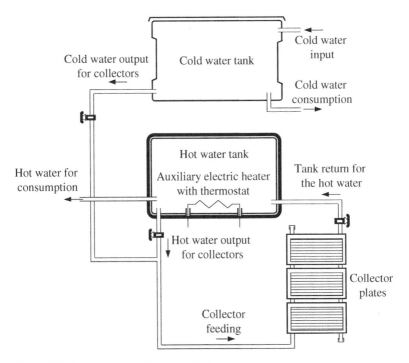

Figure 5.2 Heating water with solar collectors.

Equation (5.1) can be used for practical implementation of a flat solar plate [1, 4]:

$$Q = F \cdot A[I(a \cdot b) - U(T_i - T_a)] \tag{5.1}$$

where:

Q = Energy extracted by the plate (W)
F = Efficiency factor of the heat removal from the plate
A = Plate area (m^2)
I = Rate of the incident to absorbed solar radiation per unit of the plate area (W/m^2)
a = Coefficient of solar transmittance of the transparent covers
b = Coefficient of solar absorption of the plate sheet
U = Coefficient of energy loss of the plate (W/°C - m^2)
T_i = Temperature of the fluid (°C)
T_a = Ambient temperature (°C)

Typical dimensions (meters) of a solar collector are also shown in Figure 5.1. These collectors are carefully manufactured, taking into account the variations in the incidence of radiation, the temperature of the fluids, and the ambient temperature. The useful energy for the collector is defined by equation (5.2):

$$Q_u = A \cdot G \cdot c_p \cdot (T_i - T_o) \tag{5.2}$$

where:

G = Fluid volume per unit of the collector area
c_p = Specific heat of the collector fluid (in the case of water, 4190 J/kg $\cdot$ °C)
T_o = Output temperature of the fluid

The instantaneous efficiency of the collector η is defined by

$$\eta = \frac{Q_u}{A \cdot I} \tag{5.3}$$

An example of calculation of domestic heating energy per month can use equation (5.4):

$$E = N \cdot P \cdot 100 \cdot (T_w - T_m) \cdot \rho_w \cdot C_p \tag{5.4}$$

where:

E = Energy for hot water heating per month
N = Number of days per month (30 days)
P = Number of people using typically 100 L of hot water per person a day
T_w = Desired temperature for the water
T_m = Acceptable minimum temperature for the hot water (say, 60 °C)
ρ_w = Water density (1.0 kg/L)
C_p = Specific heat (4190 J/kg/°C)

Assuming four people using the hot water in a month time (30 days), we have

$$E = 30 \text{ days} \cdot 4 \text{ people} \cdot 100 \text{L/day} \cdot \left(60° - 11°\right) \cdot 1.0 \text{kg/L} \cdot 4190 \text{J/kg} - °\text{C}$$
$$E = 2.464 \times 10^9 \text{J} = 684.44 \text{kWh}$$

When heating water with solar energy, the use of flat collectors can convert the sun's total radiation into heat. Production of water steam through solar energy, at average temperatures between 150 and 200 °C, has numerous applications. The direct heating of air in driers of grains and seeds has the advantage that less sophisticated systems can compete with the ones using conventional fuels. The adoption of a solar process helps to avoid losses in the volume of cropped grains (of up to 50%), due to deficient storage, exposing them to humidity. In addition, solar collectors can avoid burning fossil fuels, and even help to improve some industries, such as those for food transformation, and can contribute for a reduction of electric and gas consumption in residences [4, 6, 7].

Sites with high-speed winds can unfeasibly use heat conductive solar plates due to safety. An evaluation of the site environmental conditions plus the operating conditions is recommended in order to decide if heat conductive solar plates should be deployed.

5.3 Heat Transfer Calculation of Thermally Isolated Reservoirs

Devices for the accumulation of hot water have double walls, isolating them completely against temperature losses with vacuum, liquid, or solid material. To increase their durability, accumulators should be immune to corrosion. Some commercial modules are built of stainless steel AISI 304 and special resins of polyester for hot water and can resist temperatures up to 280 °C and pressures of 7 atm. A simplified method of heat transfer calculation in thermally isolated reservoirs in steady and transient states is presented next.

5.3.1 Steady-State Thermal Calculations

Heat goes from the high-temperature region, T_{high}, to the low-temperature region, T_{low}, according to

$$P = \frac{T_{high} - T_{low}}{R_{th}} \tag{5.5}$$

where:

$P =$ The steady-state power dissipated in the reservoir (watts or joules/second)
$T =$ The temperature (°C)
$R_{rh} = R =$ The thermal resistance (°C/W), whose components depend on the manufacturing material (pipelines and reservoir), R_{ps} and the reservoir ambient, R_{ra}

The equation for the thermal calculation may be generalized as

$$T_{high} = T_{low} + RP \tag{5.6}$$

The thermal resistance R can be used to avoid the definition of what the heat distribution is exactly in a multilayer reservoir. The previous calculation obtains average values for steady-state temperatures when the loss is constant and the temperature is already stabilized at a certain value. The thermal storage capacity should be taken into account to establish the conditions for high time constants during transient and/or heat charge or discharge conditions.

5.3.2 Transient-State Thermal Calculations

At an instant, just before the measurements can be made $t < t_0$, the external temperature of the reservoir is the lowest temperature, T_{low}. At instant t_0, heat transfer begins, limited by the thermal capacity of the reservoir, which prevents the temperature from rising abruptly (exponentially). To express the instantaneous difference in temperature, it is defined as $T = Z_{th}P$, where Z_{th} is the reservoir thermal impedance, which varies with time.

The equivalent thermal circuit, including the thermal capacity C, is illustrated in Figure 5.3, with its transient response to a sudden heat step illustrated in Figure 5.4. The amount of heat Q stored in the reservoir mass m as a whole is

$$\Delta Q = c_p m \Delta T = C \Delta T = \int P_1 dt \tag{5.7}$$

where:

c_p = The specific heat of the reservoir mass m
$C = c_p m$ = The total heat capacity of the reservoir
P_1 = The supplied power necessary to keep the temperature difference ΔT across the walls

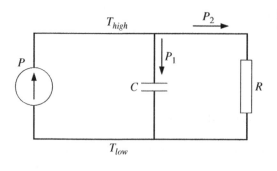

Figure 5.3 Reservoir equivalent thermal circuit.

Figure 5.4 Instantaneous temperature variations in the reservoir.

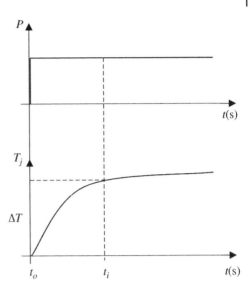

From the heat balance given in equations (5.6) and (5.7) and Figure 5.3 comes

$$\Delta T = T_{high} - T_{low} = RP_2 = \frac{1}{C}\int P_1 dt \qquad (5.8)$$

where:

P_2 = The power diverted for heat storage in the reservoir walls

Differentiating equation (5.8) with respect to time yields

$$R\frac{dP_2}{dt} = \frac{P_1}{C}$$

As $P = P_1 + P_2$, we can set $(P_1/C) = (P_2/C) - (P/C)$ and isolate P_1/C for equation (5.8) to give

$$\frac{P}{C} = R\frac{dP_2}{dt} + \frac{P_2}{C} \quad \text{or} \quad \frac{dP_2}{dt} + \frac{P_2}{RC} = \frac{P}{RC}$$

whose solution is an exponential response $P_2 = P(1 - e^{-t/RC})$. However, as $\Delta T = T_{high} - T_{low} = RP_2 = RP(1 - e^{-t/RC})$, we may use

$$\frac{\Delta T}{P} = R\left(1 - e^{-t/RC}\right) = Z_{th} \qquad (5.9)$$

This is a very simplified solution where the reservoir manufacturer supplies the value of Z_{th}, otherwise, a field evaluation has to be made. This concept of transient impedance Z_{th} can be used for applications with repetitive thermal loading and unloading conditions at high power levels, when the reservoir works under different states of heat variation. The heat path conditions to take heat out from the interior side of the reservoir and transfer it into ambient are too complex to be expressed precisely by a simple exponential form like that in Figure 5.4. This figure for the variation of Z_{th} is useful only for homogeneous materials such as copper pipes, tank walls, or similar materials where the heat spreads quickly. The same does not occur in the path of liquid in a tank where the heat goes to the tank walls, pipes, isolation, drafts, and internal layers of air, since the losses are not uniform. The larger the transient load, the smaller the mass that experiences a temperature increase; one cannot give a single value to its thermal storage capacity. In summary, the method presented previously is very simplified; even so, it is useful in daily practice.

5.3.3 Practical Approximate Measurements of the Thermal Constants *R* and *C* in Water Reservoirs

One practical way of measuring the thermal constants R and C can be performed as the following procedure: An electrothermal element is immersed in the middle of the thermal fluid in a reservoir as shown in Figure 5.2, such that the steady-state power V^2/R supplied to it can be measured. After some period of constant power to the thermal element, the internal temperature will reach a stable value to be measured simultaneously with the ambient temperature close to the tank. The value of R can be calculated by equation (5.6).

To calculate C, we continue the experiment from the previous test to evaluate R and its last measured temperature, except that now the heating active element is switched off. The temperature will begin to decrease exponentially, as in Figure 5.5, and its values at periodic intervals can be logged in a table or graph.

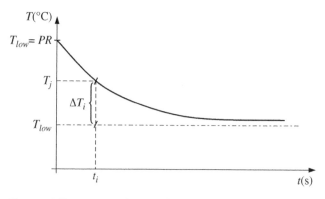

Figure 5.5 Temperature decaying for measurement of C_i.

The thermal capacity may be evaluated from equation (5.9) for $T_{i=0} = T_{high} = PR$ and put into a more convenient form:

$$C_i = -\frac{t_i}{R\ell n\left(\dfrac{\Delta T_i}{PR}\right)} \qquad (5.10)$$

There will be as many values of C_i as the instants of measurement of T_i, and a good average value could be obtained from there. A high standard thermal isolation should guarantee warm water conservation for very long periods. Large differences in temperature between the inner and outer surfaces of the reservoir will increase its losses.

5.4 Heating Domestic Water

Heating a volume of water demands a temperature rise (i.e., an increase in energy). For domestic purposes, that amount of energy is linked to the individual habits of each occupant of a residence. Recorded data show that, on average, each member of a family uses approximately 100 L of water a day. The amount of energy used to heat water each month is then estimated as

$$W = N \cdot P \cdot 100 \cdot (T_w - T_m) \cdot \rho_w \cdot C_p \qquad (5.11)$$

where:

W = Energy used for water heating per month
N = Number of days per month
P = Number of people
T_w = Acceptable minimum temperature for the hot water (say $60\,^\circ$C)
T_m = Ambient temperature of the water
ρ_w = Water density (1.0 kg/L)

Equation (5.11) can be used to predict the amount of energy utilized to heat water in any given period.

Example   To estimate the average monthly energy to heat water (joules (W/s)) for a four-person family, at an ambient temperature of 11.0 °C, equation (5.11) yields

$$W = 30 \,\text{days} \cdot 4 \,\text{people} \cdot 100\,\text{L/day} \cdot (60^\circ - 11^\circ) \cdot 1.0\,\text{kg/L} \cdot 4190\,\text{J/kg} - {}^\circ\text{C}$$

$$W = 2.464 \times 10^9 \,\text{J}$$

5.5 Thermosolar Energy

For applications of electric power generation using solar thermal energy on a large scale, the best-known technologies are the parabolic trough, parabolic dish, solar power tower, and hydrogen production. For smaller uses and remote locations, parabolic dishes are better known, due to their great development potential.

The final design of a solar thermal power plant should follow the solar power density of the place, as illustrated in the normal distribution curve described by equation (5.12) and represented in Figure 5.6 [8–10]:

$$\rho = \rho_{max}\, e^{\frac{-(t-t_0)^2}{2\sigma^2}} \tag{5.12}$$

where:

ρ_{max} = The maximum solar power density (100%)
σ = Standard deviation
t = Hour of the day using the 24 h clock
t_0 = Instant at ρ_{max} (noontime in the equator)

Solar thermal power standards use various means to generate heat, with the water being converted into steam that will be used to drive a conventional steam turbine to produce electricity (Figure 5.7). Sometimes a fossil fuel is used as a backup, so that the power plant can continue to produce energy when solar energy is not available [6, 9].

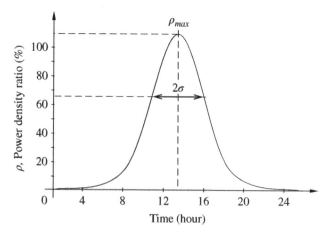

Figure 5.6 Solar density power.

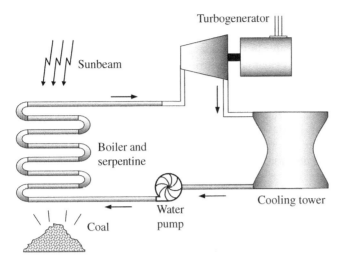

Figure 5.7 Conversion of solar heat and fossil fuel into electricity.

Solar energy is sometimes considered a land-intensive technology, due to the amount of land used in such facilities. However, the amount of energy produced by a solar thermal power plant for a given land area is larger than that produced by a large electrical thermo power plant of the same proportions [10, 11]. This is because the surface is only used in heat-generating plant, while a large electrical thermo power plant concentrates its apparatus in smaller buildings. In addition, solar thermal power plants may use desert lands that are arid or semiarid, as in the northeast Brazil, the northern Sahara, and Central America. Another positive aspect of solar thermal power is that diversifying energy reserves results in less dependence on fossil fuels, whose prices float widely on the international market and are subject to increase as their supply sources become exhausted.

The environment seems to be the biggest beneficiary of such clean technologies. Emissions of carbon dioxide in the production of energy create about 50% of the gases responsible for the greenhouse effect. The hybrid thermal solar power plants in operation in California, where fossil fuel is used only as reserve power, help in the reduction of carbon dioxide, nitrogen oxide, and sulfuric dioxide. A typical installation of 80 MW using solar troughs reduces carbon dioxide emissions into the atmosphere by 4.7 million tons and avoids using 2 million tons of coal during 25 useful years of trough life [12–16].

5.5.1 Parabolic Trough

A parabolic trough uses a type of clear oil in pipes that absorbs heat reflected off the trough. The parabolic troughs are long, trough-shaped reflectors that focus the sun's energy on a pipe running along the mirror's curve, as shown in Figure 5.8a. Heat-absorbent oil is used inside the conducting pipeline to carry the thermal energy to the water in the boiler heat exchanger, whose temperature can reach about 400 °C. Heat energy from the oil is transferred through a heat exchanger to boil water to dry steam of high pressure that drives the turbine of an electric generator. The cross section of the troughs is parabolic, to ensure conversion efficiency. In larger systems, the design of the trough includes a rotating shaft that allows each group to follow the sun from east to west. Solar farms reunite several long and parallel lines of groups of parabolic troughs to concentrate and maximize the sunbeams on heat absorbing pipelines as shown in Figure 5.8b.

The sunlight incident on the absorber is maximized by the high reflectance coefficient of the parabolic reflector, which is positioned as high as possible.

(a)

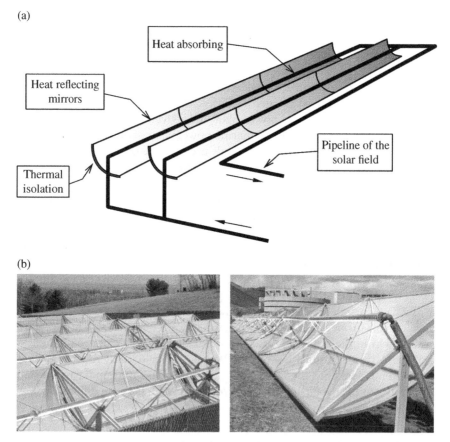

(b)

Figure 5.8 Parabolic troughs: (a) deposition of more than one solar array; (b) operational solar array. *Source*: © Jefferson County Detention Facility in Golden, Colorado.

The peak optical efficiency of a parabolic trough is in the range from 70 to 80%, but only something like 60% is practically useful since there are other losses, such as those due to heat losses in the solar field piping. Typical solar thermal power plants can supply energy at full load in 2000–3000 of the 8760 annual hours. Some losses are always present in the receiver of a trough concentrator, which is usually made of a metal absorber sometimes surrounded by glass tubes with antireflexive properties. The absorber is coated with a selective surface to filter out infrared radiance from the incoming light, to make way for light in the visible range. The intensity of the sun may be multiplied by a concentration ratio in the range from 20 to 100.

Silver reflectors have higher reflectance, but aluminum is preferred for the structures of the trough system because silver is more expensive and more difficult to protect against the corrosive effects of the outdoor environment. It is also important to keep the reflectors clean since dust and dirt will degrade the reflectance of light from the parabola. Larger systems may have to use yaw controllers (precise to a fraction of a degree), whose values are almost negligible in these cases, but as their use depends on the size of the trough collector, small systems should not use them. Typical uses of parabolic solar troughs are to generate power for hot water, space heating, air conditioning, steam generation, industrial process heating, desalination, and electrical power generation.

Solar heating design makes possible annual savings of up to 70% of the overall hot water heating bill. In addition, commercial trough systems can compete with natural gas and can deliver energy at a steady cost for more than 20 years. Little maintenance is needed since the collectors track the sun continually during the day to heat a closed-loop circulation of an antifreeze solution such as propylene glycol. Heat from the solar collectors is usually transferred through immersed copper coils to hot water storage tanks. The still-expensive manufacturing technology of trough systems is being enhanced through research on parabolic trough technology that uses new methods and designs for power plant integration of solar technology with fuel gas that result in lower investment costs, less weight, cleaner energy, higher efficiency, and lower costs for production of electricity.

5.5.2 Parabolic Dish

For remote stand-alone applications, parabolic dishes with yawing engines seem to be a very attractive solution. It is the most modular of all the options presently available for concentrating sunrays in a single point and can produce higher temperatures. It goes commercially from a few watts to several kilowatts at temperatures as high as 800 °C. For higher power capacities, several units can be combined to produce the desired output power. Figure 5.9 is an illustrative example of such a small configuration for a domestic application.

Figure 5.9 Parabolic dish. *Source*: Simões *et al.*, 2015. Reproduced with the permission of Taylor & Francis.

Solar dish–engine systems convert the energy from the sun into electricity using a mirror array formed in the shape of a large dish. The solar dish has a parabolic shape, which focuses the sun's rays onto a receiver. The receiver transmits the energy to an engine, typically a Stirling or Brayton cycle engine, to generate electric power at very high efficiency (see Appendix B). This high efficiency is due to the high concentration ratios achievable with parabolic dishes and the small size of the receiver, whose highest values are achievable at higher temperatures. Tests of prototype systems and components at locations throughout the United States have demonstrated net solar-to-electric conversion efficiencies as high as 30%, significantly higher than for any other solar technology.

Some interesting experiments with larger solar dishes have been reported in several countries, such as Australia with CSIRO in collaboration with industry partner Solar Systems Pty Ltd., to demonstrate the integration of solar thermal energy and methane gas (see Figure 5.10). It has been claimed that the system can produce a range of solar-enriched fuels and synthesis gas (CO and H_2). These gases can be used as power generation fuel gas, as metallurgical reducing gas, or as chemical feedstock (e.g., in methanol production). Dishes were built and operated by CSIRO, which constructed a demonstration facility designed to process 44 kW thermal load of natural gas. Key aspects of process chemistry, reactor design, and power generation prove the value of solar thermal technology as an economical option for large-scale energy delivery [14–17].

5.5.3 Solar Power Tower

The principle of capturing and concentrating the solar light on reception mirrors is used in an energy tower. In this case, the mirrors are distributed in a convenient form and not in arrays, as is the case with parabolic troughs (Figure 5.11). In the center of the tower circle, a receiver is placed containing a fluid that can be water, air, oil, liquid metal, molted salt, or diluted salt.

Figure 5.10 CISRO solar dish system in Australia. *Source*: Simões *et al.*, 2015. Reproduced with the permission of Taylor & Francis.

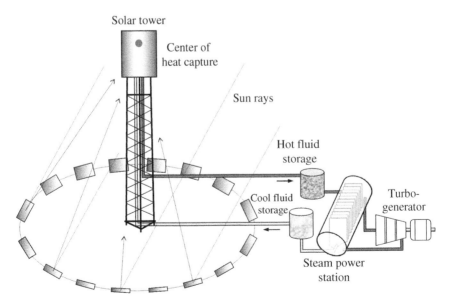

Figure 5.11 Solar power tower.

The position of the mirrors is heliostat. The warm fluid goes from the receiver to the tower block and then to a steam turbine. As the sun heats salt in the receiver, it goes to the hot storage tank. The hot salt is then pumped through a steam generator, and the steam drives a turbine generator to produce electricity. The same salt is finally returned to the cold storage tank to be used again in the concentrator and closes the cycle. This technology is not yet well established, but the tower of energy can supply temperatures higher than those supplied by any other heat concentrators [18–22].

Test designs in the United States using diluted salt as a transfer means to substitute for fossil fuel for thermal storage were built to work under complementary demand. These energy towers are considered good prospects for the long term, due to their high efficiency and low cost in the production of electricity, especially for larger units (100–200 MW). Power plants as big as 200 MW have been considered in a formal joint venture and profit-sharing agreement between SolarMission Technologies, Inc., and Sunshine Energy (Aust) Pty Ltd. In the case of small facilities, the intermediary heated fluid using these technologies (e.g., oil) can be used directly for heating poultry houses, greenhouses, and grain driers [23–25].

An interesting practical example of this type of solar heating is the Solar Two (see Figure 5.12), a retrofit of the original Solar One, located in the Mojave Desert (Daggett, California), which was operated with steam in 1982–1988. Solar Two operated with molten salt with 60% sodium nitrate and 40% potassium nitrate as the heat transfer and energy storage medium. It used 1926 heliostats with a total area of 82,750 m^2 to produce 10 MW, and it was operated as such for a few thousand hours being decommissioned in 1999 and unfortunately demolished in 2009.

5.5.4 Production of Hydrogen

In the production of hydrogen, steam reforming for natural gases and synthesized hydrogen from either gasoline or methanol are being considered for fuel distribution supply in fuel cell-powered vehicles. However, in solar belt regions, the use of an environmentally friendly option for a solar thermal-distributed hydrogen process for these vehicles will probably need to be based on the already existing infrastructure to generate the hydrogen required [16–18, 24–27].

For this purpose, sunlight can be concentrated on small solar reactors to achieve ultrahigh temperatures: 1700 °C or higher. The Solar Two power tower at Daggett, California (see Figure 5.12), is already dismantled, but it served as basis for new project as the Ivanpah solar project. It plans on using similar technology, scaled up to larger size with seven such tower receivers. The chemical reaction rates at these temperatures are enormous. These reactors are made of one outer quartz tube and two inner concentric graphite tubes. The graphite used in the concentric tubes shown in Figure 5.13 is of different consistencies.

Figure 5.12 Power tower Solar Two (Daggett, California). *Source*: © Work of the United States Department of Energy.

The external graphite tube is solid graphite; the internal graphite tube is porous graphite. Sunlight has to pass through the quartz tube to heat the solid external graphite tube, which, in turn, radiates and heats the internal porous graphite tube to the necessary reaction temperature. The natural gas to be used in these reactors must pass through a hydro generator and onto a zinc oxide bed to remove mercaptans and hydrogen sulfide from the stream. The chemical reaction is given by

$$C_x H_y + concentraded\ sunlight \Rightarrow C + 2H_2 + unreacted\ C_x H_y$$

Single-pass conversions reported by researchers from the University of Colorado say that 70% of the natural gas can be purified by this method. The hydrogen can be compressed and stored. The carbon can be sold into the carbon market, kept in storage, or used to feed a highly efficient carbon conversion fuel cell. The hydrogen that results from this process can be used in stationary fuel cells or for storage. Certainly, the cost of compression and storage of hydrogen will exceed 50% of the capital of a solar thermal plant in a distributed service station (DSS).

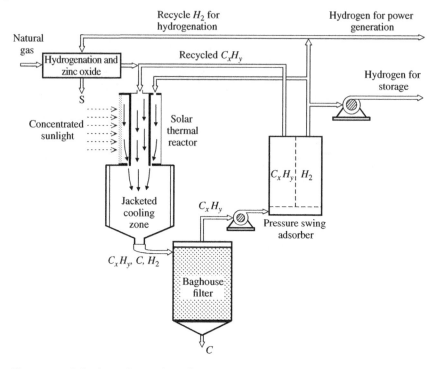

Figure 5.13 Solar thermal natural gas dissociation process.

5.6 Economics Analysis of Thermosolar Energy

There are structural problems related to financing solar energy, primarily because of the high installation costs, lack of government financial assistance, amount of time needed to gain the return of initial installation cost, competition with petroleum companies, and belief that only large power plant installations can manage the world's energy problems. However, there are real opportunities to be sought by people with long-term vision, as discussed in the 6th annual Sustainable Innovation Forum in 2015 in Stade de France, Paris (COP21).

Three general sources of capital are available for a renewable power project: equity, debt, and grant financing. An equity investment is the purchase of ownership in a project. A debt investment is a loan to the project. For example, in the purchase of a house, the person buying the house is the equity investor, and the mortgage is the debt. Since most solar power projects cannot compete with conventional fossil power technologies today, a number of organizations have grant financing to help buy down the no-economic portion of the project. These grants typically are made available to help account for environmental, developmental, and economic externalities that would not otherwise be accounted for during a normal competitive project selection [20, 23–25].

One example of the contribution of solar thermal electricity generation to climate change can be assessed by calculating the greenhouse gas costs (GGCs) incurred during the provision of services and production of materials needed for the construction and operation of solar power plants. Both process and input/output analyses are able to yield GGCs for solar-only plants with a typical uncertainty of about 25%, at a moderate level of input data detail. The uncertainty of the GGCs for hybrid systems decreases with increasing fossil share. The GGCs of utility-sized solar-only parabolic trough, central receiver, and parabolic dish plants range from 30 to 120 g CO_2 – e/kWhel. Furthermore, GGCs vary with plant size and depend on whether a fossil-fueled backup or heat storage system is chosen to increase the plant's capacity factor.

Comparisons with other renewable electricity generation technologies must be judged with care because of differences in the assessment boundary and methodology and in the power plant's location, lifetime, capacity factor, dispatchability, and other characteristics. The parabolic trough option is typically applied to grid-connected plants and is commercially available with over 9 billion kilowatt-hours operational to date; the solar collection efficiency is up to 60%, with peak solar-to-electrical conversion efficiency up to 21%. The main disadvantage is the low temperature with moderate steam quality.

The parabolic dish is typically applied to stand-alone applications or small off-grid installations. It is modular with very high conversion efficiency with peak solar-to-electrical conversion of about 30%. The drawback is the low-efficiency combustion in hybrid systems. The central receiver is for grid-connected plants with high solar collection efficiency: up to 46% at operating temperatures close to 550 °C. However, the technology is not yet commercially proven to allow precise capital cost projections.

The cost of the initial investment in thermal solar energy is high, and the solar field absorbs around 50% of the total costs of facilities. Thermal solar systems now in commercial operation are designed to integrate the solar energy in power plants fed by fossil fuel, aiming at a reduction in the cost of fuel. The factors that work against increased investment in this area are still the low cost and smaller scale of fossil fuel power plants (coal or oil), despite all the clear disadvantages (see Chapter 13).

Either indirect or direct solar energy is a reality for heating and electricity use. This technology should be widespread because it is an important source of clean energy with very minimum environmental impacts. New houses and buildings should incorporate thermal solar energy in their initial design, so they can complement and minimize electrical utilities costs. Typically they should be assembled on roofs, building fronts, backyards, and open spaces. There is still a lot of implementation and adoption to be done by builders and construction companies. It is expected that this chapter contributes to widespread fundamental knowledge and understanding on how to use thermal solar energy [26–28].

References

1 Hot water for solar energy, Energetic Power (Força Energética), Vol. 1, pp. 14–16, May–June 1992.

2 Solar energy: electricity or hot water, Energetic Power (Força Energética), Vol. 2, No. 3, pp. 25–26, March 1993.

3 W.L. Hughes, Energy in Rural Development: Renewable Resources and Alternative Technologies for Developing Countries, Advisory Committee on Technology Innovation, Board on Science and Technology for International Development, Commission on International Relations, National Academy of Sciences, Washington, DC, 1976.

4 J.F. Kreider and F. Kreith, Solar Energy Handbook, McGraw-Hill, Inc., New York, 1981.

5 D.S. Ward, S. Karaki, G.O.G. Löf, C.C. Smith, M.Z. Lowenstein, C.B. Winn, M.E. Larson, and I.E. Valentine, Solar Heating and Cooling of Residential Buildings: Sizing, Installation and Operation of Systems, U.S. Department of Commerce (EOA), Solar Heating Application Laboratory, Colorado State University, Fort Collins, CO, 1977.

6 M.R. Patel, Wind and Solar Power Systems, CRC Press, Boca Raton, FL, 1999.

7 A.W. Culp, Principles of Energy Conversion, McGraw-Hill, Toronto, ON, Canada, 1979.

8 The American Society for Testing and Materials (ASTM), http://www.astm.org/, accessed March 10, 2017.

9 National Technical Information Service (NTIS), https://www.ntis.gov/, accessed March 23, 2017.

10 KanEnergi, New Renewable Energy: Norwegian Developments, Research Council of Norway in cooperation with Norwegian Water Resources and Energy Directorate, Hanshaugen, Oslo, Norway, December 1998.

11 M. Romero, M.J. Marcos, R. Osuna, and V. Fernández, Design and implementation plan of a 10-mW solar tower power plant based on volumetric-air technology in Seville (Spain), in Proceedings of the Solar 2000 Solar Powers Life—Share the Energy, Madison, WI, June 17–22, 2000.

12 Intermediate Technology Development Group, Handson: It works, TVE-Series, Schumacher Centre for Technology and Development, Rugby, England, 1999.

13 Concentrating Solar Power Projects, National Renewable Energy Laboratory, https://www.nrel.gov/csp/solarpaces/by_country_detail.cfm/country=US, accessed March 23, 2017.

14 T. Liu, H. Temur, and G. Veser, Autothermal reforming of methane in a reverse-flow reactor, Chemical Engineering Technology, Vol. 32, No. 9, pp. 1358–1366, 2009.

15 J.K. Dahl, J. Tamburini, A.W. Weimer, A. Lewandowski, R. Pitts, and C. Bingham, Solar-thermal processing of methane to produce hydrogen and

syngas, American Chemical Society Journal on Energy and Fuels, Vol. 15, No. 5, pp. 1227–1232, 2001.

16 D.S.A. Simakov, M.M. Wright, S. Ahmed, E.M.A. Mokheimer, and Y. Román-Leshkov, Solar thermal catalytic reforming of natural gas: a review on chemistry, catalysis and system design, Catalysis Science and Technology, Vol. 5, pp. 1991–2016, 2015.

17 C. Agrafiotis, H. von Storch, M. Roeb, and C. Sattler, Solar thermal reforming of methane feedstocks for hydrogen and syngas production—a review, Renewable and Sustainable Energy Reviews, Vol. 29, pp. 656–682, 2014.

18 K. Uma Rao, Computer Techniques and Models in Power Systems, Second Edition, I K International Publishing House, New Dehli, India, 2014.

19 International Renewable Energy Agency (IRENA), Renewable Energy Technologies: Cost Analysis Series, Concentrating Solar Power, IRENA, Bonn, Germany, June 2012.

20 G. Parkinson, World's biggest solar tower storage plant to begin generation this month, 2015, http://reneweconomy.com.au/2015/worlds-biggest-solar-tower-storage-plant-to-begin-generation-this-month-22860, accessed March 10, 2017.

21 R. Kistner and H.W. Price, Financing solar thermal power plants, Paper RAES99-7727, Proceedings of the Conference on Renewable and Advanced Energy Systems for the 21st Century, Lahaina, Maui, HI, April 11–15, 1999.

22 M. Werder and A. Steinfeld, Life cycle assessment of the conventional and solar thermal production of zinc and synthesis gas, Energy, Vol. 25, pp. 395–409, 2000.

23 C. Dey and M. Lenzen, Greenhouse gas analysis of electricity generation systems, in Conference Proceedings of ANZSES Solar 2000, Grifith University, Queensland, Australia, November 29–December 1, 2000, pp. 658–668.

24 S.F. Rice and D.P. Mann, Autothermal Reforming of Natural Gás to Synthesis, Gas, SANDIA REPORT SAND2007-2331, Reference: KBR Paper #2031, Sandia National Laboratories, Albuquerque, NM, and Livermore, CA, April 2007.

25 A. Chambers, C. Park, R.T. Baker, and N.M. Rodriguez, Hydrogen storage in graphite nanofibers, Physical Chemistry, Vol. 102, pp. 4253–4256, 1998.

26 Renewable Energy Sources and Climate Change Mitigation, Edited by O. Edenhofer, R. Pichs-Madruga and Y. Sokona, Published for the Intergovernmental Panel on Climate Change, Potsdam Institute for Climate Impact Research (PIK), Brandenburg, Germany, 2012.

27 N. Hadjsaid, J.F. Canard, and F. Some, Computer applications in power systems, IEEE Power Engineering Society Magazine, Vol. 12, No. 2, pp. 22–28, 1999.

28 V. Quaschning and F. Trieb, Solar thermal power plants for hydrogen production, in Proceedings of the Hypothesis IV Symposium, Stralsund, Germany, September 9–14, 2001, pp. 198–202.

6

Photovoltaic Power Plants

6.1 Introduction

Solar irradiance decreases with the square of the distance to the sun; since the distance of the Earth to sun changes during the year, the solar irradiance outside the Earth's atmosphere varies from 1325 to 1420 W/m^2. In order to compare, in Mars, for example, the solar irradiance is under 600 W/m^2—a factor to be considered when designing photovoltaic (PV)-powered satellites. The total specific radiant power, or radiant flux per area that reaches Earth, is defined as spectrum AM 0 (air mass zero). We consider in this book that extra-terrestrial sunlight power density is 1361 W/m^2, which is filtered through the atmosphere and is received at the surface about approximately 1000 W/m^2 or maybe less. Since this is peak power value, a good average for a day is 345 W/m^2, disregarding the night hours. Such power density, decreasing from 1361 W/m^2 at space, all the way down to the average of 345 W/m^2 is mostly due to atmospheric absorption, reflection, clouds, and haze.

The amount of power given by the sun is estimated to be on the order of 3.90×10^{26} W. The Earth intercepts energy equal to a disk as large as the Earth's diameter, assuming the Earth's radius to be 3,393,000 m; the Earth's solar interception area is $(3.14) \times (3,393,000)^2 = 3.62 \times 10^{13}$ m^2. Therefore, the Earth intercepts about 5.02×10^{16} W, that is, the Earth intercepts 50 quadrillion watts of solar power each day. A BP report estimated several years ago that our current Earth energy needs are on the order of 400 EJ annually. Therefore, a simple calculation of power availability by the sun, integrated over cycles of hours over one year, shows that the Earth receives much more power from the sun to actually power all the human activity in the current society.

The air mass (AM) factor is an important factor to be considered when deciding for a PV power plant. If the sun rays are perpendicular to the Earth's surface, the sunlight will only have to pass through the AM of the atmosphere, and that condition is defined as AM 1. In all other cases, the trajectory of the

Integration of Renewable Sources of Energy, Second Edition. Felix A. Farret and M. Godoy Simões.
© 2018 John Wiley & Sons, Inc. Published 2018 by John Wiley & Sons, Inc.

solar radiation through atmosphere is longer. Through this way, depending on the sun's height, AM 2 would indicate that the path of the sunlight through the atmosphere is twice AM 1 for such a case if the sun inclination is 30° above the horizon ($\gamma_s = 30°$). In general, definition of AM is [1]

$$AM = \frac{1}{sin(\gamma_s)}$$

It has been defined that for a worldwide terrestrial average, the spectrum AM 1.5 should be used. In this chapter, irradiance is measured in W/m^2 (power) and has the symbol E. When integrating the irradiance over a given period of time, it becomes the solar irradiation (energy). Irradiation is measured in either J/m^2 or Wh/m^2 (energy), and it is represented by the symbol H. For terrestrial applications and daylight use, the visible part of the sunlight is the one considered for energy conversion.

6.2 Solar Energy

The sun is of great importance for the planet Earth and the ecosystem of our society. Rays emitted by the sun, gamma rays, reach the terrestrial orbit a few minutes after they leave the sun surface, crossing approximately 150 million kilometers. Clouds reflect about 17% of sunlight back into space, 9% is scattered backward by air molecules, and 7% is actually reflected directly off the surface back into space. Therefore, the travel through the atmosphere decreases the radiation at the Earth's surface to about 35% less than the level in the stratosphere. At noon on a clear day, the luminous power at the ground level is approximately $1000\,W/m^2$, which is defined as $1\,sun = 1000\,W/m^2$.

The PV industry, in conjunction with the American Society for Testing and Materials (ASTM) (http://www.astm.org) and US government research and industrial laboratories, developed and defined two standard terrestrial solar spectral irradiance distributions. These two spectra define (i) a standard direct normal spectral irradiance (ASTM E-891) and (ii) a standard total (global, hemispherical, within a 2π steradian field of the tilted plane view) spectral irradiance (ASTM E-892). The direct normal spectrum is the direct component contributing to the total global (hemispherical) spectrum [2–4]. The spectra represent terrestrial solar spectral irradiance on a surface of specified orientation under a set of specified atmospheric conditions (Figure 6.1). The distribution of power (watts per square meter per nanometer of bandwidth) as a function of wavelength provides a common reference for evaluating spectrally selective PV materials with respect to their performance measured under varying natural and artificial sources of light with various spectral distributions. The conditions selected by ASTM were considered a reasonable average for

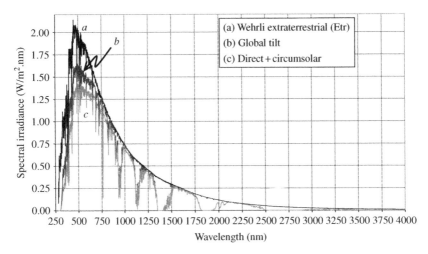

Figure 6.1 Direct and global radiation, 37 tilt (ASTM G173-03 Ref. Spectra).
© US Government.

the 48 contiguous states of the United States over a period of one year. The tilt angle selected is approximately the average latitude (40°) for the continental United States.

Global solar radiation includes their total energy, that is, thermal energy, whose use consists of domestic and industrial use of heat, either for heating fluid or for transformation to another form of energy, as discussed in Chapter 5, plus the possible direct conversion to electricity. The PV effect consists of the direct transformation of the radiant energy into electricity without the intermediate production of heat.

Several countries in the tropical regions are privileged by their solar energy potential. Therefore, there are enormous possibilities for the use of solar energy as a thermal and PV source (http://energy.gov/maps/solar-energy-potential).

Semiconductor characteristics define the electrical energy conversion efficiency of solar cells. The material energy gap, manufacturing quality of the cell material, and technology employed in the manufacturing process sum up to a final efficiency. Usually, the conversion efficiency of a commercial solar cell (in other words, the relationship between the electric power generated and the power of the incident radiation on the semiconductor) is from 7 to 15%, depending on the PV material technology. Simply stating, at noon on a clear day, a 10% efficient PV module would generate the average of $100\,W/m^2$.

Crystalline silicon and hydrogenated amorphous silicon are currently the most appropriate semiconductor materials for PV conversion. However, it is already possible in laboratories to make solar cells of crystalline silicon with up to 29% conversion efficiency, but they are still very expensive to manufacture. Hydrogenated amorphous silicon is obtained from 10 to 12% efficiency in the laboratory and 7 to 8% in modules produced industrially. Manufacturing costs are less for the crystalline silicon cell. Monocrystalline silicon cells are efficient, reliable, and durable but expensive. More appropriate for residential and commercial installations are the polycrystalline silicon types; they have a lower degree of purity or hydrogenated amorphous silicon [2, 5, 6]. The system designer must calculate if it is better to invest in more efficient PV arrays or just increase the area for less expensive ones.

6.3 Conversion of Electricity by Photovoltaic Effect

Solar light is converted directly into electricity with modules consisting of many PV solar cells. Such solar cells are usually manufactured in the form of fine films or maybe in thick wafers that are cut and assembled. They are semiconductor devices capable of converting incident solar energy into direct current (dc), with efficiencies varying from 3 to 31%, depending on their technology, the light spectrum, temperature, design, and the material of the solar cell (see Table 6.1).

A solar cell could be understood simply as a semiconductor device, such as a diode, able to produce a very low voltage (about 0.6 V), which is continually

Table 6.1 Common materials used in PV modules.

Type	Theoretical efficiency η (%)	cm^2	Practical tests η (%)	Commercial modules η (%)	cm^2	Useful life Years
Monocrystalline silicon (Si)	29	4	23	15–18	100	30
Thick monocrystalline silicon				15–20		30
Polycrystalline silicon (Si)		4	18	12–18	100	30
Thin vapor films				6–12		25
Laminate				5–10		20
Amorphous silicon (a-Si)	27	1	12	5–8	1000	20
Gallium arsenide (GaAs)	31	0.25	26			
Copper–indium–selenide (CIS)	27	3.5	17			
Cadmium telluride (CdTe)	31	1	16			

Source: Based on Ref. [1]. Osterwald *et al.*, 2012. Reproduced with the permission of John Wiley and Sons.

energized at a rate proportional to their incident irradiance. The series–parallel connections of cells allow the design of solar panels, or solar arrays, with high currents and voltages (reaching up to kilovolts). In order to complete a full system of electric power provisioning, it is also necessary to have energy storage and power electronic conditioning equipment; typically such a system is more appropriate for small-scale applications.

The most attractive features of solar panels are the nonexistence of movable parts, slow degradation of the sealed solar cells, flexibility in the association of modules (ranging from a few watts to megawatts), and their extreme simplicity of use and maintenance. In addition, solar energy is a very relevant source, with the following characteristics: it is autonomous, its operation does not pollute the atmosphere (i.e., its use does not harm any ecosystem though the fabrication has many materials that could potentially pollute the place where the cells are manufacture [6–8]); but it is an inexhaustible and renewable source with great reliability. As an example, a 6-kW PV system, instead of a thermoelectric power plant, can reduce annual pollution by 1.5 kg of NO_x (nitrogen oxides), 4.5 kg of SO_2 (sulfur dioxide), and 4.5 ton of CO_2 (carbon dioxide). Until the writing of this book, manufacturing costs have been a major impediment to their widespread use, and some governmental actions can be implemented in order to allow common people to install in their roofs and generate power for residential and small commercial applications.

6.3.1 Photovoltaic Cells

Semiconductor materials have bands of allowed and forbidden energy in their spectrum of electronic energy (the energy gap or Fermi level). The allowed bands are the valence and conduction bands, separated by the energy gap. The electrons occupy the valence band and can be excited to the conduction band by thermal energy or by absorption of photons with energy quantum in order to make them win the energy gap. The bandwidth of the energy gap is intrinsic and depends on each semiconductor characteristics and their doping.

When an electron passes from one band to another, it leaves in its place a hole that can be considered a positive charge. When voltage is applied across this semiconductor, the electrons and their holes contribute to the electrical current, since the presence of that electric field makes those particles move in opposite directions with respect to each other. Therefore, an electrostatic potential inside the material is created to separate positive from negative charges. Conversely, when the semiconductor is illuminated, it behaves like a battery; in other words, the charges accumulate in opposite areas of the chip. If an external wire connects the two areas, there will be a circulation of electric current. In solar cells made of crystalline silicon, the process of controlled and selective contamination of the semiconductor material originates in the electric field in the crystalline silicon [6–8].

6.4 Equivalent Models for Photovoltaic Panels

Figure 6.2 represents a more detailed model of a PV cell. Typical values of capacitor C are very low ($\cong 10\,\text{pF}$) and are usually neglected or may be not considered for dc analysis. The resistive parameters for a silicon PV cell are typically R_s in the range $0.01–1.0\,\Omega$ and R_p in the range $300–800\,\Omega$ [6, 7, 9–11]. The irradiance current is proportional to the radiant luminous intensity of the sun given in lumens or footcandles (luminous flow) or photocandles (luminous intensity), where 1 lumen $= 1.496\cdot10^{-10}\,\text{W}$; 1 lumen/ft$^2 = 1.609\cdot10^{-12}\text{W/m}^2$ $=1$ footcandle (fc); and 1 sun $=1000\,\text{W/m}^2$.

(a)

(b)

(c)

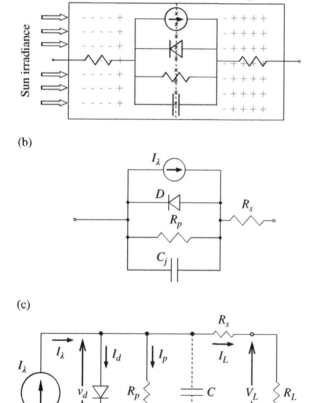

Figure 6.2 Evolutionary steps: (a) P-N junction, (b) Electrical PV current path, (c) Electrical model.

From the equivalent electrical circuit of the single cell represented in Figure 6.2, the following equation can be derived:

$$I_L = I_\lambda - I_d - I_p \tag{6.1}$$

where:

I_λ = The photon current, which depends on the light intensity and its wavelength

I_d = The Shockley temperature-dependent diode current

I_p = The PV cell leakage current

6.4.1 Dark-Current Electric Parameters of a Photovoltaic Panel

Figure 6.2 shows a classical representation of a solar cell, a parallel resistance R_p, a series resistance R_s, a diode with a saturation current I_s as a function of the ideality factor η, a sunshine irradiance source I_λ, and a capacitance C. For simplicity, this capacitance is not usually represented since the output voltage of a PV cell is essentially a dc. In the case of a connection to the load through a high-frequency power converter, this capacitance has been considered. Dark-current electric parameters of a PV panel may be measured with laboratory tests as discussed below. Characteristics such as those represented in Figure 6.3 can be used to determine the PV cell or PV panel parameters using the respective models as discussed in Sections 6.4.1 and 6.4.2 [12–15].

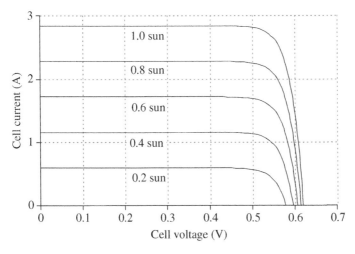

Figure 6.3 Typical I–V characteristic of a solar cell.

There are modern and advanced equipment for flash testing, that is, to maintain a nearly constant light intensity while rapidly changing the bias voltage on the cell to sweep out the I–V curve, and typically about 1000 I–V data points are acquired in less than 1 millisecond. This method does not heat the solar cell, and a controlled temperature chamber can be used to impress variable and precise steady temperature measurements. For Standard Test Conditions (STC), the temperature is 25 °C (77 °F) and the atmospheric density is $AM = 1.5$; the output of the panel under these STC is what gives the solar panel rating.

6.4.1.1 Measurement of I_λ

The photon current is proportional to the irradiance intensity and depends on the light wavelength, λ. The parameters of this current are related to the cell short-circuit current, I_{sc}, and to the cell open-circuit voltage, V_{oc}. The short-circuit current is obtained from the I–V characteristic for a given solar cell (Figure 6.3) when the cell output voltage is $V_L = 0$. In its turn, the open-circuit voltage is obtained for zero output current, $I_L = 0$ (no load). In addition, if the photon current is known for certain standard irradiance intensity, say, $G_s = 1.0$ sun and the corresponding current $I_{\lambda o}$, it is possible to obtain the photon current approximately for every other levels G through the expression

$$I_\lambda = \frac{G}{G_s} I_{\lambda o} \tag{6.2}$$

Referring to Figure 6.2, the Shockley diode current is given by a classical expression:

$$I_d = I_s \left(e^{qV_d/\eta kT} - 1 \right) = I_s \left(e^{V_d/\eta V_T} - 1 \right) \tag{6.3}$$

where:

I_s = The diode reverse saturated current, typically 100 pA for the silicon cell
$k = 1.38047 \times 10^{-23}$ J/K = The Boltzmann constant
$q = 1.60210 \times 10^{-19}$ C = The electron charge
V_d = The diode voltage in volts
η = An empirical ideality constant adjusted to around 1.2 and 1.8 for silicon
$T = 273.15 + t_C$ = The absolute temperature given as a function of the temperature t_C (°C), generally taken as $T = 298.15$K (i.e., 25 °C)
$q/kT = 38.9445$ C/J for $t_C = 25$°C or, in a more general way, for any temperature: $q/k = 11605.4677$ C
$V_T = kT/q = 26.7$ mV = The temperature equivalent

6.4.1.2 Measurement of R_p

The measurement of R_p can be performed at night or in a dark room or ambient room without light in a closed environment to ensure that the panel does

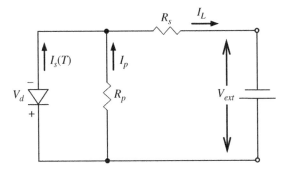

Figure 6.4 Equivalent circuit of a reverse-biased PV cell.

not receive any form of lighting to be based on the equivalent model of Figure 6.2. It is necessary to use an external voltage source V_{ext} and keep a known temperature. By experimentation the inverse polarity of the panel internal diode is determined by observing the current intensity. This external source should inversely polarize the panel as shown in the drawing in Figure 6.4, such that the main current is forced to circulate through R_p. The value of R_p is supposed to be high enough in order to avoid heating the panel cells and decreasing the overall efficiency.

Under the biasing conditions and impressed lighting for the panel for $I_s \cong 0$ and $R_p \gg R_s$,

$$R_p \cong \frac{V_{ext}}{I_L} \tag{6.4}$$

The parallel resistance R_p represents part of the internal losses together with the leakage current across the Shockley diode. In addition, there is a series resistance R_s between the photon current source and the load across the PV cell terminals, R_L.

6.4.1.3 Measurement of I_d

There is no need for other measurements of I_d for other parameters. With the forward biasing of the cell diode under controlled lighting and a short circuit across the PV cell terminals, a diode current for the calculation of other parameters is established:

$$I_{di} = I_{\lambda i} - I_{Li} - \frac{V_{di}}{R_p} \approx I_{sc} - I_{Li} - \frac{V_{Li}}{R_p} \tag{6.5}$$

where:

$$I_{\lambda i} = I_{\lambda 0} \frac{G_i}{G_s}$$

6.4.1.4 Measurement of η

Using equation (6.3), two diode currents are measured under distinct conditions 1 and 2:

$$I_{d1} = I_s\left(e^{qV_{d1}/\eta kT} - 1\right) \quad \text{and} \quad I_{d2} = I_s\left(e^{qV_{d2}/\eta kT} - 1\right)$$

The measurement conditions must use very distinct currents in order to obtain more representative values of the operating regions of the cell internal diode. One of these measurements should be in the linear region (high current) and the other around-the-knee region (lower current). Note that V_{d1} and V_{d2} can be approximated, respectively, by V_{L1} and V_{L2}, since R_s is very small.

Taking a ratio of Shockley's equations, each one for distinct measurement conditions of the diode current, we arrive at equation (6.6):

$$\frac{I_{d1}}{I_{d2}} = \frac{e^{qV_{L1}/\eta kT} - 1}{e^{qV_{L2}/\eta kT} - 1} \tag{6.6}$$

or yet

$$I_{d2}\left(e^{qV_{L1}/\eta kT} - 1\right) = I_{d1}\left(e^{qV_{L2}/\eta kT} - 1\right)$$

$$I_{d2}e^{qV_{L1}/\eta kT} - I_{d1}e^{qV_{L2}/\eta kT} = I_{d2} - I_{d1}$$

With a numerical methodology (Runge–Kutta, Euler, modified Euler), it is possible to determine η. The advantage of this solution is that it is independent of I_s. On the other hand, if the voltage used across the diode terminals is greater than 0.3 V, the exponential term of Shockley's equation at a STC gives

$$e^{qV_L/\eta kT} = e^{1.6021 \times 10^{-19} \times 0.3/1.5 \times 1.3805 \times 10^{-23} \times 298.15} = e^{7.784986} \cong 1618.2 \gg 1$$

and equation (6.6) can be approximated by

$$\frac{I_{d1}}{I_{d2}} \cong \frac{e^{qV_{d1}/\eta kT}}{e^{qV_{d2}/\eta kT}} = e^{q(V_{d1}-V_{d2})/\eta kT}$$

Isolating the exponential term and rearranging this equation results in

$$\eta \cong \frac{q\left(V_{L1} - V_{L2}\right)}{kT \ell n \dfrac{I_{d1}}{I_{d2}}} \tag{6.7}$$

It is important to remember that η changes according the position on its output characteristics, and it should be calculated for every rated value.

6.4.1.5 Measurement of I_s

With the parameters R_p and η, one can define I_s directly from Shockley's equation or as a function of T, V_{d1}, and I_{d1} as

$$I_s = \frac{I_d}{e^{qV_d/\eta kT} - 1} \tag{6.8}$$

If several points are measured, their average can be used as the final result.

6.4.1.6 Measurement of R_s

R_s can be determined by some numerical method from

$$R_s = \frac{\left(\dfrac{V_{oc}}{I_\lambda} k_4 - V_L\right)\left(I_{sc} - I_\lambda\right) - \left[V_L + \dfrac{V_{oc}}{I_\lambda}\left(I_L - I_\lambda\right)\right] k_3}{\left(I_{sc} - I_L\right) I_\lambda + I_L k_3 - I_{sc} k_4} \tag{6.9}$$

where k_3 and k_4 are defined as [8]

$$k_3 = \frac{I_\lambda}{e^{V_{oc}/\eta V_T} - 1}\left(e^{I_{sc} R_s/\eta V_T} - 1\right) \tag{6.10}$$

$$k_4 = \frac{I_\lambda}{e^{V_{oc}/\eta V_T} - 1}\left[e^{(V_L + I_L R_s)/\eta V_T} - 1\right] \tag{6.11}$$

The values of R_s reflect directly on the manufacturing quality of the PV cells.

6.4.2 Power, Utilization, and Efficiency of a PV Cell

The diode voltage V_d can be expressed by

$$V_d = I_L\left(R_s + R_L\right) = V_L\left(1 + \frac{R_s}{R_L}\right) = V_L + I_L R_s \tag{6.12}$$

Now, substituting V_d represented in Figure 6.3 and equation (6.1), more precise results for the output current can be obtained if $V_d = V_L + I_L R_s$ from

$$I_L = \frac{R_p}{R_p + R_s}\left[I_\lambda + I_s - I_s e^{q(V_L + I_L R_s)/\eta kT} - \frac{V_L}{R_p + R_s}\right] \tag{6.13}$$

For large values of the load resistance, R_L, with respect to the parallel resistance, R_p, of the cell, differences in the output current values may become remarkable. Also, it is important to use the empirical ideality factor, η, in the exponential term of either equation (6.11) or (6.13), so that they can be adjusted to the practical data from the manufacturers.

As expressed in equation (6.13), the diode voltage changes with the load current. For the very particular case where $I_L = 0$, the open-circuit voltage V_{oc} in an illuminated panel (single-cell open-circuit voltage) may be obtained as

$$V_{oc} = V_d\big|_{I_L=0} = \left[I_\lambda - I_s\left(e^{qV_{oc}/\eta kT} - 1\right)\right]R_p$$

or, yet, in the logarithmic and simplified form,

$$V_{oc} = \frac{\eta kT}{q}\ln\left(1 + \frac{I_\lambda}{I_s} - \frac{V_{oc}}{I_s R_p}\right) = \frac{\eta kT}{q}\ln\left[1 + \frac{1}{I_s}\left(I_\lambda - \frac{V_{oc}}{R_p}\right)\right]$$

This is a transcendental equation (it can be solved numerically). In practice, $I_\lambda \gg V_{oc}/R_p$ and $I_\lambda/I_s \gg 1$; then

$$V_{oc} \cong \frac{\eta kT}{q}\ln\left(\frac{I_\lambda}{I_s}\right) \tag{6.14}$$

As the series resistance of a PV cell is very small (around $0.1\,\Omega$), when the cell is short-circuited, practically the only opposition to the current through the cell is dominated by the value of R_s. So the short-circuit current can be given by

$$I_{sc} \cong I_\lambda \tag{6.15}$$

The peak power may be given from the following condition:

$$\frac{dP_L}{dV_L} = V_L + I_L\frac{dV_L}{dI_L} = 0 \tag{6.16}$$

that is,

$$\frac{dV_L}{dI_L} = -\frac{V_L}{I_L} \tag{6.17}$$

It is interesting to observe in Figures 6.3 and 6.5 that neither the maximum voltage nor the maximum current is the same for different levels of irradiance at the maximum power point. Condition of equation (6.17) is interpreted graphically in Figure 6.6.

A more convenient way to obtain the maximum power is illustrated in Figure 6.6 by doing the differentiation of equation (6.16) with respect to V_d. The derivative is equaled to zero in order to give the external load voltage V_{dm} at the maximum output power of the cell. Since the maximum load current I_L is directly dependent on the maximum internal diode voltage V_d, initially the

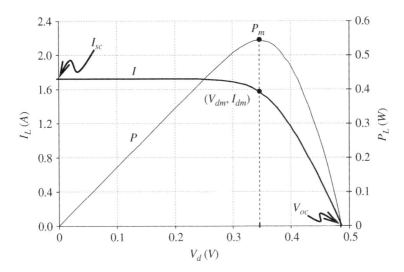

Figure 6.5 V×I and V×P output characteristics of a PV cell.

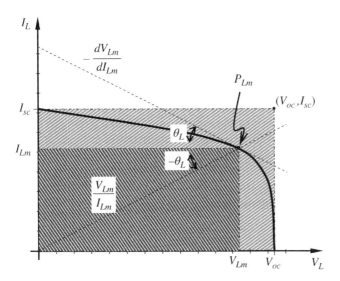

Figure 6.6 Conditions of maximum power for various irradiance levels.

PV output power under the point of view of the internal diode terminal voltage V_d is obtained, that is,

$$P = V_d I_L = V_d \left[I_\lambda - I_s \left(e^{cV_d} - 1 \right) - \frac{V_d}{R_p} \right] \tag{6.18}$$

$$f(V_d)\big|_{max} = \frac{dP}{dV_d}\bigg|_{max} = I_\lambda + I_s - I_s \frac{d(V_d e^{cV_{dm}})}{dV_{dm}} - \frac{2V_{dm}}{R_p} = 0 \tag{6.19}$$

Therefore, from the maximum of equation (6.19) equated to zero, after isolating the exponential term, we arrive at

$$e^{kV_{dm}} = \frac{\dfrac{I_\lambda + I_s}{I_s} - \dfrac{2V_{dm}}{I_s R_p}}{cV_{dm} + 1} = \frac{a - bV_{dm}}{1 + cV_{dm}}$$

where:

$$a = \frac{I_\lambda + I_s}{I_s}$$

$$b = \frac{2}{I_s R_p}$$

$$c = \frac{1}{\eta V_T}$$

A numerical solution of a function $f(V_{dm})$ will obtain V_{dm} where:

$$f(V_{dm}) = e^{kV_{dm}} - \frac{a - bV_{dm}}{1 + cV_{dm}} \tag{6.20}$$

Very often we need to know how efficient or better utilized a solar panel is (see Figure 6.6). The "fill factor," known by the abbreviation "*FF*," is a parameter that, in conjunction with V_{oc} and I_{sc}, determines the maximum power from a solar cell. The *FF* is defined as the ratio of the maximum power from the solar cell to the product of V_{oc} and I_{sc}. Graphically, the *FF* is a measure of the "squareness" of the solar cell characteristics; it is the area of the largest rectangle fitting the I–V curve. A very useful numerical figure for this is called form factor expressed as

$$FF = \frac{V_{Lm} \cdot I_{Lm}}{V_{oc} \cdot I_{sc}}$$

where:

FF = Fill factor
G = Illumination intensity
A = Cell's surface
I_{sc} = Short-circuit current
V_{oc} = Open-circuit voltage
S = Surface occupied by the panel

In this case

$$\eta = \frac{FF \cdot V_{oc} \cdot I_{sc}}{G \cdot S}$$

Example Let it be the following data from the PV cell manufacturers:

$$I_\lambda = 1 A$$
$$I_s = 2 \times 10^{-9} A$$
$$R_p = 200\Omega$$
$$R_s = 0.1\Omega$$

$$k = c = 1/\eta V_T = 1/1.5 \times 0.0267 = 24.969 \approx 25$$

From the diode voltage V_{dm} at the maximum power point is obtained:

$$a = \frac{I_\lambda}{I_s} + 1 \approx \frac{I_\lambda}{I_s} = 5 \times 10^8$$
$$b = \frac{2}{I_s R_p} = \frac{2}{2 \times 10^{-9} \times 200} = \frac{1}{10^{-9} \times 200} = 5 \times 10^6$$
$$c = 1/\eta V_T = 1/1.5 \times 0.0267 = 24.969$$

From equation (6.20), comes:

$$f(V_{dm}) = e^{kV_{dm}} - \frac{a - bV_{dm}}{1 + cV_{dm}} = e^{25V_{dm}} - \frac{5 \times 10^8 - 5 \times 10^6 \cdot V_{dm}}{1 + 25 \cdot V_{dm}}$$

Numerical solution of this equation gives $V_{dm} = 0.6865V$. With this value, it is possible to calculate the following figures:

$$I_{dm} = I_s \left(e^{cV_{dm}} - 1 \right) = 2 \times 10^{-9} \left(e^{24.969 \times 0.6865} - 1 \right) = 0.05563$$
$$I_p = \frac{V_{dm}}{R_p} = \frac{0.6865}{200} = 0.0034325$$
$$I_{Lm} = I_\lambda - I_{dm} - I_p = 1 - 0.05563 - 0.05563 = 0.9965675$$
$$V_{Lm} = V_{dm} - I_{Lm} R_s = 0.6865 - 0.9965675 \times 0.1 = 0.58684325V$$

$$I_{sc} = I_\lambda = 1 A$$
$$V_{oc} = \frac{\ln(I_\lambda / I_s)}{k} = \frac{\ln(1 / 2 \times 10^{-9})}{24.969} = 0.85772V$$

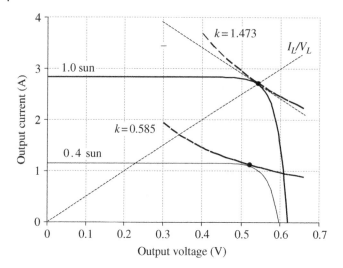

Figure 6.7 Graphical solution of the maximum power point.

$$P_{Lm} = V_{Lm}I_{Lm} = 0.58684325 \times 0.9965675 = 0.584829\,\text{W}$$

$$FF = \frac{V_{Lm}I_{Lm}}{V_{oc}I_{sc}} = \frac{0.584829}{0.85772} = 0.6818$$

$$\eta \cong \frac{\left(V_{L1} - V_{L2}\right)}{V_T \cdot ln\left(I_{d1}/I_{d2}\right)}$$

Equation (6.13) supports the fact that the output power of a PV cell increases with the irradiance level, generating a locus of maximum power. A graphical way of establishing the maximum power is by plotting equation (6.20) to obtain its maximum since it is a transcendental equation as illustrated in Figure 6.7. This locus can be obtained by hyperboles such as $k = xy$ analogous to a given $P_k = V_L I_L$ where there is a single interception point on the PV output character-istic. Figure 6.6 represents the maximum output power points of an individual PV cell for $k = 0.580\,\text{W}$ and $k = 1.475\,\text{W}$.

6.5 Solar Cell Output Characteristics

Figures 6.3, 6.5, and 6.8 show useful characteristics typically used when designing a controller of power electronic systems connected to PV arrays. They are curves of the output power versus the load voltage and the output power versus the load current at different irradiance and temperature levels. Voltage-based power control is clearly better, as suggested by Figure 6.8a, where the family of curves are smoother and better spaced [6, 8, 10, 12].

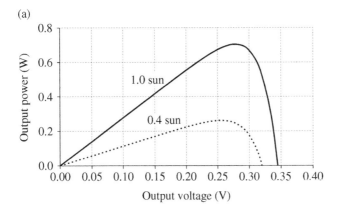

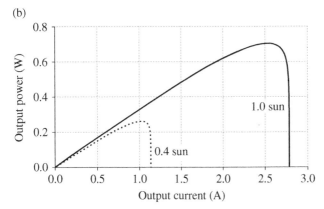

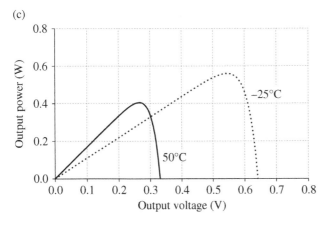

Figure 6.8 Effects of distinct sun irradiations and temperatures on the PV cell output power. (a) Effects of the radiation level on the PV power versus voltage characteristic, (b) effects of the radiation level on the PV power versus current characteristic, (c) effects of the temperature on the PV power versus voltage characteristic, and (d) effects of the temperature on the PV voltage versus current characteristic.

(d)

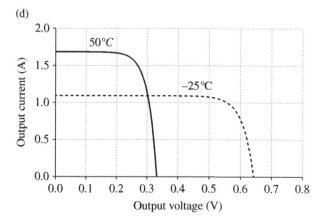

Figure 6.8 (Continued)

Cell temperature also plays an important role in the efficiency of solar panels. Figure 6.8c and d shows that for a cell illuminated by 0.6 sun under two different temperatures, $t_1 = 50°C$ and $t_2 = 25°C$, the cell short-circuit current decreases, whereas the open-circuit voltage increases.

6.5.1 Dependence of a PV Cell Characteristic on Temperature and PV Cells

Solar cells are influenced by temperature in two ways: current gain varies with irradiance; therefore reverse-biased diode current and voltage are also affected. These effects are taken into account in the terms $V_T = kT/q$, I_λ, and I_s of Shockley's equation. For the saturation current, a somehow precise expression could be given by [11, 12, 15, 16]

$$I_s(T) = KT^m e^{-V_{GO}/\eta V_T} \tag{6.21}$$

where:

K = A constant independent of temperature
$m = 2$ for germanium and $m = 1$ for silicon
$\eta = 1$ for germanium and $\eta = 1.2 \sim 1.8$ for silicon
$V_{GO} = 0.785V$ for germanium and $V_{GO} = 1.21V$ for silicon (and it is known as the equivalent voltage to the prohibited band (eV))
$V_T = kT/q = T/11,605.4677$ = An equivalent voltage to the junction temperature, usually taken as $T = 298°C$
PTC = The α positive temperature coefficient
NTC = The β negative temperature coefficient

Taking the derivative of the earlier expression in its logarithm form [3–5],

$$\frac{d(lnI_s)}{dT} = \frac{dI_s/I_s}{dT} = \frac{m}{T} + \frac{V_{GO}}{\eta TV_T} \tag{6.22}$$

It is common practice to establish the PV panel parameters with respect to standard test (STC) and normal operating (NOC) conditions (Table 6.2). Under standard conditions with the parameters given, the theoretical logarithm per unit variation dI_s/I_s with temperature is about 8%/°C for silicon and 11%/°C for germanium [1]. In commercial diodes, these variations are small because surface leakage is not an important effect for these diodes as it is for PV cells. Therefore, the theoretical values will be smaller, typically 7%/°C for both silicon and germanium diodes. These 7%/°C values may be interpreted as $(1.07)^{10}$ 2.0 (i.e., the reverse saturation current doubles for every 10 °C). With respect to I_s, some references give this value as if the diode reverse saturation current I_{s0} for temperature T_0 changes at another temperature T [14–17]:

$$I_s(T) \cong I_{s0} \times 2^{(T-T_0)/10} \tag{6.23}$$

Table 6.2 The standard (STC) and nominal (NOC) operating conditions.

Parameter	Standard conditions	Nominal conditions	Equation
Solar spectral irradiance	1000 W/m^2	800 W/m^2	—
	ASTM E-892	ASTM E-892	
Cell temperature	25 °C	42 °C	—
$I_s(T) = KT^m e^{-V_{GO}/\eta V_T}$	—	—	(6.21)
$\dfrac{dI_s/I_s}{dT} = \dfrac{m}{T} + \dfrac{V_{GO}}{\eta TV_T}$	About 7 %/°C for Si	—	(6.22)
$I_s(T) \cong I_{s0} \times 2^{(T-T_0)/10}$	—	—	(6.23)
$I_{sc} = I_{sc0}(1 + \alpha\,\Delta T)$	$\alpha \approx 500\mu$ units/K	PTC	(6.24)
$\dfrac{dV_d}{dT} = \dfrac{V_d - (V_{GO} + m\eta V_T)}{T}$	$dV_d/dT \cong -2.5$mV/°C	—	(6.26)
$V_{oc} = V_{oc0}(1 - \beta\,\Delta T)$	$\beta = -5\mu$ units/K	NTC	(6.27)

A more practical approach to this issue is the use of temperature coefficients such as a positive temperature coefficient (PTC) for the current and a negative temperature coefficient (NTC) for the voltage. For the short-circuit current,

$$I_{sc} = I_{sc0}(1 + \alpha \Delta T) \tag{6.24}$$

where:

α = The PTC of the electrical current
For a typical single-crystal silicon PV cell, α is about $500\,\mu$ units/K.

On the other hand, if temperature is increased and the voltage is kept constant, the diode current will increase. By taking the derivative from Shockley's equation, it is possible to find an expression for the variation of the diode voltage V_d with its saturation current I_s and, consequently, the variation with the temperature T as a difference of two terms as functions of two different temperatures. By balancing these two terms through reduction of voltage and increase of current under different temperatures, it may be kept a constant current for both silicon and germanium diodes. Under this condition, it is usually necessary that

$$\frac{dV_d}{dT} \cong -2.5\,\text{mV/}°\text{C} \tag{6.25}$$

However, this relationship also decreases with increasing temperature. So a more precise expression for this variation becomes necessary, such as the one given by equation (6.26) [12, 14]:

$$\frac{dV_d}{dT} = \frac{V_d - (V_{GO} + m\eta V_T)}{T} \tag{6.26}$$

With the usual values, it may be stated that
$dV_d/dT \cong -2.1\,\text{mV/}°\text{C}$ for germanium and $dV_d/dT \cong -2.3\,\text{mV/}°\text{C}$ for silicon.
The practical approach would suggest something like $\beta = -5\mu$ units/K as the NTC for the cell open-circuit voltage. Therefore, the temperature corrected for the open-circuit voltage is

$$V_{oc} = V_{oc0}(1 - \beta \Delta T) \tag{6.27}$$

Passing from the intermediary condition 1 to 2 can be done through [11]

$$I(G_2, T_2) = I(G_1, T_1)\frac{G_2}{G_1}\left[1 + \alpha(T_2 - T_1)\right]$$
$$V(G_2, T_2) = V(G_1, T_1)\cdot\left[1 + \beta(T_2 - T_1)\right]\cdot\left[1 + \delta ln\left(\frac{G_2}{G_1}\right)\right] \tag{6.28}$$

where:
α, β, and δ = Constants obtained from one known of two conditions (G_1, T_1, G_2, T_2)

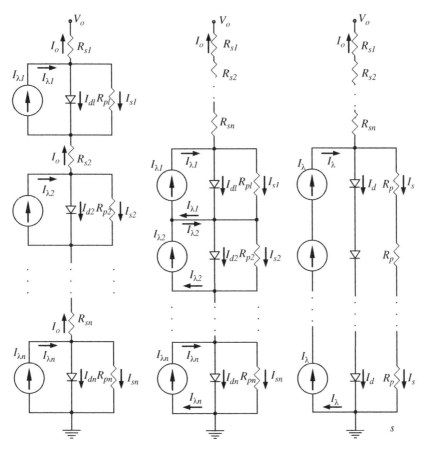

Figure 6.9 Equivalence evolution of *n* identical cells in series.

6.5.2 Model of a PV Panel Consisting of *n* Cells in Series

When stacking a group of PV cells in series, it is assumed that the cells are identical, resulting in the equivalence evolution of Figure 6.9. This assumption is usually well accepted if it is taken into account that individual cells inside the panel are manufactured with very similar characteristics, to avoid circulation of internal currents among the cells. With this assumption it is possible to consider the following:

$$I_{\lambda 1} = I_{\lambda 2} = \cdots\cdots = I_{\lambda n}$$
$$I_{d1} = I_{d2} = \cdots\cdots = I_{dn}$$
$$V_{d1} = V_{d2} = \cdots\cdots = V_{dn}$$
$$R_{s1} = R_{s2} = \cdots = R_{sn}$$
$$R_{p1} = R_{p2} = \cdots = R_{pn}$$

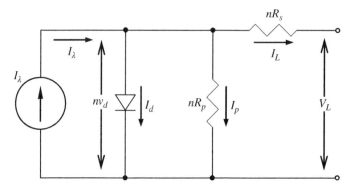

Figure 6.10 Equivalent model of a PV panel with *n* identical cells in series.

The respective output voltage and current of the group are

$$V_o = nV_{oi}$$
$$I_o = I_{oi}$$

where:

V_{oi} and I_{oi} = Voltage and current in the individual cell i, respectively

A more compact form of the equivalent circuit of Figure 6.9 is shown in Figure 6.10. As shown in Figure 6.10, there are four relevant parameters: the electrical efficiency, η_e, the series resistance, R_s, the parallel resistance, R_p, and the current saturation of the internal diode, I_s. The efficiency of a solar panel may be given for the maximum electrical power by

$$\eta_e = \frac{P_{electrical}}{P_{irradiation}} \times 100\% \qquad (6.29)$$

$$\eta_e = \frac{V_{Lm} \cdot I_{Lm}}{G \cdot S} \qquad (6.30)$$

An important aspect of the connection of PV cells in series is that during shadowing the cell producing smaller current will be the one leading the current through that chain. As an example, suppose that the manufacturer's data sheet indicates the following: effective illuminating area $S = 0.136 \text{m}^2$, $V_{mp} = 15.6 \text{V}$, and $I_{mp} = 1.4 \text{A}$ for the STC (1000W/m^2, $25\,^\circ\text{C}$, AM 1.5); then

$$\eta_e = \frac{15.6 \times 1.4}{1000 \times 0.136} \times 100\% = 15.74\%$$

Using Figure 6.9 and also equation (6.3) of Shockley, it is possible to observe that the voltage across the terminals of an individual cell can be given by

$$V_{Li} = \frac{\eta kT}{q} \ln\left(\frac{I_d}{I_s} + 1\right) - I_L R_s \tag{6.31}$$

So, the output voltage of a panel with identical cells in series will be given by

$$V_L = nV_{Li} = \frac{m\eta kT}{q} \ln\left(\frac{I_d}{I_s} + 1\right) - nI_L R_s \tag{6.32}$$

Shockley's equation for the current through the diodes then assumes the following form:

$$I_d = I_s \left[e^{q(V_L + nI_L R_s)/\eta kT} - 1 \right] \tag{6.33}$$

6.5.3 Model of a PV Panel Consisting of *n* Cells in Parallel

Assuming that the PV cells of a panel are connected in parallel across common terminals *a* and *b* as in Figure 6.11. In these conditions, there is no direct interaction among current sources or individual diode voltages because they are considered the same. Therefore, this equivalent model may assume a more compact form, as shown in Figure 6.12. Tests of this equivalent model can be seen as carried out on a single panel with individual output voltage and current for the same cell:

$$V_o = V_{oi}$$
$$I_o = nI_{oi}$$

The method just described can be extended to equivalent models of series–parallel combinations of PV panels or other arrangements of interest.

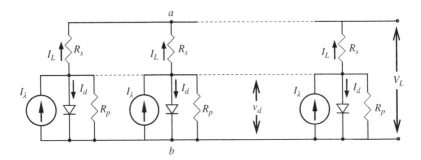

Figure 6.11 Group of *n* identical PV cells connected in parallel.

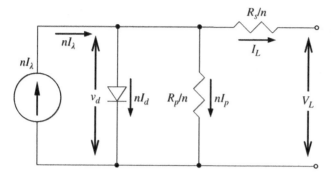

Figure 6.12 Equivalent model of *n* identical PV cells connected in parallel.

6.6 Photovoltaic Systems

A PV electric system consists of PV modules, an electronic converter, a controller, and usually a bank of batteries. Each module is made of PV cells connected in series, parallel, or series/parallel, as part of a panel supported by a mechanical structure. A power converter may be used to regulate the flux of energy from the panel to the load to assure energy quality within certain limits. The electricity converted from solar modules should be stored in banks of batteries of high capacity to make the overall system feasible economically. The bank of batteries should store enough energy to minimize the night and/or low-light-intensity periods. The public network also can be used "to store" energy after an agreement with the electricity company.

Maintenance of the modules should be careful such as wiping with a damp cloth and always using maximum care to avoid scratches and other damages to the cells. Getting leaves stuck on the panels or scratching deposits of dirty or foreign substances should be avoided, because these cells are very sensitive structures. The electric system should be installed in the following way:

1) Provide equipment grounding with a copper pole at least 2 m in length.
2) Connect the negative conductor, usually black in color with an alligator claw, to the ground bar.
3) Connect the other end of the negative cable, black in color, to the battery's negative terminal.
4) Connect the negative conductor of the solar module to the negative pole of the electric fence equipment.
5) Connect the positive conductor of the solar module to the equipment's positive pole.
6) Connect the positive cable of the electric fence to the positive terminal of the battery.
7) Connect the equipment's electric discharge conductor to the galvanized wire of the fence.

Several other aspects should also be considered at the design stage of a PV-based power supply system [18, 19]:

1) The site should be open, without constant shadows for the solar light.
2) One or more solar modules.
3) Structures or a support of aluminum.
4) A load controller.
5) A bank or set of batteries or arrangements with the electricity company.
6) An indicator of the level of battery charge.
7) Points of load consumption.

6.6.1 Irradiance Area

Solar electric power generation is based on incident sunlight rather than heat. Therefore, for satisfactory efficiency, it is necessary that the solar panels stay exposed directly to sunshine most of the day, not exposed to tree shadows or buildings or other sorts of shielding. On cloudy days, there may be some electric power generation at lower levels.

As discussed in Section 6.5, a solar panel consists of several parallel strings of many cells connected in series. These parallel arrays should match each other as closely as possible to avoid internal current circulation through the cell series. Any unbalance in these internal voltages should be avoided. Although all precautions are taken during manufacture of the units, external causes may be a source of imbalance over the string voltages.

A very common problem with PV panels is the unexpected shading over its surface. In addition to clouds, common causes of shadow over solar panels are birds flying around, planes, its own structure, people or animals passing by, stray leaves, grown trees, and wind-carried materials. Such transient effects can be analyzed using the circuit shown in Figure 6.13. With no light, the photocurrent is zero ($I_L \approx 0$), and only the internal diodes, R_s and R_p, remain as a path for the current impinged from other active cells. The voltage of the healthy cells will supply current to the shadowy string cells, dissipating internally valuable energy. In a series group of these cells under shadow, if more cells are shadowed, they will go beyond a certain critical voltage value ($I_\lambda \to 0$), and the current through the association of all these internal resistances will act as a heat source, decreasing the overall efficiency of the solar panel.

To avoid the effect of having partially shadowed cell arrays, most modern solar panels are manufactured to divide the strings into several substrings with bypass diodes. Each series group connected string current and a series diode in the parallel-connected group array, as illustrated in Figure 6.13 for a shadowed PV cell group. The diode in series with every parallel connection will avoid reverse current through that array in case of reverse current. This improves performance because some of the string power still flows through the diodes.

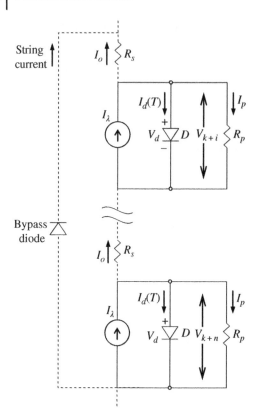

Figure 6.13 Bypass diode in a series PV cell connection.

6.6.2 Solar Modules and Panels

Solar modules are a group of cells that convert the radiant light in dc electric power, usually in standard connections of 12, 24, or 48 V. They are usually composed of monocrystalline silicon cells, protected by anti-reflexive glass and by a special synthetic material. The number of modules to be used in a system is determined by the electric power consumption needs. There are several commercial arrays for many purposes, as can be seen in Figure 6.14.

6.6.3 Aluminum Structures

Light aluminum structures may support the mechanical assembly of solar modules. Their purpose is to position the solar cells, preventing the movement of the modules and maximizing the sun's radiation.

The normal line to the panel face must be directed to the geographical south (to the north in the southern hemisphere). The latitude of the site where the solar power plant is to be installed determines the inclination of the panel.

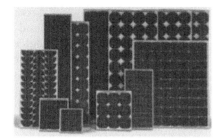

Figure 6.14 Several arrays of commercially available CSI solar panels.

Table 6.3 Panel angle based on terrestrial latitude.

Latitude (deg)	Panel angle (deg)
0–15	15
20	20
25	25
30	35
35	40

Besides, it is more advantageous to tilt the panel perpendicularly to choose the maximum radiation. The maximum energy would be obtained if the perpendicular position of the panel follows the sun in its daily and seasonal variations of inclination, although that may not be very practical because of the high cost of mechanical trackers, time of operation in case of failures, and regular maintenance.

It should also be noted that the sun's inclination varies from April to September from 5 to 23 degrees in latitude above and below the same perpendicular line of the site latitude. In industrial and other cases wherever power is needed in the winter as in the summer, it is better to set the inclination of the panel to the angle of the latitude of the site. Otherwise, the position selected for the panel should follow the seasonal power needs of the facilities. Table 6.3 indicates the suggested inclinations for solar panels based on terrestrial latitude.

A simple method for determining the best inclination for a given location is the use of a shadow diagram of a perfectly vertical stem tip (for best and easy results, use a bricklayer's plumb line) on the soil. Stem tip shadow positions are marked on the ground, let us say, from 10:00 to 15:00 hours. Then a line is drawn connecting these points. The perpendicular to this line will show the north–south direction being the rod shadow positioned to the north sides if the place is in the northern hemisphere. Note that there will be a small direction deviation with respect to magnetic north. If it is considered a ground-mounted array, the adjustment should be +15° above local latitude for winter or −15° for summer correction.

Other considerations to allow more or less inclination of a PV panel could be taken into account: (i) the typical schedule for a load is fed by PV panel, say, a daytime business machinery or a public square illumination, and (ii) the PV power used a longer time in business or illumination would be more useful in the winter or in the summer.

6.6.4 Load Controller

Chapter 15 discusses the load controller in detail. The load controller optimizes and conforms to the transfer of PV energy generated by the modules in order to match consumption needs. It also protects the battery against excessive overcharging and overdischarging [18, 20–22].

6.6.5 Battery Bank

Banks of batteries accumulate the electric power generated during the day, making it available whenever necessary. They usually consist of lead–acid deep-cycle batteries, preferably sealed and possibly with no need of water replacement (although those batteries are more expensive), with a nominal voltage rated at the dc-link voltage. Further details of battery banks, maintenance, and sizing are provided in Chapter 14. A battery bank also requires a healthy monitoring system to indicate the state of charge and to support the maintenance.

6.6.6 Array Orientation

A PV array generates energy mostly when it faces due south in the north hemisphere and due north in the south hemisphere. The azimuth angle determines the direction the array faces with respect to due south. The tilt of the array will determine how much energy the PV array will generate as well. An array will generate the most energy annually when it faces south and has an array tilt equal to the latitude of that location. For the southern hemisphere the array will have to face north.

For the installation site it is necessary to evaluate the surroundings for best orientation. A north–south cross-sectional cut may help evaluate the incidence angle and any possible block of sunlight. An east–west cross section is an open wide view in how the sun will travel from sunrise to sunset and possible shading along that path. For best results, differences between angles of incidence in winter and summer days have to be considered. The angle at which sunlight hits the solar array changes every day. An N–S sketch would help in defining possible shading throughout the year in the north hemisphere. March 20 and September 22 are the vernal and autumnal equinoxes, where the array angle should be equal to the latitude. If it is a ground-mounted array, it is possible to adjust +15° for winter or −15° for summer correction.

A flat-plate absorber aimed normal to the sun will receive energy diminishing according to the amount of atmosphere along the path (overhead $AM \equiv 1$). However, the received energy varies around the world due to local cloud attenuation. For example, in Arizona, direct normal radiation is 5.0–5.5 kWh/(m^2 - day). In the United States, daily solar energy varies from < 3.0 to 7.0 kWh/(m^2 - day).

This chapter covers basic notions of the use of solar electrical energy for direct and indirect energy production in electrical power plants.

6.7 Applications of Photovoltaic Solar Energy

There are tremendous opportunities for PV applications, such as residential electrical power, public illumination, camping, stroboscopic signaling, fluvial and marine boats, electric fences, telecommunications, telephone booths in remote areas, water supply, conservation of food and medicines, control of plagues, WiFi outdoors, cameras and surveillance, and general electric power supply [2, 3, 19–23].

6.7.1 Residential and Public Illumination

Several options are available on the market for the residential and public illumination applications. They are manufactured in a diversity of solar modules that provide illumination in a level and amount adapted for each need. There are commercial fluorescent and Dulux lighting systems ranging from 9 to 22 W and low-power incandescent illumination lamps from 45 to 110 W, in addition to LED light bulbs. Traditional manufacturers are Philips, Osram, and General Electric.

A typical home lighting commercial kit consists of a 9 W/11 W/15 W energy-efficient compact fluorescent light or LED-based luminary, solar PV module, battery, and charge controller unit. The PV panel charges the battery during the daytime, and the energy thus stored in the battery is used to power the lights whenever required. The bulb controller very often has a built-in high-efficiency electronic inverter to reduce energy losses and noise and provide more lighting time. A charge controller unit is used to control the charging of the battery. It comes with electronics for overcharging/overdischarging control of the battery, which in turn increases battery life. Solar energy can provide a good illumination solution for streets, blocks, and parks. Electric power is obtained from solar panels and stored in a system of charged batteries through a load controller. Such systems are usually installed individually on top of each illumination pole with high-efficiency compact bulbs. Solar PV street light systems use efficient 11 W/18 W compact fluorescent light, providing illumination equivalent to a 60 W/75 W/100 W/150 W incandescent light bulb. Currently there are LED bulbs with even lower electrical power with same illumination equivalence. These systems are equipped with high-efficiency, high-frequency electronic inverters to reduce energy losses and noise and provide lighting for a longer period. The system does not require any external power supply since the solar panel is inherently safe for public use and requires less frequent maintenance.

A truly portable solar lantern is also possible with an energy-efficient compact fluorescent bulb, battery, and electronics all placed in a compact housing and packaged along with a PV module. This solar lantern is equipped with a high-efficiency electronic inverter and a charge controller with deep discharge and overcharge protection to enhance battery life without need of maintenance.

In addition, solar energy is used as a practical and economical source of illumination in camping areas, maintaining the charge of the battery system or being transformed to alternative energy at compatible voltage patterns for the most varied domestic means available in the market [22–25]. For fluvial and marine embarkations and similar types of movable loads, the generated energy is indicated for those cases where there is need of a simple, compact, light, and efficient energy system without need of fuels.

6.7.2 Stroboscopic Signaling

PV systems allow electric power provisioning and signaling in remote places. They are used for warning signals in places of difficult access or where no electric grid connection is available. They are suitable for signaling air and marine traffic or highway traffic and in the signaling of towers, antennas, buildings, and so on. They have been widely employed in developed countries such as the United States and in the load supply in battery banks and in signaling buoys of ports, lights, and oceanographic rafts. Their principal advantages are their low maintenance needs, good reliability, and moderate operational cost, alleviating the need for periodic technical inspections.

The modern high towers built to integrate mobile telephone networks all around the world benefit from the use of power modules composed of solar panels, a controller, and batteries. These modules are lightweight, long lasting, autonomous, efficient, and suitable for many other applications that use high tower supports, such as for TV, radio, and satellite installations.

PV systems are also used for signaling in road infrastructure. These systems, involving a solar panel, electronic controllers, and a backup battery, are often installed in remote areas where only limited skilled labor is available for maintenance. The battery must be able to withstand daily shallow cycles and seasonal deep cycles of operating temperatures from −30 to +50 °C. There are batteries designed especially for these applications.

6.7.3 Electric Fence

Solar energy can be useful for electric fencing for cattle confinement, eliminating the need to change batteries or to install an electric network close to a pasture. Even in the absence of solar light for long periods, the simplest systems can maintain autonomy for up to a week. Twisted flexible copper wires are used in electric fences. The use of very thin wires and fence wires themselves should be avoided, because they cause unnecessary losses and can easily suffer rust or oxidation (isolation). Electric supports for a fence can be fixed directly on poles with the aid of handles. It is necessary to adjust the system for the best possible inclination angle of the panel with respect to the latitude. Batteries should be installed inside plastic, wood, or concrete cabinets to protect them from weather, humidity, and sludge. After connecting the battery terminals, they should be covered with grease or Vaseline to avoid sulfating.

6.7.4 Telecommunications

PV systems for telecommunication applications must supply energy with long autonomy to overcome long absences of sunlight. In many countries, telephone booths along highways use solar panels to avoid long passages of the feeding electric lines. The telephone signals are transmitted via radio transceptors, whose modules are fed through load controllers (see Figure 6.15). There are many installations like this in the world, for example, the electrification in Morocco with 22 control stations and the Maghreb–Europe gas pipeline from the Algeria–Morocco border to the Strait of Gibraltar. The following equipment was installed at each station: a 8.5-kWp solar generator, 4000-Ah stationary batteries, a regulation and control system, and a galvanized steel mounting structure. Another example is the Radiolink network of the Dirección General de Telecomunicaciones (Spanish Telecommunications Authority), which covers all of Spain. Installed at each station were a 6-kWp solar generator, 4000-Ah stationary batteries, a regulation and control system, a galvanized steel mounting structure integrated in the cover, and other pieces of equipment.

6.7.5 Water Supply and Micro-irrigation Systems

Solar pumping systems are designed especially for water supply and irrigation in remote areas where no reliable electricity supply is available. Rural applications of solar energy are very valuable despite the high cost [5, 6, 23, 25]. Small villages and towns far from the electric grid may invest in irrigation systems or potable water supply powered by water pump and solar modules such as the

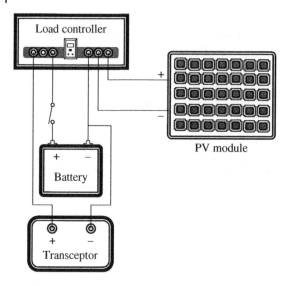

Figure 6.15 Telecommunication systems fed by solar modules.

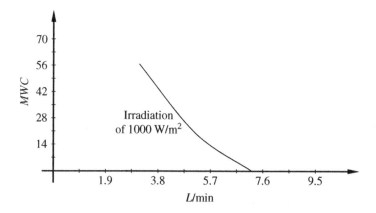

Figure 6.16 Pressure–flow characteristic for a surface water pump.

ones recommended by the California Energy Commission (http://www.consumerenergycenter.com) in the California Emerging Renewables Program. The flow characteristic is inversely proportional to the pumping head, as in any conventional system of water pumping.

For a surface pump, a performance curve such as that shown in Figure 6.16 may be used. An example is illustrated of a hydro pump reaching a flux level of 5.7 L/min for 14 m of water column (MWC) for typical days of 1000 W/m^2 radiation (i.e., noonday sun). Therefore, the maximum suction head should not

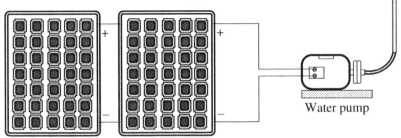

Parallel connection of solar modules

Figure 6.17 Surface water pump using solar modules.

exceed 1.5 m. The solar surface water pumping system is usually a solar PV system, with an energy-efficient dc surface centrifugal pump coupled with about a 1-kW PV panel array. This system is typically used to provide potable water, water for crop irrigation, and water for livestock. For the example shown in Figure 6.17, the connection of a surface pump to solar modules is very simple.

Submersible pumps manufactured from stainless steel and bronze with external diameters of up to 100 mm (see Figure 6.18) can be used. The flow is also inversely proportional to the pumping heads. For instance, in Figure 6.19, for a flow of 125 L/h (2.08 L/min), a head of 32 m of water column can be reached. For higher efficiency with solar panels, a current amplifier should be used.

6.7.6 Control of Plagues and Conservation of Food and Medicine

Electric, audible, or burning high-voltage traps can easily be installed at the focus points of plagues. PV energy provides a very cost-effective, clean, and silent autonomy for this kind of application.

PV refrigerators are designed especially for low consumption, fed by solar energy. The starting operation is similar to that of an ordinary refrigerator except for the power source, which is fed from 12-V batteries of 300 or 600 Ah for 75 L (from 4 to 7 °C) or 150 L (from 0 to −4 °C) for four or six solar modules, respectively, working for 24 hours of the day.

There is great interest in solar panels for conservation of food and medicine. They provide hygienic, silent, reliable, and self-operation, guaranteeing the protection of food and medicine against any shortage of electric power from the public network [7–10]. In remote areas, solar energy is associated with the reliability of batteries for emergency and security systems and for the data and information systems required in medical centers, intensive care units, and operating theaters and in similar applications where lighting interruption can

(a)

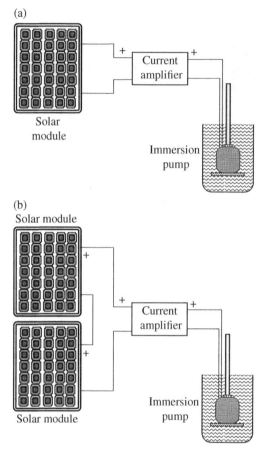

(b)

Figure 6.18 Powering a submersible pump with solar modules: (a) operation at 12 V; (b) operation at 24 V.

be a matter of life and death. In addition, the building's regular UPS system and auxiliary generator set, with its associated engine-starting battery, requires the same level of reliability. Various types of metering systems are installed to measure all types of fluids, electricity, gas, and water necessary in the conservation of food and medicines.

6.7.7 Hydrogen and Oxygen Generation by Electrolysis

Reliance on imported oil from unstable regions of the world has forced the pursuit of hydrogen as alternative to fossil energy solutions. In such a case, electrolysis by solar or wind energy can be the most practical technical means to eliminate fossil fuels from the energy production cycle, especially when the

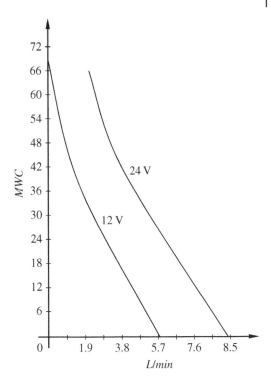

Figure 6.19 Characteristic of water pumps of the submersible type fed by solar panels.

electrolysis is powered by the sun (or maybe geothermal). High temperatures and electricity can also help in this process, enabling countries to use their own renewable electricity and water to produce emission-free hydrogen.

There are several advantages to producing hydrogen by electrolysis from solar panels: no moving parts, plug and play, installation at the point of use, virtually maintenance-free, no transportation, and direct production of high-purity hydrogen and oxygen, among others. When employing renewable energy sources for water electrolysis, it is vital to prevent electrode degradation induced by fluctuating energy input. Mechanical stability and process efficiency suffer when input power varies. The lifetime of standard electrodes in this operational mode lasts only a few hours. For this reason, electrolyzers are conventionally operated at constant rated power. During shutdown, standard electrodes will corrode without a protective voltage. This requires an electricity backup and causes energy losses. Several other losses are present in the process efficiency of water electrolysis: the low ripple controlled rectifier, electrolyzer, distribution, compression, and others. There are several companies worldwide that supply equipment for this purpose, some of them claiming reduced production costs without transportation, storage, or compressor requirements (see, e.g., http://www.electrolyser.com). In Section 7.8 we describe a detailed study regarding when electrolysis may become a convenient alternative process for electrical energy storage.

6.7.8 Electric Power Supply

Solar energy can supply alternating current (ac) power of 110, 127, or 220 V at 60 or 50 Hz (depending on the inverter) or dc power at 12, 24, or 48 V. That energy can be used for small domestic electric equipment [7, 9, 10]. The available energy is obtained from a battery that is constantly recharged by solar modules. The load is fed directly in dc or it is transformed, through switched power inverters (forced commutation), to a sinusoidal ac, rectified, and stabilized in voltage and frequency (see Figure 6.20).

Typically, the solar power available at 12:00 a.m. on a clear and sunny day is 1 sun = 1000 W/m², with an average for 365 days of 174 W/m². That means that for an electric power supply, one could expect

$$P(W) = 174\eta A \tag{6.34}$$

where:

η = The efficiency of the PV cells
A = The effective area of the panel in m²

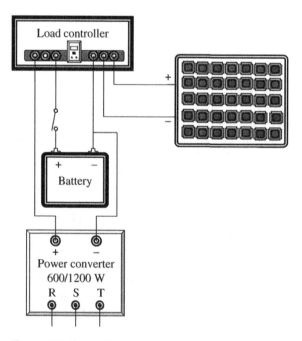

Figure 6.20 Three-phase electric supply using solar modules.

6.7.9 Security Video Cameras and Alarm Systems

Terrorist attacks have emphasized the importance of protecting water power supplies, oil fields, and defense systems against sabotage and similar attacks. The German municipal authority introduced appropriate legislation with its technical notification W1050 in March 2003. Solar power remote monitoring is used to protect property, to detect leakage, and for automatic remote reading of meters, all powered by a single solar panel.

Electronic surveillance of a plant and premises cannot replace mechanical protection and regular control. As part of an operational security concept, possible impairment of a water or gas supply can be averted by early detection through remote supervision of signaling contacts and key cylinders and the use of motion or sabotage detectors at the site.

Automatic remote reading of water meters is another application of solar panels. With manual readings, it is common for safety reasons to use two-person teams, but this method introduces reading errors, is high in labor cost, and makes few data available. Remote reading allows new data to enter the system every 15 or 20 minutes, giving an exact and instantaneous profile of consumption. An alarm signal can warn of a minimum level limit, a high consumption velocity, or faulty pipes. This system requires the attention of only one or two workers once a year.

6.8 Economics and Analysis of Solar Energy

Normalization, under study in many countries, would limit the use of electric heaters and ovens in new buildings, perhaps recommending a tariff for overuse during peak periods, and trying to induce the replacement of electric heaters and ovens by other modes of generation. The restrictions could be softened for residences endowed with demand controllers or thermal reservoirs for the storage of hot water during hours of low demand (early in the day). In that way, heating needs would not concentrate power during only certain short periods of the day.

It is known that transformation of electricity into heat represents an overall reduced efficiency (about 5%). That is due to the various processes through which electricity passes until it is available in an appropriate way to the user [5, 6, 26, 27]. Electricity has more worthy uses: for illumination, to drive motors, and in other productive processes. Without reducing the comfort level of the population, the consumption of electricity for heaters and ovens could be better used in more productive processes. However, it is difficult to compare the equipment cost that provides economic advantages of absolute reliability for the society as a whole when the user just analyzes the result in kilowatt-hours of consumption paid in the effective tariffs. Even so, growing energy prices demonstrate significant advantages for the use of direct solar energy for

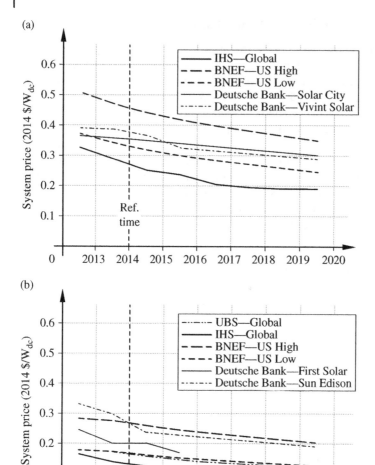

Figure 6.21 PV system prices *2014-2015 historic projections: a) residential systems; b) utility-scale systems (based on Ref. 29).*

heating of water and home appliances. These advantages are related to: easy installation, commercially modular, solar energy matches very well with daily loads, no moving parts, no noise, no wear-outs, high power capability/kg, usage without pollution, no vibrations, highly portable, and has very long useful life. There are also some disadvantages such as the need for storage of energy, fairly high costs, low efficiencies, shadowing effects with possible drastic reduction of momentary power, and pollution during its manufacturing since it is uses Si, S_2Cu-In and Te-Cd that cause greenhouse effect and acid rain [8, 22].

Table 6.4 Main standards for PV systems interconnected to the grid [26].

Indicator	IEC61727	IEEE1547	EN61000-3-2
Nominal power	10 kW	30 kW	16 A*230 V = 3.7 kW
Harmonic currents (order-h) limits	(3–9) 4.0%	(2–10) 4.0%	(3) 2.30 A
	(11–15) 2.0%	(11–16) 2.0%	(5) 1.14 A
	(17–21) 1.5%	(17–22) 1.5%	(7) 0.77 A
	(23–33) 0.6%	(23–34) 0.6%	(9) 0.40 A
		(> 35) 0.3%	(11) 0.33 A
			(13) 0.21 A
			(15–39) 2.25/h
Maximum current THD	5.0%	5.0%	—
Power factor at 50% of rated power	0.90	—	—
Dc current injection	< 1.0% of rated output power	< 0.5% of rated output power	< 0.22 A — corresponds to a 50 W half-wave rectifier
Voltage range for normal operation	85% 110% (196 V – 253 V)	88% 110% (97 V – 121 V)	—
Frequency range for normal operation	50 ± 1 Hz	59.5 to 60.5 Hz	—

The cost of electricity obviously differs from country to country. However, it has risen from an average of $49 per megawatt in 1995 to about $100 per megawatt in 2005. Such data demonstrate the great value of solar energy use in the future. Nowadays, however, the use of solar energy still demands equipment and advanced technologies, which results in a high-cost enterprise. At the time of publication of this book, the installation cost for solar power plants was about $3.50 per watt as depicted in Figure 6.21 and the operating cost was $0.20 per kilowatt-hour. The main standards for PV systems interconnected to the grid are listed in Table 6.4 [28–30].

In 2013, the California Independent System Operator (CAISO) published a chart showing the potential for "overgeneration" occurring at increased penetration of solar photovoltaics (PV). The "duck chart" shows the potential for PV to provide more energy than can be used by the system, especially considering the host of technical and institutional constraints on power system operation (see Figure 6.22). The duck curve shows sudden steep ramping needs and overgeneration risk. During overgeneration conditions, the supply of power could exceed demand, and without intervention, generators and certain motors connected to the grid would increase rotational speed, which can cause damage.

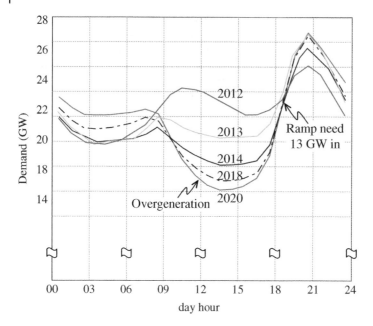

Figure 6.22 Illustration of the Californian duck curve (net load - Mar 31st).

To avoid this, system operators carefully balance supply with demand, increasing and reducing output from the conventional generation fleet. The overgeneration risk occurs when conventional dispatchable resources cannot be backed down further to accommodate the supply of variable generation (VG) [30].

Example in Sizing a PV Power Source:
Find the south-facing roof area (for latitudes at the north of the equator), say
   $6\,m \times 12\,m = 72\,m^2$;
   Assume 120 Wp solar modules are 0.66 m by 1.30 m = 0.86 m²:
      = 120 watts/0.86 m² ≈ 140 W/m²;
   Assume 90% of the roof area can be panel covered,
      = 0.9*72 m²*140 W/m² = 9072 W
   There are 5.5 effective hours of sun/day:
      = 5.5*9072 ≈ 50 kWh/day;
The south-facing modules are tilted south to the latitude angle;
Conclusion: 84 modules (= 72 m²/0.86 m²) would fit the area, but just 50 would provide a typical average home with:

   30 kWh/day and cost ~ $17600 for the modules only.

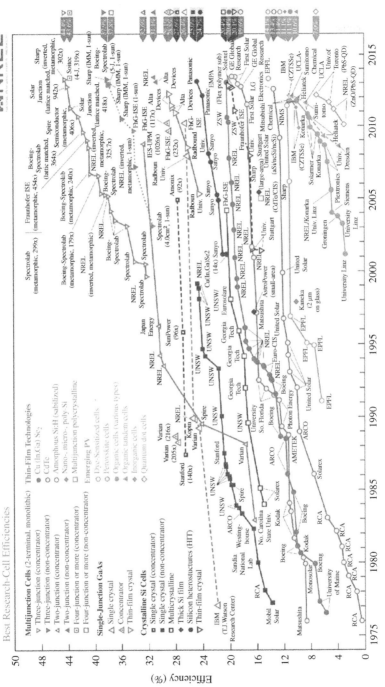

Figure 6.23 Best Research-Cell Efficiencies [29].

Calculation of Electrical Efficiency

From manufacturer's data sheet it is possible to obtain, at STC ($1000 \, \text{W/m}^2$, $25 \, ^\circ\text{C}$, AM 1.5) and an effective illuminating area $= 0.136 \, \text{m}^2$ the following:

$$V_{mp} = 15.6 \, V$$

$$I_{mp} = 1.4 \, A$$

$$\eta_e = \frac{P_{elec}}{P_{irr}} \cdot 100\% = \frac{15.6 \times 1.4}{1000 \times 0.136} 100\% = 15.74\%$$

All countries between latitudes $40 \, ^\circ$N and $40 \, ^\circ$S have potential for utilizing PV power systems. A solar panel starts producing electricity as soon as the light of around 0% of noonday sun illuminates the panel and will continue until light fades in late afternoon. Cloudy and rainy days can be accommodated by using batteries, or if the system is grid connected, the utility electrical power will supply the installation. A wind/PV hybrid system can be a solution if the location has wind resources, or diesel/PV when fuel is available for a diesel based generator. This technology can transform many villages and isolated regions, since there are no moving parts and it is relatively safe in handling the required equipment. It is expected that further manufacturing and widespread adoption will decrease the currently high installation costs and increase the interest of many countries in this very promising technology (see Figure 6.23).

References

1 C.R. Osterwald, K.A. Emery, and M. Muller, Photovoltaic module calibration value versus optical air mass: the air mass function, Progress in Photovoltaics Research and Applications, Vol. 22, p. 560, 2014.

2 KanEnergi, New Renewable Energy: Norwegian Developments, Research Council of Norway in cooperation with Norwegian Water Resources and Energy Directorate, Hanshaugen, Oslo, Norway, December 1998.

3 The American Society for Testing and Materials (ASTM), http://www.astm.org/, accessed March 10, 2017.

4 National Technical Information Service (NTIS), https://www.ntis.gov/, accessed March 23, 2017.

5 I. Chambouleyron, Solar electricity, Science Today, Vol. 9, No. 54, pp. 32–39, 1989.

6 R. Messenger and J. Ventre, Photovoltaic Systems Engineering, CRC Press, Boca Raton, FL, 2000.

7 Catalog Direct Electricity from the Sun (Eletricidade Diretamente do Sol), Soleco do Brasil, Santa Maria, Brazil, 1997.

8 D. Mulvaney, M. Cendejas, S. Davis, L. Ornelas, S. Kim, S. Mau, W. Rowan, E. Sanz, P. Satre, A. Sridhar, and D. Young, Hazardous materials used in silicon PV cell production: a primer, from the book Toward a Just and Sustainable Solar Energy Industry, Silicon Valley Toxic Coalition, San Francisco, CA, January 2009.

9 M.R. Patel, Wind and Solar Power Systems, CRC Press, Boca Raton, FL, 1999.

10 Y. Tsuno, Y. Hishikawa, and K. Kurokawa, Curves of various PV cells and modules, Conference Record of the 2006 IEEE 4th World Conference on Photovoltaic Energy Conversion, Vol. 2, 2006, pp. 2246–2249, DOI: 10.1109/WCPEC.2006.279619.

11 J. Millman and C.C. Halkias, Electronic Devices and Circuits, McGraw-Hill, New York, 1967.

12 F.T. Fernandes, L.C. Correa, C.R. De Nardin, A.J. Longo, and F.A. Farret, Improved analytical solution to obtain the MPP of PV modules, IECON 39th Annual Conference of the IEEE Industrial Electronics Society, Vienna, Austria, 10–13 November, 2013.

13 R. Boylestad and L. Nashelsky, Electronic Devices and Circuit Theory, 6th ed., Prentice Hall, Upper Saddle River, NJ, 1996.

14 F.A. Farret, J.M. Lenz, and J.G. Trapp, New methodology to determinate photovoltaic parameters of solar panels, IEEE-XI Brazilian Power Electronics Conference, Praiamar, RN, Brazil, September 2011, pp. 275–279, DOI: 10.1109/COBEP.2011.6085281.

15 F.M. Vanek and L.D. Albright, Energy Systems Engineering, Evaluation and Implementation, McGraw Hill, New York, 2008, pp. 255–257.

16 F.T. Fernandes, L.C. Correa, C. De Nardin, A. Longo, and F.A. Farret, Improved analytical solution to obtain the MPP of PV modules, Industrial Electronics Society, IECON 2013—39th Annual Conference of the IEEE, 2013, pp. 1674–1678, DOI: 10.1109/IECON.2013.6699384.

17 J. Millman and C.C. Halkias, Integrated Electronics: Analog and Digital Circuits and Systems, McGraw-Hill, New York, pp. 58, 797, 1972.

18 R.R. Spencer and M.S. Ghausi, Introduction to Electronic Circuit Design, Prentice Hall, Upper Saddle River, NJ, 2003.

19 W.H. Lee and W.G. Scott, Distributed Power Generation: Planning and Evaluation, Marcel Dekker, New York, 2000.

20 F.A. Farret, "Photovoltaics power electronics," in Power Electronics for Renewable and Distributed Energy Systems: A Sourcebook of Topologies, Control and Integration, 1st ed., Springer-Verlag, London, 2013, Chapter 3, pp. 61–109.

21 M.H.J. Bollen, Understanding Power Quality Problems: Voltage Sags and Interruptions, Series on Power Engineering, IEEE Press, Piscataway, NJ, 2000.

22 A.W. Culp, Principles of Energy Conversion, McGraw-Hill, Toronto, ON, Canada, 1979.

23 Solar Energy: The Alternative Source That Spares Energy of the Net, Catalog, Siemens, 2002.

24 W.L. Hughes, Energy in Rural Development: Renewable Resources and Alternative Technologies for Developing Countries, Advisory Committee on Technology Innovation, Board on Science and Technology for International Development, Commission on International Relations, National Academy of Sciences, Washington, DC, 1976.

25 L.B. dos Reis, Generation of Electrical Energy (Geração de Energia Elétrica), Manole, São Paulo, Brazil, 2012.

26 S.B. Kjaer, J.K. Pedersen, and F. Blaabjerg, A review of single-phase grid-connected inverters for photovoltaic modules, IEEE Transactions on Industry Applications, Vol. 41, No. 5, pp. 1292–1306, 2005.

27 L.F.L. Villa, Maximizing the power output of a partially shaded photovoltaic plant by optimizing the interconnections among its modules, M.Sc. Research 2 in Electrical Engineering, EEATS Grenoble INP/UJF, Grenoble, France, 2010.

28 D. Feldman, G. Barbose, R. Margolis, M. Bolinger, D. Chung, R. Fu, J. Seel, C. Davidson, N. Darghouth, and R. Wiser, Photovoltaic System Pricing Trends, Historical, Recent, and Near-Term Projections, 2015 Edition, NREL/PR-6A20-64898, SunShot, U.S. Department of Energy, August 2015.

29 Energy Informative, The homeowner's guide to solar panels, http://energyinformative.org/nrel-efficiency-record-two-junction-solar-cell, accessed March 10, 2017.

30 Concentrating Solar Power Projects, National Renewable Energy Laboratory, https://www.nrel.gov/csp/solarpaces/by_country_detail.cfm/country=US, accessed March 23, 2017.

31 P. Denholm, M. O'Connell, G. Brinkman, and J. Jorgenson, Overgeneration from Solar Energy in California: A Field, Guide to the Duck Chart, National Renewable Energy Laboratory, Prepared under Task No. TM13.5020, Technical Report NREL/TP-6A20-65023, November 2015.

7

Power Plants with Fuel Cells

7.1 Introduction

The operating principles of fuel cells (FCs) were demonstrated initially in 1839 at the Royal Institution of London by a Welsh barrister and physicist, Sir William Grove, when he demonstrated the reversibility of water electrolysis. The first practical application of FCs is credited to Francis T. Bacon of Cambridge University in 1950. Bacon published groundbreaking results of an alkaline cell prototype. FCs then became known worldwide when the National Aeronautics and Space Administration (NASA) decided to use them in the Apollo program during the 1950s and later in the Gemini program; the main reason was the need of drinking water in space, and they did not have budget constraints. Obviously, FCs were a very convenient technology for the space program in not being polluting, producing electricity and heat, and having as by-product potable water from hydrogen, exactly what scientists wanted for a spaceship. In the past few years, FCs have appeared as the most promising innovation in the market of alternative energies for stationary, portable, and automotive applications, as a natural energy conversion system when hydrogen is stored from electrolysis.

What appeals most about FCs is their construction. They are clean and compact and need only a few movable parts. FCs have modular technology, and they do not inflict on the environment emissions of sulfur and nitrogen oxides (SO_x and NO_x), unless of course if the hydrogen is generated from some reformation of gas [1–3].

The present interest in FCs is yet enormous. Numerous companies and research centers throughout the world have been working on many developments related to FC energy systems: Ballard Generation Systems, Global Thermoelectric, Fuel Cell Technologies (Canada), Sulzer Hexis (Switzerland), UTC Fuel Cells, Schatz Energy Research Center and Energy Partners, MC Power, General Motors, Siemens Westinghouse Corporation, GE Power

Integration of Renewable Sources of Energy, Second Edition. Felix A. Farret and M. Godoy Simões.
© 2018 John Wiley & Sons, Inc. Published 2018 by John Wiley & Sons, Inc.

Systems, Teledyne Energy Systems, H-Power, Avista, IdaTech/Northwest Power Systems, and Plug Power in the United States; Toshiba, Mitsubishi Electric Corporation, and Ebara Corporation in Japan; ECN in Holland; Nuvera Fuel Cells in Italy; Rolls-Royce in England; and MTU, Daimler-Benz, Dornier, and Buderus Heiztechnik in Germany, among others. All such interest in FCs supports the idea that direct combustion of fossil fuel might be declining in importance or maybe a breakthrough on hydrogen production will change the paradigm of electricity generation. Current efforts toward commercial and regular production of FCs are intensive, primarily to improve their performance and then to establish a worldwide distribution network and a distinct type of specialized electromechanical assistance. The space between electrodes is critical for the compactness of such energy source, as well as for the removal of sulfur and carbon monoxide, particularly in the proton exchange membrane (PEM) and solid oxide fuel cell (SOFC) types. These compounds contaminate the platinum catalysts and degrade the FC performance with time. Another critical factor is the use of more common catalysts and the design definition of a high-temperature FC versus a low-temperature one.

FC characteristics differ from those of current dominant technologies for distributed generation in electric power systems, which are based on internal combustion engines using reciprocal primary movers or vapor turbines. Such technologies are widely used and offer inexpensive and reliable energy, with satisfactory heat use and for a low system efficiency. These types of machines, engines, and thermodynamics-based power plants cannot find place in a planet concerned with their own survival. Characteristics such as noise, vibration, and emission of pollutant gases (e.g., NO_x, CO_x) are major concerns; in addition frequent maintenance should be planned, thus increasing the cost of small units, which already have low overall efficiency, on the order of 30%. FCs seem to be a good option despite not reaching a mature technology as a feasible solution for the world market [3–6].

7.2 The Fuel Cell

FCs are electrochemical devices that are frequently compared with conventional automobile batteries. However, there is a fundamental difference: fuel and oxidizer are supplied to FCs continually so that they can generate continuous electric power, while a battery is charged or discharged through a chemical reaction. Therefore, an FC is considered as an *energy vector*, combined with the stored fuel as a potential energy source. FCs are based purely on the electrochemical reaction of hydrogen or other fuel gases, while batteries either have their reactants consumed or must be regenerated through electric recharge, such as in automotive lead–acid batteries [1, 2]. In an FC, the hydrogen combines with the oxygen inside a combustion-free process, liberating electric power in a chemical reaction that is very sensitive to the operating temperature. After their initial use in aerospace applications, new materials have been contributing significantly to their improved use, efficiency, and decreasing cost of fuels for terrestrial applications. A very significant figure that shows a lot of improvement is the amount of platinum catalyst needed in the electrodes. It has decreased in the last few years from 28 mg/cm^2 of electrode area to about 0.1 mg/cm^2 [3].

There are different types of FCs classified in accordance with their operation at low or high temperatures. Low-temperature proton exchange membrane fuel cells (PEMFCs) (about 120 °C) are available and ready for mass production. They are used in applications where high-temperature cells would not be suitable, such as in commercial and residential power sources as well as in electric vehicles (EVs). The most compact systems are appropriate for powered EVs and are available from 5 to 100 kW. High-temperature (about 1000 °C) SOFCs are typically used for industrial and large commercial applications and operate as decentralized stationary units of electric power generation. Because of such high operating temperatures, great heat dissipation is expected to be recovered, and integration with gas-powered microturbines can be very successful or any combined heat and power (CHP) solution. A 70% overall efficiency is reported in such applications, representing a great improvement with respect to the energy generation in coal power plants, whose efficiency is between 35 and 40%.

Figure 7.1 shows the operating principles of an FC. They are powered by the flow of a fuel gas such as hydrogen, which could be in pure or derived form (e.g., methanol, coal gas, natural gas, gasoline, naphtha) plus an oxidizer (oxygen or regular air). After the hydrogen passes through a porous anode separated by an electrolyte of a porous cathode, the resulting ions meet the oxygen and form water. Because of this reaction, the accumulated electrons in the anode form an electrical field and find their way through an external conductive connection between anode and cathode. An electrical current is established through an external load. As in any other chemical reaction, heat is produced, which has to be dissipated in some way. Because of the gas reaction, there is by-product generation of heat and water, in contrast with the conventional mixture and burning of fossil fuel. Therefore, two electrochemical reactions occur in an FC: (i) one is the oxidation of the fuel, typically familiar combustible gaseous or liquid materials such as hydrocarbons, natural gas, alcohols, carbon monoxide, or hydrogen, and (ii) another one is the reduction of oxygen to absorb the resulting hydrogen ions. These two reactions look like the overall combustion of a fuel and the corresponding electrical energy generated related to the combustion energy. Therefore, from a physicochemical point of view, FCs are the electrolysis antipode process, in the sense that the first consumes H_2 and O_2 producing electricity, heat, and water, whereas the

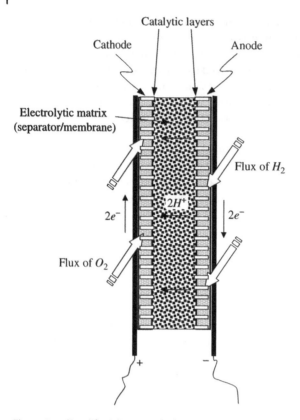

Figure 7.1 Simplified diagram of a fuel cell.

second consumes electricity and water and produces heat, H_2, and O_2. This duality widely motivates practical possibilities for fully clean and sustainable systems, as discussed in Chapter 2.

7.3 Commercial Technologies for the Generation of Electricity

Hydrogen and pure oxygen are not commercially available in the quantities required for cost-effective use in FCs. There is abundant oxygen in the air and abundant hydrogen in several organic materials, particularly natural gas. Hydrogen can be available at low temperature but demanding a fuel processor, or reformer, before reaching the cell. The reformer converts, methane, for example (added in a steam), to a reformed mixture of hydrogen, sulfur, ammonia, carbon dioxide, and carbon monoxide. The fuel processor still has to accomplish a catalytic endothermic steam reform, that is, the reaction of water steam with fuel hydrocarbon to produce hydrogen, thereby increasing complexity and losses, leading to an efficiency of around 35–48%. Purification of hydrogen is still required because particulates may degrade the FC membrane life.

Natural gas is a fossil fuel: on one hand it can bridge the current economic status toward sustainable production of hydrogen, and on the other hand within a couple of decades, it may get into exhaustion. One issue that needs to be evaluated is the cost comparison of microturbines powered by natural gas with FCs running from reformed hydrogen. The FC choice makes sense only when combined with other applications that are interconnected to hydrogen storage and production. Therefore, FCs will be feasible under an economics point of a view when (i) combined with hydrogen-powered EVs, (ii) electrolyzers, and (iii) small-scale renewable energy sources.

One of the peculiarities of FCs to be used as commercial sources of energy is the need to consider the benefits and restrictions of a very complex market concerned with the following four points:

1) **Cost**: Projected at $2–$3/W, the cost of FCs is much higher than the cost of generators, batteries, and other forms of alternative energy. A recent very positive trend motivated by the manufacture of FC-powered cars allows a projection of $0.79/W in full-scale production. Competitiveness is a little more difficult with river and maritime applications, equipment for telecommunications, and recreation vehicles, which already have a reasonable market share [4–7].
2) **Product availability**: Customers want to have products to try out and compare with other options; otherwise, it **becomes** very difficult to convince people to use them.
3) **Fuel**: With the exception of large boats that use diesel or gasoline, several other industry segments use **propane**. However, if hydrogen is available for automobiles, which is a much larger market, this will easily be extended to boats.
4) **Safety**: A certain apprehension exists on behalf of potential buyers of recreation vehicles and small boats **about** replacing propane with hydrogen.

Power generation faces a highly competitive market under deregulation policies. Therefore, economic factors must convey a proper decision as to the type of generator to be used for each application. The main economical factor is maximization of the commercial power made available to consumers with minimum maintenance at the highest levels of efficiency and minimum capital investment [4–7]. The only item difficult to analyze in FC applications is still a possibly lower capital investment. Many countries, including Germany and England, have been investing effectively in clean generation policies by subsidizing initial production and use of FCs. For a market with such diversified demands, several FC technologies are in development [8–11]. The main research lines being considered at the time of publication of this book are:

- PEMFC or solid polymer fuel cells (SPFCs)
- Phosphoric acid fuel cells (PAFCs)
- Alkaline fuel cells (AFCs)
- Molten carbonate fuel cells (MCFCs)
- SOFCs
- Direct methanol fuel cells (DMFCs)
- Reversible fuel cells (RFCs)

Common application ranges for these FCs are shown in Figure 7.2.

The designation of an FC type is related to the electrolyte for the reactant ions. For example, in a PEM, a solid polymer (SPFC) such as Nafion (a registered trademark of DuPont) is used to enable conversion of hydrogen chemical energy (H_2) and an oxidizer, oxygen (air or O_2), into electrical energy. Nafion operates at temperatures between 45 and 100 °C with a typical efficiency of 40%. This solid polymer membrane is the basis for the strongest interest in its use with automotive applications. The main function of such a polymer is to provide good electronic insulation and an efficient gas barrier between the two electrodes while allowing rapid proton transport and high current densities. When in operation, the solid electrolyte does not move with respect to the electrodes. It does not diffuse or evaporate but occupies a moderate space with appreciable weight. These cells are promising for automotive applications and are expected to have some use in CHP schemes and automotive applications.

Typical applications	Vehicles, portable, and electronic equipment.			Cars, boats, and domestic CHP		Distributed power generation, CHP, buses		
Power (W)	1	10	100	1k	10k	100k	1M	10M
Main advantages	Higher energy density than batteries. Faster recharging			Potential for zero emissions, higher efficiency		Higher efficiency, less pollution, quiet		
Range of application of the different types of FC								

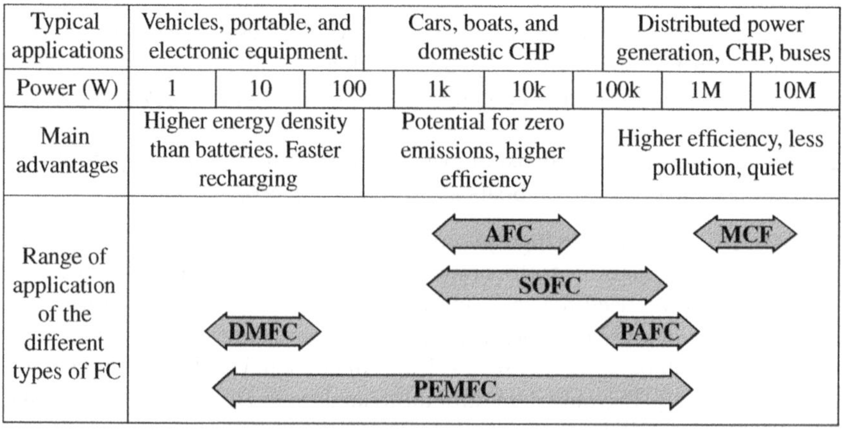

Figure 7.2 Applications of fuel cells.

In order to understand the chemical reactions and unique features of each FC technology, the basic operation may help. This subject can be started with the PEMFC. Their chemical reaction at the anode is [3, 5]

$$2H^+ + 2e^- \rightarrow H_2$$

and their cathode reaction of the fuel ionized with oxygen is

$$O^2 + 4H^+ + 4e^- \rightarrow 2H_2O$$

To make FCs as light and small as possible, one has to construct compact modules for easy and efficient setup. This is the case for the PEM polymeric membrane, bonded on each side by catalyzed porous electrodes. Such an electrode–electrolyte–anode or membrane electrode assembly (MEA) is made with bipolar electrode plates so that the positive side of one electrode plate is the negative side of the next plate. Therefore, the MEA is extremely thin, compact, and ready to be associated with others. Furthermore, the electrode plate pair is typically machined with flow fields for an even distribution of fuel and air or oxygen to anode and cathode. It must also be provided with a compatible path for the cooling air or water at the back of the reactant flow field of the metallic electrodes. The humidity must be kept in the approximate range 85–100%, as the membrane hydration assists in proton conduction through the electrodes. Thus, a small section of the MEA has to be set aside for humidification of the reactant gases. High standards for the MEA ohmic, mass transport, and kinetic characteristics are essential for its commercial application.

The PEMFC is favored for transportation vehicles because of its high power density and moderate operating temperature. Nevertheless, it demands additional means for a quick start-up and rapid response to changes in system demand. The leading developer is Ballard Power Systems Inc., since it is estimated that Ballard has 98% of the transportation-based market. Ballard's brochures describe a 68-kW light-duty engine for the automobile market and a 190-kW engine for buses and trucks, as well as a 1.2-kW module for small applications. Smaller PEMFCs up to 250 kW capacity are being developed for the portable market field trialed for stationary power generation. Electric potential of an PEMFC is $0.6V_{dc} \times 0.7 A/cm^2 \simeq 0.42 W/cm^2/cell$.

The PAFC has a certain similarity with the PEMFC, but the acid H_3PO_4 is the electrolyte and the cell is operated with hydrogen and air. The power density produced by this FC ranges from about 0.20 to 0.35 W/cm^2. The temperature range is about 220 °C, with a typical efficiency of 40%. This cell is tolerant to poisoning up to 1% of carbon monoxide, although good water management is essential in its operation to improve reagent kinetics as dependent on temperatures varying from 150 to 220 °C. The internal environment becomes very corrosive, due to the high electrode potential of its operation, demanding the use of strong, corrosion-resistant materials in its construction. This type of FC achieves only moderate current density. Its construction and chemical reactions are quite similar to those of the PEMFC, from which most of its features were derived. Because the temperature range of the PAFC is higher than that of the PEMFC, the PAFC is recommended for CHP applications. There are many 200-kW units installed all over the world based on this technology, which have a strong record of operation. The chemical reaction at the anode of a PEMFC is similar to that in a PAFC [3, 5]:

$$2H^+ + 2e^- \rightarrow H_2$$

and the cathode reaction of the so ionized fuel with oxygen is

$$O^2 + 4H^+ + 4e^- \rightarrow 2H_2O$$

The PAFC is the most commercially advanced technology when compared with medium-scale power generation. Presently, UTC Fuel Cells offers a 200-kW-rated version of a PAFC system that has been available for over a decade. They claim worldwide installation of more than 250 units, accumulating over 6 million hours of operational experience. PAFC commercialization was given a kick start in the mid-1990s when the US government offered a one-third subsidy to stimulate sales and lower production costs.

The AFC was the first workable power unit used by NASA in the manned Apollo spaceship mission. It exploits the high conductivity and boiling point of a concentrated alkali solution (potassium hydroxide). It operates in temperatures between 150 and 120 °C with a typical efficiency of 50%. Nevertheless, it was soon found that cost, reliability, ruggedness, safety, and ease of AFC operation could never compete with some of the other technologies known at that time, although it had the least expensive construction technology. It uses nickel and silver electrodes. As the electrolyte is alkaline (carbonates), the carbon dioxide degrades the electrolyte and must be completely eliminated. So, its major problem is that strongly alkaline electrolytes such as *KOH* and *NaOH* adsorb CO_2 and so reduce its electrolyte conductivity. It runs at 150–200 °C with only highly purified hydrogen and oxygen. This feature limits its current densities and imposes a severe limitation to applications other than airspace. The chemical reaction at the anode of an AFC is

$$2H_2 + 4OH^- \rightarrow 4H_2O + 4e^-$$

and the cathode reaction of the so ionized fuel with oxygen is

$$O_2 + 4e^- + 4OH^-$$

The electrolyte of an MCFC is a mixture of alkali carbonates of potassium and lithium. A carbonate ion moves from cathode to anode, where in combination with hydrogen it forms water and carbon dioxide plus the electrical charge. The power density produced by this FC ranges to about $0.10\,W/cm^2$, limited by ohmic losses. The high operating temperatures (around $650\,°C$) and the high corrosivity of the molten carbonate salts demand special types of manufacturing materials, justified by the noticeable improvement in reactant kinetics and reduced needs for expensive noble catalysts. The typical efficiency is above 50%. The bipolar electrodes are made from high-quality stainless steel protected by additional coatings of nickel or chrome. The high temperatures allow for internal fuel steam reforming, such as the gas methane. Its high operating temperatures recommend MCFC primarily for stationary power applications in the megawatt range. The chemical reaction at the anode of an MCFC is [3, 5]

$$H_2 + CO_3^{2-} \rightarrow H_2O + CO_2 + 2e^-$$

and the cathode reaction of the ionized fuel with oxygen is

$$\frac{1}{2}O^2 + CO_2 + 2e^- \rightarrow CO_3^{2-}$$

One of the main developers of the MCFC, FuelCell Energy Inc., describes models of $250\,kW$, $1\,MW$, and $2\,MW$ capacities. An operating life of 16,000 hours for a recent demonstration unit has been reported.

The SOFC uses yttrium and zirconia as oxides to operate at high temperatures (around $1000\,°C$). Under such conditions, hydrogen and ionized oxygen are combined to form water in the anode and to liberate a pair of electrons with excellent reactant kinetics, although the cell's reversible potential is a little lower for lower-temperature cells. There is no water management problem because this FC has a solid-state construction and a single side tube sealing. This cell may use hydrogen or carbon monoxide as fuel with efficiencies higher than 50%. Its main drawback for high power generation is that of the materials required in its construction. Recent research developments involve three main types of reactant core design for this cell: (i) planar (similar to that of other FCs), which has sealing problems; (ii) tubular, to improve the electronic conductivity and overcome the sealing limitations of the planar design; and (iii) flattened tube, with better air guidance, to improve the packing densities of the tubular design. In any case, the tube itself forms the air electrode. The power density produced by the tubular design cells ranges from about 0.18 to $0.20\,W/cm^2$. The planar design cell may reach a power density of about $0.35\,W/cm^2$. SOFC use is restricted almost entirely for stand-alone and CHP applications. The chemical reaction at the anode of an SOFC is [5]

$$H_2 + O^{2-} \rightarrow H_2O + 2e^-$$

and the cathode reaction of the ionized fuel with oxygen is

$$\frac{1}{2}O^2 + 2e^- \rightarrow O^{2-}$$

The SOFC has more than 40 years' development history, started when Siemens Westinghouse first produced its tubular design, aimed at avoiding the problems of sealing a conventional battery-type structure that would comprise arrays of flat plates. The tubular concept has been demonstrated with 100- and 220-kW units. Australia's Ceramic Fuel Cells Ltd. is one of the developers of a flat-plate SOFC design.

A reversible solid oxide fuel cell (RSOFC) is a device that can operate efficiently in both FC and electrolysis operating modes. In the FC mode, an RSOFC functions as an SOFC, generating electricity by electrochemical combination of a fuel (hydrogen, hydrocarbons, alcohols, etc.) with air (oxygen in the air). In the electrolysis mode, an RSOFC functions in an electrolyzer mode (in this case, referred to as a solid oxide electrolysis cell (SOEC)), producing hydrogen (from water) or chemicals such as syngas (from mixtures of water and carbon dioxide) when coupled with energy source (fossil, nuclear, and renewable).

The RSOFC has all the desired characteristics to serve as a technology base for green, flexible, and efficient energy systems in the future. It unifies the following attractive demonstrated or potential features: environmentally compatible with reduced CO_2 emissions in power generation mode, fuel flexible and suitable for integration with any type of energy sources, capability for different functions, suitable for a variety of applications and adaptable to local energy needs, and competitive in costs.

High operating temperatures and slow start-up mean that both the MCFC and SOFC should be used to power generation in larger, continuously operating installations, in the range of tens of kW to several MW power.

In the DMFC, methanol is oxidized electrochemically by water at the anode to produce carbon dioxide and positive and negative ions, in contrast to what happens in a PEMFC, where the positive ions from the hydrogen are supplied directly to the anode. To improve rejection of the carbon dioxide in a DMFC, an acid electrolyte is used; otherwise, insoluble carbonates may form in the alkaline electrolyte. Similar to what is depicted in Figure 7.1, the hydrogen ions produced at the anode permeate the polymer electrolyte in the direction of the cathode, where they react with oxygen from the air to produce water. The external circuit provides a path for the flow of electrons accumulated in the anode. The most attractive feature of the DMFC for the transportation and portable-use industries is that a fuel reformer is not required. As may be inferred from this explanation, DMFC have poorer performance in the anode, where finding more efficient catalysts is still a major problem. In other parts of the world, other types of alcohol are being experimented with, such as ethanol and benzol. The power density produced by direct methanol cells as reported by Newcastle University researchers in England is about 0.20 W/cm^2. Its overall efficiency does not go above 50%. The chemical reaction at the anode of a DMFC is [2, 5]

$$CH_3OH + H_2O \rightarrow 6H^+ + CO_2 + 6e^- \quad \text{for } E_{anode} = 0.046\,\text{V}$$

and the cathode reaction of the so ionized fuel with oxygen is

$$\frac{3}{2}O_2 + 6H^+ + 6e^- \rightarrow 3H_2O \quad \text{for } E_{cathode} = 1.23\,\text{V}$$

and for the cell voltage

$$CH_3OH + \frac{3}{2}O_2 + H_2O \rightarrow CO_2 + 3H_2O \quad \text{for } E_{cell} = 1.18\,\text{V}$$

The DMFC has made a comeback because of the market needs in consumer electronics. The high energy content of methanol is the attraction for loads like notebook computer because it can run for 10 hours on 100 mL of methanol fuel. Technical challenges include yet lowering catalyst usage as it uses platinum-based catalysts and prevents methanol and water crossing over through the membrane, and it is not in market yet.

The reversible fuel cell (RFC) (also known as a regenerative or unitized fuel cell) is a special class of FCs that produces electricity from hydrogen and oxygen but can be reversed and powered with electricity to produce hydrogen and oxygen, as depicted in Figure 7.2. The three key concepts of the reversible fuel cell are output power, run time, and recharge rate. The number and size of FCs in the stack determine the output power, which is primarily dependent on the effective area of each electrode–membrane–electrode assembly. The run time is determined by the capacity of the hydrogen storage tank available. The recharge rate is determined by the output rate of the electrolyzer used to produce the hydrogen. An example of this is a PEM-fed uninterruptible power supply under development by Unigen as part of the US Department of Energy's State Energy Programs for developing ready-made RFC modules [8].

Reverse fuel cells are also under development in many parts of the world. They can convert hydrogen directly from water using photovoltaic, hydro, or wind power. It has been observed that RFCs are capable of an energy density of about 450 Wh/kg, which is 10 times that of lead–acid batteries and more than twice that of the modern chemical batteries. GreenVolt Power Corp. typically reports one of these systems, in which their FC splits water into its components hydrogen and oxygen for use in a variety of industrial and transportation applications. The unit uses 20% less energy in a significantly smaller unit size

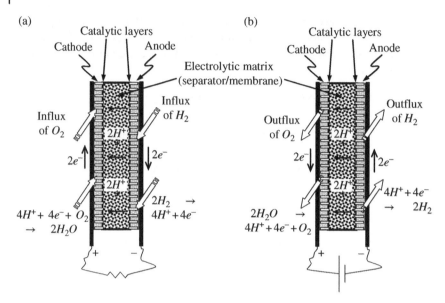

Figure 7.3 Simplified diagram of a PEM reversible fuel cell: (a) fuel cell mode and (b) electrolyzer mode.

than that of water–alkali electrolyzer devices and requires only distilled water to produce 99.5% pure hydrogen and oxygen. When used to power hydrogen FCs, the water by-product created can be reused by the RFC, potentially lowering operating costs. On-site hydrogen gas generation can also improve the economical constraint of hydrogen transportation by pipelines and bottles. It is possible to fabricate ceramic microtubes by EPD, although further research into certain areas of this investigation is in progress.

Figure 7.3 shows the chemical reactions in a reversible fuel cell. In the FC mode, a PEMFC combines hydrogen and oxygen to create electricity and water. When the cell reverses its operation to act as an electrolyzer, electricity and water are combined to create oxygen and hydrogen. RFCs are expected to be useful mostly in passenger cars, solar-powered aircraft, energy storage schemes, required propulsion of satellites for orbit correction, micro spacecraft, and power systems, as in load leveling in remote sources of wind turbines and solar cells.

Another type of FC is the zinc–air fuel cell (ZAFC), which has a gas diffusion electrode (GDE), an anode separated by electrolyte, and mechanical separators. The GDE is a permeable membrane that allows atmospheric oxygen to pass through. After the oxygen has been converted into ions and water, the ions travel through an electrolyte and reach the zinc anode. There it reacts with the zinc, forming zinc oxide, creating an electrical potential. This electrochemical process is very similar to that of the PEMFC described previously, but the refueling is very different, although it has some of the characteristics of batteries. ZAFCs are best suited for battery replacement but have another wide range of potential applications. Some developers of ZAFCs are Cinergy and Metallic Power and Electric Fuel Co.

A more recent technology is the regenerative fuel cells (RFC) that can be seen as a closed-loop power generation. The water is separated into hydrogen and oxygen by a solar electrolyzer. Therefore, after hydrogen and oxygen were fed into the FC to generate electricity, heat, and water, the resulting water is then recirculated back to the electrolyzer, powered with solar energy, and the process begins all over again. Worldwide speaking, NASA seems to be the most interested in this technology.

In the ZAFC, there is a GDE; one separates zinc anode in an electrolyte and separators by mechanical means. The GDE is a permeable membrane that allows the passage of air. Once the air oxygen has been converted into hydroxyl ions and water, these ions travel through the membrane and reach the zinc anode reacting with zinc to form zinc oxide. As a result, it generated electricity. Connecting several ZAFCs, higher voltage levels can be obtained to form a power source. The difference between this electrochemical process and the one present in a PEMFC is the refueling process that presents some similarities with batteries. In this closed-loop system electricity is generated as zinc and oxygen are mixed in the presence of an electrolyte, creating zinc oxide. Once the fuel has ended, the system is connected to the network and the process is reversed, again leaving a pure zinc-pelletized fuel. The key is that the rollback process only takes about 5 minutes to complete and time for battery recharge is not a problem in this system.

The useful life of the zinc–air technology when compared with other batteries is their specific energy, that is, with respect to its weight. This is an important aspect to power EVs since it allows longer time between refueling stops. Another point is the cost of zinc used in ZAFCs, making the zinc–air batteries very low.

Perhaps the newest type of FC is the proton ceramic fuel cells (PCFCs), which are based on a protonic ceramic electrolyte material with high protonic conductivity at elevated temperatures. The PCFCs share similar thermal and kinetic advantages of operation at high temperature of 700 °C with the more conventional types like MCFC and SOFC. This PAFC has the benefits of proton conduction in polymer cell electrolyte and phosphoric acid. The higher the operating temperature, the higher the electrical efficiency of the hydrocarbon-based fuels. One relevant point under the environmental point of view is that PCFCs can operate at high temperatures and electrochemically oxidize fossil fuels directly to the anode, making the hydrogen production cheaper since it dispenses reformers. The process uses gaseous molecules of hydrocarbon fuel absorbed on the surface of the anode in the presence of water vapor and hydrogen atoms. These atoms are efficiently wrenched to be absorbed into the electrolyte, producing carbon dioxide as main product reaction. Additionally, PCFCs have a solid electrolyte different from that of the PEMFCs and PAFCs. In Table 7.1 some types of FCs with their main characteristics are listed.

Table 7.1 Types of fuel cells with their main characteristics.

Characteristic	PEMFC	DMFC	AFC	PAFC	MCFC	SOFC
Electrolyte	Proton exchange membrane	Proton exchange membrane	Potassium hydroxide	Phosphoric acid	Molten carbonate (Li, K, Na)	Solid oxide ($ZrO_2 - Y_2O_3$)
Temperature (°C)	50–90 °C	50–130 °C	50–250 °C	180–200 °C	650 °C	750–1050 °C
Charge carrier	H^+	H^+	OH^-	H^+	CO_2^{3}	O^2
Catalyst	Pt	Pt	Pt, Ni	Pt	Ni, LiNi	Ni
Fuel	H_2 (pure or reformed)	CH_3OH	H_2 (pure)	H_2 (reformed)	H_2 and CO reformed and CH_4	H_2 and CO reformed + CH_4
Poison	$CO > 10$ppm	Adsorbed intermediates	CO, CO_2	$CO > 1\%$ $H_2S > 50$ppm	$H_2S > 0.5$ppm	$H_2S > 1$ppm
Main applications	Transportation, portable	Transportation, portable	Space and spaceships	Cogeneration, transportation	Electrical power cogeneration	Electrical power cogeneration

7.4   Practical Issues Related to Fuel Cell Stacking

Figure 7.4 shows all different reactions for main types of FC technology, with a range of their operating temperatures.

7.4.1   Low- and High-Temperature Fuel Cells

There are two broad categories of FCs: low temperature or first generation, which include AFCs and SPFCs (or PEMFCs) and the phosphoric acid type (PAFCs). The latter has been the main objective of resource investment to obtain a commercial model with compatible costs for terrestrial applications. High-temperature cells such as MCFCs and SOFCs are considered second generation, and their commercial development is at an advanced stage, being considered for use in large-scale electric power generation [1–5].

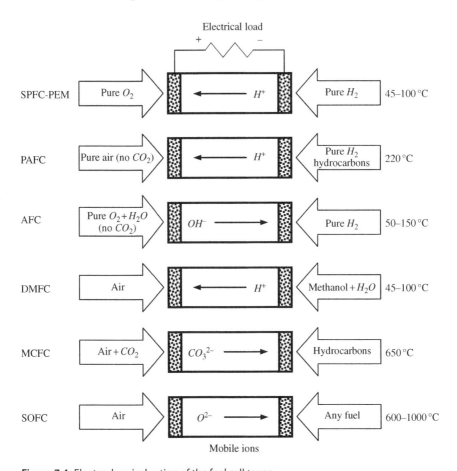

Figure 7.4 Electrochemical action of the fuel cell types.

Second-generation cells have been developed for operation at high temperatures to have an easier coupling with reforming of the hydrogen inside the cell itself. Such a procedure makes the overall system very simple and increases the efficiency up to 60%, as demonstrated under laboratory conditions. A single FC can produce just a few watts. Therefore, they are stacked up and connected to other cells, adding their effects electrically by series and/or parallel connection, very similar as ordinary batteries or photovoltaic cells. A stack will then form autonomous modules of energy supply in accordance with the end purpose. There are currently very large power stacks, up to 1.5 MW.

The electric power generated by an FC stack is always direct current (dc), just as ordinary batteries. The dc power generated in the cells should be conditioned to the types of contemporary applications, as explained in Chapter 12. For all FC types there is a lot of packaging research aimed at turning them into useful commercial and practical products. The following characteristics are also the subject of intense research:

- Pretreatment of fuel for cogeneration systems
- Heat recovery by preheating fuel and oxidizer
- Injection and recirculation of water or steam
- Energy administration, conditioning, conversion, and optimization
- Separation and recycling of CO_2 in MCFCs by blowing air and careful administration of the water in SPFCs to assure that the membrane stays damp for optimal conditions

7.4.2 Commercial and Manufacturing Issues

The Carnot cycle can be used to compare FCs with internal combustion engine (ICE) motors as generators of energy. This cycle imposes limits on the performance and construction of any energy transformation unit by high pressures and temperatures according to the upper efficiency limit $\eta_c = (T_{high} - T_{low})/T_{high}$. The dimensions of the machine also influence the overall efficiency directly, hopefully by about 30%. In the same way, theoretical limitations exist for FCs, although they may not be as narrow. The purer the oxygen and hydrogen used, the higher the overall FC efficiency rates, which are around 70%. Such rates would be justified only in an aerospace mission, or in oceans or similar environments, where the costs of hydrogen and oxygen are not as relevant as in ordinary power plants. Using methanol, for example, to obtain the fuel gas, the efficiency falls to around 44%; however, the cost reduction is dramatic. A practical alternative that is perhaps already feasible is the use of SPFCs, as in the Gemini spaceship mission and as used for many years for the production of oxygen in submarines.

One problem with FCs, which could not be completely solved by the time this book was written, is related to how to obtain the largest possible contact area among electrolytes and gaseous reactants, called the *notable surface action* of the electrocatalytic conductor. This conducting contact path involves the electrolyte itself, the flooded pores of the electrolyte, the platinum on the catalytic carbon particles, the polytetrafluoroethylene (PTFE), and the gas through the pores.

Power systems with fuel cells for remote applications (FCPS-RA) seem to be the great opportunity for FCs in the current market. The powers range between 0.1 and 5 kW and include loads such as recreation vehicles, communication stations, yachts, residences isolated from the public network, and canalizations of natural gas and oil. For these applications, the current retail prices in for FCPSs are $0.13 (per kilowatt-hour) (generator), $0.29 (per kilowatt-hour) (battery), and $0.50 (per kilowatt-hour) (photovoltaic panels) in the consuming markets. Current FCPSs in the wholesale can produce electricity for $0.44 (per kilowatt-hour). In the best of hypotheses with a reduction of 50% in stack cost, an improvement of 100% in the performance, and a hydrogen generation cost of $0.12/GJ, FCPSs could supply energy at $0.23 (per kilowatt-hour).

Commercial and manufacturing investment has been concentrated in three areas: (i) the development of a 5-kW reformer to be integrated with an FC stack tolerant to fuel reforming, (ii) commercially available FCs that operate with pure hydrogen and are integrated with EVs, and (iii) diffusion and training in the manufacture, production, and use of FCs.

For small- and medium-energy systems, the present state of the art seems to point toward the establishment to two already advanced technologies: PEMFC and SOFC. This chapter therefore concentrates on those two technologies.

7.5 Constructional Features of Proton Exchange Membrane Fuel Cells

Porous separation between fuel and oxidizer is achieved by the use of specialized conventional materials (e.g., tissues made of carbon or carbon paper obtained from Teflon) and is the most expensive part of the FC stack. The main characteristic of these materials is that they should allow the flow of ions generated in the fuel and oxidizer reaction when going from anode to cathode, preventing the passage of electrons, so they have to circulate through an external circuit.

Figure 7.1 is a simplified diagram of an FC. At the anode side of the PEMFC, fuel is supplied under a certain pressure, usually enough to make it go through the flow-field channels of the electrodes and cross the electrolyte. It is assumed for the sake of discussion that the fuel is the pure gas H_2, although other gas compositions may be used, as discussed in Section 7.3. In these cases, the hydrogen concentration should be determined in the mixture. The fuel spreads through the electrode until it reaches the catalytic layer of the anode, where it reacts to form protons and electrons according to the reaction

$$H_{2(g)} \rightarrow 2H^+ + 2e^- \tag{7.1}$$

Protons are transferred to the catalytic layer of the cathode through the electrolyte (a solid membrane for a PEMFC). On the other side of the cell, the oxidizer flows through the channels of the plate and spreads itself through the electrode until it reaches the catalytic cathode layer. The oxidizer used in this model may be air or pure O_2. The oxygen is consumed by reacting with the H_2 protons and electrons, and water is produced with some residual heat on the surface of the catalytic particles. The electrochemical reaction in the cathode is

$$2H^+ + 2e^- + \frac{1}{2}O_2 \rightarrow H_2O + heat \tag{7.2}$$

Then, the full FC chemical reaction becomes

$$H_2 + \frac{O_2}{2} \rightarrow H_2O + heat + electrical\ energy \tag{7.3}$$

Figure 7.5 is a schematic diagram of a complete electrical energy generation system using an FC stack with PEMFCs. The most vital part of such a system is a stack with 70 PEMFCs and $300\,cm^2$ per cell of active area, supplying $3\,kW$ of alternating current (ac) through a dc–ac inverter. The diagram shows the stack fed with hydrogen and oxygen (air) as well as water for refrigeration and output products. Hot water and electricity are both available for the user. The overall stack output voltage is represented by V_s. Table 7.2 gives the main nominal characteristics of this system. The reformer for the cells to obtain hydrogen from a fuel with hydrocarbon and water steam is also represented. The capacity of the components in the system will depend primarily on the total output power of the stack.

The system comprises of a low-pressure, high-volume air blower to supply the oxidizer. A water circulation system is used to cool the stack reactions and to provide humidification of the input flows of air and reformer fluids. The reformer is of the partial oxidation type to generate fuel rich in hydrogen for the stack from the tank of natural gas or propane.

A typical system setup consists of an FC stack conditioned by a dc–dc converter of 48 V to feed a blower, water pump, 4-kW power inverter, and another inverter to generate a 120 V ac output voltage. Auxiliary power can typically be fed from dc–dc converters operating from the main FC dc-link bus for microcontrollers, relays, solenoid valves, and so on.

A microcontroller can be used to manage the FC operation, providing acquisition of operating data and to inform the user about the stack operating conditions. A start-up battery must be included to bootstrap the beginning of operations, the microcontroller, and other load controls. After FC starting

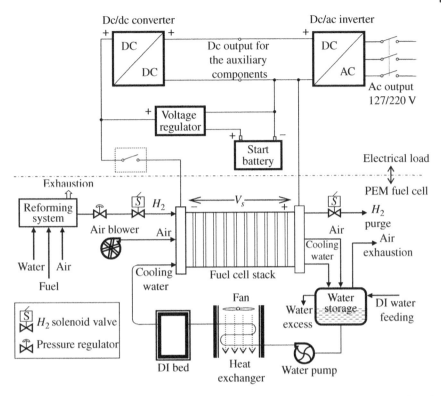

Figure 7.5 Practical outline of a small fuel cell power plant. *Source*: Hoogers [5]. Reproduced with the permission of CRC Press LLC.

Table 7.2 Rated values of a typical commercial fuel cell stack.

Description	Specification
Ac output power	3 kW
Ac output voltage	127/220 V
Dc voltage	42–70 V
Temperature of the membrane	60 °C
Normal maximum flow of the reformatted H_2	100 L/min
CO contents of the reformatted output	<50 ppm

procedures (warm-up time), the system will be fully running only with power from the FC stack. The PEMFC is recommended for either stationary or vehicle power applications, and there are many 5- to 250-kW units based on this technology worldwide that have had many hours of operation.

7.6 Constructional Features of Solid Oxide Fuel Cells

SOFCs use an entirely solid oxide ion-conducting ceramic of zirconia (zirconium oxide) stabilized with yttria (yttrium oxide) without any liquid-state interacting product. It works at very high temperatures, typically between 800 and 1100 °C. Therefore, there is no need for electrocatalysts, the most complex item commonly associated with research in ceramics and membranes in all the other FCs. It may operate with hydrogen with some level of carbon monoxide, and as a result, CO_2 recycling is not necessary. As a contrast to low- and medium temperature FCs, the ions crossing the electrolyte from the cathode to the anode are the oxygen and by-product water formed at the anode side. At about 800 °C, zirconia allows conduction of oxygen ions starting up the energy production process. The open-circuit voltage of SOFCs is usually lower than of MCFCs, but in compensation, it has lower internal resistance, thinner electrolytes, and therefore reduced losses. For these reasons, SOFCs may operate at higher current densities (around $1000\,mA/cm^2$).

A zirconia mixture of ceramic and metal (cermet) is widely used to construct a highly resistant and stable anode for the high SOFC temperature environment. The metal used in this mixture is nickel because of its good electrical conductivity and catalyst properties, widening the operating range of this FC since the fuel-reforming process can take place at lower temperatures. On the other hand, the cathode composition is still a complicated matter because of the cost of effective conducting materials at high temperatures. Some materials presently used for this purpose are based on strontium-doped lanthanum manganite.

The most challenging issues of such a slowly maturing technology are related to high-temperature-resistant materials, combined applications with other FCs, heat, and power management. At this stage of development, it is difficult to predict which FC technology is going to be the most suitable or which is going to become a successful commercial version that takes the best possibilities of CHP and the construction of hybrid systems with various other FC types. Figure 7.6 displays a possible system for SOFC cogeneration, including the electric conversion oxidizer input and fuel gases, heat exchanger, and catalytic burner [11–14].

There are other configurations considered seriously for a high-temperature combination of SOFCs with steam or gas turbines (combined cycle system) where the exhaustion gases of the FC would feed the gas turbine, which would power an alternator. SOFCs of size 200 kW are being widely considered for large CHP generation units in shopping centers, hospitals, military headquarters, residential condominiums, public buildings, and stand-alone villages. In all these applications, the natural gas has to be desulfurized before feeding the anode, and air is admitted into the FC through preheaters using exhausted anode and cathode hot gases.

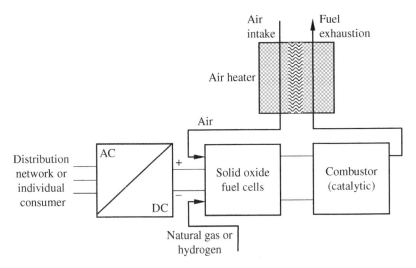

Figure 7.6 Cogeneration system using SOFCs.

7.7 Reformers, Electrolyzer Systems, and Related Precautions

The main objective of using a reformer is to supply hydrogen locally to the FC stack, a relatively complex system that demands automated control. The system is sensitive to several aspects, such as control failure, the presence of inflammable gases, carbon monoxide poisoning gas, and over-temperature operation. An appropriate design should foresee several redundant sensors to guarantee safe operation, monitoring the components, and controlling the stack operation. Although such complex systems are usually encountered in industrial and automotive environments, a lot of research is still required for its use in efficient, safe, and reliable FC systems.

There are two major fluids present in the reformer: the fuel input and the hydrogen generated. Care should be taken for transportation, storage, distribution, and safety standards of the input fuel (methanol, propane, or natural gas). Safety requirements for hydrogen have already been adopted as related to small-scale storage and proper pipelining, but it is expected that the FC industry will instill new practices with the growth of FC applications [15–17].

During reforming, small amounts of rusted carbon monoxide can be produced, but as long as there is no leakage, it does not represent a danger. However, if there is any leakage before the oxidation process, it is not safe and is a potential system safety concern. Reformers usually work at a temperature of 1650 °C and must be thermally insulated to maintain a bearable exterior temperature. Internally, water steam is used to cool the hydrogen. High internal temperatures can overheat and rupture the equipment. Leakage in the cooling area can produce steam jets at high temperatures with eminent danger.

Another way to obtain hydrogen is through renewable sources of energy. The process to be used for that is the electrolysis of water or direct concentration of sunlight on reactors. This process can produce almost pure hydrogen and oxygen fed only by secondary sources of energy, such as the surplus in power system installations in the early hours of the day; Earth heat, such as geothermal, deserts, and ocean temperature gradients; and renewable sources of energy. Among renewable sources of energy with most promise for these purposes are hydroelectric power plants such as those in Hamburg and Vancouver; biomass, such as sugarcane in Brazil; solar, geothermal, and wind, as in the Middle East; and sea tidal and sea wave energy in almost every corner of the planet. Other means of hydrogen production are:

- Photolysis, electrolysis of vapor, thermochemical decomposition, or photoelectrochemical processing of water
- Reform or gasification of petrol or coal
- Reform of ethanol or methanol
- Catalytically reformed or thermal cracking of natural gas
- Gasification of biomass
- Fermentation or photodecomposition of organic compounds
- Partial oxidation of hydrocarbon

7.8 Advantages and Disadvantages of Fuel Cells

FCs are very appropriate for power plants using cogeneration systems in circumstances where environmental issues such as a clean atmosphere, silence, and absence of vibrations are of particular concern. This is because of the absence of movable parts other than the circulation pump and gaseous fluid blowers. PAFCs or cells using fuel reduction are not very convenient for power plant purposes because they are complex and necessitate specialized engineering services. These features maintain them safe and reliable, despite being smaller than the ones required for power plants with rotating machines.

The modular nature of FCs allows their use in virtually any application that allows flexible expansion of a power plant and gradual investment to follow evolution of the loads. The energetic conversion factors are aided very favorably by flexibility in the use of fuel, high efficiency (it may get near to 70%), and low volume/power ratio. In addition, the fuel hydrogen can be homemade using water electrolysis with solar energy or reforming technology-producing hydrogen from hydrocarbonate fuels.

FC systems adapt easily for energy injection applications into the utility grid because they have energy density (Wh/m^3) higher than that of standard batteries, relatively fast response to load fluctuation, high reliability, low cost of operation, and very low maintenance cost. Spilling of hydrogen will never be a major safety concern as this lighter-than-air gas flows up and away. When refueling FCs in vehicles, a very fast time is possible, about the same as that to fuel an ordinary gas-driven car, a very good advantage compared with recharging a battery.

The disadvantages are the typical ones in any new technology that requires time to mature and to be widely adopted. FC energy systems are unfortunately not characterized by historical data of reliability and continuous operation under harsh conditions. Another drawback of FC systems is the need for expensive noble materials and susceptibilities to the contaminants present in the fuel.

Hydrogen is considered to be the most efficient FC fuel, but it is still not freely available in nature. It has to be manufactured, and the full cycle of efficiency and cost must make sense. Therefore, the market for FCs is currently limited to special applications. On the other hand, hydrogen will not be widely available until there is an increase in demand with a corresponding economy of scale where the full cycle of efficiency sums up to a steady market. Although there are still some pessimist objections to adoption of a hydrogen economy, it is expected that eventually it will become a reality.

7.9 Fuel Cell Equivalent Circuit

Evaluation of the dynamic performance of FCs for studies of electrical energy generation systems is important to reduce cost and time at the design and testing stages. An electrical model may be derived from the electrochemical equations to enable determination of the open-circuit voltage and voltage drops of cells for a specified operating point [11–16]. In power generation systems, the dynamic response is of extreme importance for the control planner and system management, especially when energy is injected into the grid. So special attention has to be given to the dynamic response of FCs [1, 4, 5]. For energy injection into the grid, the generation control has to set the amount of power the FC will supply as a function of the load demand. As such, the dynamic FC response should be compatible with a fast variation in the random load curve, which is not always the case, as we discuss next [11–14]. From the reaction outlined in equations (7.1), (7.2), and (7.3), it is possible to obtain the electrical voltage generated in the electrochemical process in the cell if one recalls that two electrons pass through the external circuit for each water molecule produced and each molecule of hydrogen present in the process. So electrical work to move these charges is expressed as

$$G = \int 2qde = -2FE \tag{7.4}$$

where:

F = The Faraday constant (or the electron charge in every molecules)
E = The FC voltage

If the system is reversible (no losses), the electrical work is equal to the Gibbs free energy released, which is the "energy available to do external work, neglecting any work done by changes in pressure and/or volume." All these forms of chemical energy are rather like ordinary potential energy with respect to the zero-point energy and energy variation with respect to this point. The Gibbs energy is listed in the literature [1] for the reaction of water formation from $2H_2$ and O_2 as in equation (7.3). For example, for a hydrogen cell operating at 200 °C, the Gibbs energy is 220 kJ, and from equation (7.24), $E = 1.14$ V.

The activity of the reactants and products changes the Gibbs free energy of a reaction. Balmer [17] showed in 1990 that temperature and pressure do affect the reaction activity, resulting in an electromotive force given in terms of the product and/or reactant activity, called the Nernst reversible voltage, E_{Nernst}.

Therefore, reversible cell voltage is the cell potential obtained in an open-circuit thermodynamic balance (without load). In this section, E_{Nernst} is calculated from a modified version of Nernst's equation, with two extra terms to account for changes in temperature with respect to the standard reference temperature, 25 °C, and 100 kPa or 1.00 atm pressure [13–15], respectively. This is all given by

$$E_{Nernst} = \frac{\Delta G}{2F} + \frac{\Delta S}{2F}(T - T_{ref}) + \frac{RT}{2F}\left[ln\left(p_{H_2}^{\bullet}\right) + \frac{1}{2}ln\left(p_{O_2}^{\bullet}\right) \right] \tag{7.5}$$

where:

ΔG = The change in the free Gibbs energy (J/mol)
F = The constant of Faraday (96,487 C)
ΔS = The change of entropy (J/mol)
R = The universal constant of gases (8314 J/K· mol)
$p_{H_2}^{\bullet}$ and $p_{O_2}^{\bullet}$ = The partial pressures (atm) of the hydrogen and oxygen, respectively
T = The absolute temperature of the operating cell (K)
T_{ref} = The reference absolute temperature (K)

Using the previous standard temperature and pressure values for ΔG, ΔS, and T_{ref}, equation (7.5) can be simplified to [13–17]

$$E_{Nernst} = 1.229 - 0.85 \times 10^{-3}(T - 298.15) + 4.31 \times 10^{-5} T\left[ln\left(p_{H_2}^{\bullet}\right) + \frac{1}{2}ln\left(p_{O_2}^{\bullet}\right) \right] \tag{7.6}$$

The ohmic overpotential results from the resistance to the electron transfer in the collecting plates and carbon electrodes and the resistance to the proton transfer in the solid membrane. This resistance is essentially linear and dependent on the membrane resistivity and temperature. In this model, a general expression for the resistance was defined to include all the important parameters of the membrane. The equivalent resistance of the membrane is then calculated by

$$R_m = \rho_m \frac{\ell}{A} \qquad (7.7)$$

where:

ρ_m = The specific resistivity of the membrane for the flow of electrons ($\Omega \cdot$ m)
ℓ = The thickness of the membrane serving as cell electrolyte (μm)
A = The cell active area (cm^2 = 10^{-4} m^2)

The Nafion membrane type is a registered trademark of DuPont and broadly used in PEMFC technology. This book considers the available parameters for a Nafion membrane. DuPont uses the product designations shown in Table 7.3 to denote the Nafion membrane thickness. The following numeric expression for the resistivity of the Nafion membranes can be used:

$$\rho_m = \frac{181.6 \left[1 + 0.03 \left(\dfrac{i_{fc}}{A} \right) + 0.062 \left(\dfrac{T}{303.15} \right)^2 \left(\dfrac{i_{fc}}{A} \right)^{2.5} \right]}{\left[\psi - 0.634 - 3 \left(\dfrac{i_{fc}}{A} \right) \right] \cdot e^{\left[4.18(T-303.15)/T \right]}} \qquad (7.8)$$

where:

$181.6/(\psi - 0.634)$ = The specific resistivity ($\Omega \cdot$ cm) at no load current and 30 °C
T = The absolute temperature of the cell (K)
ψ = The average water content, which is an adjustable parameter with a possible maximum value of 23
$e^{[4.18(T-303.15)/T]}$ = A temperature factor correction if the cell is not at 30 °C

The parameter ψ is influenced by the membrane preparation procedure and is a function of the relative humidity and stoichiometric rate of the anode gas. It can have a value of about 14 under the ideal conditions of 100% relative humidity, and values approximately 22 and 23 have been reported under over-saturated conditions.

Table 7.3 DuPont designations for Nafion membrane thickness.

DuPont designation	Thickness
Nafion 117	7 mils (178 µm)
Nafion 115	5 mils (127 µm)
Nafion 112	2 mils (51 µm)

Equation (7.8) may be extended to obtain the ohmic overpotential of the membrane, electrodes, and contact resistances as

$$V_r = i_{fc}\left(R_m + R_c\right) \tag{7.9}$$

where:

R_c = The resistance of the electrodes and contacts to the ion transfer through the membrane (electrolyte), usually considered constant

The concentration or mass transport affects the hydrogen and oxygen concentrations. This, in turn, causes a decrease in the partial pressures of these gases. Reduction in the pressure of oxygen and hydrogen depends on the electrical current and physical characteristics of the system. To determine an equation for this voltage drop, it is defined as the maximum current density, J_{max}, under which the fuel is being used at the same maximum supply rate. The current density cannot surpass this limit because the fuel cannot be supplied at a greater rate. Typical values for J_{max} are in the range 1000–1500 mA/cm². Thus, the voltage drop due to mass transport is

$$V_{con} = -B \cdot ln\left(1 - \frac{J}{J_{max}}\right) \tag{7.10}$$

where:

B = A constant depending on the cell and its operating state (V)
J = The actual current density of the cell electrode (A/cm²)

As shown in [15–18], the electrode activation overpotential, including anode and cathode, can be calculated by

$$V_{act} = -\left[\xi_1 + \xi_2 T + \xi_3 Tln\left(c_{O_2}^*\right) + \xi_4 Tln\left(i_{fc}\right)\right] \tag{7.11}$$

where:

ξ_i's = The parametric coefficients for each cell model ($i = 1,2,3,4$)
i_{fc} = The cell operating current (A)
$c_{O_2}^* = p_{O_2}^* / 5.08 \times e^{-498/T}$ = The oxygen concentration on the cathode catalytic interface (mol/cm³)

The values used in equation (7.11) are set by theoretical equations with kinetic, thermodynamic, and electrochemical foundations [15–18].

The resistance R_a is determined from steady-state evaluation of the cell current and the activation and concentration voltages. That is, after the end of the transient, there will be stable final values of voltage and current, which are used to calculate the resistance through [18–21]

$$R_a = \frac{V_{act} + V_{con}}{i_{fc}} \tag{7.12}$$

Before setting up an equivalent circuit to represent the cell dynamics, it is interesting to examine in detail the phenomenon known as the charge double layer. Such a phenomenon normally exists on every contact between two different materials due to a charge accumulation on the opposite surfaces or charge transfer from one to the other. The charge layer on both electrode–electrolyte interfaces (or close to the interface) is the storage of electrical charges and energy; in this way, it behaves as an electrical capacitor. If the current changes, there will be some elapsed time for the load (and its associated voltage) to decay (if the current decreases) or to increase (if the current increases). Such a delay affects the activation and concentration potentials. It is important to point out that the ohmic overpotential is not affected, since this has a linear relationship with the cell current through Ohm's law. Thus, a change in the current causes an immediate change in the ohmic voltage drop. In this way, it can be considered that a delay of first order exists due to the activation and concentration voltages only. The associated time delay t(s) is the product [18–21]

$$\tau = CR_a \tag{7.13}$$

where:

C = The equivalent capacitance of the FC (farads)
R_a = The equivalent resistance to the activation and concentration losses (ohms)

The value of the capacitance C is some few farads. In broad terms, this capacitive effect assures the good dynamic performance of the cell, because the voltage moves smoothly to a new value in response to a change in the load current. The electrical output energy of the cell is linked to a certain load, such as the load represented in Figure 7.5. There is no restriction with respect to load type as long as the power supplied by the stack is capable of feeding it or if it does not represent starting motors and faster transient response loads. For example, in systems for injection of energy into a distribution network, the load can be a dc–dc boost converter, followed by a dc–ac inverter, connected to the public network through a transformer. In stand-alone systems, it can be a pure resistive load (heating) or a resistive-inductive load (motor) [22–25]. In any case, the average current density J (A/cm^2) through the cell cross-sectional area is defined as

$$J = \frac{i_{cell}}{A} \tag{7.14}$$

Table 7.4 Typical parameters of a Ballard-Mark-V fuel cell.

Parameter	Value	Parameter	Value
T	343.15 K	ξ_1	−0.948
A	50.6 cm^2	ξ_2	$0.00286 + 0.0002 \ln A + (4.3 \times 10^{-5}) \ln C_H^*$
ℓ	178 μm	ξ_3	$7.6 \cdot 10^{-5}$
$\dot{c}_{O_2}$	1.10^{-4} mol/cm^3	ξ_4	$-1.93 \cdot 10^{-4}$
$\dot{c}_{H_2}$	1.10^{-4} mol/cm^3	λ	23.0
$\dot{p}_{O_2}$	1 atm	B	0.016 V
$\dot{p}_{H_2}$	1 atm	J_{max}	1500 mA/cm^2

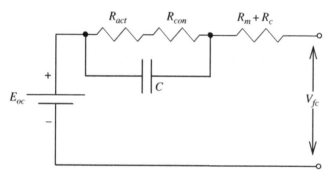

Figure 7.7 Equivalent circuit of the dynamic behavior of a fuel cell.

The instantaneous electrical power supplied by the cell to the load can be determined by

$$P_{cell} = V_{cell} I_{cell} \tag{7.15}$$

where:

V_{cell} = The cell output voltage for each operating condition (volts)
P_{cell} = The corresponding power (watts)

In Table 7.4 values of the parameters discussed in this section for a typical Ballard-Mark-V FC are listed.

The equations derived in this section allow the construction of an equivalent circuit model as shown in Figure 7.7 to represent FC dynamic behavior as given in many references [2, 13, 14, 16–18]. Equation (7.6) represents the active source E_{Nernst}. The membrane and contact ohmic losses are represented by $R_r = R_m + R_c$ (equation (7.9)). The time delay given in equation (7.8) includes the capacitor, C, to represent the double-charge-layer effects on the output voltage and current. This capacitance is very large (a few farads), since it is directly proportional to the electrode area and the manufacturing material and it is inversely proportional to the electrode thickness, which is numerically very small (a few nanometers).

The activation and concentration losses (mass transport) cause a voltage drop across the resistor represented by R_a (equation (7.12)). The fuel crossover and internal currents are essentially small, equivalent, and less important than other losses in terms of the cell operating efficiency. However, it has a very notable effect on the open-circuit voltage of low-temperature cells. Therefore, from Figure 7.7, the output voltage of a single cell can be defined as the result of the expression

$$V_{fc} = E_{Nernst} - V_{act} - V_{con} - V_r \tag{7.16}$$

Typical values for the components used in the circuit model of Figure 7.8 may vary widely with the size, construction, and type of FC, but as an example, the following parameters could be used: $E_{Nernst} = 1.48V$ and $C = 3F$; R_a is a function of i_{cell} as given in equations (7.10) and (7.16).

The first term appearing in equation (7.36) represents the open-circuit voltage of the FC (operating voltage without load); the last three terms represent voltage drops to the net voltage of the cell, V_{fc}, at a certain operating current. The term E_{Nernst} is the thermodynamic potential of the cell and represents its reversible voltage. V_{act} is the voltage drop due to the anode and cathode activation overpotential, a measure of the voltage drop associated with the electrodes. V_{con} represents the voltage drop resulting from the concentration or mass transportation of oxygen and hydrogen (concentration overpotential). Finally, V_r is the ohmic voltage drop (ohmic overpotential), a measure of the ohmic losses associated with the conduction of protons through the solid electrolyte and internal electronic resistances [8]. Many other models can be found in the literature [14, 15, 19–22].

Figure 7.8 includes performance curves for a typical membrane used in FCs, which may be superimposed on the theoretical curves described by equations (7.15) and (7.16), respectively, for P_{fc} and V_{fc}. The efficiency η is defined as the relationship between the electric output power and the energy corresponding to the fuel input [3]. For a water product in liquid form, it is usually given as

$$\eta = \mu_f \frac{V_c}{1.48} \tag{7.17}$$

For a water product in vapor form, the efficiency is generally given as

$$\eta = \mu_f \frac{V_c}{1.25} \tag{7.18}$$

where:

μ_f = The fuel utilization factor, generally about 95% and 1.48 and 1.25 V are the maximum voltages that can be obtained using the largest and smallest values, respectively, of the cell enthalpy

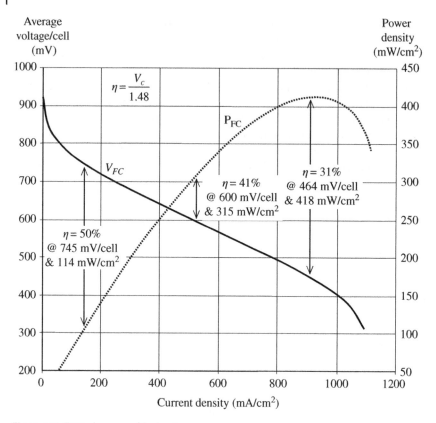

Figure 7.8 Typical curves of fuel cells.

7.10 Water, Air, and Heat Management

Optimum operation management of FCs is vital for maximum profits in this energy system in financial and energy use terms. The products to deal with are water, air, and heat, which are closely related to FCs in three aspects: (i) as a by-product of chemical reactions, (ii) in the cooling circulation fluid, and (iii) for heat recovery. The FC heat affects especially the speed of electrochemical reactions because it eases hydrogen extraction from hydrocarbonates and increases the overall efficiency of the FC by means of heat recycling for both ambient and water heat and/or for heating boilers used as primary energy sources for turbine drivers of electrical generators and for hot water storage. As the heat is produced locally, the FC generating units may increase their overall efficiency using internally produced electricity and heat. Local production of energy with heat integration decreases the transportation and distribution losses and alleviates the public electric grid from problems associated with concentrated power plants; those are usually located at very long distances from consumers [3, 5].

The water resulting from chemical reactions in FCs must be removed only partially since its presence increases the conductivity of the electrolyte. Conversely, if there is too much water in the electrodes bonding the electrolyte, pore flooding will cause difficulties for fuel permeation, and as a result, the concentration losses might be higher. A good relative humidity should be maintained between 85 and 100% for a reasonable operating balance. In some small FCs the air going through the cell may be the same for both oxygen reactions and moisture balance (stoichiometric feeding is usually at a rate higher than 2). For larger cells there are usually two independent air supplies, and the moisture may have to be complemented by an external vapor source. Use of these quantities may be expressed as given in Ref. [3]:

$$Air\ rate = 3.6 \times 10^{-7} \lambda \frac{P_e}{V_{fc}} \quad (kg/s) \tag{7.19}$$

$$H_2\ rate = 1.05 \times 10^{-7} \frac{P_e}{V_{fc}} \quad (kg/s) \tag{7.20}$$

$$H_2O\ production\ rate = 1.33 \times 10^{-4} \quad (kg/s) \tag{7.21}$$

The membrane relative humidity may be given in terms of the partial pressure of the water and saturated water vapor pressures as

$$RH(\%) = \frac{P_w}{P_{sat}}$$

7.10.1 Fuel Cells and Their Thermal Energy Evaluation

The chemical reactions in an FC produce heat as a result of the conversion enthalpy of a fuel into electricity. Therefore, the electrical efficiency can be defined as

$$\eta_e = \frac{P_e}{P_t} \tag{7.22}$$

where:

P_t = The total power converted by the FC stack into electric power P_e and heating rate $\dot{Q}_{fc}$

The total power P_t is obtained from

$$P_t = P_e + \dot{Q}_{fc} \tag{7.23}$$

where:

$\dot{Q}_{fc}$ = The thermal energy rate produced in the stack

The total amount of heat generated by the chemical reactions is expressed in equation (7.17) by

$$\dot{Q}_{fc} = P_t - P_e = \frac{P_e}{\eta_e} - P_e = P_e \left(\frac{1}{\eta_e} - 1 \right)$$

that is,

$$\dot{Q}_{fc} = P_e \left(\frac{1.48}{V_c} - 1 \right) \tag{7.24}$$

This heat has to be removed from the FC by a heat sink:

$$\dot{Q}_{rem} = \eta_{hs} \dot{Q}_{fc}$$

where:

η_{hs} = The heat sink efficiency

As a result, the temperature in the stack will drop at a rate:

$$\frac{dT}{dt} = \frac{\dot{Q}_{fc} - \dot{Q}_{rem}}{c \cdot m} = \frac{\dot{Q}_{fc} - \dot{Q}_{rem}}{C}$$

where:

c = The specific heat
m = The considered mass

Only part of this heat is useful and expressed as

$$\dot{Q}_u = \eta_{th} \dot{Q}_{fc}$$

The total FC efficiency then is given by

$$\eta_t = \frac{P_e + \dot{Q}_u}{P_e + \dot{Q}_{fc}} = \frac{P_e + \eta_{th} \dot{Q}_{fc}}{P_e + \dot{Q}_{fc}} \tag{7.25}$$

Again, from equation (7.24), the thermal energy produced by the FC with power equal to P_{fc} (kW) working during h_{fc} hours may be expressed as

$$Q_{fc} = P_e \left(\frac{1.48}{V_s} - 1 \right) h_{fc} \quad (\text{kWh}) \tag{7.26}$$

The stack heat losses can be separated into three categories: the rate of heat removed by the water cooling system Q_w, the rate of heat loss in the air flowing out of the stack Q_a, and other losses Q_{other} (mostly stack heating and surface heat exchange with the surroundings). The total heat loss is given then by

$$Q_{fc} = Q_w + Q_a + Q_{others} \tag{7.27}$$

It is common practice to consider the amount of heat removed by the air in a stack to be about 2% of P_t and in ambient losses, 18% [16]:

$$Q_w = P_e\left(\frac{1.48}{V_s} - 1\right)nh_{fc} - Q_a - Q_{others} \tag{7.28}$$

An interesting case to be considered is water storage during the early hours of the day using the heat dissipated by the FC between midnight and 06:00 p.m. An FC operating in this period allows enough heat dissipation in water for household applications, such as those in the kitchen and bathroom in addition to air heating. The needs of a family can be accounted for on a per-person basis, that is, assuming that each person uses an average of 100 L/day of hot water, so a family with n_p people will need an average flow of $100n_p$ L/day [16]. Based on these data, the necessary energy to heat up the required amount of water is calculated as

$$Q = cm\Delta T \tag{7.29}$$

where:

$c = $ The specific heat of water (1.163×10^{-3} kWh/kg-°C)
$m = $ The mass of water in ℓ
$\Delta T = $ The difference (°C) or absolute temperature between the initial and final temperature in the process

The thermal energy needed to heat up mn_p liters of water for n_p people is

$$Q = mn_p c\Delta T \tag{7.30}$$

The output power of an FC to heat up an amount of water (liters) during h_{fc} hours is obtained from equations (7.26), (7.27), and (7.30):

$$P_{fc} = \frac{\left(mn_p c\Delta T + Q_a + Q_{other}\right)}{\left(1.48 - V_{fc}\right)h_{fc}}V_{fc} \quad (\text{kW}) \tag{7.31}$$

Based on the equations described in this section, it is possible to control the stack electric efficiency in order to obtain more thermal or electrical energy according to the appliance demand. Therefore, two operating modes may be selected for the FC output power: variable power or constant power. Variable power aims at a constant load, and constant power aims at an optimal FC operating point for best operation and hydrogen consumption. In the general case, different operation regimes can be established between the storage and generation of energy. The electrolysis process can be used for these purposes in any off period or along shorter periods, in accord with the needs of the utility or the consumer. To evaluate operation at variable power, an iterative process has to be established. As already mentioned, due to losses in the electrolysis and in power generation with FCs, the total daily consumption of energy will increase, although now shifted to a more convenient period.

If the water product is in a liquid form (without the latent water heat) from equation (7.17), the heat production may be approximated by

$$Heating\ rate = P_e\left(\frac{1.48}{V_s} - 1\right) \quad \left(J/s\ or\ W\right) \tag{7.32}$$

If the water product is in vapor form, the heat production may be approximated by

$$Heating\ rate = P_{fc}\left(\frac{1.25}{V_{fc}} - 1\right) \quad \left(J/s\ or\ W\right) \tag{7.33}$$

Either air heat from the cooling process or hot water from the chemical reactions may be used in heat exchangers for commercial, industrial, and residential purposes and warming up vapor and/or air conditioning in domestic homes.

7.11 Experimental Evaluation of the Fuel Cell Equivalent Model Parameters

Several techniques have been used for practical parameter determination of the FC equivalent circuit models. A possible method is the electrical impedance spectroscopy in which a variable frequency ac is applied through the cell, causing a voltage drop across its terminals. The cell equivalent impedance is derived from the relationship between this voltage and the applied ac. The frequencies used in these tests may be as low as $10\,mHz$.

Another very simple method to obtain the equivalent model parameters is the current interruption technique. In this method, the concentration or mass transport overvoltage has to be neglected. Assume that a load is being fed across the FC terminals with V_L volts and I_L amperes. An interruption in the FC load current will cause a step voltage increase V_r across its terminals. This is measured through a data acquisition system or a storage oscilloscope to show the step of voltage followed by a capacitor charge-like response, as illustrated in Figure 7.9. Therefore, there is a need not only for instantaneous

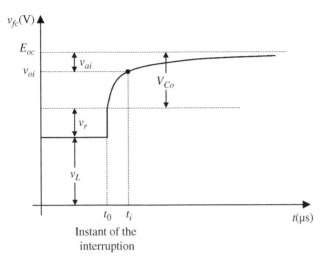

Figure 7.9 Typical FC characteristic for the current interruption test.

switching but also for fast sampling of the voltage increase to enable a clear separation of the activation overvoltage and the overvoltage due to the ohmic losses. This can be interpreted as if at the exact instant of the current interruption, the voltage across the charge double layer capacitance does not immediately change [21–24]. After some time, the FC output voltage would tend to $E_{oc} = V_a + V_r + V_L$. Data extrapolation may be necessary to intercept the vertical line of the exact instant when there was the current interruption. As zero current is assumed soon after the load current interruption, there will solely happen a capacitor discharge through R_a. The following expressions could be used to express these changes:

$$R_r = \frac{V_r}{I_L} \tag{7.34}$$

$$R_a + R_r = \frac{E_{oc} - V_L}{I_L} \tag{7.35}$$

Notice that for a given electrode area, R_a is dependent on the load current, temperature, and specific resistivity. So, the initial voltage across the double layer capacitance $V_{Co} = I_L R_a = E_{oc} - V_L - V_r$ at the very instant the load current was interrupted. This voltage may be considered the initial discharge voltage of a capacitor through the variable resistance R_a. Therefore, the initial activation voltage drop may be approximated as

$$v_{ai}(t) = V_{Co} e^{-(t_i - t_0)/R_a C} = E_{oc} - v_{Li} \quad \text{for } t_i \geq t_0 \tag{7.36}$$

This expression can be used to create a generic time delay at the instant t_i soon after the load current was interrupted:

$$R_{ai}C = \frac{t_0 - t_i}{\ell n\left(\dfrac{v_{ai}}{V_{Co}}\right)} \tag{7.37}$$

where:

V_{Co} = The initial charge voltage across capacitance C

From equation (7.36), the tangent slope at the exact current interruption instant $t_1 \to t_0$ is

$$\left.\frac{dv_a}{dt}\right|_{t \to t_0} = -\frac{V_{Co}}{R_{a0}C} \tag{7.38}$$

From the incremental value of equation (7.38) rearranged, then

$$C = \frac{V_{Co}}{R_{a0}\left.\dfrac{\Delta v_{ai}}{\Delta t}\right|_{\Delta t \to 0}} \tag{7.39}$$

An interactive calculation of successive values of C will give a better approximated average for this constant capacitance.

Taking Figure 7.7 as the equivalent dynamic model of an FC, the following steps may be used to determine these circuit parameters. The best representation for it depends on the precise detection of the voltage jump V_r at the exact instant of the current interruption, which demands for a high-speed data acquisition system or storage oscilloscope. Calculations of the previous parameters can be made through the following procedure:

1) Make the current interruption test and plot the current interruption characteristic.
2) Obtain on this plot the following values at the current interruption instant: E_{oc}, V_{Co}, V_r, V_L, and measured I_L.
3) At the interruption test, the membrane R_m plus contact resistance R_c is $R_r = V_r/I_L$, which does have current through it and has to be calculated apart at every instant t_i of the voltage discharge by

$$R_{ri} = R_c + R_{mi} = \frac{E_{oc} - V_{Ci} - V_{Li}}{I_{Li}}$$

In a way, the value of R_{ri} will be constant for a given load current or even going to vary according to the load current.

4) At this instant the activation resistance becomes

$$R_{a0} = \frac{E_{oc} - V_L}{I_L} - R_r$$

5) The initial capacitance voltage is $V_{Co} = I_L R_{a0}$.
6) From equation (7.36), $v_{ai} = E_{oc} - v_{Li}$.
   Assuming a constant and large value of C, for a time instant as close as possible to the current interruption instant, from equation (7.38),

$$-\frac{dv_{ai}}{dt}\bigg|_{t \to t_0} = \frac{V_{Co}}{R_{a0}C} = \frac{I_L}{C} \cong -\frac{\Delta v_{ai}}{\Delta t}\bigg|_{t \to t_0} = \frac{\Delta v_{Li}}{\Delta t}\bigg|_{t \to t_0} \tag{7.40}$$

and from equation (7.39)

$$C \cong \frac{I_L}{\dfrac{\Delta v_{Li}}{\Delta t}\bigg|_{t \to t_0}} \quad \text{for } i \to 1 \tag{7.41}$$

From equation (7.37),

$$R_{ai} = \frac{t_0 - t_i}{C \cdot \ell n\left(\dfrac{E_{oc} - v_{Li}}{V_{Co}}\right)} \quad \text{for } i \geq 1 \tag{7.42}$$

7.11.1 Determination of FC Parameters

Using a 32-cell PEM stack with an effective membrane area $A = 100\text{cm}^2$, whose current interruption characteristic is given in Figure 7.10 and tested to obtain the stack parameters at $J_L = 520\,\text{mA/cm}^2$.

Step 1. Using a digital storage oscilloscope, the current interruption characteristic of the FC was plotted as indicated in Figure 7.10.
Step 2. With the plot in Figure 7.11, obtain E_{oc}, V_{Co}, V_r, V_L, and measured I_L:

$$E_{oc} = 29.148\text{V}$$
$$V_{Co} = E_{oc} - v_{o1} = 29.148 - 22.273 = 6.875\text{V}$$
$$V_r = 22.273 - 16.269 = 6.004\text{V}$$
$$V_L = 16.269\text{V}$$
$$I_L = 100\text{cm}^2 \cdot 520\text{mA/cm}^2 = 52\text{A}$$

Step 3. Assuming uniformly distributed current through the FC membrane, $R_r = V_r/I_L = 6.004/52 = 0.1455\Omega$. This value can be compared with the one obtained from calculation with equations (7.7) and (7.8) not taking into

$t(s)$	$v_{oi}(V)$
0.0000	16.269
0.2000	16.269
0.4000	16.269
0.6000	16.269
0.7356	16.269
0.7357	–
0.7405	24.035
0.7455	24.348
0.7520	24.753
0.7670	25.601
0.7990	26.976
0.8740	28.758
1.0000	29.757
1.2000	30.016
1.6000	30.040

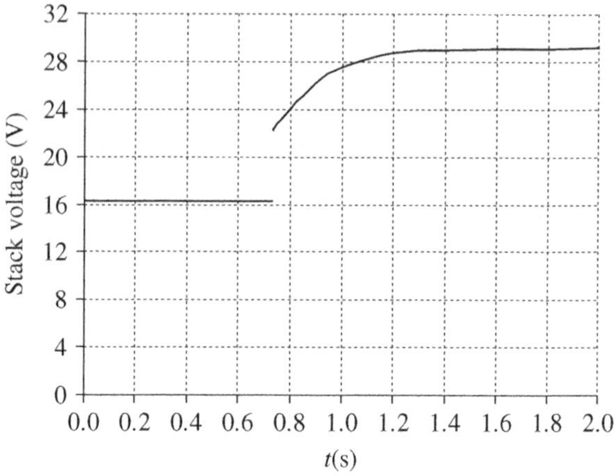

Figure 7.10 Plotting of the FC current interruption test for the circuit in Figure 7.8.

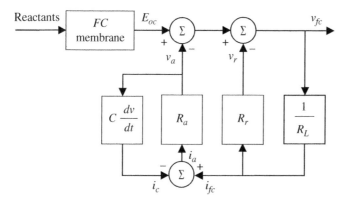

Figure 7.11 Flow diagram calculation of FC output voltage.

account the contact resistance. The small differences to be corrected are due to the membrane thickness ($\ell = 0.178\mu m$), unevenly distributed current through the membrane, and effects of the test temperature conditions ($T = 30°C$).

Step 4. $R_{a0} = \dfrac{E_{oc} - V_L}{I_L} - R_r = \dfrac{29.1480 - 16.2690}{52} - 0.1455 = 0.1028\Omega$

Step 5. The current interruption happened at $t_0 = 0.7334s$ when it was registered:

$$V_{Co} = I_L R_{a0} = 52 \times 0.1028 = 5.3456V$$

Step 6. Assuming a constant and large value of C for an instant very close to the current interruption time (next instant), from equation (7.41),

$$C \cong \frac{I_L}{\left.\dfrac{\Delta v_{Li}}{\Delta t}\right|_{t \to t_0}} = \frac{52}{\left.\dfrac{22.390 - 22.273}{0.7336 - 0.7335}\right|_{t \to t_0}} = 44.44 \, \text{mF}$$

Step 7. From the digital oscilloscope data, at $t_1 = 0.7335$s and $t_2 = 0.7336$s from the beginning of the test, the respective values of the output voltage were $v_{L1} = 22.273$V and $v_{L2} = 22.390$V. Therefore, the activation voltages $v_{ai} = E_{oc} - v_{oi}$ were

$$v_{a1} = E_{oc} - v_{L1} = 29.148 - 22.273 = 6.875\text{V}$$
$$v_{a2} = E_{oc} - v_{L2} = 29.148 - 22.390 = 6.758\text{V}$$

From equation (7.42) it is possible to calculate the variation of $R_{ai}(t)$ as a function of time:

$t_i(s)$	$v_{Li}(V)$	R_{ai}
0.7335	22.27300	0.015229
0.7336	22.27582	0.007971
0.7337	22.27800	0.007611
0.7400	22.45100	0.005541
0.8000	24.05100	0.004914
0.9000	26.29990	0.004146
1.0000	27.59000	0.003894
1.2000	28.79500	0.003164
1.4000	29.00510	0.002849
1.6000	29.03400	0.002781
1.8000	29.04000	0.002158
2.0000	29.14800	—

From equations (7.7) and (7.8),

$$\rho_m = \frac{181.6 \left[1 + 0.03(0.52) + 0.062 \left(\dfrac{353.15}{303.15} \right)^2 (0.52)^{2.5} \right]}{\left[23 - 0.634 - 3(0.52) \right] \cdot e^{\left[4.18(353.15 - 303.15)/353.15 \right]}} = 0.0017533 \Omega \text{m}$$

$$R_m = 4.9841 \times \frac{178 \times 10^{-8}}{0.00506} = 0.0017533 \Omega$$

Finally, from equation (7.9) and Step 3,

$$R_c = R_r - R_m = 0.145500 - 0.001753 = 0.143745 \Omega$$

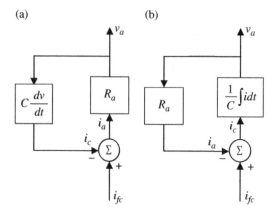

Figure 7.12 Options of equivalent circuits to represent a charge double layer: (a) solution by derivative and (b) solution by integration.

Several computer softwares can be used for solving or simulating FC electrical equivalent circuits, such as MATLAB®/Simulink®, PSIM, or PSpice using a computation flow as in the graph illustrated in Figure 7.11. In this calculation, the internal voltage drops are calculated from the FC internal parameters measured in laboratory tests. Either solutions suggested in Figure 7.12 could be used to obtain the voltage across the equivalent capacitance of the charge double layer [18, 19, 26–28]. The solution in Figure 7.12a is preferred since it is very easy to calculate the initial value of i_a assumed soon after any constant state condition prior to the desired calculation interval:

$$i_{ao} = I_L = \frac{V_{Co}}{R_{ao}} = \frac{V_{oc} - (V_L + V_r)}{R_{ao}}$$

7.12 Aspects of Hydrogen as Fuel

Hydrogen is the lightest and most buoyant element, so if it is released into an open space, it disperses quickly, reducing dramatically the chance of ignition. In general, for every 96 parts of air, at least four parts must be hydrogen before there is a threat of combustion. This is actually quite a high concentration relative to other commonly used fuels. For this reason, it is good to know the conditions under which this happens based on such properties as ignition energy, ease of flotation, diffusivity, flammability limits in the air, and combustion energy.

The combustion energy of the hydrogen is very low, which makes it even easier to catch fire, being in the ratio 1:10 for gasoline and 1:15 for natural gas or propane. All these gases are in fact of very low ignition energy, such that the probability of ignition of a mixture of any of them is relatively high even for a weak ignition source. Comparable examples include mixtures of 4–75% of hydrogen in the air, 5–16% of natural gas in the air, and 1.4–7.6% of gasoline steam in the air.

Hydrogen possesses 2.4 times more combustion energy stored per unit of mass than natural gas or gasoline. In volumetric terms, hydrogen has much less energy: it has 25% of the explosion energy of natural gas and 0.3% of the liquid gasoline per unit of volume under normal conditions of temperature and pressure. The amount of energy stored in small systems of hydrogen is usually less than that corresponding to 4 L of gasoline.

Advantages of hydrogen include its high diffusivity and flotation capacity. Such properties help to avoid fuel mixtures, and when this happens, they last for shorter times. Hydrogen is four times more diffusive than natural gas and eight times more diffusive than gasoline.

The first strategy used to prevent hydrogen from starting a fire is based on reducing the possibility of forming a combustible mixture, which can be accomplished using very well-sealed canalization to avoid leakage. When one of these happens, the hydrogen will be dispersed quickly unless it is contained. The second strategy would be an ambient well ventilated to reduce or, in some cases, to eliminate the area in which a combustible mixture may develop. In the case of leaks, this reduces the exposure time to the possibility of developing an eventual combustible mixture. The third strategy is the minimization of near-ignition sources such as static discharges, open fires, hot surfaces (temperatures higher than 585 °C), and other equipment able to produce sparks.

The flammability of the hydrogen is the greatest danger in its use. However, the problem may be limited by storage of only small volumes. The most dangerous and susceptible parts refer to canalization for the gas, electrical equipment, the FC itself, the control system, and the system reformer. A control system should be used to establish safe operation of the power plant. If it extrapolates the safety limits, the control system should turn off the power plant, switching alarms on and possibly informing operators regarding details of the dangerous conditions. The control system should also minimize dangerous situations due to single, multiple, or simultaneous failure of components.

The efficiency of an FC is a function of the operating voltage of the entire stack. Higher voltages produce higher efficiency and therefore less consumption of hydrogen per kilowatt-hour, but less output power is supplied. For a better trade-off, the choices for a given output power are between (i) operating at higher voltages, increasing the number of cells in the stack and thus the capital costs, and (ii) operating at higher densities of current with fewer cells, higher fuel costs, and a shorter useful life.

FC and hydrogen costs, together with stack efficiency, will determine the final costs ($/kWh). The current technology, shown in Figure 7.4, displays the performance of an individual cell belonging to a PEM stack with 70 cells, providing of 33% efficiency in generating electric power at a cost of $0.23/kWh based on a hydrogen cost of $12/GJ. A higher fuel cost would require a stack with more cells operating at higher efficiency to minimize the total cost.

7.13 Load Curve Peak Shaving with Fuel Cells

The strategies presented in this section allow specification of the power rating for an FC stack system under the approach of imposing a load curve to be as flat as possible. FC sizing can be based on the maximal load curve flatness, limiting the maximum load peak or the amount of the necessary thermal energy. Other methodologies may also be used for FC sizing, depending on operating mode and how hydrogen is going to be produced. In principle, the FC can operate at constant or variable power during a selected period, depending on the amount of hydrogen available. For each situation, a dedicated procedure should be applied [15–17, 29, 30].

7.13.1 Maximal Load Curve Flatness at Constant Output Power

The methodology of maximum load curve flatness at constant output power allows calculation under a set of constraints. The major consumer interest is to decrease demand during the peak period so as to reduce the cost of the energy consumed. On the other side, from a utility company point of view, there is, in addition to reduction in the maximum demand, a possibility of increasing the efficiency of the distribution network, due, for example, to a decrease in overall losses. The optimization rationale is based on limitations of load curve changes. Load can only be decreased during the load peak period (say, 3 hours) and can increase with the use of an electrolyzer only during nightly runs (say, a 6-hour period). With this purpose, flatness of the load curve can be quantified by a form factor defined as equation (7.43) [16]:

$$k_f = \frac{P_{qav}}{P_{av}} \tag{7.43}$$

where:

P_{qav} = The quadratic average of the instantaneous powers, defined as

$$P_{qav} = \frac{1}{n}\sqrt{\sum_{t=1}^{n} P_t^2}, \quad P_{av} = \frac{1}{n}\sum_{t=1}^{n} P_t$$

where:

P_t = The transformer load at the time interval t
n = The number of the intervals considered

The factor k_f is always higher than 1. If $k_f = 1$, the load curve would have an ideal flat shape. For the case of FC usage,

$$P_{qav} = \frac{1}{n}\sqrt{\sum_{t=1}^{n}\left(P_t - P_{fc}\right)^2} \qquad (7.44)$$

where:

P_e = The output power of the FC stack

The objective, in this case, is to determine P_{fc} to minimize the function $z = k_f - 1$, aiming at determination of the FC power output to make the load curve as flat as possible. The value of P_e can be determined by the solution of

$$\frac{\partial z}{\partial P_e(t)} = 0 \qquad (7.45)$$

Current converters to feed electrolyzers are of the current source-controlled converter (CSCC) types. However, they present special characteristics of voltage and current. The rectifier current capacity depends on the amount of gas flow required. In the same way, it is possible to determine the power of the electrolyzer for production of hydrogen out of the peak period to obtain a factor k_f as close as possible to 1. The specific electrical power that can be obtained from a hydrogen-fed FC is $26.6 \cdot V_c$ kWh/kg, where V_c is the output voltage across the terminals of every single cell belonging to the FC stack [3, 16]. This calculation uses the standard FC electrochemical model discussed in Section 7.11. The hydrogen density in the FC is 0.084 kg/m^3 (NTP), so the amount of H_2 (m^3) necessary to generate the right amount of energy is determined from

$$Volume \ of \ H_2 = 0.44755 \cdot \frac{P_e h_{fc}}{V_c} \quad \left(Nm^3\right) \qquad (7.46)$$

where:

P_e = The FC rated power
h_{fc} = The number of operating hours within the peak period

One Faraday ($96,489$ C) produces 1 g of hydrogen (ideal case), that is, $8,105,076$ As/m^3 or 2251.41 Ah/m^3. Based on Ref. [2], it can be also stated that $96,489$ Ah can produce 42.84 Nm^3 of hydrogen. So, 1 Nm^3 demands 2251.41 Ah. Taking into consideration that the voltage across each electrolyzer cell is V_{elec} (volts), the power demand to produce 1 Nm^3/h of H_2 would produce an ideal value of $2251.41 \times V_{elec}$ watts (neglecting losses). Now, considering the electrolyzer efficiency equal to η_{elec}, the requisite energy to produce any amount of hydrogen is given by

$$W_{elec} = \frac{1.0135 \times P_{fc} h_{fc} V_{elec}}{V_c \eta_{elec}} \tag{7.47}$$

To incorporate the actual losses, the following equation for the electrolyzer power holds:

$$P_{e(t)} = \frac{1.0135 P_{fc} h_{fc} V_{elec}}{V_c \eta_{elec} h_e} \tag{7.48}$$

where:

$P_{e(t)}$ = The electrolyzer power at the time instant t
$P_{fc(t)}$ = The FC power at the same instant t
h_e = The number of operating hours of the electrolyzer

So, P_{qav} can be now calculated by

$$P_{qav} = \frac{1}{n} \sqrt{\sum_{t=1}^{n} \left(P_t - P_{fc(t)} + P_{e(t)} \right)^2} \tag{7.49}$$

Figure 7.13a represents an example of the daily load curve of a small village served by an energy system with both an FC and an electrolyzer operating under constant full power during both peak hours and off-peak hours. The difference between the two curves in the figure is the energy diverted to the electrolyzer from midnight to 6 p.m. and the energy received by the load from the FC power plant between 6 and 9 p.m. The difference is due to the efficiency of the entire conversion loss.

There are other possible modes of peak shaving (Figure 7.13b) when the period of hydrogen production could be restricted to the morning hours. The FC may operate at constant power or at variable power to track as closely as possible consumer demand to keep the load curve at constant maximum demand.

7.14 Future Trends

FC principles were discovered more than one and half centuries ago by a layperson. Those foundations continue to be studied today in sophisticated laboratories. Therefore, FC technology is still going to become mature and to change the way our society handles the production, storage, and delivery of energy and hydrogen. Enormous investments are being committed to the development of a reliable and workable energy system able to replace the fossil

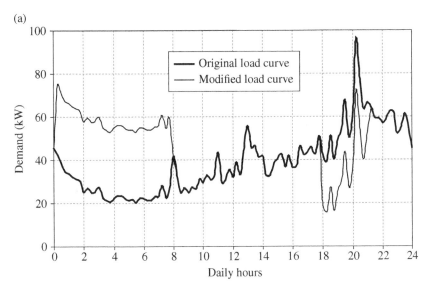

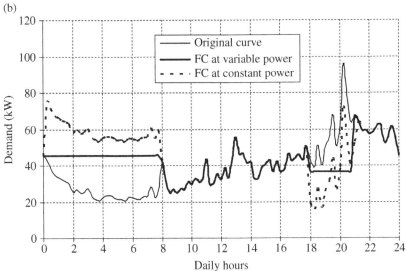

Figure 7.13 Typical residential load curves, including production of H_2: (a) hydrogen production during off-peak hours of a weekday (from 9 p.m. to 6 a.m. the next day) in a private residence; (b) hydrogen production during the early hours of a weekday (midnight to 8 a.m.) in a condominium.

fuel model in current use. The benefits of such a breakthrough would be to decrease pollution in urban areas, reduce greenhouse gas emissions, and increase the energy independence of oil-consuming countries. However, hydrogen cannot be compared directly with fossil energies, because it is only an energy vector, not an energy source. As such, it simply makes it possible to transmit a given quantity of energy from the place of production to the place of consumption.

Concerns about global warming and greenhouse gas abatement have dominated the FC and energy vicissitudes. Their reputation for efficiency, being environmentally friendly, and energy savings has helped FCs gain prominence for applications such as transportation, distributed power generation, and easy hydrogen conversion energy to electricity and envisaged for portable consumer electronics such as notebook computers and mobile phones have emerged [31–33].

FC applications have been changing very slowly in these last few years [34–38]. Because of current social and economic conflicts, such as the continuing profitability of fossil fuel providers, sea transportation, fuel distribution networks, and port installations, it is not perhaps the lack of technological knowledge that causes the greatest difficulty but selection of the right moment to begin the transforming process. Automobile manufacturers have been very aggressive and investing massively in developing not only FC systems but also all the required peripherals for easy integration to household use.

The most important barriers to development of FC applications are the present cost, the absence of production infrastructure and distribution networks, and the difficulties encountered in developing hydrogen storage technologies. Therefore, use of hydrogen in the transport sector should remain still relatively limited in the short run. On the other hand, the stationary FCs and the storage of energy for mobile equipment (domestic utilities, cell phones, laptops, smartphones, etc.) are expected to be commercially available at competitive costs in the very near future. Literature has reported several storage alternatives [39]. The trends of all that can be reunited in:

- New catalytic materials and concepts to increase the efficiency and reduce at the same time the costs and increase the durability of FCs
- Investments in basic science research in areas such as photoelectrolysis, high-temperature water splitting, biological production of hydrogen, new materials for hydrogen storage and FC electricity generators, and nanotechnologies
- Secure, compact, durable, reliable, and economic systems for hydrogen on-board storage in FC vehicles
- Cost-effective production of hydrogen that meets public consumption, environmental paradigms, and quality standards
- New concepts and technologies for reducing the cost and energy consumption of hydrogen transportation, security, storage, and distribution

There are expected worldwide shifts in the political balance, new commercial developments, and environmental concerns. Our modern society is working toward energetic solutions, and a new renewable and alternative energy paradigm is being consolidated. A new equilibrium point will eventually be found for a future era that will provide us with a great level of comfort, lifestyle, and power strategies or better than those of our present society under a sustainable approach. Maybe hydrogen and FCs will be part of this future since they are silent, flexible, safe, clean, nonseasonal, relatively long lasting, easily deployable, and expensive, but the technology/market is not mature enough for intense use yet. Above all, hydrogen and oxygen are strategic and can be encountered in any part of the world.

References

1 M.R. Fry, Environmental impacts of electricity generation: fuel cells, IEE Proceedings, Vol. 140, No. 1, pp. 40–46, 1993.

2 J.G. Seo, J.T. Kwon, J. Kim, W.S. Kim, and J.T. Jung, Impurity effect on Proton Exchange Membrane Fuel Cell, 2007 International Forum on Strategic Technology, Ulaanbaatar, Mongolia, pp. 484–487, October 3–6, 2007.

3 J.E. Larminie and A. Dicks, Fuel Cell Systems Explained, 2nd ed., Wiley, Chichester, England, 2003.

4 Teledyne Brown Engineering, Energy Systems Using Fuel Cells for Remote Applications, Final Report, Teledyne Brown Engineering Energy Systems, Schatz Energy Research Center Humboldt State University, Arcata, CA, and Palm Desert City Canada, February 1998.

5 G. Hoogers, Fuel Cell Technology Handbook, CRC Press, Boca Raton, FL, 2003.

6 Advanced Alternative Energy Corp., Business opportunities, http://www.aaecorp.com, accessed June 1999.

7 U.R. Prasanna and K. Rajashekara, Fuel Cell based Hybrid Power Generation Strategies for Microgrid Applications, 2015 IEEE Industry Applications Society Annual Meeting, Addison, TX, USA, pp. 1–7, October 18–22, 2015.

8 W.L. Hughes, Energy in Rural Development: Renewable Resources and Alternative Technologies for Developing Countries, Advisory Committee on Technology Innovation, Board on Science and Technology for International Development, Commission on International Relations, National Academy of Sciences, Washington, DC, 1976.

9 C. Von D. Gesellschaft, Sustainable Germany: The Contribution to Sustainable Global Development, Wuppertal Institute for Climate, Environment, Energy in the North Rhine-Westphalian Science Centers, Bünde, Germany, 1995.

10 R. Hill and A.E. Baumann, Environmental costs of photovoltaic energy, IEE Proceedings, Vol. 140, No. 1, pp. 76–80, 1993.

11 R.F. Mann, J.C. Amphlett, M.A.I. Hooper, H.M. Jensen, B.A. Peppley, and P.R. Roberge, Development and application of a generalized steady-state electrochemical model for a PEM fuel cell, Journal of Power Sources, Vol. 86, pp. 173–180, 2000.

12 J.J. Baschuck and X. Li, Modeling of polymer electrolyte membrane fuel cells with variable degrees of water flooding, Journal of Power Sources, Vol. 86, pp. 181–196, 2000.

13 J.C. Amphlett, R.F. Mann, B.A. Peppley, P.R. Roberge, and A. Rodrigues, A model predicting transient responses of proton exchange membrane fuel cells, Journal of Power Sources, Vol. 61, pp. 183–188, 1996.

14 G.R. Ault and J.R. McDonald, An integrated SOFC plant dynamic model for power systems simulation, Journal of Power Sources, Vol. 86, pp. 495–500, 2000.

15 D. Chu and R. Jiang, Performance of polymer electrolyte membrane fuel cell (PEMFC) stacks, part I: Evaluation and simulation of an air-breathing PEMFC stack, Journal of Power Sources, Vol. 83, pp. 128–133, 1999.

16 L.N. Canha, V.A. Popov, A.R. Abaide, F.A. Farret, A.L. König, D.P. Bernardon, and L. Comassetto, Multicriterial analysis for optimal location of distributed energy sources considering the power system, presented at the 9th Symposium of Specialists in Electric, Operational and Expansion Planning, Rio de Janeiro, Brazil, May 2004.

17 R. Balmer, Thermodynamics, West Publishing, St. Paul, MN, 1990.

18 R.M. Nelms, D.R. Cahela, and B.J. Tatarchuk, Modeling double-layer capacitor behavior using ladder circuits, IEEE Transactions on Aerospace and Electronic Systems, Vol. 39, No. 2, April 2003.

19 A. Ebadighajari, J. DeVaal, and F. Golnaraghi, Multivariable Control of Hydrogen Concentration and Fuel Over-Pressure in a Polymer Electrolyte Membrane Fuel Cell with Anode Re-circulation, 2016 American Control Conference (ACC),. Boston, MA, July 6–8, 2016.

20 M.G. Godoy Simões and F.A. Farret, Modeling power electronics and interfacing energy conversion systems, IEEE Press-John Wiley, Hoboken, New Jersey, 2017.

21 J.M. Corrêa, F.A. Farret, L.N. Canha, and M.G. Simões, An electrochemical-based fuel-cell model suitable for electrical engineering automation approach, IEEE Transactions on Power Delivery, Vol. 17, No. 2, pp. 467–476, 2002.

22 W. Friede, S. Raël, and B. Davat, Mathematical model and characterization of the transient behavior of a PEM fuel cell, IEEE Transactions on Power Electronics, Vol 19, No. 5, pp. 1234–1241, September 2004.

23 J.C. Amphlett, R.F. Mann, B.A. Peppley, P.R. Roberge, and A. Rodrigues, A practical PEM fuel cell model for simulating vehicle power, Proceedings of the Tenth Annual Battery Conference on Applications and Advances, Royal Military College of Canada, Kingston, ON, Canada, 1995, pp. 221, 226, DOI: 10.1109/BCAA.1995.398535.

24 F.A. Farret, J.R. Gomes, and A.S. Padilha, Comparison of the hill climbing methods used in micro power plants, Proceedings of INDUSCON 2000, IEEE Brazil-South, Porto Alegre, RS, Brazil, 2000, pp. 756–760.

25 D. Yu and S. Yuvarajan, A novel circuit model of PEM fuel cells, Proceedings of the Nineteenth Annual IEEE Conference and Exposition on Applied Power Electronics, APEC, CA, USA, Vol. 1, 2004, pp. 362–366.

26 C.J. Hatziadoniu, A.A. Lobo, F. Pourboghrat, M. Daneshdoost, J.T. Pukrushpan, A.G. Stefanopoulou, and H. Peng, A simplified dynamic model of grid-connected fuel-cell generators, IEEE Transactions on Industrial Electronics, Vol. 51, No. 5, pp. 1103–1109, 2004.

27 Y. Kim and S. Kim, An electrical modeling and fuzzy logic control of a fuel cell generation system, IEEE Transactions on Energy Conversion, Vol. 14, No. 2, pp. 239–244, 1999.

28 F. Gonzatti, F.Z. Ferrigolo, V.N. Kuhn, D. Franchi, and F.A. Farret, "Automation of thermal exchanges of metal hydrides cylinders integrated with energy storage," in 11th IEEE/IAS International Conference on Industry Applications INDUSCON 2014, Juiz de Fora, MG, Brazil, IEEE, Piscataway, NJ, 2014.

29 F.N. Biichi and G.G. Scherer, In-situ resistance measurements of Nation® 117 membranes in polymer electrolyte fuel cells, Journal of Electroanalytical Chemistry, Vol. 404, pp. 37–43, 1996.

30 F. Gonzatti, V.N. Kuhn, F.Z. Ferrigolo, M. Maicon, and F.A. Farret, "Theoretical and practical analysis of the fuel cell integration of an energy storage plant using hydrogen," in 11th IEEE/IAS International Conference on Industry Applications INDUSCON 2014, Juiz de Fora, MG, Brazil, IEEE, Piscataway, NJ, 2014.

31 R. Carnieletto, J.B. Parizzi, A.C. Schittler, and F.A. Farret, "Evaluation of the use of secondary energy for hydrogen generation," in 2010 IEEE International Conference on Industrial Technology, Valparaiso, Chile, IEEE, Piscataway, NJ, 2010.

32 C. Xu, P.M. Follmann, L.T. Biegler and M.S. Jhon, Numerical Simulation and Optimization of a Direct Methanol Fuel Cell, Computers and Chemical Engineering, Elsevier, Vol. 29, No. 8, pp. 1849–1860, July 15, 2005.

33 J.T.S. Irvine and P. Connor (eds.), Solid Oxide Fuels Cells: Facts and Figures: Past Present and Future Perspectives for SOFC Technologies, Springer, Berlin, Germany, 2013.

34 N.Q. Minh and M.B. Mogensen, Reversible Solid Oxide Fuel Cell Technology for Green Fuel and Power Production, The Electrochemical Society, Pennington, NJ, 2013, pp. 55–62.

35 M. Hinaje, S. Raël, P. Noiying, D. Nguyen, and B. Davat, An equivalent electrical circuit model of proton exchange membrane fuel cells based on mthematical modelling, Energies, Vol. 5, pp. 2724–2744, 2012.

36 I. Khazaee, Analytical investigation and improvement of performance of a proton exchange membrane (PEM) fuel cell in mobile applications, International Journal of Applied Mechanics and Engineering, Vol. 20, No. 2, pp. 319–328, 2015.

37 F. Gonzatti, V.N. Kuhn, M. Miotto, F.Z. Ferrigolo, and F.A. Farret, Distinct renewable energy systems maximized by P&O algorithm, Journal of Control, Automation and Electrical Systems, Vol. 27, pp. 310–316, 2016.

38 F. Gonzatti, V.N. Kuhn, F.Z. Ferrigolo, F.A. Farret, and M.A.S. De Mello, Experimental hydrogen plant with metal hydrides to store and generate electrical power, International Journal of Emerging Electric Power Systems, Vol. 17, pp. 59–67, 2016.

39 Y.-X. Wang, K. Ou, and Y.-B. Kim, Modeling and experimental validation of hybrid proton exchange membrane fuel cell/battery system for power management control, International Journal of Hydrogen Energy, Vol. 40, pp. 1713–1721, 2015.

8

Biomass-Powered Microplants

8.1 Introduction

Biomass energy (or bioenergy) is converted from energy derived from organic matter related to residential waste, particularly related to municipal trash programs, plus commercial and industrial waste, agricultural and forestry residues, general human waste, and the use of landfills [1–3]. Organic components from municipal and industrial waste, plants, agricultural and forestry residues, home waste, and landfills can be used very efficiently in our society to reduce worldwide greenhouse gas emissions. Table 8.1 illustrates the biomass potential of several sources. Dedicated energy plantations are spread throughout the world as in the United States, Brazil, India, and China [1–3]. For example, there is a program in China targeted at 13.5 million hectares of wood for fuel applications since 2010. Three million hectares of eucalyptus are used for charcoal in Brazil, 16,000 ha of willow plantations is used for generation of heat and power in Sweden, and 50,000 ha of agricultural land has been converted to woody plantations and may rise to as much as 4 million hectares (10 million acres) in the United States. The US Department of Energy road map for biomass is illustrated in Figure 8.1.

Biomass generates the same amount of carbon dioxide as do fossil fuels (when burned). However, from thermodynamics and a chemical balance point of view, every time a new plant grows, carbon dioxide is removed from the atmosphere. The net emission of carbon dioxide will be zero as long as plants continue to be replenished for biomass energy purposes. The overall idea is that biomass does not have fossil origin, so it belongs already to the balance of the ecosystem. If the biomass is converted through gasification or pyrolysis, the net balance can even result in the removal of carbon dioxide. Energy crops such as fast-growing trees and grasses are called biomass feed stocks. The use of biomass feed stocks can help increase profits for the agricultural industry.

Combustion derived from biomass, such as burning wood, has been used from prehistoric times to the present. However, it is not very efficient, because

Integration of Renewable Sources of Energy, Second Edition. Felix A. Farret and M. Godoy Simões.

Table 8.1 Biomass potential (data for United States).

Classification	Biomass type	Amount
Municipal solid waste/ landfills	Quantity of raw material	167 million tonnes
	Direct use from combustion	217,722 TJ
	Electricity generation capacity	2,862,000 kW
	Electricity generating	71,405 TJ
	Total energy production	**289,127 TJ**
Forestry/wood processing	Electricity generating capacity	6,726,000 kW
	Electricity generation	124,712 TJ
	Direct use from combustion	2,306,026 TJ
	Total energy production	**2,430,738 TJ**
Agricultural residues—corn	Quantity of raw material	13.5 million tonnes
	Ethanol fuel production capacity	152,376 TJ/year
	Yield of ethanol	8. 8 GJ/tonne
	Ethanol fuel production	**118,010 TJ**
Agricultural residues—soy bean and waste food oils	Biodiesel production capacity	6,708 TJ/year
	Yield of biodiesel	40 GJ/tonne
	Biodiesel production	**671 TJ**
Wood pellets	Quantity of raw material	0.582 million tonnes
	Direct use from combustion	8,872 TJ
Other biomass	Electricity generating capacity	10,602,000 kW
	Electricity generation	11,328 TJ
	Direct use from combustion	102,084 TJ
	Total energy production	**113,412 TJ**

Source: www.worldenergy.org. © The World Energy Council, London, UK.

it is an external combustion engine conversion. Converting solid biomass to a gaseous or liquid fuel by heating it with limited oxygen prior to combustion greatly increases the heating value and overall efficiency, making it possible to convert the biomass to other valuable chemicals or materials. The gasification of biomass is a developing energy technology among various systems for the energetic utilization of biomass. Biomass gasification has the following advantages over conventional combustion technologies: The CHP generation

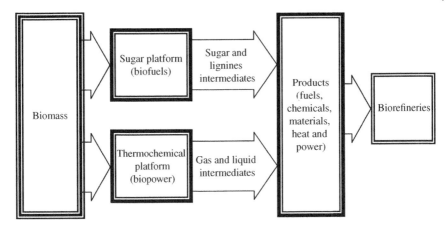

Figure 8.1 Biomass energy conversion. *Source*: From Refs. [1, 2]. Hackett [2]. Reproduced with the permission of M.E. Sharpe, Inc.

(via biomass gasification techniques connected to gas-fired engines or gas turbines) can achieve significantly higher electrical efficiencies, between 22 and 37%, than those of biomass combustion technologies with steam generation and steam turbine, 15–18%. If the gas produced is used in fuel cells for power generation, an even higher overall electrical efficiency can be attained, in the range of 25–50%, even in small-scale biomass gasification plants and under partial load operation.

The gasification allows improved electrical efficiency of the energy conversion. Therefore, the potential reduction in CO_2 is greater than with regular combustion. The formation of NO_x compounds can also be largely prevented, and the removal of pollutants is easier for various substances. The NO_x advantage, might be partly lost if the gas is subsequently used in gas-fired engines or gas turbines. The two primary NO_x formation mechanisms in gas turbines are thermal and fuel NO_x. In each case, nitrogen and oxygen present in the combustion process combine to form NO_x. Thermal NO_x is formed by the dissociation of atmospheric nitrogen (N_2) and oxygen (O_2) in the turbine combustor and the subsequent formation of NO_x. When fuels containing nitrogen are combusted, this additional source of nitrogen results in the formation of fuel NO_x. When fuels have little or no nitrogen content, thermal NO_x is the dominant source of NO_x emissions. The formation rate of thermal gas NO_x increases exponentially with temperature. Because the flame temperature of oil fuel is higher than that of natural gas, NO_x emissions are higher for operations using oil fuel than natural gas. Significantly, lower emissions of NO_x, CO, and hydrocarbons can be expected when the gas produced is used in fuel cells rather than in gas-fired engines or gas turbines.

Researchers at the US Department of Energy Biomass Program are conducting efforts for developing thermochemical technologies to tap the enormous energy potential of lignocellulosic biomass more efficiently [4, 5]. In addition to gasification, pyrolysis, and other thermal processing, program research focuses on cleaning up and conditioning the converted fuel. This is a key step in the effective commercial use of thermochemical platform chemicals.

Biomass pyrolysis refers to a process where biomass is exposed to high temperatures in the absence of air, causing the biomass to decompose. The final product of pyrolysis is a mixture of solids (char), liquids (oxygenated oils), and gases (methane, carbon monoxide, and carbon dioxide). Flash pyrolysis gives high oil yields, but because of the technical effort needed to process pyrolytic oils, this energy-generating system does not seem very promising at the present stage of development. However, pyrolysis as a first stage in a two-stage gasification plant for straw and other agricultural feed stocks that pose technical difficulties in gasification does deserve consideration.

Biomass gasification technologies have been a subject of commercial interest for several decades. Gasification operates by heating biomass in an environment where the solid biomass breaks down to form a flammable, low-caloric gas. The biogas is then cleaned and filtered to remove chemical compounds. The gas is used in more efficient power generation systems called combined cycles, which combine gas turbines and steam turbines to produce electricity. The efficiency of these systems can reach 60%. Gasification systems may also be coupled with fuel cell systems using a reformer to produce hydrogen and then convert hydrogen gas to electricity (and heat) using an electrochemical process. Interest in biomass gasification increased substantially in the 1970s because of uncertainties in petroleum supplies. Most development activities were concentrated on small-scale systems. Low-energy gasifiers are now commercially available, and dozens of small-scale facilities are in operation.

During the 1980s, government and private industries sponsored research and development (R&D) for gasifier systems primarily to gain a better understanding of reaction chemistry and economies-of-scale issues. In the 1990s, CHP was identified as a potential near-term opportunity for biomass gasification because of incentives and favorable power market drivers. R&D concentrated on integrated gasification combined cycle (IGCC) and gasification cofiring demonstrations, which culminated in a number of commercial-scale systems. In US projects, most processes are very recalcitrant feeds such as bagasse and alfalfa.

The use of biogas for energy generation represents a high-efficiency alternative source to replace fossil fuels. The annual world bioenergy potential is about 2900 EJ, although only about 10% of that could be considered available on a sustainable basis and at competitive prices. In urban areas, another

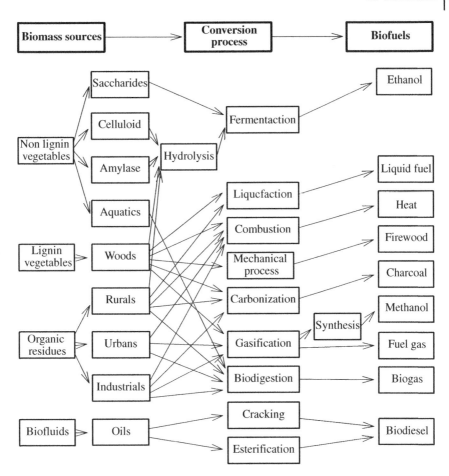

Figure 8.2 Processes of biomass conversion into biofuels.

important benefit is the use of the enormous amounts of organic garbage or liquid effluents such as industrial residues, sewers, and trash cans. Transformation of these items into industrial or automotive fuels can avoid environmental damage. Figure 8.2 depicts the main processes of biomass conversion into biofuel [6, 7].

The use of biomass and biodigesters introduces several advantages for rural applications, where leftover cultural and animal residues can be used to obtain biofertilizer (i.e., the organic material processed in biodigesters can be used as fertilizer). Furthermore, biomass and biodigesters can be used to provide necessary energy for illumination and heating and to drive motors. The market potential and future prospects of small-scale electricity generation from RES are covered in an OPET report [5, 8] and summarized in Figure 8.3.

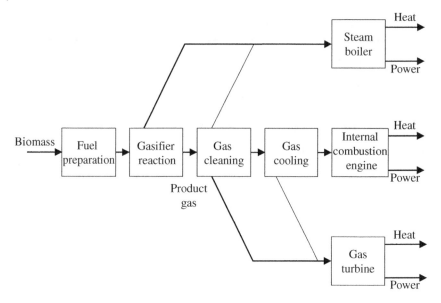

Figure 8.3 Gasification paths to heat and power. *Source*: From Ref. [9].

8.2 Fuel from Biomass

Biomass seems to be a very attractive and renewable option for liquid transportation fuel. Biomass use strengthens rural economies, decreases US dependence on imported oil, avoids use of methyl tertiary butyl ether (MTBE) or other highly toxic fuel additives, reduces air and water pollution, and reduces greenhouse gas emissions. Fuel from biomass is related to ethanol, biodiesel, biomass power, and industrial process energy. To expand the role of biomass in our modern society, the Department of Energy's Office of the Biomass Program is fostering biomass technologies with a balanced portfolio of R&D [9].

It is important to consider that when using biomass, most of the electricity, heat, and steam produced by the industry are consumed on site. However, some manufacturers sell excess power to the grid. Wider use of biomass resources will directly benefit many companies whose growth generates more residues, such as wood or animal wastes, than they can use internally. New markets for these excess materials will also support business expansion.

The pulp and paper industry is a very important stakeholder in the forest industry, because it is estimated that 85% of wood waste is used for energy in the United States, with a tremendous potential for market growth for other wood waste products and biomass renewable sources of fuel. Gasification technologies using biomass pulp by-products has aided the pulp industry, and the paper industry has been improving chemical recovery and generating process steam and electricity at higher efficiencies and with lower capital costs than have conventional technologies. Pulp and paper industry by-products that can be gasified include hogged wood, bark, and spent black liquor.

For biomass gasification, organic materials are heated with less oxygen than that needed for efficient combustion. Since combustion is a function of the mixture of oxygen with hydrocarbon fuel, there is a stoichiometric amount of oxygen and other conditions to determine if biomass gasifies or pyrolysis to a mixture of carbon monoxide and hydrogen, known as synthesis gas or syngas. Gaseous fuels mix with oxygen more easily than do liquid fuels, which in turn mix more easily than solid fuels. Therefore, inherently, syngas burns more efficiently and cleanly than the solid biomass from which it is made.

Biomass gasification can improve the efficiency of large-scale biomass power facilities. Forest industry residues and specialized facilities such as black liquor recovery boilers used in pulp and paper industry can be major sources of biomass power. Syngas can also be burned in gas turbines, a more efficient electrical generation technology than steam boilers, since its use of solid biomass and fossil fuels is limited.

Most electrical generation systems are relatively inefficient, because half to two-thirds of the energy is lost as waste heat. If such heat is used for an industrial process, space heating, or another purpose, the efficiency can be greatly increased. Small, modular biopower systems are more easily used for such cogeneration than are most large-scale electrical generation.

Syngas mixes more readily with oxygen for combustion and with chemical catalysts than solid fuels do, thus enhancing its ability to be converted to other valuable fuels, chemicals, and materials. The Fischer–Tropsch process converts syngas to liquid fuels, which aids in transporting such a fuel [9, 10]. The process is named after two German coal researchers, who discovered the method in 1923. A variety of other catalytic processes can turn syngas into a myriad of chemicals or other potential fuels or products, and it is used for the synthesis of hydrocarbons and other aliphatic compounds. Syngas, a mixture of hydrogen and carbon monoxide, reacts in the presence of an iron or cobalt catalyst; much heat is involved, and such products as methane, synthetic gasoline and waxes, and alcohols are made, with water or carbon dioxide produced as a by-product. In other words, the water–gas shift process converts syngas to more concentrated hydrogen for fuel cells. This important source of the hydrogen–carbon monoxide gas mixture is the gasification of coal.

Biomass gasification involves thermal conversion to simple chemical building blocks that can be transformed into fuels, products, power, and hydrogen. Components include feed preparation, the biomass gasifier, and a gas treatment and cleaning train. Syngas initially contains particulates and other contaminants and must be cleaned and conditioned prior to use in fuels or chemical or power conversion systems (e.g., catalyst beds, fuel cells). The gasification process readily converts all major components of biomass, including lignin, which is resistant to biological conversion, to intermediate building blocks. Utilization of the lignin, which is typically 25–30% of the biomass, is essential to achieve high efficiencies in the biorefinery. The gasification process can convert most biomass feedstocks or residues to a clean syngas. Once such a gas is obtained, it is possible to access and leverage the process technology developed in the petroleum and chemical industries to produce a wide range of liquid fuels and chemicals.

There are several widely used process designs for biomass gasification: (i) staged reformation with a fluidized bed gasifier, (ii) staged reformation with a screw auger gasifier, (iii) entrained flow reformation, or (iv) partial oxidation. In staged steam reformation with a fluidized bed reactor, the biomass is first pyrolyzed in the absence of oxygen. Then the pyrolysis vapors are reformed to syngas with steam, providing added hydrogen as well as the proper amount of oxygen and process heat that comes from burning the char. With a screw auger reactor, moisture (and oxygen) is introduced at the pyrolysis stage, and process heat comes from burning some of the gas produced in the latter. In entrained flow reformation, external steam and air are introduced in a single-stage gasification reactor. Partial oxidation gasification uses pure oxygen with no steam, to provide the proper amount of oxygen. Using air instead of oxygen, as in small modular uses, yields produce gas (including nitrogen oxides) rather than syngas.

Biomass gasification is also important for providing a fuel source for electricity and heat generation for the integrated biorefinery. Virtually all other conversion processes, whether physical or biological, produce residue that cannot be converted to primary products. To avoid a waste stream from the refinery, and to maximize the overall efficiency, these residues can be used for combined heat and power (CHP) production. In existing facilities, these residues are combusted to produce steam for power generation. Gasification offers the potential to utilize higher-efficiency power generation technologies, such as combined cycle gas turbines or fuel cells. Gas turbine systems offer potential electrical conversion efficiencies approximately double those of steam cycle processes, with fuel cells being nearly three times as efficient.

A workable gasification process requires the development of some technology: for example, feed processing and handling, gasification performance improvement, syngas cleanup and conditioning, development of sensors, analytical instruments and controls, process integration, and materials used for the systems.

8.3 Biogas

Biogas, also termed methane or gobar gas, is a clean unpolluted and cheap source of energy in rural areas. It consists of 55–70% methane that is inflammable. Biogas is produced from cattle dung in a plant that allows digestion, and

Table 8.2 Probable composition of some combustible gases.

Methane (CH_4)	60–70%	
Carbonic gas (CO_2)	30–40%	
Nitrogen (N)	Traces hydrogen (H)	Traces
Gas sulfidric (H_2S)		Traces

at the end gobar comprises a mixture of gases. It is a fuel of high caloric value resulting from anaerobic fermentation of organic matter called biomass. Composition of this gas varies with the type of organic material used. Its basic composition is listed in Table 8.2.

Methane is a colorless, odorless gas with a wide distribution in nature. It is the principal component of natural gas, a mixture containing about 75% methane (CH_4), 15% ethane (C_2H_6), and 5% other hydrocarbons, such as propane (C_3H_8) and butane (C_4H_{10}). The firedamp of coal mines is chiefly methane [5, 6], and methane is the main component of biogas. It is colorless and odorless, highly flammable, and in combustion presents a lilac blue flame and small red stains. It does not leave soot and produces minimal pollution. The caloric power of biogas depends on the amount of methane in its composition, which could reach 5000–6000 kcal/m^3. Biogas can be used for stove heating, campaniles, water heaters, torches, motors, and other equipment.

8.4 Biomass for Biogas

Biomass can be considered as all materials that have the property of being decomposed by biological effects, that is, by the action of bacteria. Biomass can be decomposed by methanogenic bacteria to produce biogas, in a process depending on factors such as temperature, pH, carbon/nitrogen ratio, and the quality of each. Usable and accessible organic matter includes animal residue, agricultural residue, water hyacinth (*Eichornia crassipes*), industrial residue, urban garbage, and marine algae.

There exists a potential biomass use for animal feces produced in cattle farming, which is easily mixable with water. In building a biodigester, the first step is to determine its capacity, observing the amount of biomass available locally. For this calculation, one simply multiplies the weight of a live animal by 0.019 to find the approximate daily production of manure [10, 11].

Agricultural residue produces an average of seven times more biogas than animal waste but is not very practical with respect to harnessing, provisioning, and discharging in biodigesters. For this biomass, intermittent biodigesters are most appropriate. Agricultural remains are not mixable with water because

Table 8.3 Biogas volume produced by organic residues.

Biomass	Production of biogas (m³/ton)	Methane (%)
Sunflower leaves	300	58
Rice straws	300	Variable
Wheat straws	300	Variable
Bean straws	380	59
Soy straws	300	57
Linen stem	359	59
Grapevine leaves	270	Variable
Potatoes leaves	270	Variable
Dry leaves of trees	245	58

they may be lighter; they emerge only after initial decomposition, when the production of gas begins. If triturated, they can be used in any type of biodigester, although that represents greater energy and labor. Provisioning is made once every 15 or 20 days, and the fermentation period and production of biogas take from 60 to 120 days. Residues containing agricultural chlorinated pesticides or herbicides are not used because chlorine does not allow the development of methanogenic bacterium. Table 8.3 lists the approximate volume of biogas produced by agricultural waste.

Under favorable conditions, water hyacinth produces up to 600 kg of dry matter per hectare a day, offering excellent biogas production. According to research data, 350–410 L of biogas per kilogram of dry water hyacinth was captured. Each square meter of a plantation can produce, on average, 18 L/day, or 30,000 m³/ha. The amount of methane reaches 80% in volume, greater than that of other types of biomass.

The use of industrial residues is minimal. Except for the industrialization of fruits, meats, cereals, and alcohol, such biomass can be used when the residue has been dumped in rivers or elsewhere outdoors and is polluting the environment. Urban garbage can be transformed into sources of energy if processed in biodigesters. Biodigestion could be considered a powerful instrument for the disposition of garbage if associated with noble products such as electricity.

Marine algae offer a strong option as a prime matter in the production of biogas. Scientific studies show that the biogas produced from marine algae is of good quality, characterized by the lack of an odor of sulfur, and provides a clear to bluish flame. On average, this biogas produces 300 L/kg of dry raw material and is 60–70% gas methane. Its caloric value reaches about 6600–7200 calories.

8.5 Biological Formation of Biogas

Animal manure is a biomass found in considerable quantities on rural properties [4, 12]. Anaerobic compost is formed from a mixture of organic manure with animal–water residues in three stages. In phase 1 (the solid stage), substances such as carbohydrates, lipids, and proteins are attacked by ordinary fermentative bacteria for the production of fatty acids, glucose, and amino acids. In phase 2 (the liquid stage), the substances formed previously are attacked by different bacteria like the propionibacteria, acetogens, and acidogenic bacteria, forming organic acids, mainly propionic and acetic, forming carbon dioxide, acetates, and H_2. In phase 3 (the gaseous stage), the methanogenic bacteria act on the organic acids to produce primarily methane CH_4 and carbon dioxide CO_2 (biogas). The third stage is the most important, because bacteria demand special care; they are responsible for limiting the speed of a chain of reactions. This is due primarily to the formation of microbulbs of methane and carbon dioxide around the methanogenic bacteria, which isolates them from the mixture in digestion. Furthermore, they require temperature and acidity suited for their reproduction. Methanogenic fermentation is a highly sensitive biological process involving many microorganisms, such as the thermophilic and mesophilic methanogenic bacteria, which reproduce between 45 and 50 °C and 20 and 45 °C, respectively.

8.6 Factors Affecting Biodigestion

Fermentation is more intense when the temperature of the material is between 30 and 35 °C. At these conditions, the biogas production per kilogram of raw material is higher and faster. Therefore, the biodigester is usually buried because underground temperatures are higher and more constant. Additionally, a buried biodigester is easier to handle. Ideal temperatures for the material used to produce methanogenic bacteria are as follows: psychrophilic, mesophilic, and thermophilic with the highest growth at 20, 35, and 55 °C, respectively. In general, the biomass inside a digester should be at a temperature of approximately 35 °C. Table 8.4 lists the amount of waste per semistabled animal and the corresponding production of gas.

The biodigestion acidity measured by the hydrogenionic potential (pH) should remain between 6 and 8. If the pH is below those values, the environment becomes better adapted to the acid bacteria that will prevent the growth of methanogenic bacteria.

For any biodigester system, it is imperative that agitation of the mixture achieves a uniform distribution of temperature, thus improving the distribution of the intermediary and final products in the biodigester and reducing top mud layer growth [4, 12].

Table 8.4 Production of gas from manure from semistabled animals.

Manure-producing animal (kg/day)		m^3 of gas/kg of manure	m^3 of gas/ animal/day
Birds	0.09	0.055	0.0049
Bovines	10.00	0.040	0.4000
Equines	6.50	0.048	0.3100
Oviparous	0.77	0.070	0.0500
Suidae	2.25	0.064	0.1400

Source: Refs. [6, 12]. © Tamera – Peace Research Center.

The concentration of nutrients in a mixture is another important factor. Nutrients for the development of bacteria include carbon, nitrogen, nitrate, phosphorus, sulfur, and sulfates. Animal urine (urea) or chemical fertilizers such as ammonia sulfate can be used to cause fermentation. For such, the following points should be observed:

1) The carbon/nitrogen ratio: Nitrogen is a nutrient that contributes to the formation and multiplication of bacteria, but in excess, it induces ammonia formation, which can increase and even stop production of biogas. Carbon is responsible for the production of energy. As most animal solid waste has a low carbon/nitrogen ratio, those elements should be corrected with vegetable residues to reach the ideal point.
2) The concentration of solids should be between 7 and 9%, for the following reasons:
   • To avoid mud formation on the top of the digester
   • To ease movement of the material in the digester
   • To obtain good fermentation
   • To allow good digestion

Tables 8.5 and 8.6 list the average dry material content used to establish the correct amount of water in the mixture [6, 12].

The retention period of the substratum in the biodigester is the digestion period or the time that elapses between the input and output of the biomass in the digester and it varies as a function of the pH, volume of material used, agitation inside the biodigester, and internal digester temperature. Generally, this period ranges from 20 to 50 days, lengthening the digestion time when the biofertilizer is not smelly at the output, detected by the presence of flies.

Digestion can produce some toxic substances. Thus, as the biogas is not a decomposition product of sterilized organic residues, it should be mixed

Table 8.5 Average percent of dry matter in selected animal manure.

Producing animal	Dry matter in the manure (%)
Bovine	16.5
Equines	24.2
Oviparous	34.5
Capra	34.8
Suidae	19.0
Birds	18.0

Source: Refs. [6, 12]. © Tamera – Peace Research Center.

Table 8.6 Manure/water ratio for 7, 8, and 9% solid contents.

	Solids in the biomass preparation		
Producing animal	7%	8%	9%
Bovine	0.74:1	0.94:1	1.20:1
Equines	0.41:1	0.49:1	0.59:1
Oviparous	0.26:1	0.30:1	0.35:1
Capra	0.25:1	0.30:1	0.35:1
Suidae	0.58:1	0.73:1	0.90:1
Birds	0.64:1	0.80:1	1.00:1

Source: Refs. [6, 12]. © Tamera – Peace Research Center.

without chlorinated water. At the same time, the presence of disinfectants and excess antibiotics and/or copper in animals should be watched for, as these are fatal to methanogenic bacteria and thus prevent the production of gas.

8.7 Characteristics of Biodigesters

Several biodigester models are available for assembling and construction, with some open-source designs. For this study, a typical biodigester of average capacity of the Indian type will be referenced; this type has been used to reference most facilities worldwide [5, 12]. In the Indian model, the system of biogas production consists primarily of a digester and a gasometer (see Figure 8.4). The digester is a reservoir built of bricks or concrete below ground level. A wall

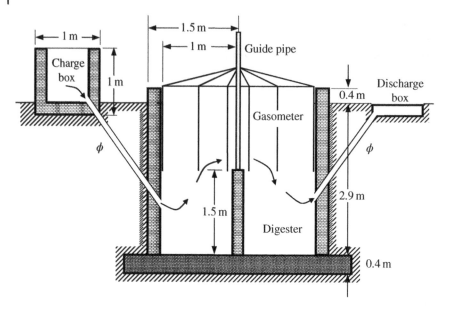

Figure 8.4 Production of biogas in digesters. *Source*: From Ref. [12]. © Tamera – Peace Research Center.

divides the biodigester into two semicylindrical parts for the purpose of retaining and providing circulation for the biomass loaded in a biofertilization process. The biodigester is loaded through the input box, which also serves as a prefermenter. The input box communicates with the digester through a pipe going down to the bottom. The output of the biofertilizer is through another pipe at a level that assures that the amount of biomass entering the biodigesters is the same as that leaving it in biofertilizer form. It should also have a discharge box, tank, or dam to pump and/or deliver the biofertilizer directly to the consumer.

The gasometer is responsible for storing the biogas produced at an almost constant pressure equivalent to its weight. Sometimes, counterweights are placed in the gasometer to supply gas at higher pressure at the output; they move up or down in accordance with the volume of the stored biogas. The counterweights are often bags of sand or cement blocks. The gasometer is generally made of steel foil 14, which should be welded in a metallic structure made of 3/4-in. carbon steel. It is usually cylindrical with an arched cover (in cone form) to avoid deposition of sludge and water in its exterior parts. The gasometer can move up or down as guided by a galvanized guide tube 2.5–3.0 in. in diameter.

The biogas output is a device on the top of the gasometer through which the biogas leaves the interior of the gasometer for the consumption point. It should be a flexible hose, to ease gasometer movement. The gas is transported in a 1-inch-diameter tube.

In the dimensions shown in Figure 8.4, the biodigester produces $6 \, m^3$ of gas per day, corresponding to $8568 \, kWh$ ($1 \, m^3/day = 1428 \, kWh$). The tank will be charged with $240 \, L$ of biomass a day, of which $120 \, L$ is water and $120 \, L$ is bovine manure. Twelve or thirteen adult semistabled bovine animals are necessary for this level of production.

8.8 Construction of a Biodigester

The biodigester capacity to be installed on a rural property is determined by the number of people and their minimum energy needs. The biodigester should be placed between the point of consumption (kitchen, refectory, and drying deposits) of the gas and the site of manure collection, at a point 20–25 m from the consumption point. Sloped grounds, to ease withdrawal and distribution of the biofertilizer, are preferable [6, 12].

The digester well should be cylindrical, with a diameter a little larger than the external diameter of the digester and a depth 20–30 cm smaller than its useful height. During excavation of the well, two rifts should be constructed in opposite directions, with an approximate slope of 180 for placement of the charge and discharge pipes. If the land is flat, a shallower excavation should be made, using the soil extracted to form a mound around the digester, which will simplify charging and discharging.

The floor of the digester should support the entire hydraulic load that it holds, in one of following ways: (i) in wells with a rock or very firm dirt bottom on which a brick floor is set on fine sand or (ii) in wells without a firm base, in which case the floor should be of concrete 10–15-cm thick.

The external and dividing walls of the digester can be built of concrete or of ordinary bricks settled firmly with no gaps through the walls, to avoid undesired leaks or cracks in the digester. The dividing walls have the double purpose of driving the mass flow to be digested and serving as a support for the gasometer guides. The latter function requires very resistant supports to withstand the gasometer movement along the guides; building double walls or concrete dividing walls allowing apertures of 40–50 cm where the pipe guides of the gasometer will be fixed is recommended.

The digester charges and discharges through pipes, allowing mass circulation during the digesting process. The pipes can be made of 100-mm-diameter PVC and placed in rifts in the digester well, fixed on the bottom and fastened on the top of the wall.

The charge tank can be used either as the prefermentation tank or for mixing the material. Therefore, its volume should be a little larger than the daily charge and a little bit above the liquid level of the digester. The discharge tank is constructed primarily to protect the output flow.

The cladding used for the digester is of fundamental importance to its construction and operation. The cladding should be made of a 1:6 ratio of cement for fine sand and should be at least 10 mm thick. Finally, to guarantee the accuracy of the diameter measurements, the use of a standard scale fixed on the guide pipes of the plant to allow free rotation around it is recommended.

8.8.1 Typical Size for a Biodigester

Figure 8.4 is a schematic diagram of a typical biodigester with a production capacity of 6 m^3/day and a retention time of 50 days. For the daily production of gas, the following is required:

Amount of biomass: 240 L/day
Retention time: 50 days
Total useful volume: (240 L/day) × (50 days) × (1 m^3/103 L) = 12 m^3
Diameter of the chamber: 2.5 m
Total height: 2.85 m
Useful height: 2.45 m
Total volume: 14 m^3
Volume of charge: 12 m^3
Volume of the gasometer: 6.3 m^3
External diameter of the gasometer: 2.40 m
Plant height of the gasometer: 1.85 m
Lateral height of the gasometer: 1.40 m
Height of the dividing wall: 1.52 m
The charge tank dimensions to carry 300 L of biomass: 1 m × 1 m × 0.8 m

8.9 Generation of Electricity Using Biogas

For the installation of a biogas microplant as an alternative source of electric energy for small power generators, it is necessary to use gas motors coupled with turbines and electrical generators, as illustrated in Figure 8.5. Many manufactured generator–turbines work with gas. Because the majority of the smaller commercial units are in the high-power range (1.6–216 MW) [7, 8], some adaptations are required. For micropower plants, alcohol and gasoline motors can be made to operate with methane without affecting their operational integrity. This adaptation is made by installing a bottle of biogas in place of using conventional fuel. For gas flow regulation, a reducer is placed close to

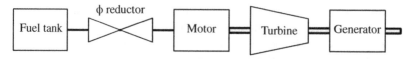

Figure 8.5 Generating group of electric energy using biogas.

Table 8.7 Average consumption of some biogas motors.

Motor power (hp)	Average consumption (m³/h)
1.0	0.45
2.0	0.92
5.5	2.24
9.0	3.16

Table 8.8 Equivalent energy per cubic meter of biogas to obtain 6148.98 kcal of heat.

Fuel	Equivalent amount
Gasoline	0.98 L
Alcohol	1.34 L
Crude oil	0.72 L
Natural gas	1.50 m³
Coal	1.51 m³
Electricity	2.21 kWh

Source: Adapted from Ref. [12]. © Tamera – Peace Research Center.

the motor. As the biogas flow operates at low pressures, it demands a larger aperture in the injector (with a diameter of 1.5–2 mm). The adaptation of equipment to work with PLG gases as biogas involves simply a replacement of the injector or an increase in the diameter of biogas flow to the motor. Table 8.7 shows the average consumption of biogas for motors of several power levels.

For a better understanding of these quantities, the equivalent energy of a cubic meter of biogas to obtain 6148.98 kcal of heat in accordance with Table 8.8 can be considered. Table 8.9 shows the daily amounts of biogas used for domestic purposes by a family of five or six people.

Observing that a biogas flow of $1 m^3/h = 1.428 kWh$, it is sufficient to use a simple rule of thumb to determine the power consumption in kilowatts of each motor, as in the following example of a 9-hp motor:

$$x = 1.428 kWh \times (3.16 m^3/h) / \left(1 m^3/h\right)$$
$$x = 4.51 kWh$$

Figure 8.6 shows a renewable fuel gas generator where a gas production module converts biomass to a clean gas for power and heat, biofuels research, and hydrogen generation. Depending on a number of variables, the fuel cost for the product gas measured in dollars per million Btu is at least 50% less than that of nonrenewable natural gas and propane. Figure 8.6 shows integration with a gas turbine generator, but it can typically be converted to operate with microturbines, thermoelectric power generators, and fuel cells.

Table 8.9 Daily domestic amounts of gas use by a family of five or six people.

Use purpose	Biogas
Kitchen	$1.960\,m^3$
Bathtub	$0.588\,m^3$
Shower	$0.336\,m^3$
Refrigeration of victuals	$2.800\,m^3$
Illumination	$0.140\,m^3$
Gasoline (fuel)	$0.5\,m^3/hp/h$

Source: Adapted from Ref. [12]. © Tamera – Peace Research Center.

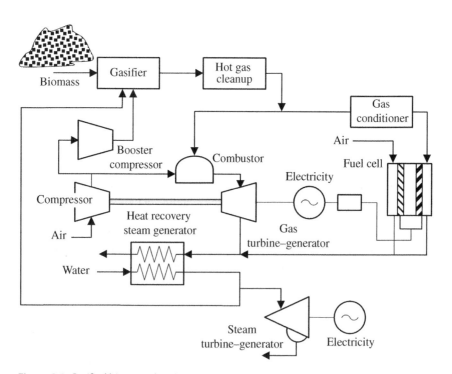

Figure 8.6 Gasified biomass electric generator system.

Table 8.10 Brazilian installed capacity of electric power according to the biofuel.

Source	Biofuel	Number of power plants	Capacity (MW)
Agricultural waste	Sugarcane bagasse	394	10,594.4
	Rice husk	12	45.33
	Elephant grass	3	65.7
	Biogas	2	1.72
	Vegetable oil	2	4.35
Forestry	Black liquor	17	1,978.1
	Forest residues	51	389.52
	Blast furnace gas	10	114.26
	Charcoal	7	51.39
	Firewood	1	11.5
Urban solid waste	Biogas	14	83.69
Animal waste	Biogas	10	1.92
	Total	523	13.352 MW

Source: Data Bank of Generation (BIG) (© ANEEL, Brazil, 2016).

Another application related to biomass is the utilization of digester gas derived from anaerobic sludge digestion in wastewater treatment plants. Many wastewater treatment plants in the United States employ anaerobic sludge digestion as a component of their treatment process. Microturbines are being used to transform the energy contained in the methane-based biogas produced as a by-product of the sludge digestion process into useful forms of electricity and thermal energy rather than disposing of the biogas as a waste product. There are varieties of technical and financial factors that must be considered and various barriers to the adoption of the microturbine technology that must be overcome before a microturbine installation can be deemed practical for a given site.

For example, in Brazil the share of biomass in the national energy matrix is approximately 8.8%, equivalent to 13,352 MW of installed capacity in 2015. In addition, their share is 33.7% of thermal generation capacity of the country, having 523 plants, as presented in Table 8.10. Furthermore, the most representative biofuel derived from agricultural sources and mills in that country, 411 of them represent a generation capacity of 10,711 MW—equivalent to 80.3% of the biomass generation [13].

Several interesting HOMER and other simulations describe the use of hybrid energy systems consisting of biomass/PV/wind for energy requirement of

Table 8.11 HOMER hybrid system components [10, 11] [13].

Diesel generator set	Wind turbine
Size: 250 kW	Size: 5 kW
Biodiesel generator set	Batteries
Size: 250 kW	Commercial models
PV module	Converters
Size: 150 kW	Size: 200 kW

Source: © ANEEL, Brazil, 2017.

residential and remote areas with some small-scale industries [10–12]. The biomass generator is used as the main source of power, whereas using photovoltaic panels and wind generators as supporting additional sources. This study concluded that the system could manage to supply a total area load with a cost of energy at approximately $0.04/kWh, which is very low as compared with the local domestic tariffs rated at about $0.1/kWh. Typical types of equipment used in such setups are listed in Table 8.11. Therefore this analysis leads to an appreciable cost reduction to make biomass a considerable solution for remote areas.

References

1 F.X. Aguilar (ed.), Wood Energy in Developed Countries, Resource Management, Economics and Policy, CRC Press, Taylor and Francis Group, Boca Raton, FL, 2014.

2 S.C. Hackett, Environmental and Natural Resources Economics, Theory, Policy and the Sustainable Society, 4th ed., CRC Press, Taylor and Francis Group, Boca Raton, FL, 2011.

3 R. Kulothungan, Dedicated energy plantation for sustainable biomass power generation, National Workshops on Technological Developments in Efficient Operation and Maintenance on Boilers, Pune, India, February 2014, pp. 46–51.

4 D.Y. Goswami and F. Kreith, Energy Efficiency and Renewable Energy Handbook, 2nd ed., CRC Press, Taylor and Francis Group, Boca Raton, FL, 2016.

5 T. Lensu and A. Alakangas, Small Scale Electricity Generation from Renewable Energy Sources: A Glance at Selected Technologies, Their Market Potential and Future Prospects, OPET Report 13, VTT Processes, Jyväskylä, Finland, May 2004.

6 Renewable Energy Sources and Climate Change Mitigation, Edited by O. Edenhofer, R. Pichs-Madruga and Y. Sokona, Published for the Intergovernmental Panel on Climate Change, Potsdam Institute for Climate Impact Research (PIK), Brandenburg, Germany, 2012.

7 U.R. Prasanna and K. Rajashekara, Fuel Cell based Hybrid Power Generation Strategies for Microgrid Applications, 2015 IEEE Industry Applications Society Annual Meeting, Addison, TX, USA, pp. 1–7, October 18–22, 2015.

8 J.K. Maherchandani, C. Agarwal, and M. Sahi, Economic feasibility of hybrid biomass/PV/wind system for remote villages using HOMER, International, Journal of Advanced Research in Electrical, Electronics and Instrumentation Engineering, Vol. 1, No. 2, pp. 49–53, 2012.

9 U.S. Department of Energy, Energy Efficiency and Renewable Energy, Bioenergy Technologies Office (BETO), Multi-Year Program Plan (MYPP), DOE/EE-1385, U.S. Department of Energy, Energy Efficiency and Renewable Energy, Washington, DC, March 2016.

10 G. Liu, M.G. Rasul, M.T.O. Amanullah, and M.M.K. Khan, "Feasibility study of stand-alone PV-wind-biomass hybrid energy system in Australia," in IEEE Power and Energy Engineering Conference (APPEEC), Wuhan, Asia-Pacific, IEEE, Piscataway, NJ, March 2011, pp. 1–6.

11 A. González, J.R. Riba, and A. Rius, Optimal sizing of a hybrid grid-connected photovoltaic-wind-biomass power system, Sustainability, Vol. 7, pp. 12787–12806, 2015.

12 T.H. Culhane, Biogas Digester, SolarVillage, Tamera – Peace Research Center, Monte Cerro (Valerio Marazzi – ARCO), Portugal, 2016.

13 Brazilian Energy Balance, Final Report, National Agency of Electrical Energy (ANEEL), Ministry of Mines and Energy, 2016.

9

Microturbines

9.1 Introduction

Utility gas turbine (GT) generators ranging from 500 kW to 250 MW are too large for distributed power applications. Microturbines were developed by the industry through improvements in auxiliary power units originally designed for aircraft and helicopters and customized for customer-site electric user applications. Therefore, microturbines from 30 to 400 kW were developed for small-scale distributed power either for electrical power generation alone, in distributed electrical power generation, or in combined cooling or heat and power (CCHP) systems.

Microturbines can burn a variety of fuels, including natural gas, gasoline, diesel, kerosene, naphtha, alcohol, methane, propane, and digester gas. The majority of commercial devices presently available use natural gas as the primary fuel. Modern microturbines have evolved dramatically, with advanced components such as inverters, heat exchangers (recuperators), power electronics, communications, and control. Microturbines have several inherent advantages over conventional fossil-fueled power systems:

- Microturbines have very low weight per horsepower ratio, resulting in light generator sets.
- Microturbines demonstrate pure rotary motion as opposed to stroking, resulting in less vibration, low noise compared to diesel generators, high mechanical performance, lower mechanical stress, and very high reliability.
- A liquid cooling system is not required.
- Some microturbines run on air bearings with very low maintenance.
- Microturbines exhibit a very fast response to load variation (high momentum), since they do not need to build up pressure as in steam turbines or have high torque as in reciprocating engines.
- Microturbines operate on a variety of fuels.
- Combustion usually runs on an excess of air, resulting in very low emissions.

Integration of Renewable Sources of Energy, Second Edition. Felix A. Farret and M. Godoy Simões.
© 2018 John Wiley & Sons, Inc. Published 2018 by John Wiley & Sons, Inc.

- Even low-power microturbines can provide recoverable heat for water and space heating, and the larger units can be used for industrial purposes or in a combined cycle with other turbines.

The disadvantages are:

- Difficulties in finding new commercial spare parts
- Difficulties to find an adequate model it in the market
- Lower efficiencies (from 28 to 32%) compared to reciprocating engines (over 38%)
- Very sensitive to the environment (temperature, pressure, and humidity needing derating for these conditions)
- Relatively new technology
- Need for better skilled technicians for repairing

Reciprocating internal combustion engines have some advantages over microturbines, such as efficiencies over 38%, whereas microturbines have efficiencies ranging from 28 to 32%. Microturbines are very sensitive to ambient air temperature, pressure, and humidity, thus requiring derating for those ambient variables. A major drawback of microturbines is the requirement for better skilled technicians to repair and do maintenance.

The US Department of Energy (DOE), the industry, and the National Fuel Cell Research Center made analysis of a hybrid power system that combines a GT with a high-temperature fuel cell (HTFC). These efforts have demonstrated that such a combination provides remarkably high efficiencies [1].

Microturbines are ideally suited for distributed generation applications [2] because of their flexibility in connection methods, ability to be stacked in parallel to serve larger loads, ability to provide stable and reliable power, and low emissions. The most typical applications include:

- Peak shaving and base load power (grid parallel)
- Combined heat and power
- Stand-alone power
- Backup/standby power
- Ride-through connection
- Primary power with grid as backup
- Microgrids (DG and mobile units)

Customers for microturbines include banks, hospital, financial services, data processing, telecommunications, restaurant, lodging, retail, office building, and other commercial sectors. Microturbines are currently operating in resource recovery operations at oil and gas production fields, wellheads, coal mines, and landfill operations, where by-product gases serve as essentially free fuel. Unattended operation is important since these locations may be remote from the grid and even when served by the grid may experience costly downtime when electric service is lost due to weather, fire, or animals. In CHP applications, the waste heat from microturbines can be used to produce hot water, to heat building space, to drive absorption cooling or desiccant dehumidification equipment, and to supply other thermal energy needs in a building or industrial process.

9.2 Principles of Operation

Figure 9.1 illustrates the mechanical features of a small GT engine. From left to right there are a compressor, a combustor, a turbine, and a generator, all sharing an only shaft. In general, an auxiliary machine starts the entire process initially by spinning the turbine shaft to introduce air compressed by the blades indicated on the left side of the shaft in Figure 9.1. The air is then mixed with fuel gas and passed into the combustion chamber of the turbine. This mixture is compressed to a point of continuous combustion, which drives the set of blades indicated on the right side of the shaft, increasing the speed of the shaft, which in turn admits more incoming air and fuel gas. A high-speed generator is connected on the shaft, which converts this mechanical energy into electricity. The operation is based on the Brayton cycle as presented in Chapter 2. At the end of the process, the hot residual gases from combustion are expelled.

Figure 9.1 shows air coming in through the inlet. The shape of the inlet must be designed carefully to prevent particulates and outside objects from being drawn into the engine. The air passes through a fanlike assembly of rotating impellers fitted with curved vanes at the intake area, which guides the air in, pulling it at higher speed and pressure.

There are several blades in succession in a turbine, each row taking the air shoved by the preceding wheel and providing an even greater push back into the combustion area, gradually accumulating significant differential pressure. The pressure of the air is raised about 10 times the atmospheric pressure and eventually enters the combustion chamber.

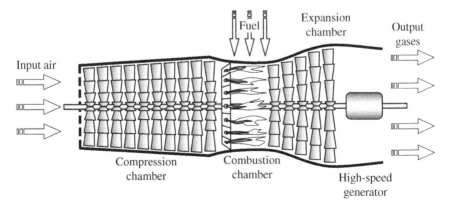

Figure 9.1 Principles of gas-microturbine power conversion.

The combustion assembly might include a single or multiple combustors. The combustion chamber is a reacting system (i.e., a chemical reaction takes place). In the combustion chamber, heat is added to the gas at constant pressure. The density decreases, while the specific volume and temperature are increased. Entropy is also increased, since combustion is not a reversible process. In the turbine, the situation is the opposite of the compressor, where the pressure and temperature decreases (for ideal expansion the entropy is constant). The compressor and other components are considered nonreacting. The combustion chamber takes the internal energy of the fuel, mixes the burning fuel with air at constant pressure, and increases the temperature. Usually, natural gas is used as fuel source. The best natural gas input should be at 80 psig or higher pressure. As the gas pipeline gets closer to customers, a regulator station reduces the pressure in the pipeline to about 60 psi or less. Therefore, pre-compression of natural gas can be required for microturbines.

The combustion chamber is of a lean premix emission type, achieving low emissions of NO_x, CO, and unburned hydrocarbons in the exhaust gases. The hot high-speed mix of air with burned fuel expands rapidly as it enters the expansion turbine stage. The expansion turbine is the power-producing element of the overall set. It consists of turbine wheels, each set on increasing greater diameters than the preceding one. Work is extracted from the hot gas by letting it expand. The turbine has a carefully designed cross section, wheel diameters, and vane angles, which force the gas to slow, cool, and expand at an optimal rate to collect the most energy out of the gas as it passes from one blade to another. Eventually, less than one-third of the power extracted from the hot gases is available to drive the generator, while the other two-thirds is used to drive the compressor and auxiliary equipment such as a fuel booster, ventilation fan, oil pump, water cooling pump, oil separator, and buffer air pump.

The generator is typically a high-speed permanent-magnet machine. The air leaves the turbine through the exhaust outlet. Although the operation seems simple, it must be designed and tailored to the turbine to have the correct speed and pressure combination to avoid backpressure and optimize overall performance.

The exhaust gases can be used in process heating applications or can be used indirectly with a heat exchanger to produce hot water. Typically, microturbines for distributed generation do not operate at very high temperatures, so no steam is produced, although large GTs can produce steam and may be coupled and combined with other cycles.

The shaft design used by some manufacturers is supported by airfoil bearings requiring no lubrication. If heat recovery is implemented, a careful study must be done to minimize piping and loss of heat. The heat exchangers (or recuperators) can heat water or glycol but may not be able to generate steam. A good combination is to use microturbines to heat boiler feedwater, effectively reducing boiler fuel needs. However, recuperators also cool the exhaust gas and hence limit the residual thermal energy available for use. Some microturbine manufacturers include a recuperator bypass valve to reduce the electrical efficiency but to increase the overall system efficiency by increasing the recoverable heat available. This option also provides increased flexibility to balance the electricity and heat production demands on site.

9.3 Microturbine Fuel

All GTs use the same energy source (i.e., an expanding high-pressure gas produced by the combustion of mixed fossil fuel and air). This mixture has the same physical properties regardless of the size of the GT. Therefore, the tips of the turbine blades must move at a speed appropriate to capture the energy from this expanding gas. This means that large utility turbines with 2.5 m wheels will spin at 1800 or 3600 rpm, while smaller turbine with wheels only 0.15 m in diameter will have a much higher rotational speed, up to 100,000 rpm. Therefore, large GTs use the same philosophy as that of central-station steam turbines or diesel technology driven at constant speed. On the other hand, microturbines have special generators running at very high speed, which must be able to supply stand-alone loads. Therefore, a variable-speed generator is typically used as an intermediate power electronics stage to convert the variable voltage and frequency to a fixed grid frequency (i.e., the output of the high-speed generator is rectified and coupled to a dc/ac power converter).

Most GTs are equipped to burn natural gas. A typical range of heating values of gaseous fuels acceptable for GTs is 900 to 1100 Btu per standard cubic foot (SCF), which covers the range of pipeline-quality natural gas. Clean liquid fuels are also suitable for use in GTs. Larger GTs may have special combustors to handle cleaned gasified solid and liquid fuels. Fuels are permitted to have low levels of specified contaminants (typically less than 10-ppm total alkalis and single-digit ppm amounts of sulfur).

Liquid fuels require their own pumps, flow control, nozzles, and mixing systems. Many GTs are available with either gas- or liquid-firing capability. In general, such GTs can be converted from one fuel to another. Several GTs are equipped for dual-fuel firing and can switch fuels with minimal or no interruption. Lean-burn/dry low NO_x combustors can generate NO_x emission levels as low as 9 ppm (at 15% O_2), and those with liquid fuel combustors have NO_x emissions as low as approximately 25 ppm (at 15% O_2). Although there is no substantial difference in the general performance with either fuel, the different heat levels of combustion result in slightly higher mass flows through the expansion turbine when liquid fuels are used, thus a very small increase in power and efficiency performance is obtained. Fuel pump power requirements for liquid fuels are less power demanding than fuel gas booster compressors, further increasing net performance with liquid fuels.

A fuel gas booster compressor can be required to ensure adequate fuel pressure for GT flow control and combustion systems. GT combustors operate at pressure levels of 75–350 psi. Although the pressure of pipeline natural gas is higher in interstate transmission lines, the pressure is typically reduced at the city gate metering station before it flows into the local distribution piping system. The cost of the fuel gas booster compressors adds to the installation capital cost. High reliability will often require redundant booster compressors. Liquid-fueled GTs use pumps to deliver the fuel to the combustors.

GTs are among the cleanest fossil-fueled power-generation equipment commercially available. The primary pollutants from GTs are oxides of nitrogen (NO_x), which can be thermal and fuel NO_x (thermal NO_x is formed by the dissociation of atmospheric nitrogen (N_2) and oxygen (O_2) in the turbine combustor and the subsequent formation of NO_x, while fuel NO_x is present in the fuel itself). In addition, there are carbon monoxide (CO) and volatile organic compounds (VOCs). Other pollutants, such as oxides of sulfur (SO_x) particulate matters, depend primarily on the fuel. Emissions of sulfur compounds, primarily SO_2, reflect the sulfur content of the fuel. GTs operating on natural gas or distillate oil that has been desulfurized in a refinery emit insignificant levels of SO_x. In general, SO_x emissions are significant only if heavy oils are fired in the turbine, and SO_x control is a fuel purchasing issue rather than a turbine technology issue. Particulate matter is a marginally significant pollutant for GTs using liquid fuels. Ash and metallic additives in the fuel may contribute to particulate matter in the exhaust.

It is important to note that the GT operating load has a significant effect on the emissions levels of the primary pollutants (NO_x, CO, and VOCs). GTs are typically operated at high loads. Therefore, they are designed to achieve maximum efficiency and optimum combustion at these high loads.

9.4 Control of Microturbine

Modeling of microturbines is a very complex matter. Although there are several modeling efforts [3–7], there is no complete electrical engineering-based model available for feedback control and automation development purposes. To support the modeling and control approach taken in this book, several assumptions and simplifications had to be made. For example, constant physical and chemical properties are assumed, such as specific heat at constant pressure and volume and specific gas constant. Heat loss is neglected and there is no internal energy storage. Through a thermodynamic evaluation, a microturbine can be considered a topping cycle (i.e., electricity or mechanical power is produced first and heat is recovered to meet the thermal loads). Topping cycle systems are generally used when there is no requirement for high process temperatures. Modern commercial microturbines operate on the nonideal Brayton open cycle with heat recovery.

9.4.1 Mechanical-Side Structure

The control structure of a microturbine will depend on whether the unit has a single- or two-shaft structure. The physical arrangement of a microturbine can be used for classification, for example, single shaft or two shaft, simple cycle or recovered, and intercooled or reheated. Unrecovered turbines have the compressed air mixed with fuel and burned under constant-pressure conditions. The resulting hot gas is allowed to expand through a turbine to perform work. Simple-cycle microturbines have lower efficiencies, lower capital costs, higher reliability, and more heat available for cogeneration applications than recuperated units do. Recuperated units use a metal sheet heat exchanger that recovers some of the heat from an exhaust stream and transfers it to the incoming airstreams, boosting the temperature of the airstreams supplied to the combustor. Further exhaust heat recovery can be achieved in a cogeneration configuration. Advanced materials such as ceramics and thermal barrier coatings are some of the key enabling technologies used to improve microturbines further. Efficiency gains can be achieved with materials like ceramics, which allow a significant increase in engine operating temperature.

A lot of knowledge of large steam turbines and large synchronous generators built a very solid understanding of the steady state and dynamic behavior for such systems. The basic operation has two modes: steady state and load transient. At steady state, the power of the steam rate into the turbine is equal to the electrical power removed from the generator. The speed of the generator and turbine is considered synchronous, implying that the output electrical sinusoidal voltage is in phase with the grid. This operation requires good speed control of the turbine. During a load transient, the change in power is taken from the rotor speed of the large turbine generator set. Because these devices are enormous, there is considerable stored energy in the rotating masses. The speed control of the turbines sees this speed change and corrects the rate at which the steam is supplied to the turbine, correcting the speed until the set point is achieved. In this manner, the turbine generator set is capable of nearly instantaneous load tracking.

Unfortunately, the same base of knowledge is not yet fully developed for microturbines and the correspondent generators. However, the following similar principles are important in establishing the control approach for a microturbine:

- At steady state, the power of the natural gas combustion and air into the turbine is ideally equal to the electrical power removed from the generator. The speed of the generator and turbine is not critical, as the output sinusoids from the generator are rectified. The dc-link voltage needs to be supported to ensure that the power requirements are conserved. This operation requires good speed control of the turbine.

- During a load transient, the change in power is taken from the rotor speed of the microturbine. However, because these devices are small, there is very little stored energy in the rotating masses, and the rotor speed changes very quickly. The speed control of the microturbine sees this speed change and corrects the rate at which the fuel is supplied to the microturbine, correcting the speed until the set point is achieved. The speed of a microturbine needs to be changed quickly to ensure that the generator does not stall. In this manner, the turbine generator set is capable of load tracking.

The shaft construction defines many important characteristics that will eventually influence the control system. There are essentially two types of shaft construction. One is a high-speed single-shaft design with the compressor and turbine mounted on the same shaft as the alternator rotates at speeds of 90,000 to 120,000 rpm. The split-shaft design has a power wheel on a separate shaft and transfers its output to a conventional generator via a gear reducer [8, 9], as indicated in Figure 9.2. This system philosophy employs the same technology as that for two-pole generator sets running at 3600 rpm or may use an induction generator with a static VAR compensator to assist as a source of reactive power. This is a proven robust scheme, and although manufacturers claim that it does not need power electronics, it requires synchronizing equipment and relays for connection to the electric grid. The gear reducer requires maintenance along with the lubrication system. There are two turbines, one a gasifier turbine driving a compressor and another free power turbine driving a generator at a rotating speed of 3600 rpm. Therefore, there is one combustor and one gasifier compressor. In conventional power plants, a two-pole permanent-magnet generator is driven via a gearbox, running at constant speed, since it is synchronized to the electric network.

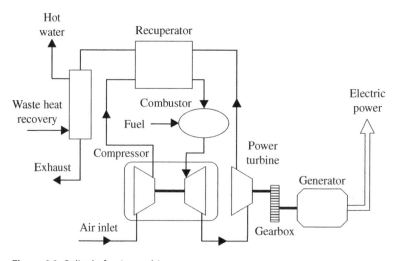

Figure 9.2 Split-shaft microturbine system.

Figure 9.3 Capstone single-shaft microturbine. *Source*: Courtesy of National Renewable Energy Laboratory.

The intake gas is controlled through a nozzle, where its velocity is increased and the pressure decreases. The high-speed gas exerts a force on the turbine blades, and the geometry of the blades causes the turbine to rotate, thus producing mechanical work. After the turbine rotor, the gas passes through the diffuser of the turbine, where the velocity is again decreased and the pressure is increased, but not as much as after the compressor. Most of the steam power-based technology is used to control a split-shaft GT system [8, 9].

A commercial single-shaft microturbine system is portrayed in Figure 9.3. It is an integrated design where the rotor spins on a thin film of pressured air rather than oil inside its bearings. The high-speed generator is integrated to the shaft as well as the recuperator. The compressor and turbine blades are designed to shape the airflow to go through the combustion chamber and exhaust the hot gas at optimal pressure.

9.4.2 Electrical-Side Structure

One advantage of a high-speed generator is that the size of the machine decreases almost in direct proportion to the increase in speed, leading to a very small unit that can be integrated with the GT. In the split-shaft design, a starter is needed to bring the turbine up to operational speed, whereas in the single-shaft design, the generator may operate in motoring mode through a bidirectional power electronic system for start-up. The generator is a rare-earth permanent-magnet alternator. The stator core is made with thin laminations of low-loss electrical steel, and coils are made of Litz wire to achieve good high-frequency characteristics. A four-pole design contributes to short end windings that make possible a short distance between bearings.

Water cooling might be used to keep the temperature low in the windings. Vacuum-pressurized impregnation and insulation thicker than standards ensures a long lifetime for the windings. An important issue is to ensure that the rotor never reaches a temperature that would demagnetize the magnets. This is done by reducing the rotor losses through providing efficient cooling in the air gap.

Power electronic inverters are also integrated to the generator system, and high switching transistors with customized monitoring system control the operation of the generator. As opposed to the split-shaft design, the use of an inverter provides several advantages, such as elimination of a gearbox and start-up features. The power electronics provides most of the protection and interconnection functionalities in addition to power factor correction and control flexibilities impossible with the split-shaft concept.

Figure 9.4a shows a bidirectional power electronics converter that allows motoring and regeneration control of the high-frequency permanent-magnet generator [10–12]. Figure 9.4b shows a more simplified scheme where the generator voltage is rectified and delivered to an inverter tied to the grid. The scheme provided by Figure 9.4a is more versatile; it requires very fast devices and very fast signal processing. This is the current challenge for power electronics engineers. On the other hand, the scheme in Figure 9.4b requires standard power electronic devices at the expense of an auxiliary mechanical start-up on the shaft for bringing the turbine to the optimal generation speed range. Most commercial turbines are constructed with a front-end rectifier such as Figure 9.4b for lower cost and less complexity.

The overall single-shaft microturbine system is very intricate, as depicted by Figure 9.5. The main input to the regulator is an electric power reference (i.e., the amount of electric power the generator should produce). The main output of a regulator is the fuel rate (i.e., how much fuel will be injected in the combustion chamber). There is a maximum value, with the fuel valves fully open, and a minimum value, when the valves are open just to keep the flame alive.

9.4.3 Control-Side Structure

There are measurements in the feedback loop for the speed, the power currently produced in the generator, and the outlet temperature at the turbine. When the microturbine is connected to the grid, the GT provides heat and power at a constant level to a local load. If the power demand in the local power grid increases to a higher value than the microturbine maximum output, the

(a)

(b)

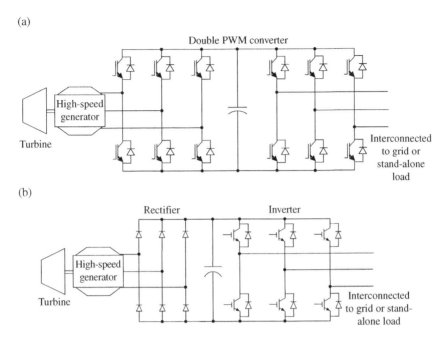

Figure 9.4 Power electronic topologies for permanent-magnet generators: (a) bidirectional; (b) unidirectional.

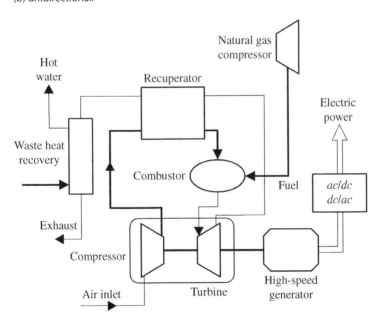

Figure 9.5 Single-shaft microturbine overall system.

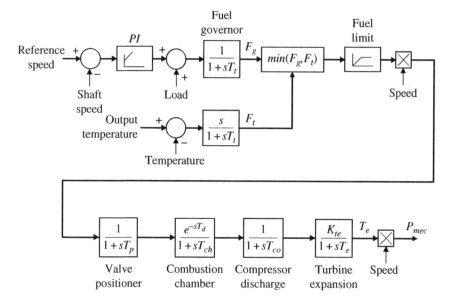

Figure 9.6 Modeling a microturbine system from speed to mechanical power delivery.

extra power is taken from the outer power grid. The microturbine usually runs at its optimum, full load. Since the control system does not need to meet any sudden changes, all changes can be done slowly, and the thermal fluctuations inside the GT are minimized.

The life-span of a GT is closely related to the size and speed of the temperature changes it experiences. These changes can be slowed by including a rate limiter function. The total efficiency is optimal at a certain temperature and speed, which makes the microturbine more economical if always run at the optimum point. The purpose of the control system is to produce the power demanded while maintaining optimal temperature and speed characteristics.

In stand-alone mode, the control system must maintain a constant voltage and frequency insensitive to variations in the load power. Therefore, a very fast control loop is required for stand-alone systems. The faster control is achieved at the expense of higher thermal variations in the machine, thus reducing its life span. Figure 9.6 depicts a simulation block diagram of a microturbine system from speed control to mechanical power delivery [13]. There are several time constants and coefficients to be determined, but it is not easy to obtain this information from a manufacturer's data sheet. System identification procedures [14–16] must be conducted to define:

- Fuel governor response
- Temperature rise time
- Valve positioner

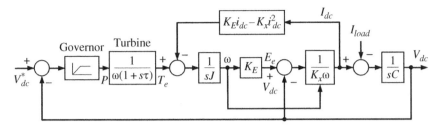

Figure 9.7 Combined electromechanical system model for a microturbine system.

- Combustion chamber response
- Time delay of the combustion chamber
- Compressor discharge
- Turbine expansion

The control system consists of a normal speed regulator that tries to maintain a predefined reference speed (e.g., at 65,000 rpm). Such speed may not be optimal for lower power loads, but the regulator tries to maintain the speed by controlling the fuel input to the turbine. If the load is increased suddenly, it is very difficult to increase the speed without reducing the power load for a moment, and a very complex control must consider all those variable. The upper performance limit here is the limitation of the fuel injectors. The maximum amount of fuel that can be injected is based on temperature limitations.

The full system of Figure 9.6 can be simplified, and a first-order dominant pole can be assumed to represent the torque response to the fuel governor. In this case, Figure 9.7 represents a possible control system diagram that integrates a microturbine generator with a front-end rectifier with control of the dc-link voltage [17, 18].

Some parameter and system identification procedures are still required to determine the proportional integral parameters for the governor and turbine time constant t, system inertia J, and apparent turbine impedance to the dc-link system by constants K_e and K_x. The ac load I_{load} will eventually draw power from the dc link. Therefore, this model considers an equivalent current at the dc link, which is reflected by the ac load.

The power–speed-operating regions are constrained by three variables, as indicated in Figure 9.8. A minimum dc-link voltage is required for maintaining a linear pulse width modulation. Therefore, such a constraint imposes a minimum shaft speed. Maximum power is determined by the fuel injection rate to the combustors, limiting the maximum power over the operating range. Bearings, vibrations, and overall mechanical design constraints will limit the maximum speed on the right side of the power–speed envelope of Figure 9.8.

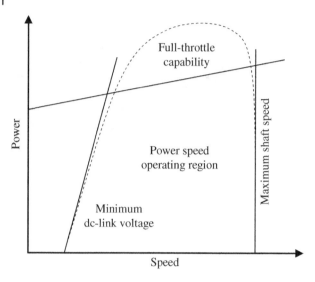

Figure 9.8 Constraints on power and speed of a microturbine.

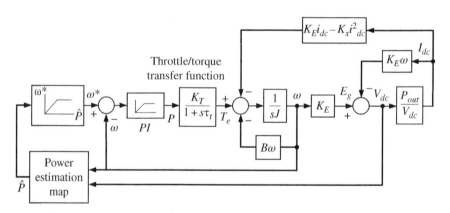

Figure 9.9 Direct self-control approach of a microturbine with power-speed mapping.

Inside the power–speed-operating region of Figure 9.8, a mapping of power to speed can be defined either by system identification or by nonlinear methods. If a family of curves defines such a power–speed mapping, a very robust direct self-control system can be implemented as indicated by Figure 9.9. For the sake of simplicity, the dc bus capacitance response is neglected in Figure 9.9. As dc-link voltage has a trend to increase or decrease, the power estimation map will define the estimated output power and will generate the speed reference. The shaft speed will accommodate the microturbine within the power–speed-operating region by a self-tracking procedure [19]. Figure 9.10

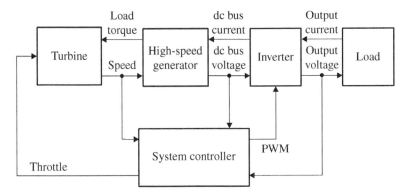

Figure 9.10 Control system and output inverter for a microturbine.

depicts the signal flow of all the interconnecting modules (i.e., the system controller controls the fuel injector by direct self-control after reading the dc bus output voltage, output voltage, current, and shaft speed) [20–22]. The output inverter is controlled by sinusoidal pulse width modulation (SPWM) to deliver clean output voltage.

9.5 Efficiency and Power of Microturbines

The work produced by the turbine can be calculated by an equation using the inlet and outlet pressures (p_1 and p_2) and input temperature (T_1) at isentropic conditions. Those conditions are assuming ideal conditions (the process can be reversed without losing energy) and no heat flow in or out of the system:

$$W = \left(\frac{\kappa}{\kappa - 1} \right) R T_1 \left[\left(\frac{p_2}{p_1} \right)^{\frac{\kappa - 1}{\kappa}} - 1 \right] \tag{9.1}$$

where:

k = The ratio of specific heat at constant pressure and constant volume
R = The gas constant for the gas specified, in this case, air

In addition to simplifying ideal isentropic conditions, friction, turbulent eddies behind the vanes, and efficiency of the compressor, all of them decrease the output power. Therefore, the thermodynamic variable called the isentropic efficiency η_{is}, defined as the ratio of the enthalpy change at isentropic conditions and the actual enthalpy change, is used to compensate for the actual conditions. The value of the isentropic efficiency can be determined through experiments and tabulated for variable temperature and pressure ratio. The work multiplied by the mass flow rate $\dot{m}$, the isentropic efficiency, and the mechanical efficiency due to the bearing and windage losses, η_{mec}, gives the output power, as indicated by

$$P_{turbine} = \dot{m} \cdot \eta_{is} \cdot \eta_{mec} \cdot \left(\frac{\kappa}{\kappa-1}\right) RT_1 \left[\left(\frac{p_2}{p_1}\right)^{\frac{\kappa-1}{\kappa}} - 1\right] \tag{9.2}$$

Formulation of equations (9.1) and (9.2) embeds the heating value of the natural gas in the outlet pressure increase in the air. However, for economical evaluation, it is very important to define the turbine efficiency from fuel to the output power. The natural gas input power in Btu/hr or kJ/s can be evaluated by

$$P_{fuel} = HV_s \cdot Q \cdot \frac{P}{P_s} \cdot \frac{T_s}{T} \tag{9.3}$$

where:

HV_s = The heating value of the natural gas at standard temperature and pressure (energy/volume)
Q = The natural gas volume flow rate
P = The natural gas pressure
P_s = The standard pressure
T = The natural gas temperature
T_s = The standard temperature.

For example, for natural gas at $T_s = 20\,°C$ and $P_s = 14.7\,psi$, the heating value is $HV_s = 1042\,Btu/ft^3$.

As depicted in the power equations previously mentioned, one can consider a microturbine as a mass flow machine where changes in the air density affect the performance significantly. The industry has set the inlet air temperature at 15 °C for determining performance. Therefore, if a microturbine is rated at 60 kWh output, it is for 15 °C inlet air temperature. Manufacturers provide information on the derating of power and efficiency for variable inlet air temperature (see Table 9.1). An example is given in Figure 9.11.

The elevation is also very relevant technical information, and a manufacturer data sheet is usually available to provide the relative sensitivity with altitude. A rule of thumb is that for every 300 m above sea level, the net power should be derated by 3%. However, a combination of altitude and inlet temperature must be used to determine output power [23]. Since exhaust gases are usually ducted for either heat recovery or use as a drying furnace, it is important to calculate the decay in power and efficiency, as this backpressure affects microturbine

Table 9.1 Typical characteristics of gas turbines.

Specification	Micro	Mini	Large
Turbine speed	110,000 rpm	5,000–15,000 rpm	1,800–3,600 rpm
Generator speed (rpm)	110,000 rpm	3,600 rpm	1,800–3,600 rpm
Generator type	dc generator + dc/ac	ac synchronism	ac synchronism
Gearbox	None	3 : 1	None
η	20–25%	30–31%	45%
Power	15–250 kW	500 kW to 10 MW	>25 MW
Starter	Generator as motor		

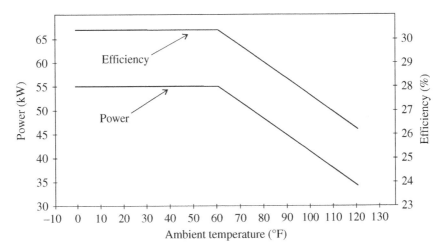

Figure 9.11 Power and efficiency in terms of ambient temperature.

performance. Losses as a function of the duct diameter, duct length, number of bends in the duct, and type of material that determines the roughness, and friction factors must be accounted for as the total head loss and must be limited to the level specified by the microturbine manufacturer.

9.6 Site Assessment for Installation of Microturbines

Microturbines can be used for standby power, power quality and reliability, peak shaving, and cogeneration applications. In addition, because microturbines are being developed to utilize a variety of fuels, they are being used for resource recovery such as landfill gas applications. Sitting and sizing a microturbine require a comprehensive economical evaluation and a study of safety,

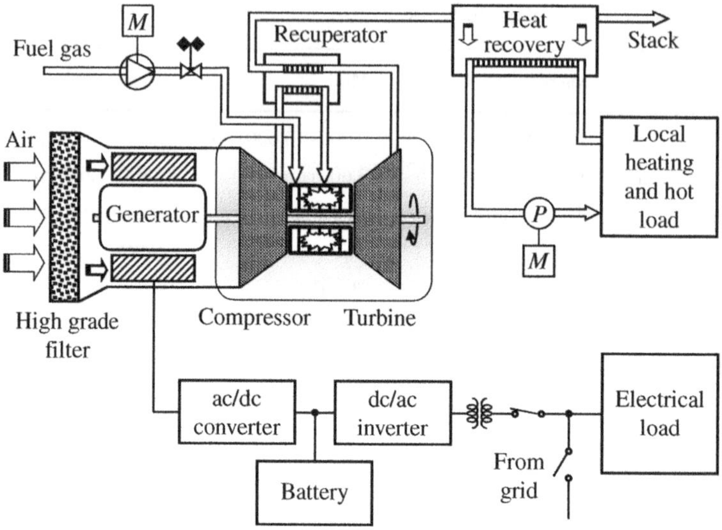

Figure 9.12 Microturbine heat recovery system.

building, and electrical codes. Microturbines are well suited for small commercial building establishments such as restaurants, hotels/motels, small offices, retail stores, and many others.

Microturbines provide an aesthetic power source, improving sightlines and views with off-the-grid systems that eliminate the need for overhead power lines and enable cost savings by reducing the peak demand at a facility. As a result, it lowers demand charges and provides better power reliability and quality, especially for those in areas where brownouts and surges are common or utility power is less dependable. Microturbines can provide power to remote applications where traditional transmission/distribution lines are not an option, such as construction sites and offshore facilities, and possess combined heat and power capabilities for sensitive applications.

Figure 9.12 illustrates a microturbine heat recovery system. Typical microturbine power and heat output efficiencies for power generation are at around 30% now. Notice that a typical microturbine speed can be quite above 100,000 rpm, and therefore, it produces high-frequency power voltages that has to be converted into dc voltage for battery charge and then to the standard 60 Hz grid. The electrical efficiency falls to about half of that when the recuperator bypass is engaged (about 30%). In this case, the overall thermal efficiency may rise to about 80%. The exhaust gas temperature is typically about 260 °C while using the recuperator and 870 °C with the recuperator bypassed. Not all of this heat can be transformed effectively into useful energy. For example, the Capstone microturbine can produce hot water at about 90 °C when joined with a CHP heat recovery unit [22, 23].

Placement of microturbines is critical to avoid distribution losses, piping, and wiring costs. Therefore, ideal locations are near already existing emergency diesel generators and boiler rooms with available connectivity to natural gas, electricity, and hot water. Often, outdoor installations are more cost-effective because audible noise is typically on the level of 70 dBA at 100 m, and it is probably convenient to find an outdoor place in the building with easy connections to exhaust and fresh air ducting and close to utility connections.

Microturbine capital costs currently range from $700 to $1100/kW, including hardware, software, and initial training. Adding heat recovery increases the cost by $75–$350/kW. Installation costs vary significantly by location but generally add 30–50% to the total installed cost. Microturbine manufacturers are targeting a future cost below $650/kW. This appears to be feasible if the market expands and sales volumes increase. With fewer moving parts, microturbine vendors hope the units can provide higher reliability than that for conventional reciprocating engine generator technologies. Most manufacturers are targeting maintenance intervals of 5000–8000 hours. Maintenance costs for microturbine units are still based on forecasts with minimal real-life situations. Estimates range from $0.005 to $0.016/kWh, which would be comparable to that for small reciprocating engine systems [24, 25].

References

1 S. Samuelsen, Fuel cell/gas turbine hybrid systems, NFCRC report, ASME, National Fuel Cell Research Center, University of California, Irvine, CA, 2004.

2 W.G. Scott, Micro-turbine generators for distribution systems, IEEE Industry Applications, Vol. 4, No. 3, pp. 57–62, 1998.

3 P. Ailer, Modeling and nonlinear control of a low-power gas turbine, M.Sc. thesis, Department of Aircraft and Ships, Budapest University of Technology and Economics, Budapest, Hungary, 2002.

4 S. Haugwitz, Modeling of micro-turbine systems, M.Sc. thesis, Department of Automatic Control, Lund Institute of Technology, Lund, Sweden, 2002.

5 Improving the operating efficiency of microturbine-based distributed generation at an affordable price, U.S. Department of Energy, Advanced Manufacturing Office, Energy Efficiency and Renewable Energy, DOE-EE-0432, June 2014.

6 A.A. Pérez Gómez, Modeling of a gas turbine with Modelica, M.Sc. thesis, Department of Automatic Control, Lund Institute of Technology, Lund, Sweden, May 2001.

7 Y. Zhu and K. Tomsovic, Development of models for analyzing the load following performance of micro-turbines and fuel cells, Journal of Electric Power Systems Research, Vol. 62, No. 1, pp. 1–11, 2002.

8 W.I. Rowen, Simplified mathematical representations of heavy-duty gas turbines, ASME Journal of Engineering for Power, Vol. 105, pp. 865–869, 1983.

9 L.N. Hannett, G. Jee, and B. Fardanesh, A governor/turbine model for a twin-shaft combustion turbine, IEEE Transactions on Power Systems, Vol. 10, No. 1, pp. 133–140, 1995.

10 A. Al-Hinai, K. Schoder, and A. Feliachi, Control of grid-connected split-shaft micro-turbine distributed generator, Proceedings of the 35th Southeastern Symposium on System Theory, Morgantown, WV, March 16–18, 2003.

11 R.H. Staunton and B. Ozpineci, Micro-turbine Power Conversion Technology Review, Report ORNL/TM-2003/74, Oak Ridge National Laboratory, Oak Ridge, TN, April 2003.

12 E. Mollerstedt and A. Stothert, A model of a micro-turbine line side converter, Proceedings of the International Power System Technology Conference, Perth, Western Australia, Vol. 2, December 4–7, 2000, pp. 909–914.

13 O. Aglen, Back-to-back tests of a high-speed generator, Proceedings of the IEEE International Electric Machines and Drives Conference, IEMDC'03, Madison, Wisconsin, USA, Vol. 2, pp. 1084–1090, June 1–4, 2003.

14 J. Padulles, G.W. Ault, and J.R. McDonald, An integrated SOFC plant dynamic model for power systems simulation, Journal of Power Sources, Vol. 86, pp. 495–500, 2000.

15 Y. Tanaka, Alpha parameter method and its application to a gas turbine, Proceedings of the 35th Conference on Decision and Control, Kobe, Japan, December 1995, pp. 2792–2793.

16 P.D. Fairchild, S.D. Labinov, A. Zaltash, and B.D. Rizy, Experimental and theoretical study of micro-turbine-based BCHP system, presented at the 2001 ASME International Congress and Exposition, New York, November 11–16, 2001.

17 A. Cano and F. Jurado, Modeling micro-turbines on the distribution system using identification algorithms, Proceedings of the IEE Conference on Emerging Technologies and Factory Automation, ETFA'03, Vol. 2, pp. 717–723, September 16–19, 2003, Lisbon, Portugal.

18 R. Lasseter, Dynamic models for micro-turbines and fuel cells, Proceedings of the IEEE Power Engineering Society Summer Meeting, Vol. 2, pp. 761–766, July 15–19, 2001, Juan les Pins, France.

19 M. Etezadi-Amoli and K. Choma, Electrical performance characteristics of a new micro-turbine generator, Proceedings of the IEEE Power Engineering Society Winter Meeting, Vol. 2, pp. 736–740, January 28–February 1, 2001, Ohio, USA.

20 A.M. Azmy and I. Erlich, Dynamic simulation of fuel cells and micro-turbines integrated with a multimachine network, Proceedings of the IEEE Power Tech Conference, Bologna, Italy, Vol. 2, June 23–26, 2003.

21 A. Al-Hinai and A. Feliachi, Dynamic model of a micro-turbine used as a distributed generator, Proceedings of the 34th Southeastern Symposium on System Theory, March 18–19, pp. 209–213, 2002, Huntsville, Alabama, USA.

22 Q. Zhang and P.L. So, Dynamic modeling of a combined cycle plant for power system stability studies, Proceedings of the IEEE Power Engineering Society 2000 Winter Meeting, Singapore, January 23–27, 2000.

23 G. Venkataramanan, M.S. Illindala, C. Houle, and R.H. Lasseter, Hardware Development of a Laboratory-Scale Microgrid Phase 1-Single Inverter in Island Mode Operation, Report NREL/SR-560-32527, National Renewable Energy Laboratory, Golden, CO, November 2002.

24 B.F. Kolanowski, Guide to Micro-turbines, Fairmont Press, Lilburn, GA, and Marcel Dekker, New York, 2004.

25 H.L. Willis and W.G. Scott, Distributed Power Generation: Planning and Evaluation, Marcel and Dekker, Inc., New York, 2000, p. 195.

10

Earth Core and Solar Heated Geothermal Energy Plants

10.1 Introduction

Geothermal energy is heat derived from the Earth that can be used either directly or indirectly. There are many areas in the western United States and across the developing world for all sizes of geothermal projects. Geothermal energy has been around for as long as the Earth has existed. "Geo" means Earth and "thermal" means heat, so geothermal represents Earth heat. It is so named because the core of the Earth is a very hot mixture of incandescent matter under its thin, crust-like mantle or is heat accumulated by the sunrays over the Earth's crust. This is why such energy has to be distinguished according to its origin as the Earth core and solar geothermal energy, respectively. Long-term solar geothermic is fully sustainable but the Earth core geothermic may not because we are extracting hot water from the Earth's core and injecting back cold water.

The lithosphere is the rigid outermost shell of a planet (crust and upper mantle) broken up into plate tectonics. Plate tectonics are the probable outer shell divisions of several plates gliding over the external mantle. Below the Earth's crust, the top layer of the mantle is hot liquid rock called magma on which the oval crust of the Earth floats. When magma breaks through the surface of the Earth in a volcano, it is called lava (see Figure 10.1).

The temperature underground raises about $3\,°C$ for every $100\,m$ of depth. Such approximation is valid for underground temperature still close to the crust, far and away from plate tectonics boundaries. The superficial underground, which is considered to be "surface geothermal," down to less than $100\,m$ from the soil is pretty much maintained in average temperature, impressed by the amount of the sun's heat above the surface. Therefore, in summary, the first $100\,m$ of depth, we have an underground with constant temperature, close to the average temperature of the open-air soil above. Then, for every $100\,m$ below, the temperature raises approximately $3\,°C$. Therefore, about $3000\,m$ underground, the temperature will be high enough to

Integration of Renewable Sources of Energy, Second Edition. Felix A. Farret and M. Godoy Simões.
© 2018 John Wiley & Sons, Inc. Published 2018 by John Wiley & Sons, Inc.

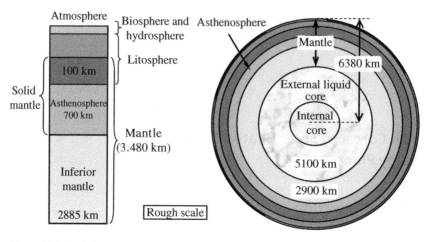

Figure 10.1 Earth layers.

boil water. If the steam cannot escape, the mix of water and steam can make hot water to reach very high temperature. Therefore, for about 4–5 km of depth, any trapped water there can easily get as hot as 150 °C (this water does not turn into steam because it is trapped, without contact with air, and not allowed to expand). When such hot water comes through a crack, that phenomenon is called as a "hot spring," for example, Emerald Pool and geyser called Old Faithful, both in the Yellowstone National Park. Hot springs have been harnessed and used by human beings for many centuries. In North America, several Native American tribes built settlements close to hot springs. In Japan, they are plentiful and still attracting a lot of tourists. The Romans built buildings to enjoy their hot baths, and several European cities still use those installations in modernized ways. For deeper depths, more than 5 km, the temperature is not linearly increasing, but it depends on the influence of the Earth's core, and most of the time this high temperature reservoir is not harnessed for industrial or commercial applications, but oil exploration sometimes reaches such heat reservoirs. For the upper layer of the Earth, close to the surface, that is not very hot. The sunshine irradiation makes it much hotter just below the ground, working like an isolation layer. That is the solar geothermal energy. Almost everywhere across the planet, the upper 6 m below ground level stays at the same temperature (the local average ambient temperature), something between 10 and 16 °C. This can be felt in a basement of a building or in a cavern underground, where the temperature of the area is usually cooler in hot places and warmer in cold places.

The Earth's core and/or solar geothermal energy are advantageous because they are renewable, reliable, and efficient, and they do not need any fuels. These plants are seldom offline for maintenance or repairs and have the highest-capacity factors of all power plants [1]. In general terms, geothermal energy is very attractive because it is renewable (solar), reliable, efficient, easy to produce, and water recyclable (Earth core); it generates 95% of the time; it depends on the location (Earth's core); it rarely needs any maintenance; and it has the highest-capacity factor among all, with operating costs of $0.05–0.08/kWh.

10.2 Earth Core Geothermal as a Source of Energy

The Earth core geothermal heat flows outward from the Earth's interior, causing a convective motion in the mantle usually as brine and vapor. This, in turn, drives tectonic plates. Tectonic movement of plates within the Earth refers to plates that collide periodically usually causing other events to take place. When two plates collide, one plate is subducted beneath the other, and the subducted plate slides downward, reaching pressures and temperatures that cause the plate to melt, forming magma. The magma ascends on buoyant forces, pushing plates into the crust of the Earth and causing large amounts of heat to be produced. At the surface, magma forms volcanoes; below the surface, magma creates subterranean regions of hot rock. Faults and cracks at the Earth's surface allow water seepage into the ground. The water is then heated by the hot rock and can naturally circulate back to the surface, forming hot springs, geysers, and mud pots. Assuming that the water is not allowed to recirculate but is trapped by impermeable rock, the water fills the rock pores and cracks, forming a geothermal reservoir.

The direct use of Earth core geothermal energy is achieved primarily by using shallower reservoirs of lower temperatures. Hot springs and spas are very popular around the world and are utilized when the sulfur content is relatively low. Other applications involve the use of the reservoirs for agricultural production, aquacultures, industrial applications, and heating.

Earth core geothermal resources can be used for either power generation or direct consumption. There are four methods of power generation: direct steam, flash steam, binary cycle, and combined/hybrid.

Direct steam power generation is the least common type of generation and is characterized by vapor-dominated and dry steam reservoirs. Most common are shallower reservoirs of lower temperatures for bathing and spas, agriculture, aquaculture, industrial applications, and district heating. The industrial use is related to drying of fish, fruits, vegetables, and timber; washing of wool; dyeing of cloth; manufacturing of paper; pasteurizing of milk; and piping under sidewalks and roads to keep them from icing over in freezing weather.

Flash steam generation (above 180 °C) can be used in flash plants to make electricity, and it is the most common type of generation used in high temperature applications. Fluid is sprayed onto a tank at a much lower pressure than the sprayed fluid, causing some of it to rapidly vaporize, or "flash" over the tank. The vapor then drives a turbine, which drives a generator. An example is

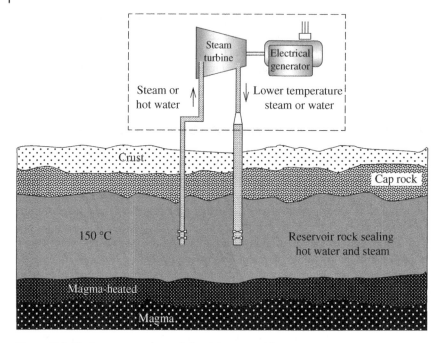

Figure 10.2 Flash steam geothermal electricity power plant.

shown in Figure 10.2. It is characterized by liquid-dominated reservoirs, flashing the water into steam to turn turbines.

Binary cycle power generation is also very common but is the most expensive, as it utilizes a working fluid. Most geothermal areas contain moderate-temperature water (below 200 °C). Energy is extracted from these fluids in binary cycle power plants where hot geothermal fluid and a secondary binary fluid with a much lower boiling point than water pass through a heat exchanger (see Figure 10.3). Heat from the geothermal fluid causes the secondary fluid to flash to vapor, which then drives the turbines. The binary cycle can be used with lower-temperature reservoirs, as the water heats the working fluid, and the working fluid steam actually turns the turbine.

Finally, combined or hybrid generation is not used very widely and is currently being researched to combine the foregoing techniques to obtain the most efficient use of the Earth core geothermal energy.

10.2.1 Earth Core Geothermal Economics

Today, people use geothermally heated water in swimming pools and health spas. The hot water from underground can also warm buildings for growing plants. In Earth core geothermic energy, there is an increase of roughly 1 °C/km. This temperature may go well above 150 °C and a typical plant needs from

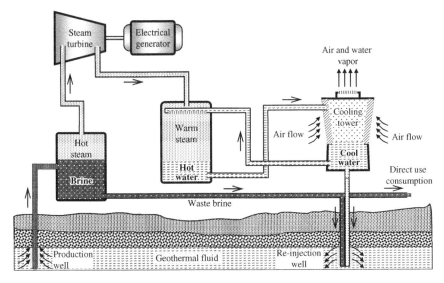

Figure 10.3 Binary-cycle electrical power generation.

6 to 20 ha (±3 ha/MW). In San Bernardino in southern California, hot water from below ground is used to heat buildings during the winter; the hot water runs through miles of insulated pipes to dozens of public buildings. City hall, animal shelters, retirement homes, state agencies, a hotel, and a convention center are some of the San Bernardino buildings heated this way. In Iceland, many of the buildings and swimming pools in the capital of Reykjavik and elsewhere are heated with geothermal hot water. Iceland has at least 25 active volcanoes and many hot springs and geysers.

There are many advantages of Earth core geothermal energy, but there are also some disadvantages. To determine whether the advantages outweigh the disadvantages, it is important to conduct a feasibility study for the area in question. Generally, the advantages of Earth core geothermal power plants are that the energy is renewable, reliable, and efficient; the power plants are online about 95% of the time; and they have the highest-capacity factors of all power plants. On the other hand, the disadvantages are environmental impacts due to drilling, the land required (the power plants require 1–8 acres of land per megawatt), limited use due to site availability, and high costs associated with start-up and consumption (see Table 10.1).

The implementation cost of Earth core geothermal power depends on the depth and temperature of the resource, well productivity, environmental compliance, project infrastructure, and economic factors such as the scale of development and project financing costs. Tables 10.2 and 10.3 illustrate the costs associated with geothermal applications.

Table 10.1 Geothermal resources cost ($/tonne).

	Steam	Hot water
High temperature (>150 °C)	3.5–6.0	n/a
Average temperature (100–150 °C)	3.0–4.5	0.2–0.4
Low temperature (<100 °C)	n/a	0.1–0.2

Table 10.2 Geothermal plant unit cost (cents/kWh).

	High-quality resource	Average-quality resource	Low-quality resource
Small plants, <5 MW	5.0–7.0	5.5–8.5	6.0–10.5
Medium plants, 5–30 MW	4.0–6.0	4.5–7.0	Normally not suitable
Large plants, >30 MW	2.5–5.0	4.0–6.0	Normally not suitable

Table 10.3 Direct capital costs of large plant exploration ($/kW installed capacity).

Large plants (>30 MW)	100–200	100–400
Steam field	$300–450	$400–700
Power plant	$750–1100	$850–1100
Total	$1150–1750	$1350–2200

As an example, the overall cost of a geothermal plant in South Fork, Colorado, with a capacity of about 300 MW, is approximately $39 million [2]. This cost does not include the exploration costs, as the initial feasibility study/exploration has already been completed. This cost would remain virtually unchanged for 25–30 years, as there is little maintenance and repair associated with these geothermal systems.

10.2.2 Examples of Earth Core Geothermal Electricity

Some areas have enough steam and hot water to generate electricity, and hot water or steam from underground can also be used to make electricity in an Earth core geothermal power plant. Steam carries no condensable gases of variable concentration and composition. In California, there are 14 areas where geothermal energy is used to make electricity. Some are not yet used because the resources are too small or too isolated, or the water temperatures are not hot enough to make electricity [2–4]. California's main sources are:

- The Geysers area north of San Francisco
- The northwest corner of the state near Lassen Volcanic National Park
- The Mammoth Lakes area, the site of a huge ancient volcano
- The Coso Hot Springs area in Inyo County
- The Imperial Valley in the south

To extract hot steam from underground, holes are drilled into the ground and pipes lowered into the hot water, like a drinking straw in a soda. The hot steam or water comes up through these pipes from underground, as the Geysers Unit 18 located in the Geysers geothermal area of California. California's geothermal power plants produce about one-half of the world's geothermally generated electricity. These geothermal power plants produce enough electricity for about 2 million homes.

An Earth core geothermal power plant is like a regular electrical power plant except that no fuel is burned to heat water into steam. The Earth heats the steam or hot water in a geothermal power plant. It goes into a special turbine whose blades spin, and the shaft from the turbine is connected to a generator to make electricity. The steam is then cooled in a cooling tower. The white "smoke" rising from the plants is steam given off in the cooling process. The cooled water can then be pumped back underground to be reheated by the Earth.

Figure 10.2 gives a glimpse inside a power plant using Earth core geothermal energy. Hot water flows into and out of the turbine, and this hot water circulation turns the generator. The electricity goes out to the transformer and then to huge transmission lines that link the power plants to homes, schools, and businesses.

10.3 Solar Heat Stored Underground as a Source of Energy

The Earth ground is composed of a very wide variety of chemical compounds that confer it special physicochemical properties. Among these are insulating properties (thermal resistance) and heat storage (thermal capacity) of the heat received from the sun [4–6]. With this, an important part of the energy coming from the sun and reaching the Earth is absorbed and stored underground since the Earth's crust works as a thermal insulation for the underground heat. The insulating properties of the Earth's crust and its great mass cause the temperature of the underground. From about 2 m from the surface, the soil temperature remains virtually constant throughout the year. This temperature varies depending on the terrain, the depth of the basement, and even the solar radiation in the region. In Brazil, a country with high levels of solar radiation in its whole territory, the temperature of the Earth to a depth of over 2 m is relatively high (between 13 and 27 °C). Therefore, shallow ground depths store energy from the sun, which is a renewable and inexhaustible source of heat.

Under the power generation point of view, geothermal energy is characterized with respect to the environment, represents savings in costs of heating/cooling environments, does not use fuels, and can make life more comfortable to everyone in the planet. At a lower thermal level, an air conditioner can extract heat from the ground for winter heating or insert energy into the ground to gain a more efficient cooling sink.

The shallow geothermal potential comes from the heat storage capacity of the first meters below the Earth's surface, say, something around 15 m. This underground temperature was accumulated over time and keeps quite stable, so depending on the ambient average temperature just above the Earth's surface, as represented in Figure 10.4. A universal average for the Earth's inhabited regions, the stored temperature goes up somewhere around about 10–27 °C.

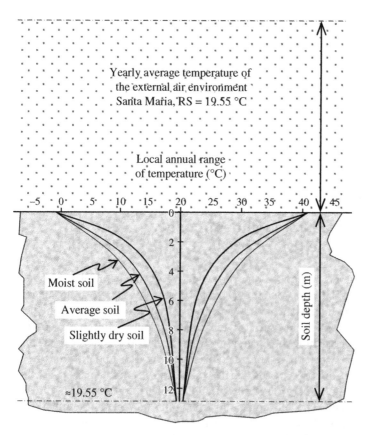

Figure 10.4 Soil temperature estimated by the annual average of the external environment.

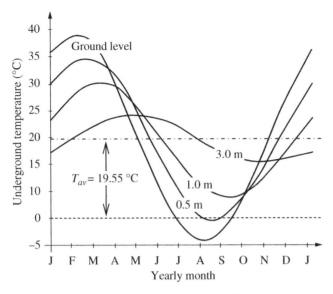

Figure 10.5 Time delay of the soil temperature along annual seasons (south hemisphere).

As we go deeper and deeper into the underground, soil there is a time delay along the year due to the time constant of the temperature penetration into the ground (see Figure 10.5).

Notice that for depths somewhere between 10 and 15 m, the tendency is to have a stable temperature in the soil (Figure 10.4). This temperature is set by the annual average temperature in the local, which over the years has remained stable. The low temperature of the moist soil with respect to the dry soil is due to the higher thermal capacity of the soil and the evaporation, so it has to do with weather conditions of the location including wind, humidity, soil composition, and temperature.

10.3.1 Heat Exchange with Nature

The solar geothermal or ground source heat pump system can use the constant temperature under the ground warmed by the sun radiation to heat or cool a building or drying crops. To do that, pipes are buried in the ground near the building to be conditioned as depicted in Figure 10.6. Inside these pipes, a fluid is circulated to harness the heat and move it to the interior of the ambient to be warmed. In winter, heat is taken from the warmer underground going through the heat exchanger of the heat pump, which sends warm air into the home or business. During hot weather, the process is reversed. Hot air from inside the building goes through the heat exchanger, and the heat is passed into the relatively cooler ground. Heat removed during the summer can also be used to preheat water [7–9].

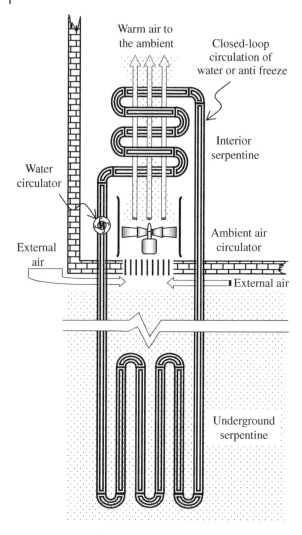

Figure 10.6 Direct extraction of solar geothermal heat.

Indoors residences, farms, or businesses can exchange heat with nature in different ways, each selected according to the convenience of the location and type of room. The most common ways for heat exchange are (i) use of streams of water, sea, or near rivers; (ii) circulation of thermal fluid between two sources of heat, one represented by the subsoil; and (iii) pressurized circulation of heat transfer fluid done through a heat pump.

Solar geothermal heat can be easily extracted a few meters from the ground surface and used in air conditioners or other heat exchangers increasing their efficiency. The example illustrated in Figure 10.6 is an application of a direct heat exchange with the environment immediately underground, with or without a heat pump. The exchange efficiency increases because instead of heat being exchanged with the ambient temperature, it can be exchanged with the underground heat, land and rivers, wells, and ocean. Countries such as Brazil and the United States have coasts bathed by the ocean over thousands of kilometers, usually with temperatures quite different from that of the seafront cities, warmer in the winter and, in reverse, cooler in summer.

The operation of an air-conditioning system can be completely powered by geothermal energy with no other external energy input and no heat pump. In the example of direct extraction of the solar heated geothermal depicted in Figure 10.6, a test room with geothermal air conditioning and a reference room under natural conditions were compared. The temperature limits imposed by a hill climbing control were set in between 20 and 25 °C. This type of geothermal air conditioning can contribute to the reduction of electricity consumption in residences and small businesses, as illustrated in Figures 10.7 and 10.8, and fits any small installation since it is a universal surface geothermal energy. This system was designed to circulate water in a serpentine driven by a small hydraulic pump. In the room ambient, there is a part of the underground serpentine exposed to a fan coil to dissipate the conditioning heat. That is possible because soil temperature, to a certain depth has a constant temperature throughout the year since the Earth's soil has a large mass and heat

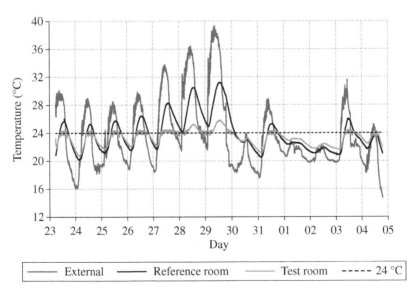

Figure 10.7 Conditioned temperatures from October 23 to November 11, 2014 in Santa Maria-RS, Brazil. *Source*: Todesco *et al.* [10]. Reproduced with the permission of Elsevier.

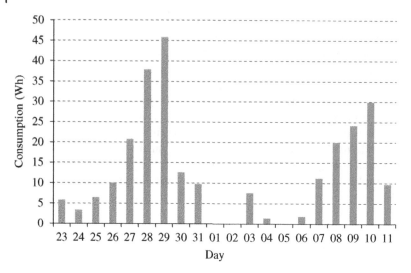

Figure 10.8 Power consumption from October 23 to November 11, 2014. *Source*: Todesco *et al.* [10]. Reproduced with the permission of Elsevier.

capacity [11–14]. If the temperature inside the room is lower or higher than the soil temperature, there is heat exchange between the room ambient and the soil. Therefore, the surface geothermal energy can be used to either warm up or cool down a room ambient.

10.3.2 Heat Exchange with Surface Water

Surface water for heat exchanges refer to water streams, sea, wells, or rivers that are present near the surface or the circulating water to the facilities consumption themselves. This heat exchange is based on the thermal contact surface or exchange efficiency between the water sources, as one wishes to absorb heat with the geothermal exchange unit. It is the cheapest way but requires the availability of water circulation nearby in reasonable quantities.

10.3.3 Heat Exchange with Circulating Fluid

The heat exchange with fluid circulation occurs through two serpentines or radiators: one buried in the ground and the other placed in the environment to be conditioned. With proper uptake, this heat can be directly used just by circulation of a thermal fluid (air, water, transformer oil, etc.) between radiators buried about 3–10 m, receiving heat and another one placed within the environment to be acclimated. The energy efficiency is high, since it does not use a high pressure conduit and does not work with high temperatures. In this type, heat exchangers are serpentines that can be arranged horizontally, vertically, or in a mixed arrangement.

10.4 Solar Geothermal Heat Exchangers

Depending on the weather and soil conditions of the location where the solar geothermal conditioner is to be installed, heat exchange can be performed using only a simple hose of flexible material like high density polyethylene (HDPE) having a thermal coefficient lower than the soil [10, 15, 16]. For this, there is no need to have an extremely low coefficient of heat exchange as in the case of the metal pipe (copper or aluminum). The inconvenience of metal is the cost, difficulty in handling and shaping, and being subject to corrosion and thermal deformation.

Note in Table 10.4 the exchange coefficients of some types of soil compared with other materials. These coefficients vary greatly with soil moisture as exemplified in Figure 10.3. Because the thermal coefficient of the Earth is much higher than the hose material, the exchange rate becomes irrelevant. This should be compensated with a longer length of hose. Table 10.5 shows the thermal conductivity of some materials at $27\,°C$ commonly encountered in design or manufacture of geothermal installations [10, 16, 17].

The soil density is directly related to their ability to store heat (mass m) in a smaller volume V of terrain. The soil density can be given by

$$\rho = \frac{m}{V}$$

and that there is therefore a relationship between the specific gravity ρ and volumetric capacity as depicted in Table 10.5 and given by

$$c_v = \rho c_p$$

where:

ρ = The soil-specific density
c_p = The specific heat
c_v = The volumetric thermal capacity

Table 10.4 Geothermal soil conditions.

Material	ρ (g/cm^3)	c_p (cal/g-°C)	c_v (cal/cm^3-°C)
Quartz	2.65	0.175	0.4637
Water	1.00	1.000	1.00
Air	0.00129	0.240	0.000309
Dry mineral soil	2.65	0.18–0.20	0.477–0.53
Dry organic soil	1.30	0.46	0.598

Source: Edgar Ricardo Schöffel—Agrometeorology, UFSM, Santa Maria-RS, Brazil.

Table 10.5 Thermal conductivity of some conducting materials (27 °C).

Material	Thermal conductivity (Wm^{-1}K^{-1})
Silver	426
Copper	398
Aluminum	237
Tungsten	178
Iron	80.3
Carbon steel	38.0
Glass	0.72–0.86
Water	0.61
Brick	0.4–0.8
Pine wood	0.11–0.14
Fiberglass	0.046
Polystyrene foam	0.033
Air	0.026
Polyurethane foam	0.020
Polypropylene	0.25
Polypropylene HDPE 3408	0.38
Epoxy (with silica)	0.30
Epoxy	0.120–0.177

10.4.1 Horizontal Serpentines

The horizontal serpentines are buried as shown in Figure 10.9 as close as possible to the environment being conditioned to a typical economic depth between 4 and 10 m. The required land area to bury the whole set is roughly around 1.5 m of serpentine pipe length for every 1.0 m^2 of air-conditioning environment. Accidents in the near ground, such as trees, loose stones, building with shadows, chemicals, garbage, large temperature differences between winter and summer, and landslides should be avoided as they can cause damage to the facilities. The difficulty installing this system is related to the size of the land area required to have an acceptable result. The cost is the lowest among the heat exchange systems with nature, but the free area to bury the pipeline is the unwanted counterpart.

Figure 10.10 shows an example of a horizontal solar geothermal serpentine installed in the Center of Excellence in Energy and Power Systems (CEESP) in the Federal University of Santa Maria, Santa Maria-RS, Brazil. They are a 100-m ordinary plastic tube buried in a 5-m deep hole, long enough to keep

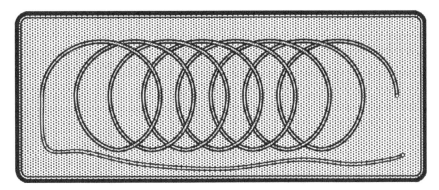

Figure 10.9 Top view of a horizontal solar geothermal serpentine.

Figure 10.10 Practical setup of horizontal solar geothermal serpentine.

nice and warm a $2.5 \times 5.0\,\mathrm{m}^2$-geothermic case study, capable to supply a room in a building. Using the principle of the communicating vessels in these tubes, the electrical power to drive a fan and the water pump is used only to win the pipe losses, which is almost negligible compared with the traditional air-conditioning systems.

10.4.2 Vertical Serpentines

A vertical serpentine has a form of helicoidal U or double U buried in separate wells apart about 4 or 6 m, typically as in Figure 10.11. The depth of a single well should be of a numerical factor of about 1.3 times larger than the surface to be conditioned. For example, if the surface to be heated is 10 m², the depth of the well should be about 13 m. If used over a pit, the depth of them can be reduced in inverse proportion to the number of wells. The type and soil moisture could contribute favorably, or not, depending on its heat transfer capability. An advantage of the vertical serpentine is a smaller ground area needed for installation of the serpentines, but the efficiency is lower than that of the horizontal serpentine. The cost is higher and the rocky terrain makes it almost unviable.

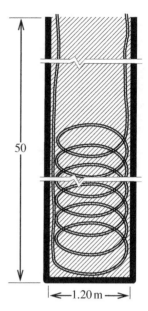

Figure 10.11 Solar geothermal vertical serpentine.

10.4.3 Mixed Serpentines

The use of mixed sets of horizontal and vertical serpentines is actually a good solution for reducing the land area in such projects. Those hybrid systems use the most intense and stable heat from the ground depths and of the available surface area closer to the solar radiation. The dimensioning of the serpentine must follow the same procedures used for the horizontal and vertical serpentines [1, 4].

10.4.4 Pressurized Serpentines Heat Pump

When there is no much difference between the average annual ambient temperature and the average subsurface temperature, or the limited available area or depth for solar geothermal energy, one can improve the heat exchange with the nature using a geothermal heat pump. These machines can be used to transfer heat from underground, say, from an environment at about 12 °C, and the water circuit accumulator at about 40 °C. The heat thus transferred may be used in ambient heating or as hot water or yet to preheat water for showers and sinks as described in the following text.

The heat transfer efficiency between two environments (power delivered/absorbed energy) depends on the temperature difference between the sources, which provide heat to the condenser and the radiator's ability to absorb the difference. Conventional air-conditioning systems absorb heat from the atmosphere, represented by the temperature in winter down a few degrees Celsius. At these temperatures, the condenser cannot capture virtually any heat and pump efficiency becomes too low. In summer, with the highest temperatures, the pump must give heat to the atmosphere, which can be about 40 °C. In this situation, the performance is not very good as well. A solar geothermal capture system extracts heat from underground with a temperature more or less constant, the performance will always be higher, not being so relevant with the local atmospheric temperature conditions.

For the purpose of this chapter and briefly, a heat pump is a device to extract heat from underground soil, air, or water to heat water in a confined ambient as the ones used in air conditioners or fridges. Even being driven by electricity, this device is an extremely efficient way for heating water or home ambient. In particular, geothermal heat pump returns to almost five times more energy in the output compared with the input. It works with the help of a gas compressor and two heat exchangers placed in each of the environments with which heat has to be exchange. The operation of this device is based on the Carnot cycle and the gas properties. When a gas is compressed, it emanates heat (exothermic exchange). On the other hand, when gas is expanded it absorbs heat (endothermic exchange). Thus, the same heat pump can absorb heat from the surroundings say at 40 °C and transfer it to the subsurface using the same heat transfer machine, that is, to function as a heater and a cooler as shown in Figure 10.12.

The heat exchange can be reversed in the heat pump by use of suitable bypass valves that allow gas expansion in the warm environment and the gas compression in the cooler environment. They are called "splitters," devices that can distribute heat or cold in the environment, using the same circuit of a heat pump.

In terms of energy efficiency, using electrical power to move the conditioning compressor using an example of a basement at 12 °C as the heat source, it is possible to have around 4.5 times more energy for heating and 6 times more for cooling. Moreover, this heat exchange is environmentally sound and friendly because it does not use any combustion and does not generate any toxic gases. Moreover, it can also be extended to large clusters and homes [1, 4].

The heat exchange condenser/radiator of the nature and the conditioned environment occur directly between a copper and an aluminum serpentine operating as an extension of the heat pump. For a conventional house, it requires a 100–150-m^2 horizontal serpentine whose cost is higher than that of a pipe without pressure, whose efficiency is much higher because it is a forced heat exchange and uses metal pipes. Among the disadvantages of this method is a high pressure conduit with gas leakage possibilities probably underground and the consequent difficulties to spot the leak, continuous heat loss, and reduced efficiency.

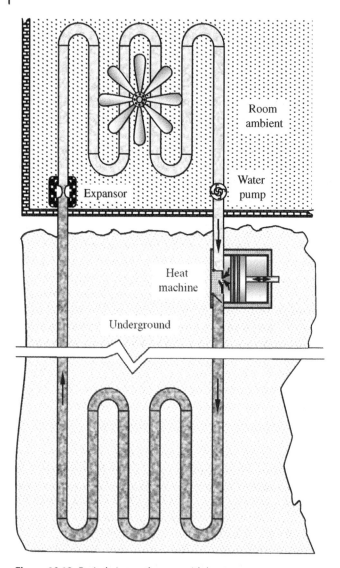

Figure 10.12 Buried pipe exchanger with heat pump.

10.5 Heat Exchange with a Room

For heat exchange in a room using a heat pump, air must circulate through the surface of the room to distribute cold/heat throughout it using thermally insulated pipeline and splitters, as mentioned earlier. The main advantages of this system are the affordability and simplicity of installation. The disadvantages are low efficiency, moderate comfort only, requirement of works prior to construction, and better suitability for smaller homes.

To make an exchange of heat, it is necessary to pass water through a closed circuit of hot or cold water and distribute it over the room surface. These exchanges are limited by the temperature of the water that can be heated only up to about $10\,^{\circ}$C in winter and to about $45\,^{\circ}$C in summer, depending on the actual weather conditions of each region. For such an exchange, it is possible to use hydraulic convectors with air, earth, or radiating walls or yet ordinary radiators. A convector is a partially closed surface directly heated by some form of energy from which the heated air is circulated by convection.

Hydraulic convectors are unique air-conditioning systems. In the case of radiating ground, it still needs an air serpentine to cool the environment. As the radiant soil has better performance and gives more comfort heating, this implies the need for another cooling system (see details in Figure 10.6). Its advantages are higher comfort degree and high efficiency, and as disadvantage it has to duplicate the equipment for full comfort.

In recent years, solar heated geothermal has awaken an interest such that modeling heat exchangers with nature has become one of the main focuses of research on clean energy efficiency and use of solar heat. The purpose of these studies has been the development of new tools for engineering and a better understanding of the coupling between the flow of fluids in hydraulic conduits or circuits, heat transfer, chemical reactions, and mechanical formation of rocks and soil in general [10, 14–18].

References

1 Energy Quest, Geothermal energy, http://www.energyquest.ca.gov/story/chapter11.html, accessed March 10, 2017.
2 K. Tweed, Geothermal Utility Launches in Colorado, https://www.greentechmedia.com/articles/read/geothermal-utility-launches-in-colorado, accessed March 2017.
3 A. Kleidon, Thermodynamic Foundations of the Earth System, Cambridge University Press, Cambridge, UK, 2016.
4 F.A. Farret, Use of Small Sources of Electrical Energy (Aproveitamento de Pequenas Fontes de Energia Elétrica), Editora UFSM, Santa Maria-RS, Brazil, 2014.
5 Geotics, Energetic Efficiency, http://www.geotics.net/nweb/esp/index.php, accessed March 10, 2017.
6 GS Geothermal Solar, http://www.geothermal-solar.ie/, accessed March 10, 2017.
7 K. Ooi, P.X.W. Zou, and M.O. Abdullah, A simulation study of passively heated residential buildings, Procedia Engineering, 9th International Symposium on Heating, Ventilation and Air Conditioning (ISHVAC) and the 3rd International Conference on Building Energy and Environment (COBEE), Vol. 121, pp. 749–756, 2015.

8 C. De Nardin, F.T. Fernandes, S. Cunha, A. Longo, L.P. Lima, F.A. Farret, and E.M. Ferranti, Reduction of electrical load for air conditioning by electronically controlled geothermal energy, Proceedings of the 39th IEEE Industrial Electronics Conference, Vienna, Austria, November 2013.

9 A.J. Longo, F.A. Farret, F.T. Fernandes, and C.R. De Nardin, Instrumentation for surface geothermal data acquisition aiming at sustainable heat exchangers, Industrial Electronics Society, IECON 2014—40th Annual Conference of the IEEE, 2014, pp. 2164–2138, DOI: 10.1109/IECON.2014.7048797.

10 M. Todesco, J. Rutqvist, G. Chiodini, K. Pruess, and C. Oldenburg, Modeling of recent volcanic episodes at Phlegrean Fields (Italy): geochemical variations and ground deformation, Geothermics, Vol. 33, No. 4, pp. 531–547, 2004.

11 C.R. Sommer, M.J. Kuby, and G. Bloomquist, The spatial economics of geothermal district energy in a small, low-density town: a case study of Mammoth Lakes, California, Geothermics, Vol. 32, pp. 3–19, 2003.

12 J. Rutqvist and C.F. Tsang, TOUGH-FLAC, a numerical simulator for analysis of coupled thermal-hydrological-mechanical processes in fractured and porous geological media under multi-phase flow conditions, Proceedings, TOUGH Symposium 2003, Lawrence Berkeley National Laboratory, Berkeley, May 2003, pp. 12–14.

13 C. Shan, K. Pruess, and K. Eosn, A new TOUGH2 module for simulating transport of noble gases in the subsurface, Geothermics, Vol. 33, No. 4, pp. 521–529, 2004.

14 F.T. Fernandes, F.A. Farret, A.J. Longo, C.R. De Nardin, and J.G. Trapp, PV efficiency improvement by underground heat exchanging and heat storage, Renewable Power Generation Conference (RPG 2014), Naples, Italy, 2014, pp. 1–6, DOI: 10.1049/cp.2014.0875.

15 C. Shan and K. Pruess, Sorbing tracers—a potential tool for determining effective heat transfer area in hot fractured rock systems, Stanford Geothermal Workshop, Stanford, California, USA, 2005.

16 N. Todaka, C. Akasaka, T. Xu, and K. Pruess, Reactive geothermal transport simulation to study the formation mechanism of impermeable barrier between acidic and neutral fluid zones in the Onikobe Geothermal Field, Japan, Journal of Geophysical Research: Solid Earth, Vol. 109, No. B5, p. 5209, 2004.

17 C.R. De Nardin, F.T. Fernandes, A.J. Longo, F.A. Farret, A.C. De Nardin, L.P. Lima, and B.R. Cruz, Underground geothermal energy to reduce the residential electricity consumption, Renewable Power Generation Conference (RPG 2014), Naples, Italy, 2014, pp. 1–6, DOI: 10.1049/cp.2014.0933.

18 C.R. De Nardin, F.T. Fernandes, F.A. Farret, L.P. Lima, A.J. Longo, and M. Godoy Simões, Surface geothermal energy applied to low cost and low power consumption residential air conditioning, Industrial Electronics Conference, Florence, Italy, October 23–26, 2016.

11

Thermocouple, Sea Waves, Tide, MHD, and Piezoelectric Power Plants

11.1 Introduction

There are several other options of power generation in addition to the traditional alternative sources of energy, previously discussed in Chapters 1 to 10. Most of the time, these sources meet very specific purposes such as spaceships, favorable local conditions, need of low vibrations or noises, backup, and others. Among these sources, there are thermocouples, sea tides, sea waves, piezoelectric, and the magnetohydrodynamic (MHD) generators.

The alternative sources discussed in this chapter are still subject for further research and may not have yet been used in common practices or large scale. Technical and economic feasibilities of these sources still depend on possible depletion of other sources. There is also a possibility of combining them to take advantage of heat and by-products for a greener system solution. There is also a lot of research and development of new or more suitable materials that may bring some of these new energy sources to be widely deployed in the future.

11.2 Thermocouple Electric Power Generation

Thermocouple devices consist of a junction of two different metals, joined at one end. When the junction of the two metals is heated or cooled, a voltage is produced that can be correlated to a temperature measurement. On the other hand, it can produce a voltage by the temperature difference at the extremes of the junction elements. There will be an electric current flowing through the junction every time there is heat flowing from the environment to the junction. Conversely, when passing a current through the junction, there will be a flow of heat to the environment. Therefore, there exists a conversion of electrical current into heat or vice versa. The typical voltage variation of a thermoelectric junction of two metals with temperature is $50\,\mu\text{V}/^\circ\text{C}$. This value is very small.

Integration of Renewable Sources of Energy, Second Edition. Felix A. Farret and M. Godoy Simões.
© 2018 John Wiley & Sons, Inc. Published 2018 by John Wiley & Sons, Inc.

Therefore, the junction should be associated with high temperatures, in addition to a careful element selection and a combination of many junctions to increase its electrical output power.

11.2.1 Thermocouples

The electricity generation with thermocouples is based on three physical effects: Peltier, Thompson, and Seebeck. The first, Peltier effect (due to Jean Charles Peltier), explained in 1834 how a junction (pair) of different materials can diffuse electrons on each other by an electrical current [1–3]. Depending on the types of materials involved, there will be a temperature rise on the end of one of the junction elements and a corresponding reduction on the other end. In other words, a junction (pair) of different materials diffuses electrons into each other by passing electric current with a temperature rise on the edge of the junction elements and corresponding reduction at the other end. If there is a stack of these materials, this effect can be used for cooling or electricity generation.

The temperature difference due to a Peltier element is limited by the heat dissipation, because the Peltier current is also very limited. Such constraint is due to the temperature difference of the Peltier coefficient, which related to the heat flow, the electrical current, the electrical resistance, and the thermal resistance between the hot and cold sides of the material.

The second physical effect is the Thompson effect, which explains how a differently heated rod (temperature gradient) at both ends can develop an emf by density differences in the electron concentration.

The Seebeck effect is called negative Thomson effect. Metals such as zinc and copper have a positive Thomson effect. The Seebeck effect, sometimes known as Peltier–Seebeck thermoelectric effect, is the association of the Peltier and Thompson effects manifesting an emf by heat interaction as in the thermocouple as represented in Figure 11.1a [4–6]. Differently from the Joule heating (I^2R), the Peltier effect is reversible depending on the direction of the current.

Though the phenomena of the Seebeck and Peltier effects were first discovered in circuit loops using two dissimilar metals, the fundamental phenomenon actually exists in a single conductor as was later pointed out by Kelvin and named the Thomson effect in his honor. It is thus not the dissimilar metals that generate the voltage but the temperature gradient. It just happened that it was the use of dissimilar metals with different thermal properties in the circuits, which enabled the discovery of the phenomena. For example, metals such as cobalt, nickel, and iron, with the cold terminal, when connected to the higher electric potential terminal and the hot terminal connected to a lower electric potential, an electric current will flow from the low thermal potential point to the highest point of thermal potential. Electric current flows from the cold end to the hot end; in this condition there is heat absorption.

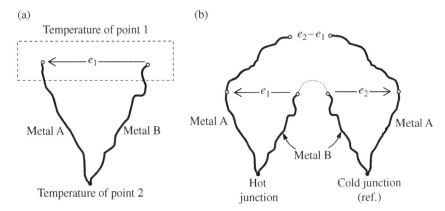

Figure 11.1 Thermocouples. (a) Thermocouple junction and (b) pair of thermocouples.

The thermoelectric effect is a composed result of the Peltier effect (current flowing through a junction of two different materials causing a heat flow from the environment to the junction), plus Thompson (current flowing in an element when subjected to a temperature gradient) and the Seebeck (carrier rearrangement at the junction of two different materials at the same temperature). The thermoelectric emf effect when used for thermoelectric generation can be approximated by

$$E = s\Delta T$$

where:

$E =$ The emf produced (V)
$\Delta T =$ The temperature difference between the junctions (°C or K)
$s =$ The sensitivity of the thermocouple temperature (V/°C)

In fact, there is a nonlinear relationship with the emf and the junction temperature, T, expressed as

$$E = a_1T + a_2T^2 + a_3T^3 + \cdots \tag{11.1}$$

Taking the derivative of equation (11.1) with respect to temperature, we can obtain the Seebeck coefficient of the thermoelectric pair. The values for some materials are listed in Table 11.1 [2]. Note that for many pairs of metals, the terms a_2, a_3, and so on are sufficiently small and can be ignored. There are other metals that can be inserted in the circuit that will not have effect on the emf if their junctions are at the same temperature, that is, the first and last metal temperatures are the ones that make the difference, as shown in Figure 11.1b.

Table 11.1 Thermoelectric Seebeck coefficient.

Material	Power (μV/°C)
Antimony	47.0
Bismuth	−72.0
Carbon	5.0
Copper	6.5
Constantan	−35.0
Nichrome	25.0
Platinum	0.0
Silver	6.5
Rhodium	6.0
Tungsten	7.5

Table 11.2 shows the main characteristics of thermocouples. Examples of alloys can be used in thermocouples such as constantan (alloy made of 55% copper to 45% nickel), chromel (alloy of 90 to 10% nickel chrome), and alumel (alloy of 96% nickel, 2% manganese, and 2% aluminum).

When there is a need to increase the output emf of the thermocouples, one can put them in series. Such cells are called thermocouple stacks. If the temperature increases to near the maximum, it reduces the lifetime of the thermocouple. There are other materials better qualified currently being considered for power production on a larger scale as discussed in the following text.

11.2.2 Power Conversion Using Thermocouples

Electrical power converted from heat flux passing through an ordinary conductor is called as "Seebeck effect." Most thermocouples today are used as temperature sensors with the purpose of measurements [7–9]. But they can be also used as electric generators in some peculiar places such as deep caves, sea bottom, or space where the solar intensity is very low and electricity may not be generated in photovoltaic panels to supply instruments and devices (such as a far space mission from the sun). Thermocouples have been used in outer space to generate small amounts of electricity in satellites. There are also some other applications of installation in ground stations or in the polar areas and submerged in the deep ocean.

Thermoelectric generators have assembled with burning propane or kerosene and use in camping. There are also applications for ocean buoys and

Table 11.2 Main types of thermocouples.

Type	Material of the pair (material$_1$–material$_2$)	Temperature range (°C)	Sensitivity (μV/°C)	Output 100 °C (mV)	Output 1000 °C (mV)	R (Ω/m)
B	Platinum/94% rhodium/6% –70%/30%	0 to 1800	0.00	0.033	4.83	3.15
E	Chromel–constantan	0 to 1000	60.48	6.32	76.36	12
J	Iron–constantan	–180 to 760	51.45	5.27	—	6
K	Chromel–alumel	–180 to 1370	40.28	4.10	41.27	10
R	Platinum–platinum/87% rhodium/13%	0 to 1750	5.80	0.65	10.50	3.17
S	Platinum–platinum/90% rhodium/10%	0 to 1750	5.88	0.64	9.58	3.17
T	Copper–constantan	–180 to 370	40.28	4.28	—	5

Source: Adapted from Horowitz and Hill, *The Art of Electronics* [2, 3].

remote places requiring to power loads of only a few watts. These places are considered radioisotope thermoelectric generators (RTGs) that generate energy using radioactive elements such as plutonium (PU), which in turn generates heat for the thermocouples, and that energy is finally converted into electricity. The RTGs are reliable and more durable than most alternative sources because they do not have moving parts. Such devices can produce electricity at a virtually constant voltage, no moving parts, and minimal maintenance during operation. They are suitable for places where frequent maintenance is not possible, such as in space probes and pacemakers.

Thermocouples may also be passive devices using the reverse effect of the traditional thermocouples. In such operation, the electricity passing through the thermocouple generates a heat difference; those are the thermoelectric or Peltier coolers. Refrigerators can be made of simple thermocouples based on the Peltier effect and used to cool down moving parts. Such devices can also be used in computers where the cooling effect can be used to directly cool electronic parts, making them more effective than traditional fans [7–10].

The first known thermoelectric power generator heated by nuclear energy was constructed in 1954 in Mound Laboratory, Ohio, USA. A set of thermocouples of metal wires was used, with efficiency below 1%. Other thermocouples, the SNAP 3 series, made of thick cylindrical semiconductor lead telluride (PbTe), were also tested with length of 5 cm and efficiency higher than 5%. This percentage seems small when compared with the 35–40% obtained in a modern thermal power plant, but the SNAP 3 generators have the advantage of operating in remote places without any technical supervision unlike conventional power plants.

The characteristics of the junctions of thermoelectric couples with heated metal plates have many difficulties: the board should be strong enough to withstand high temperatures and have small electrical resistance in order to have low losses. Thermocouples are surrounded by mica cases to prevent surrounding gases to contact with the junction and change the balance of the semiconductor material. One important problem to be solved is the fragility of all the elements involved in constructing and assembling thermocouples.

11.2.3 Principle of Semiconductor Thermocouples

Semiconductor thermocouples use both n-type and p-type to allow different electron densities. The semiconductors are placed thermally in parallel to each other, and electrically in series, joined by a heat conducting plate placed on each side (see Figure 11.2). When a voltage is applied across the free ends of the two semiconductors, there is a flow of DC current through the semiconductor junction, causing a temperature difference. The side with the cooling plate

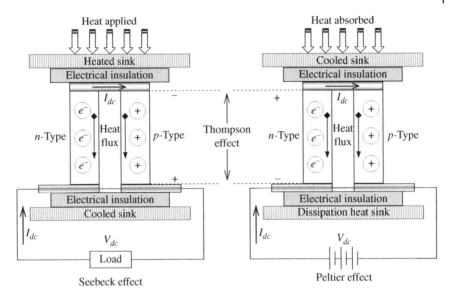

Figure 11.2 A semiconductor thermocouple.

absorbs heat, which is then moved to the other side of the device where the heat sink was placed. A thermoelectric cooler (TEC) is usually connected side by side and sandwiched between two ceramic plates.

In a semiconductor thermocouple, the heat is applied to one side of the device causing charge carriers to be released into the conduction band, which in turn, produces electrons in the n-type material and holes in the p-type material. The charge carriers concentrated at the hot side of the device will repel each other so that they tend to migrate toward the device cold side. In the n-type material, such electron flow constitutes a current from the cold to the hot side, and the electron movement causes a negative charge at the cold side against a corresponding positive charge on the hot side. In the p-type material, the migration of the holes constitutes a current flowing in the opposite direction. At the cold side there will be a positive charge corresponding to the negative charge at the hot side. With metallic interconnections, these junctions will produce a current flowing in an external circuit proportional to the temperature gradient between the hot and cold junctions, and the voltage is proportional to the temperature difference. Heat must be removed from the cold junction since otherwise the migration of the charge carriers will equalize their distribution in the semiconductor, eliminating the temperature difference across the device causing the migration, which can make the current to stop.

When a voltage is applied across a semiconductor thermocouple, the surplus charge carriers present in the semiconductor will be attracted toward the terminal with the opposite polarity. Thus, electrons in the *n*-type material migrate toward the positive terminal, causing a surplus to accumulate in the region of the semiconductor next to the terminal, leaving a deficit at the negative side of the device. Similarly, holes in the *p*-type material migrate toward the negative terminal. In other words, as the charge carriers flow through the material due to the electric field created by the voltage between the terminals of the device, their increased kinetic energy manifests as heat. The total heating capacity of the device is then proportional to the number of TECs [7–10].

The temperature within the device will depend on the number of charge carriers and their kinetic energy. The temperature will be higher in the region where the charge carriers are concentrated and lower in the region behind where charge density is consequently lower. Thus a temperature gradient, proportional to the magnitude of the applied current, builds up across the device [8–10].

11.2.4 A Stack of Semiconductor Thermocouples

Semiconductor thermocouples can be joined in a rod arrangement to allow heat transfer and increase power production. A scheme illustrating this arrangement is illustrated in Figure 11.3 with the positive and negative carriers moving in the semiconductors. A hot fin and a cold fin bar are used for heat conduction. These two bars hold together by metallic plates to let the current flow through each thermocouple and the load.

11.2.5 A Plate of Semiconductor Thermocouples

Several semiconductor thermocouple stacks can be used to form a Peltier generating plate as depicted in Figure 11.4. Typically, power may go on the order of 300 W for this application.

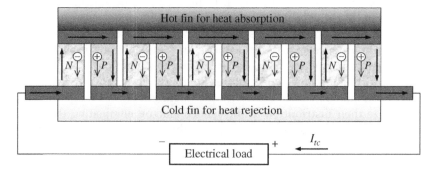

Figure 11.3 A stack of semiconductor thermocouples.

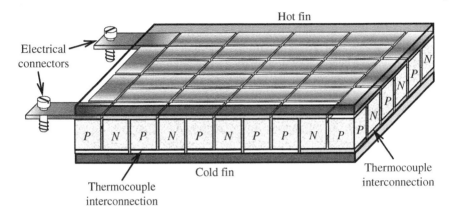

Figure 11.4 A plate of semiconductor thermocouples.

11.2.6 Advantages and Disadvantages of the Semiconductor Thermocouples

The advantages of the semiconductor thermocouples with respect to other types of power generation are as follows:

- They have no moving parts and therefore less maintenance.
- They are very robust devices for power generation.
- It is possible to keep the temperature control in degree of fractions.
- They can have a flexible size (form factor) and, in particular, a very small size.
- They can be used in tight environments where it is not possible to have conventional cooling.
- They do not use chlorofluorocarbons.
- They have a long life, with a mean time between failures (MTBF) greater than 100,000 hours.
- An input voltage controls them.

Disadvantages of semiconductor thermocouples to generate electrical power are as follows:

- Limitations in amount of heat flux that can be dissipated by each plate.
- They are not as efficient in terms of performance coefficient as in wind systems and vapor compression.
- Used only for applications with low heat flow.

11.3 Power Plants with Ocean Waves

Ocean waves are caused by the wind. When the wind blows consistently, they may force continuous waves along the shoreline. When directly generated and affected by local winds, a wind wave system is called a wind sea. Since waves in the oceans can travel thousands of kilometers before reaching land, their size will range from small ripples to waves over 30 m high. A swell consists of wind-generated waves that are not significantly affected by the local wind at that time. They have been generated further away or some time ago.

Wind waves in the ocean are called ocean surface waves. There is randomness in the wind waves; subsequent waves differ in height, duration, and shape with limited predictability. They can be described as a stochastic process, in combination with the physics governing their generation, growth, propagation, and decay. Ocean waves contain tremendous energy potential. Wave power devices are able to extract energy from the surface motion of ocean waves or from pressure fluctuations below the surface.

Just to compare wind resources with wave and tidal, the data for wind potential is usually in the order of gigawatts (GW), whereas the potential of wave and tidal are usually given in terawatt-hours/year (TWh/year). Plenty of wave energy is available all over the world [10, 11]. However, they may not be possible to utilize because it is necessary to have suitable conditions that may conflict with established beaches, shipping, commercial fishing, and naval operations or simply due to environmental and touristy concerns. Wave and tidal resource potential are typically given in terawatt-hours/year (TWh/year). The Electric Power Research Institute (EPRI) concluded in a recent analysis of potential energy resources of waves in the United States and estimated the total resource in wave energy along the outer continental shelf of the country is 2640 TWh/year. This is a huge potential, considering that only 1.0 TWh/year of energy can supply annually about 100,000 homes of the middle class. The EPRI estimates that total recoverable resources along the US coasts are 1170 TWh/year. This is almost a third of the 4000 TWh of electricity used by them a year. While an abundance of wave energy is available, it cannot be fully harnessed everywhere for a variety of reasons, such as other competing uses of the ocean (i.e., traveling, shipping, commercial fishing, and naval operations) or environmental concerns in sensitive areas. Therefore, it is important to consider how much resource is recoverable in a given region. EPRI estimates that the total recoverable resource along the US shelf edge is 1170 TWh/year, which is almost one-third of the 4000 TWh of electricity used in the United States every year [10].

The energy of the waves has been mostly used in desalination plants, power plants, and water pumps. Power production from sea wave is determined by their height, speed, length, and density of ocean water. Ocean wave devices capture wave energy and convert it into electricity. There are several energy wave conversion devices: the hydro–piezoelectric system, the oscillating water column, and the wave moving upward in a tapered channel and marine snail. The "snails" particularly involve the action of waves forcing air between blades located on the perimeter of a barge type circular structure. The air is then directed onto the air turbines coupled to electrical generators.

There are energy-generating devices from sea waves that can produce electricity at a cost of less than US$0.10/kWh (about the cost at which electricity becomes economically viable); some of these projects have been deployed in Europe, particularly in the United Kingdom. The most efficient of these devices, the Salter Duck, can produce electricity for less than $0.05/kWh. The "Salter Duck" was developed in the 1970s by Professor Stephen Salter from the University of Edinburgh, Scotland, and generates electricity with the rise and fall of the waves [11, 12]. While it may produce energy with an extreme efficiency, it disappeared in the mid-1980s when the European Union reported the results with a factor of 10 to estimate the cost of other produced electricity. In recent years, the error was noticed, and interest in Duck is becoming intense [12].

There are some obvious disadvantages of wave power plants in comparison with wind energy: (i) a wave power unit will probably have no available parts; (ii) it costs more than three times more to produce wave power devices than producing an equivalent same power wind turbine; and (iii) the construction costs are likely to be much larger because of complexity of the anchorage problems, volume, and corrosion compared with the whole structure and location of the base in water, in addition to investment requirements in studying the right and unique assembling and care with the environmental impact.

Wave energy is successfully used in small-scale applications, such as power navigation lights and buoys. Prospects for use on a large scale appear to be economically not viable at the current state of the art. Therefore, there is a very low probability that medium-term tidal waves will play an important role in providing energy to the industrialized nations. In addition, a project like this must take into account the cost of maintaining the current rate of carbon dioxide emissions within acceptable limits for construction of new plants and developing new environmental and economic policies, for electrical generation to make sea energy less dependent of the volatile markets of fossil fuel.

11.3.1 Sea Wave Energy Extraction Technology

In many areas of the world, the wind blows with consistency and force to continuously provide waves along the coast. Those waves contain a lot of energy that is extracted from mounting surface movement of the waves or by pressure fluctuations below the water, caused by wind blowing over the sea surface [13, 14]. Roll waves are swells; they travel long distances through the ocean, are not generated by the local wind but by distant storms, are typically smooth waves (not choppy-like wind waves), and are measured from the crest (top) to the trough (bottom).

There are four basic applications of roll waves: point absorbers, dampers, transposition devices, and terminals. There are major differences in the technical concept and design itself depending on how the wave energy technologies will be deployed: if on the water surface or shallow. The wave technology has been mostly designed for coastal sites but more at the sea and far away from the coast. For example, they may differ in their orientation to the waves or in the form of converting energy.

There are many devices for extracting energy from ocean waves. However, all of them are still in the phase of development and testing; they are not yet widely available. There is no predominant technology, but some promising projects are in demonstration and prototyping. Some of these projects are in advanced commercial scaling such as the Scottish Pelamis Project 2, the Andritz Hydro Hammerfest, and the OIG Giant II [15, 16].

Converting energy from swells or roll waves is possible by using a point absorber, that is, a floating structure with components moving relative to each other due to the action of the waves (e.g., a buoy floating within a fixed cylinder). The point absorbers appear for people who lot at them as floating buoys. They use the rise and fall of the wave height at a single point for energy conversion. Such an up-and-down movement caused by passing waves is used to drive electromechanical or hydraulic converters to generate power, which can be used mechanically, for example, with pressurized air systems, or for water desalination and/or for electrical power generation.

Attenuators are long and multi-segmented floating structures that are installed in a targeted manner parallel to the direction of the waves. They practically surf the waves, like a ship, extracting energy from them by the arc installed along the device length. The difference between the heights of the waves along the length of the device causes bending segments where the pickups are connected. Each segment is connected to a hydraulic pump or other converters that generate energy by the waves. Their connection to the electrical transmission lines has a power transformer for high-voltage offshore transmission. The electricity generated feeds a common feeder in a junction box on the seabed, connecting it to the shore with other machines over a common undersea cable.

Some floating platforms are built to generate electricity by channeling sea waves to turbines on land and then back to the sea. The transposition devices have reservoirs, which are filled by incoming waves, causing a slight rise of the water movement by air pressure such as a dam, called an oscillating water column (OWC). As the water is released, gravity causes it to flow back into the ocean. The energy of this falling water is used to rotate electric turbines (see Figure 11.5) [16, 17].

Another solution for energy conversion using the movement of sea waves is shown in Figure 11.6. The waves generate a sequence of compression and decompression in the air chamber through a turbine, maintaining a unidirectional flow of air over the blades of such a turbine, with synchronized valves that open and close in accordance with the direction of the pressure p in the air chamber. In order to understand this using an equivalent circuit,

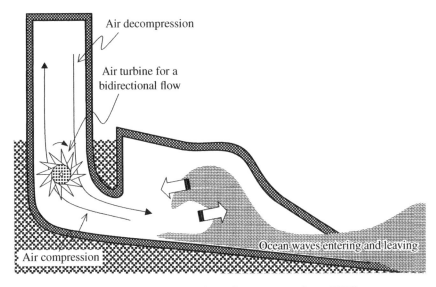

Figure 11.5 Power generation system with oscillating water column (OWC).

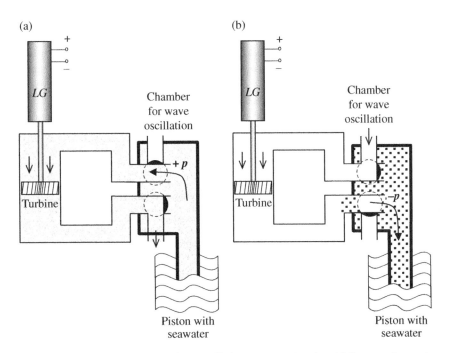

Figure 11.6 Power generation with an oscillating pressure chamber. (a) Compression and (b) decompression.

it is possible to imagine a kind of a "rectifying hydraulic diode." In general, the turbine might be connected to an electrical linear generator (LG), which can be a permanent-magnet-based linear machine or a linear induction-based linear machine [16, 18].

Terminal devices extend perpendicularly to the direction of the wave in order to capture or reflect the wave energy. These devices are usually set on beachfront or near the waterfront. Several versions of floating terminals have been designed for these applications. The OWC has a way for the water to go through an opening that is mounted in an underground chamber, trapping the air inside it. Wave action causes the captured water column to move up and down like a piston, forcing air to pass through an opening connected to a turbine, which drives an electrical generator. These devices have a rated power range from 500 kW to 2 MW, depending on the location and dimensions of the installed device. The electrical machines connected to this piston must be designed to operate in very low speed, and they are usually huge and very difficult to design and implement.

11.3.2 Energy Content in Sea Waves

Sea wave energy depends on the length and frequency of the waves. The dispersion in the length and frequency values of the waves can be studied under a statistical approach. The phase velocity of a wave may be different from the rate at which the energy wave propagates. Distraction is the time that takes consecutive wave crests to pass a defined, maybe imaginary, line. Waves in deep water are more distracting because longer wavelengths travel faster than shorter wavelengths, where on the other hand, waves in shallow water are not as much as distracting as the deep ones. Long-term strong winds generate higher waves [19–24].

According to the shape of the sea surface, there may be a complex superposition of waves with all possible lengths or frequencies traveling in all directions without a net result for generating usable energy. Thus, the wave frequency length or range contributes to the variance on the surface displacements. The wave energy is proportional to the variation of surface displacements; when measurements are taken, it may occur that digital data spectra has limited range, and it may not contain useful information on wave frequencies higher, because of the Nyquist frequency constraints. Therefore, the minimum data sampling frequency must be twice or more the frequency of the wave to be measured [22–24].

In order to have wave data analysis useful for power generation, several methodologies have been proposed to define an idealized wave spectrum, usually assuming homogeneous and stable wind. There are two major formulations: one is the statistical equation Pierson–Moskowitz and the other, the spectrum called JONSWAP [25]. Data is supplied from empirical observations of sailors on ships and altimeter satellites. Such data has been widely used to make global maps of the wave heights. Waveguides are also used in shallow water rigs and offshore. Measurements can be taken using immersed pressure gauges on the edge of beaches for recording wave's data. Synthetic aperture radars are used for information on the wave directions. With all these data, it is possible to have a good expectation on the return of invested capital in generating plants with ocean waves. Better knowledge of the behavior could be useful for safety procedures of the generating plant. If the dispersion of the waves is measured accurately, they can be used to track distant storms that may propagate some low frequency waves. This could prevent damage to the generating station power [9].

11.4 Tide-Based Small Power Plants

Tidal energy is an attractive source for electricity generation near coastal regions. It is an inexhaustible source of energy and environmentally friendly and does not produce greenhouse gases or other pollutants. As 71% of the Earth's surface is covered by water, there is a lot of potential of using tidal energy in large or scale [26, 27].

Tides are the alternating pattern of rising and falling sea level with respect to land. As the Earth spins on their axis, if we consider the ocean water to be at equal levels around the planet, the gravity will pull inward, and the inertia will provoke an equivalent centrifugal force pushing outward. The rising sea level is called high tide, and the falling is called low tide or low seas. Such rise and fall happen twice a day (12-hour cycle) and cause a huge movement of water. The movement is so powerful that has caused many accidents in the past, resulting in the flooding of coastal areas and in sinking of boats and ships. Thus, tidal power constitutes a great potential as an energy source, and that can be harnessed in some coastal areas in many parts of the world [27–29].

Dams for tidal energy utilization are built near the coast. During the high tide, the water flows into the dam through a turbine. At low tide, the tank is emptied, and the water coming out of the tank again passes through the turbine, producing electricity. An outline of underwater tidal power energy conversion is shown in Figure 11.7. In this scheme, the movement of tides produces a seawater flow in a turbine linked to an electric power generator of many MW.

As it turns out, tidal power is very similar to hydroelectric power using a reservoir but uses a tank with seawater. This solution has been used in the United States, Japan, and England. Brazil has large amplitudes of tides, for example, in São Luís, in the Bay of San Marcos with an almost 7-m gap. However, the topography of the coastline in that region makes it economically unfeasible due to the difficulties to build reservoirs with the present technology.

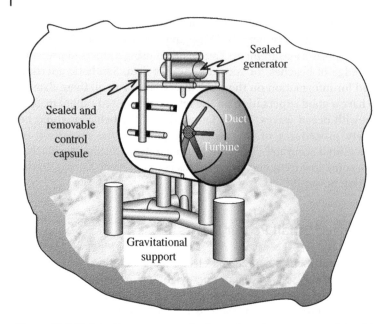

Figure 11.7 Sketch of a power generator for underwater movement of the tides.

The tides are predictable and renewable forms of cyclic power accumulation/dissipation. They are clean, do not require a lot of land, and can be exploited on a small scale to solve problems located along the coasts. The efficiency in converting tidal energy can reach up to 80%, which is much higher than coal, solar, or wind energy. Likewise, the energy density is relatively higher than other sources of renewable energy, despite the fact that the cost of construction of the dam is still high. Maintenance costs are relatively low, constituting only in cleaning the tank and pipelines, not requiring any kind of transport or storage of fuel to function. The lifetime of a tidal power plant is very long.

There are several excellent advantages for using tidal power, but many issues must still have to be resolved concerning the construction costs. Moreover, there are very few suitable places for the construction of these plants, which must be naturally located in coastal regions. The intensity of the waves is very variable demanding a construction suitable for the worst-case scenario. As the tide occurs only twice a day, the useful power generation may occur only during this short period, and it must be converted at the same time it is generated, or it must be stored, or it must be interconnected with the public network with utility scheduling of that energy in the transmission or sub-transmission power flow system. Moreover, the central facilities must be well designed by environmental engineers to avoid adversely affecting aquatic life and disrupting the natural migration of fishes [10, 12].

The driving tide generation is not yet a profitable solution. Further technological advances are needed to make it commercially viable. There are problems associated with sea ice; low or weak tides, straight coast, rises, and low
current fall are still strong obstacles to allow the widespread adoption of tidal
power plants. In order to make things more complicate, the places where tidal
energy can be deployed are far from the load markets, and it is a necessary
infrastructure of transmission lines in addition to substations and maybe other
utility-oriented facilities.

11.5 Small Central Magnetohydrodynamic

An MHD device transforms thermal energy and kinetic energy into electricity. It works with a heavily ionized gas or through liquid circulation with high
electrical conductivity in the presence of internal or external magnetic fields.
The Faraday's law explains the operation of an MHD. The induction that
relates the electric current with a magnetic field in a closed circuit in proportion to the number of lines flux crossing the involved area of the circuit per
time unit will support that an induced voltage e_{ind} occurs when a conductor
moves at a speed v between the poles of a magnetic field B, as illustrated in
Figure 11.8 [28, 29]:

$$e_{ind} = B\ell v \tag{11.2}$$

One application of an MHD is when warming gases or liquids have high electrical conductivity, but with low ionization energy, such as potassium, cesium,
or argon. Such elements are obtained when burning fuel gas or coal simply
(e.g., wood lignin). The temperature required to achieve this process should be
high enough in order to cause gas or liquid ionization. Therefore, after passing
it between metal plates subjected to an intense magnetic field, there is an interaction between the fluid conductor moving and the intense magnetic field,
generating an electromotive force. The electromotive-induced force gives an
electrical current, which is directed to drive
loads, as shown in Figure 11.9 [29, 30].

A very good application of MHD is in ship
propulsion systems. In these applications, an
electric current is applied to create the MHD
propulsion force. This electric current passes
through the seawater in the presence of a
strong magnetic field. The seawater is conductive as in a wire movement of an electric
motor. By pushing out water, this motor vehicle accelerates backwards. Mitsubishi made a

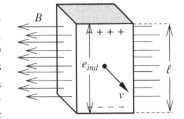

Figure 11.8 Induced
electromotive force.

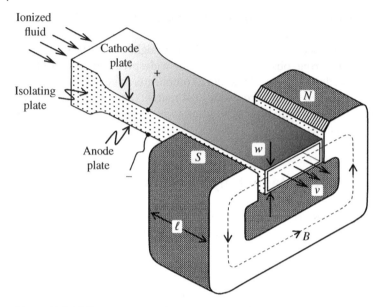

Figure 11.9 MHD power generation.

prototype of an MHD propulsion ship, the Yamato I, in Japan in 1991. Then, a ship was successfully driven in 1992. Yamato had two MHD thrusters that work without any moving parts. In addition, MHD powered some Mitsubishi-built prototype ships, during the 1990s. These ships were able to reach maximum speed of 15 km/h [30, 31].

The main efficiency limitations of the MHD energy are due to the high temperatures (2400–2800 °C) that the materials must withstand. This technology is at a very experimental stage yet, and Japanese expectations for future, larger prototypes will probably reach higher propulsion operating velocity.

The importance of generating MHD energy is related to the development of fusion reactors. These reactors can use a hot deuterium highly ionized gas and perform the controlled nuclear fusion of MHD energy generators. Such generators are suitable for ride-through short-term demand peaks from the utility grid, allowing an increase in thermal efficiency of steam generators for thermal turbines of about 40–50%.

A potassium-enriched gas can power an MHD generation system. Such system can be built with coal combustion. It has a high potential efficiency of the conversion of thermal energy into electric energy, particularly because there are no solid moving parts, and it is possible to construct a system capable to operate at very high temperatures. Maybe in the near future, we will see the use of silicon carbide or wide-bandgap devices (WBGD) associated with MHD high temperature energy conversion systems. However, there are not still many unknowns in deploying MHD generators because of the prohibitive costs and various technical and economic constraints [29–31].

11.6 Small Piezoelectric Power Plant

Piezoelectric (PZ) crystals can generate emf impressed by pressure on their axis transverse to which such pressure is applied. The principle is based on the idea that a crystal structure under pressure moves ions from their normal positions generating an electric potential difference.

The husband and wife Curie discovered the PZ effect in 1880, and Paul Langevin who made a first attempt to practical application used it in the development of sonar during the First World War. Langevin engaged metal masses of quartz crystals (coincidentally inventing the Langevin type transducer) to generate ultrasound in the range of a few tens of kHz's [32–34].

Until the First World War, it was difficult to excite the transducers built with quartz crystals, as they demanded for high-voltage generators. From there the decades of 40 and 50 began the development of synthetic PZ materials because of the improvement of PZ ceramics made with barium titanate by the USSR and Japan, and ceramic PZ of lead zirconate titanate (PZT) made by US researchers. The development of PZ ceramics was revolutionary. After correct bias it showed better properties than crystals with more flexible geometries and sizes. These PZ materials are manufactured by sintering ceramic powders shaped via pressing or extrusion.

Currently PZT ceramic type (and many variations) is the one prevailing in the market. It is possible also to find other materials, such as the PT ($PbTiO_3$) and PMN [$Pb(Mg1/3Nb2/3)O_3$], used in devices that require very specific and special properties, such as high temperature transducers or for image and nondestructive testing. Various ceramic materials have been described exhibiting a PZ effect. These include PZT, lead titanate ($PbTiO_2$), lead zirconate ($PbZrO_3$), and barium titanate ($BaTiO_3$). The best known are quartz, tourmaline, Rochelle salt, and some PZ ceramics as the crystal titanate and zirconate. These ceramics are not really the PZ type but have a polarized electrostrictive effect. The material deformed by a force F must be composed of a single crystal to be truly PZ.

A ceramic substrate may have a large number of crystal grains randomly oriented forming a crystalline structure. The random orientation of the grains results in a net cancelation of the PZ effect. To this end, the ceramic must be polarized in order to align the majority of the effects of the individual grains. The term PZ became interchangeable with the term "biased electrostrictive effect" in most scientific texts on this issue [35–38].

11.6.1 Piezoelectric Energy Conversion

PZ crystals have an interrelationship (electromechanical coupling) between mechanical variables such as tension T or strain S and electrical quantities such as electrical field E, electrical displacement (flux density) D, or polarization P. When a force F (given in Newtons) per unit area is applied to the initial length x_0 of a crystal structure, it shifts the balance center of the positive and negative charges, causing a polarization of the material, and, as a result, a displacement will happen (see Figure 11.10) [39–41]. The proportionality of the applied longitudinal voltage is given by Hooke's Law $F = -kx$ given by $\sigma = -ce$, where σ is the material strain, c stiffness or elasticity tensor or yet the strain sensor is $e = -s\sigma$ where the tensor s, is called the compliance tensor.

Such PZ transformations can be interpreted as approximately proportional to the output electrical energy due to the compression in such materials. Therefore, the PZ effect must be understood as a linear electromechanical interaction between the mechanical strength and the electrical state (Coulomb force) in crystalline materials (ceramics, polymers, etc.).

The reverse PZ effect is also possible: by applying a voltage on the faces of the crystal, it will contract or expand, which makes it an excellent transducer to produce ultrasound waves. Figure 11.11 shows the electric and mechanical equivalents of the reactions offered by PZ bodies. In the electrical equivalent, E_o is the open-circuit voltage produced across the PZ material due to the applied force F, C_{stack} is the cell capacitance, R_{loss} is the loss resistance, and $R_{leakage}$ is the leakage resistance.

In the mechanical equivalent circuit F is the compressive force on the stack formed by the PZ body of transversal area A and the fixed reference structure, m_{pz} is the mass PZ, C_{pz} is a damping constant, k_{pz} is the spring constant, and x_{pz} is the length of the PZ material compression. If the applied voltage and the

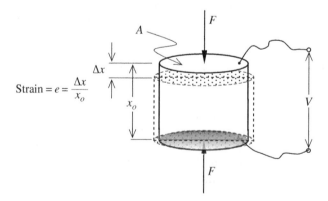

Figure 11.10 Generation of piezoelectric voltage.

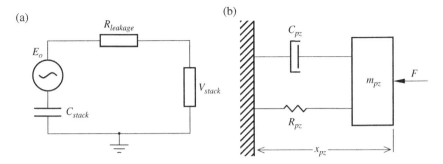

Figure 11.11 Equivalent models of a piezoelectric generator. (a) Electrical equivalent circuit resistance capacitor and (b) Mechanical equivalent circuit mass spring.

generated electric field are not great, the displacement magnitude x proportional to F and the dielectric displacement D can be represented by the electrical (E, D) and mechanical (σ, T) properties related as [35–37]

$$D = \varepsilon \cdot E + d \cdot \sigma \qquad (11.3)$$

and

$$e = d \cdot E + s \cdot \sigma \qquad (11.4)$$

where:

$D = $ The electric displacement charge density (C/m^2)
$\varepsilon = $ The free space permittivity (F/m)
$E = $ The electric field generated or applied (V/m)
$d = $ The PZ coefficient, deformation or fatigue (stress) of the material $(C/N$ or $M/V)$ (Hooke's Law)
$\sigma = $ Applied longitudinal tension (tensile stress) (N/m^2)
$e = $ The elongation (strain) (m/m)
$s = $ The crystal compliance or elastic stiffness (m^2/N)

In addition, it can be said that the Young's modulus Y, or module elasticity, will be given by [14, 15, 17, 40]

$$Y = \frac{1}{s} = \frac{F/A}{\Delta x / x_o} = \frac{\sigma}{e} \qquad (11.5)$$

where:

$F/A = $ The PZ pressure in N/m^2
$\Delta x = $ Elongation (strain) (m/m)
$x_0 = $ Initial length of the body subjected to stress (m)

11.6.2 Piezoelectric-Based Energy Applications

The PZ effect can be used either in two ways, to generate electricity or as a sensor to measure acceleration movements, vibration, pressure, influence on the surface, and floating forces. The most common applications are ultrasound, weighing scales, eco measurement of river depths and streams, PZ potential transformer, and PZ motor. However, the PZ phenomena may have other useful applications such as the production and sound detection, generation of high voltages, electronic generation of stable frequency, micro-ultrafine concentration of scales, and optical assemblies. It is also the basis for a series of instrumental scientific techniques with atomic resolution (scanning probe microscopy) and other uses for the daily life such as ignition source for electric lighters, crystal microphones, and capsules of acoustic guitar, basses, cellos, and other instruments acting as a kind of microphone. More recently, PZ generation has been designed to produce electricity from the natural human activity. Examples are the use of PZ materials in nightclub floors, tours, streets, bridges, and roads, where the pressure caused by the movement of people, machinery vibration, or passage of vehicles can be used to generate electricity cheaply, clean, and self-sustained. The use of the energy generated by these high electrical impedance sources would naturally be wasted, as it could be recovered with efficiency higher than 50%.

In order to make useful such a random and virtual pulse generation, this power must be smoothed, stored, and made compatible with the electrical loads for a proper PZ energy conversion. Storage methodologies that can be employed are discussed in Chapter 14 of this book. In order to optimize the power transfer, it requires impedance matching, which can be achieved using electronic power converters. PZ generation reacts to the generated cause causing some damping in the generating structure or the machine; this can reduce the maximum amplitude of the vibration and the vibration time after a transient excitation because there is use of kinetic energy. Electronic re-injection of PZ energy can be used for better generating quality [38, 39, 41].

References

1 P.P.L. Regtien, Electronic Instrumentation, VSSD Publishers, Norway, 2005, p. 109.

2 R.A. Freedman, F.W. Sears, and M.W. Zemansky, Sears and Zemansky's University Physics, 12th ed., Pearson, São Paulo, Brazil, 2010.

3 P. Horowitz and W. Hill, The Art of Electronics, Cambridge University Press, London, England, 2000, 716 p.

4 A.J. Diefenderfer, Principles of Electronic Instrumentation, W.B. Saunders Co., 1972, 667 p.

5 W. Bolton, Industrial Control and Measurement, Longman Scientific and Technical, Essex, England, 1991, 203 p.

6 A.D. Helfrick and W.D. Cooper, Modern Electronic Instrumentation and Measurement Techniques, Prentice-Hall, New York, 1994, 324 p.

7 Peltier as a semiconductor, http://en.wikipedia.org/wiki/Thermoelectric_cooling, accessed March 10, 2017.

8 L.B. Reis, Electrical Power Generation (Geração de Energia Elétrica), 2nd ed., Manole, São Paulo, Brazil, 2011, 460 p.

9 J.L. Acioli, Power Sources (Fontes de Energia), UnB, Brasília, Brazil, 1994, 146 p.

10 M. Previsic, R. Bedard, and G. Hagerman, E2I EPRI Assessment Offshore Wave Energy Conversion Devices, Report: E2I EPRI WP–004–US–Rev 1, Electricity Innovation Institute, Washington, D.C., USA, June 2004.

11 Salters Duck, http://science.howstuffworks.com/environmental/green-science/salters-duck.htm, accessed March 10, 2017.

12 R. Sæther, Wave energy conversion: Salter duck wave energy converter with direct electrical power conversion, Master Thesis, Institutt for elkraftteknikk, Norwegian University of Science and Technology (NTNU), Norway, 2010.

13 N.W. Bellamy, The Circular Sea Clam Wave Energy Converter Hydrodynamics of Ocean Wave-Energy Utilization, part of the Series International Union of Theoretical and Applied Mechanics, Springer-Verlag, Berlin, Germany, 1986, pp. 69–79.

14 M. Greenhow, J.H. Rosen, and M. Reed, Control strategies for the clam wave energy device, Applied Ocean Research, Vol. 6, pp. 197–206, 1984.

15 T.W. Thorpe, A Brief Review of Wave Energy, A Report Produced for The UK Department of Trade and Industry, The UK Department of Trade and Industry, Harwell, UK, May 1999.

16 R. Bhattacharyya and M.E. McCormick, Wave Energy Conversion, Elsevier Ocean Engineering Book Series, Vol. 6, Elsevier Science Ltd, Oxford, UK, 2003.

17 N.W. Bellamy, A. Bucchi, and G.E. Hearn, Analysis of the SEA-OWC-Clam wave energy device—Part A: historical development, hydrodynamic and motion response formulations & solutions, Renewable Energy, Vol. 88, pp. 220–235, 2016.

18 G.E. Hearn and J.R. Chaplin, The fluid structure interaction of wave energy devices: some old and some new theoretical and experimental challenges, The Royal Institution of Naval Architects Conference on Marine Renewable Energy, London, 2008.

19 T.V. Heatha, Review of oscillating water columns, Philosophical Transactions Royal Society, Vol. 370, pp. 235–245, 2012.

20 L.A. Vega, P. Tharakan, C.L. Torregosa and M. Lillis, Wave Energy Conversion and Ocean Thermal Energy Conversion Potential in Developing Member Countries, Asian Development Bank (ADB), Mandaluyong City, Philippines, 2014.

21 L.F. Yeung, P. Hodgson, and R. Bradbeer, Generating Electricity Using Ocean Waves as Renewable Source Of Energy, report for the Hong Kong Electric Company Ltd, Hong Kong, July 2007.

22 R.H. Stewart, Introduction to Physical Oceanography, Department of Oceanography Report, Texas A&M University, College Station, TX, September 2006.

23 M.E. McCormick, Ocean Engineering Mechanics: With Applications, Cambridge University Press, Cambridge, UK, 2010.

24 J. Taylor, Edinburgh Wave Power Group, http://www.homepages.ed.ac.uk/v1ewaveg/index.htm, accessed March 10, 2017.

25 The JONSWAP spectrum, http://www.codecogs.com/library/engineering/fluid_mechanics/waves/spectra/jonswap.php, accessed March 10, 2017.

26 European Marine Energy Centre (EMEC) Ltd, Old Academy Business Centre, http://www.emec.org.uk/about-us/our-tidal-clients/andritz-hydro-hammerfest/, accessed March 10, 2017.

27 J.H.G.M. Alves, M.L. Banner, and I.R. Young, Revisiting the Pierson–Moskowitz asymptotic limits for fully developed wind waves, Journal of Physical Oceanography, Vol. 33, pp. 1301–1323, 2003.

28 M. Mitchner and C.H. Kruger Jr., "Magnetohydrodynamic (MHD) power generation," in Physical Gas Dynamics and Partially Ionized Gases, Mechanical Engineering Department, Stanford University, Stanford, CA, November 1973, Chapter 9, pp. 214–230.

29 S. Smolentsev, N. Morley, and M. Abdou, MHD and thermal issues of the SiCf/SiC flow channel insert, Fusion Science and Technology, Vol. 50, pp. 107–119, 2006.

30 S. Smolentsev and R. Moreau, Modeling quasi-two-dimensional turbulence in MHD duct flows, CTR, Stanford University, Proceedings of the Summer Program, Stanford, CA, 2006.

31 H. Goedbloed and S. Poedts, Principles of Magnetohydrodynamics with Applications to Laboratory and Astrophysical Plasmas, Cambridge University Press, University of Cambridge, Cambridge, UK, 2004, Chapter IV, pp. 214–232.

32 C. Keawboonchuay and T.G. Engel, Electrical power generation characteristics of piezoelectric generator under quasi-static and dynamic stress conditions, IEEE Transactions on Ultrasonics, Ferroelectrics, and Frequency Control, Vol. 50, No. 10, pp. 1377–1382, 2003.

33 H.-B. Fang, J.-Q. Liu, Z.-Y. Xu, L. Dong, L. Wang, D. Chen, B.-C. Cai, and Y. Liu, Fabrication and performance of MEMS-based piezoelectric power generator for vibration energy harvesting, Microelectronics Journal, Vol. 37, No. 11, pp. 1280–1284, 2006.

34 J.R. Phillips, Piezelectric Technology Primer, Report on CTS Wireless Components, Albuquerque, New Mexico, EUA, 2010.

35 J. Sirohi and I. Chopra, Fundamental understanding of piezoelectric strain sensors, Journal of Intelligent Material Systems and Structures, Vol. 11, pp. 246–257, 2000.

36 S. Sherrit and B.K. Mukherjee, Characterization of Piezoelectric Materials for Transducers, Report Jet Propulsion Laboratory and NASA, California Institute of Technology, Pasadena, CA, 2008, pp. 1–45.

37 H.A. Sodano, D.J. Inman, and G. Park, A review of power harvesting from vibration using piezoelectric materials, The Shock and Vibration Digest, Vol. 36, No. 3, pp. 197–205, 2004.

38 K. Uchino, Introduction to Piezoelectric Actuators and Transducers, Report of the International Center for Actuators and Transducers, Penn State University, University Park, PA, June 2003.

39 Vibration Energy Harvesting with Piezoelectrics, http://www.mide.com/collections/vibration-energy-harvesting-with-protected-piezos, accessed March 10, 2017.

40 OIG_GIANT_II, http://www.marinetraffic.com/pt/ais/details/ships/shipid:364879/mmsi:305508000/imo:9509970/vessel:OIG_GIANT_II, accessed March 10, 2017.

41 Catalog of Piezo Systems, Inc., Piezoceramic Application Data, Woburn, MA, No. 8, pp. 59–61, 2011.

12

Induction Generators

12.1   Introduction

Rotating generators for renewable energies might be of any kind, as long as their shaft mechanical power is fully converted to the electrical power, and such a generator will respond to torque and speed range impressed on the mechanical shaft by the renewable source. It is possible to use any type of construction, electrical windings, number of phases, dc, or ac, where the ac can be either synchronous or asynchronous (induction) types. Nowadays, dc machines are only justified for very small power plants because they are bulky, require frequent maintenance, are more expensive, and are relatively inefficient than ac machines. There are some other contemporary permanent magnet-based design machines, but they are also used for low-power ranges.

Typically, small renewable energy power plants mostly use induction machines, because they are widely available and very inexpensive. It is also very easy to operate them in parallel with large power systems, because the utility grid will keep control of voltage and frequency, while static and reactive compensating capacitors can be used for correction of the power-factor and harmonic reduction [1–3]. Since induction generators are important for applications in small hydropower, diesel, and wind systems, in this chapter we briefly discuss details that will enable the reader to understand how to use such electrical machine.

Although the induction generator is suitable for hydro- and wind power plants, it can be used efficiently in prime movers driven by diesel, biogas, natural gas, gasoline, and alcohol motors. Induction generators have outstanding operation as either motor or generator; they have very robust construction features, providing natural protection against short circuits, and have lowest cost among generators. In addition, they do not need expensive permanent magnets. Abrupt speed changes (that usually happens in small renewable power sources) due to load variations are easily absorbed because

Integration of Renewable Sources of Energy, Second Edition. Felix A. Farret and M. Godoy Simões.
© 2018 John Wiley & Sons, Inc. Published 2018 by John Wiley & Sons, Inc.

their shaft and rotor construction are rugged and solid. Any current surge is damped by the magnetization path of its iron core without fear of demagnetization, as opposed to permanent magnet-based generators. In this chapter, we consider an analysis of the induction generator in both stand-alone and grid-connected modes, further details of the induction generator can be studied in Refs. [3–5]. The term asynchronous generator can be used for standard induction generators, but it may also mean generators that do not have strict synchronism with the utility frequency as described in Chapters 15 and 17. These are, for example, the case for photovoltaic-, fuel cell-, and battery-powered systems, which have no synchronism with the utility grid, because power electronic inverters are used to convert dc into ac. The lack of synchronism may be responsible for frequency oscillations and instabilities in the power systems to which they are connected, and system controllers must be able to synchronize the inverters.

Utility companies have policies to constrain the amount of injected asynchronous energy something about 15% of their total capacity, which is proportional to their installed reactive power (VAR) management. Therefore, in the future, widespread and deep penetration of distributed generation will only be possible with sophisticated controls and interaction with the utility grid. Induction generators can be used with or without power electronics, making them very flexible and reliable to use as it is discussed in this chapter.

12.2 Principles of Operation

The induction generator has same construction as induction motors, with some possible improvements in efficiency [1, 5], but usually any induction machine can operate as an induction generator. There is an important operating difference; the rotor speed of the generator is advanced with respect to the stator magnetic field rotation. In quantitative terms, if n_s is the synchronous mechanical speed (rpm) at the synchronous frequency f_s (Hz), the resulting output power comes from the induced voltage proportional to the relative speed difference between the electrical synchronous rotation and the mechanical rotation within a speed-slip factor range given by

$$s = \frac{n_s - n_r}{n_s} \tag{12.1}$$

where:

s = The slip factor
n_r = The rotor speed in rpm

Calling the electrical stator frequency f_s, the rotor frequency f_r can be related to the electric frequency through the expression

$$f_r = \frac{p}{120}(n_s - n_r)\frac{n_s}{n_s} = sf_s \qquad (12.2)$$

where:

$s =$ The slip factor
$n_r =$ The rotor speed (rpm)

When there is no relative movement of rotation between rotor and stator $\omega_r = \omega_s$, the rotor frequency is dc. For other angular speeds, in quantitative terms, the voltage induced on the rotor is directly proportional to the slip factor s and, from there, transferred to the stator windings [1, 5]. From this definition, the voltage induced on the rotor is established for any speed E_r with respect to the blocked rotor E_{r0}, that is,

$$E_2 = E_r = sE_{r0}$$

and the stator voltage comes from

$$E_1 = a_{rms}\frac{E_r}{s} \qquad (12.3)$$

Similarly, the rotor current can be established as

$$I_2 = \frac{I_r}{a_{rms}}$$

and

$$Z_r = \frac{E_r}{I_r} = \frac{sE_{r0}}{I_r} = R_r + jsX_{r0} \qquad (12.4)$$

From this definition, the rotor impedance for any rotor speed can be established from a blocked rotor test as

$$Z_{r0} = \frac{E_{r0}}{I_r} = \frac{R_r}{s} + jX_{r0} = \frac{Z_r}{s} \qquad (12.5)$$

where:

$X_{r0} =$ The blocked rotor reactance
$R_r =$ The rotor winding resistance

Then, the rotor impedance reflected on the primary can be obtained from equations (12.3), (12.4) and (12.5) as

$$Z_2 = a_{rms}^2\left(\frac{R_r}{s} + jX_{r0}\right) \qquad (12.6)$$

In terms of rotor values referred to stator values, comes

$$Z_2 = \frac{E_1}{I_2} = \frac{R_2}{s} + jX_2 \tag{12.7}$$

and the rotor current is

$$I_2 = \frac{E_1 s}{R_2 + jX_2 s} \tag{12.8}$$

It is possible to understand the voltage induction in the rotor of induction motors as similar to an electric transformer, except that the transformer secondary windings (in the induction motor) are rotating. Because there is an air gap in electric machines, the magnetic coupling between the primary and secondary windings makes the B–H characteristic slope under the magnetomotive force (mmf) curve of these machines much less accentuated than that of a good quality transformer. The larger air gap will help the machine core to be less prone to magnetic saturation, an effect well studied in magnetic circuits. Saturation of the induction generator is decreased significantly, and as a consequence, the necessary magnetizing current is a lot higher with respect to a transformer (X_m is much lower).

The apparent rotor power (in complex variables) can also be established from equations (12.5) and (12.6) as

$$S_2 = E_1 I_2 = I_r^2 \left(\frac{R_r}{s} + jX_{r0} \right)$$

or

$$S_2 = \frac{E_1^2 s}{R_2 + jX_2 s} \tag{12.9}$$

From these expressions, it is clear that the rotor power factor of the induction generator depends on the slip factor and other parameters in addition to the load. The leading current component (with respect to voltage) is almost constant for all voltages across the output terminals and fixed frequencies. Therefore, the reactive power absorbed by this circuit must be supplied by an external source, which can be the utility grid (since any established utility grid will have a synchronous generator providing reactive power), or a capacitor bank, or a power electronics compensator (also called STATCOM).

12.3 Representation of Steady-State Operation

In Section 12.2, a formulation was developed toward an equivalent circuit of the induction machine, similar with a transformer model. Such model is the one used by the IEEE Std. 112—2004, and it is illustrated in Figure 12.1.

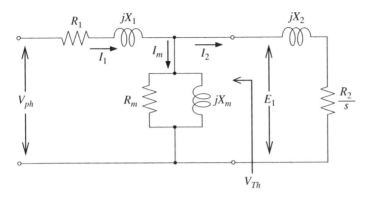

Figure 12.1 Equivalent model of the induction generator.

This model is also known as a per-phase equivalent model, valid only for perfect sinusoidal excitation and balanced conditions [3–6]. For a more representative approach, every element of the induction generator model has to be excited by the grid voltage, considering any distortions and harmonics. However, the per-phase equivalent model in Figure 12.1 is usually considered for analysis, where R_1 and X_1 are the resistance and leakage reactance of the stator, R_2 and X_2 are the resistance and reactance of the rotor, and R_m and X_m are the equivalent core loss resistance and magnetizing reactance.

The only equivalent resistance in the rotor model R_2/s depends on the rotor speed (i.e., on the slip factor). This is a very important difference with respect to a transformer model, because rotor voltage is subjected to a variable frequency, causing E_r, R_r, and X_r to be variable. In addition, it must be emphasized that the circulating current in the rotor will also depend on the impedance, whose resistance and inductance change slightly because of the skin effect. Nevertheless, the inductance is indeed affected in a more intricate way. For very large slip factor, the rotor impedance and skin effect will impose nonlinear torque and air gap power relationships. For small slip factor, we can assume that $s \rightarrow 0$, the rotor impedance becomes predominantly resistive, and the rotor current can be considered primarily to vary linearly with s.

For voltage excitation with frequency corresponding to a speed below the synchronous speed, the slip factor becomes positive, and the current is delayed with respect to the voltage. For current excitation (e.g., with vector-controlled drives), the slip factor can be either positive or negative. Beyond the synchronous speed, the rotor moves faster than the magnetic field. For voltage excitation, the slip factor s becomes negative and the current I_2 leads to the E_2 voltage. At first glance, this fact seems contradictory since the rotor circuit is essentially inductive. However, the reason for such phase inversion is the reversal of the relative move of the rotating magnetic field with respect to the rotor with changes in s when both real and imaginary parts of the current will be reversed.

The equivalent per-phase impedance seen by across the output terminals represented in Figure 12.1 is [4, 5]

$$Z = R_1 + jX_1 + \cfrac{1}{\cfrac{1}{jX_m} + \cfrac{1}{R_m} + \cfrac{1}{\cfrac{R_2}{s} + jX_2}} \tag{12.10}$$

In order to simplify the analysis and improve the formulation of the performance calculation, it is convenient to use a Thévenin circuit. Thus, the impedances of the induction generator can be taken with the following notation:

$$Z_1 = R_1 + jX_1 \tag{12.11a}$$

$$Z_2 = \frac{R_2}{s} + jX_2 \tag{12.11b}$$

$$Z_m = \cfrac{1}{\cfrac{1}{R_m} + \cfrac{1}{jX_m}} \tag{12.11c}$$

12.4 Power and Losses Generated

A power balance in an induction generator is expressed as

$$P_{out} = P_{in} - P_{losses} \tag{12.12}$$

The machine losses can be expressed by their individual contributions as

$$P_{losses} = P_{copper} + P_{iron} + P_{f\&w} + P_{stray} \tag{12.13}$$

Copper losses in the stator are obtained from $P_{stCu} = I_1^2 R_1$. Iron losses are caused by hysteresis currents (magnetizing) and Foucault currents (current induced in the iron). It should be emphasized that stator and rotor iron losses are conveniently grouped together, in accordance to experimental procedures because it is very difficult to separate their contributions. Therefore, the loss resistance R_m in the equivalent circuit of the induction generator practically represents all those losses combined with the mechanical losses. These include friction losses plus those resulting from the rotor movement against the air (windage) and spurious losses. The spurious losses can be reunited, among other additional losses, they are high-frequency losses in the stator and rotor dents, mainly through the parasite currents produced by fast flux pulsation when dents and grooves move from their relative positions [4, 5]. Subtracting these losses from input power makes the air gap power transferred from stator to the rotor. Then in such power received by the rotor, one must subtract the rotor copper losses, $P_{rCu} = I_2^2 R_2$, as well. Such final value remains for transformation from mechanical into electrical power (in motoring mode). Finally, the output power is the net power in the shaft (called by shaft useful power).

For the generating mode, the reasoning must be reversed, therefore, starting from the rotor and moving the power conversion toward the machine terminals. The increase in shaft rotation also increases losses by friction, windage, and spurious effects. In compensation, the losses in the core are reduced at the synchronous speed. These rotating losses are approximately constant, since some of them increase and others decrease with rotation. In order to quantify losses, a current I_1 will produce losses in the stator winding of the three phases by

$$P_{stCu} = 3I_1^2 R_1 \tag{12.14}$$

The iron core losses are given by

$$P_{iron} = \frac{3E_1^{\,2}}{R_m} \tag{12.15}$$

The only possible power dissipation (active power) in the secondary part of the circuit of Figure 12.1 is through the rotor resistance, therefore,

$$P_g = 3I_2^2 \frac{R_2}{s} \tag{12.16}$$

However, from the rotor equivalent circuit shown in Figure 12.1, the resistive losses are given by

$$P_{rCu} = 3I_r^2 R_r \tag{12.17}$$

In any ideal lossless transformer, the rotor power would remain unchanged when referred to the stator and, therefore, from equations (12.16) and (12.17),

$$P_{rCu} = 3I_2^2 R_2 = sP_g \tag{12.18}$$

The losses following the electrically converted power are in the mechanical parts, in the copper, in the core, and in other spurious losses. They can be all expressed as

$$P_{losses} = P_{stCu} + P_{rCu} + P_{iron} + P_{frict+air} + P_{stray}$$

With equations (12.14) and (12.18), the copper losses in the stator and rotor can be regrouped as

$$P_{copper} = 3I_1^2 \left(R_1 + R_2 \right) = \frac{3V_{ph}^2 \left(R_1 + R_2 \right)}{\left(R_1 + \dfrac{R_2}{s} \right)^2 + F^2 \left(X_1 + X_2 \right)^2} \tag{12.19}$$

The mechanical power converted into electricity, or the power developed in the shaft for a negative s, is the difference between the power going through the air gap and that dissipated in the rotor [4–6]. Then, from equations (12.16) and (12.18) comes

$$P_{mech} = 3I_2^2 \frac{R_2}{s} - 3I_2^2 R_2$$

After simplification, the mechanical power becomes

$$P_{mec} = 3I_2^2 R_2 \frac{1-s}{s} = (1-s)P_g \tag{12.20}$$

With the machine losses expressed by their individual contributions as given in equation (12.13), it is possible to establish a routine to calculate the relationships given in the flowchart of Figure 12.2, where voltage, current, and power are represented in the equivalent circuit of Figure 12.1. Notice that if $Z_m \gg Z_1$ and $Z_m \gg Z_2$, as is usually the case, the stator Thévenin voltages become $V_{Th} \approx V_{ph}$ and $Z_{Th} \approx Z_1$. Conversely, assuming that $R_m \to \infty$, Z_1 and V_{ph} will have to be replaced by their Thévenin equivalent.

It is possible to take into account how the frequency affects the parameters of the induction generator. These differences are included in the per unit (p.u.) frequency related to the frequency used in the parameter measurement usually, $\omega_{meas} = 2\pi \cdot 60$, defined as

$$F = \frac{f}{f_{meas}} = \frac{\omega}{\omega_{meas}} = \frac{pn_r}{7200} \tag{12.21}$$

where:

f, ω = The actual operating frequencies in Hz and rad/s, respectively
f_{meas}, ω_{meas} = The frequencies of the parameter measurement in Hz and rad/s, respectively
p = The number of poles of the machine
n_r = The rotor speed in rotations per minute

Corrections using factor F for parameter values are extremely important when using an induction generator model in frequencies other than that used to obtain the machine parameters in laboratory tests.

12.5 Self-Excited Induction Generator

The induction generator needs a reasonable amount of reactive power, which must be fed externally to establish the magnetic field necessary to convert the mechanical power from its shaft into electrical power [5, 6]. From the point of

Figure 12.2 Calculation for the induction generator.

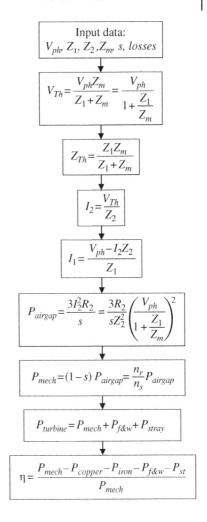

Input data:
$V_{ph}, Z_1, Z_2, Z_m, s, losses$

$$V_{Th} = \frac{V_{ph}Z_m}{Z_1+Z_m} = \frac{V_{ph}}{1+\dfrac{Z_1}{Z_m}}$$

$$Z_{Th} = \frac{Z_1 Z_m}{Z_1+Z_m}$$

$$I_2 = \frac{V_{Th}}{Z_2}$$

$$I_1 = \frac{V_{ph}-I_2 Z_2}{Z_1}$$

$$P_{airgap} = \frac{3I_2^2 R_2}{s} = \frac{3R_2}{sZ_2^2}\left(\frac{V_{ph}}{1+\dfrac{Z_1}{Z_m}}\right)^2$$

$$P_{mech}=(1-s)\,P_{airgap}=\frac{n_r}{n_s}P_{airgap}$$

$$P_{turbine}=P_{mech}+P_{f\&w}+P_{stray}$$

$$\eta = \frac{P_{mech}-P_{copper}-P_{iron}-P_{f\&w}-P_{st}}{P_{mech}}$$

view of reactive power, we can distinguish inductive receivers that take reactive power from the system to produce their magnetic field (mainly asynchronous motors and transformers) and capacitive receivers that give reactive power to the system (static capacitors and overexcited synchronous machines). The power absorbed from the supply network is considered positive ($P_c > 0$ and $Q_c > 0$), and that one from the network is considered negative ($P_c > 0$ and $Q_c > 0$). The required reactive power Q_c of these receivers comprises the magnetization reactive power as the main component of reactive power, which depends directly proportional to the amount of iron and the air gap volume in magnetic components, such as any induction machine.

In utility-connected induction generator applications, the synchronous network utility grid will supply such reactive power. In stand-alone applications, a bank of capacitors connected across its terminals or an electronic inverter must supply the reactive power. For this reason, the external reactive source must remain permanently connected to the stator windings responsible for the output voltage control. When capacitors are connected to induction generators and provide reactive power, the system is usually called a self-excited induction generator (SEIG). Self-excitation of the induction generator is usually recommended only for small power plants because the size of the capacitor in respect to the size of the machine has electrical constraints and maximum cost for such application [6–11].

The self-excitation process occurs when an external capacitance is associated with an ordinary induction motor. When the shaft is rotated externally, such movement interacts through a residual magnetic field and induces a voltage across the external capacitor, resulting in a current in the parallel circuit, which, in turn, reinforces the magnetic field and the system will build up an increasing excitation. This process is illustrated in Figure 12.3. Notice that in practical schemes, it is advisable to connect each excitation bank of capacitors directly across each motor winding phase in either Δ–Δ or Y–Y connection [5–11].

Each induction machine will require a certain capacitance bank that will depend on the primary power, the magnetization curve of the core, and the instantaneous load. Therefore, interaction among the operational state of the primary source of energy, the induction generator, the self-excitation parameters, and the load will define the overall performance of the power plant [12–14]. The performance is greatly affected by the variance of many of the parameters related to the availability of primary energy and load variations.

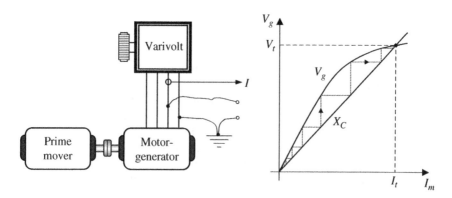

Figure 12.3 Self-excitation process of the induction generator.

The equivalent circuit given in Figure 12.1 is conveniently transformed in the circuit of Figure 12.4 to include the self-excitation capacitor and the load in p.u. frequency values F as given in equation (12.21), representing a more generic form of the stand-alone generator [2, 3]. From the definition of secondary resistance (rotor resistance) shown in Figure 12.1 and from equation (12.6), the following modification can be used to correct R_2/s to take into account changes in the stator and rotor p.u. frequencies

$$\frac{R_2}{Fs} = \frac{R_2}{F\left(1 - \dfrac{n_r}{n_s}\right)} = \frac{R_2}{F - v}$$

where:

n_r = The rotor speed in p.u. referred to the test speed used in the rotor.

Most transient modeling studies consider magnetizing reactance under magnetic saturation, but the other parameters are typically considered constant [15–18]. However, the use of uncorrected parameters may lead to wrong evaluation of overall machine performance.

Unfortunately, when working in stand-alone mode, a SEIG may collapse and stop producing any terminal voltage and current. This can happen with high output electric load or maybe because of short circuit through their terminals, resulting in complete loss of residual magnetism. Although the machine is protected, because such collapse is an intrinsic protection feature, it may affect the restarting of the overall system. Four methods for its re-magnetization are possible: (i) maintaining a spare capacitor always charged and, when necessary, discharging it across one of the generator phases; (ii) using a charged battery for the same; (iii) using a rectifier fed by the distribution network for the same; and (iv) keeping it running without load, for a possible long time, until the residual magnetism build ups, naturally, and the system restarts again.

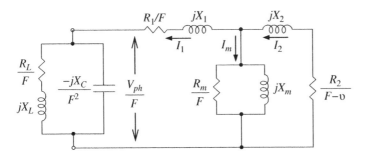

Figure 12.4 Equivalent circuit of the self-excited induction generator. Simões and Farret [5]. Reproduced with the permission of Taylor and Francis.

12.6 Magnetizing Curves and Self-Excitation

The equivalent impedance seen by the voltage across the terminals of the stator reactance can be evaluated with Figure 12.1 at the synchronous speed ($F = 1$) as

$$Z = R_1 + jX_1 + \cfrac{1}{\cfrac{1}{jX_m} + \cfrac{1}{\cfrac{R_2}{s} + jX_2}} \tag{12.22}$$

The impedance given by equation (12.22) is useful in determination of load limits, in both amplitude and amount of reactive power required for the induction generator when interconnected to the utility grid [5, 19–23].

It is often more convenient to assume the electric frequency supplied to the load and then obtain the mechanical rotation. Therefore, the equation used for calculation of these variables should be altered conveniently [3, 5], in order to accommodate the electrical frequency as

$$\omega_r = \frac{p}{2}\left(2\pi\frac{n_r}{60}\right) = \frac{p}{2}\omega_r' \tag{12.23}$$

where:

$\omega_r' = 2\pi n_r/60 =$ The angular mechanical speed of the rotor (rd/s).

The required capacitance connected across the induction generator terminals can be determined by the intersection of the magnetizing curve $V_g = f(I_m)$ of the machine and the straight line represented by the capacitive reactance given by $1/\omega C$ of the self-exciting bank, as depicted in Figure 12.3. This intersection point gives the terminal voltage V_t and current I_t of the induction generator at no-load conditions. At this point, the magnetizing current is lagging with respect to the terminal voltage (about 90°), that is, depending on the losses of the motor at no-load conditions, the current through the capacitor is advanced by approximately 90°. This means that the intersection of these two lines is the point on which the necessary reactive power of the generator is supplied only by the capacitors and can be understood as the resonant conditions for the reactive power flow. Location of this point should be in between the rated current of the machine and the end of the straightest portion of the air gap line.

The magnetizing characteristic shown in Figure 12.3 can be determined with the help of a secondary driving machine coupled with the shaft of the induction motor generator, compensating for any shaft mechanical loading during the experimental procedure for obtaining the magnetizing curve. With a constant rotation, say at 1800 rpm for a 60-Hz four-pole machine, an ac voltage is imposed across the terminals of the motor generator by a variac that induces the nominal current. From this point, several decreasing steps of voltage have to be recorded with the corresponding current until the zero voltage. A constant rotation must always be guaranteed during the tests, because any mismatch will change the frequency and terminal voltage of the rotor. The mutual inductance is essentially nonlinear, and for best results in numerical computations, it should be determined for each value of the instantaneous magnetizing current, I_m [24–27].

From the discussion earlier, it can be observed that the induction generator accepts constant and variable loads; it starts either loaded or without load, it is capable of continuous or intermittent operation, and it has a natural protection against short circuits and over currents through its terminals. When the load current goes above certain limits, the residual magnetism falls to zero and the machine is de-excited (or the system collapses) [5].

12.7 Mathematical Description of the Self-Excitation Process

A mathematical description of the self-excitation process can be explained based on the Doxey model [5, 20–25], which is derived from the steady-state model depicted in Figure 12.4. Since the excitation impedance is much larger than the winding impedance, it is usual to represent the induction generator model with a separate resistance for losses. Thus a RL branch is included to represent a typical load connected across the generator terminals [5, 26, 27].

By observing Figure 12.4, we see that the inductive load will decrease the effective value of self-exciting capacitance because some of the reactive power will be deviated to the load branch. Equation (12.24) quantifies the effective capacitance due to such an effect:

$$C_{eff} = C - \frac{L}{R_L^2 + \left(\omega_s L\right)^2} \tag{12.24}$$

Equation (12.24) leads to the conclusion that for a small L, the inductive reactance of the load does not have much influence on C_{eff}. However, for high values of L, this inductance can be decisive for the rated output voltage, because it approximates the straight line of the capacitive reactance slope to the almost linear portion of the excitation curve slope (air gap line), drastically reducing the terminal voltage until total collapse. The benefit is that for a very small load resistance, the discharge of the self-excitation capacitor is very quick, establishing a natural protection against high currents and short circuits. On the other hand, iron saturation and winding currents will impress limitation for high values of the self-excitation capacitance. The reason is that for heavy iron saturation, the intersection point on the straight line of the capacitive reactance matches very high current values for the generator, causing the winding to carry high circulation current, with possible damage in the insulation, wiring, and in the iron magnetic properties.

By applying power balance principles, Figure 12.4 helps to obtain equations (12.25) and (12.26), eventually equating the active and reactive power as [5, 21]

$$\Sigma P = I_2^2 R_2 \frac{1-s}{s} + I_2^2(R_1 + R_2) + \frac{V_f^2}{R_m} + \frac{V_f^2}{R_{Lp}} = 0 \tag{12.25}$$

$$\Sigma Q = \frac{V_f^2}{X_p} + \frac{V_f^2}{X_m} - \frac{V_f^2}{X_c} + \frac{V_f^2}{X_{Lp}} = 0 \tag{12.26}$$

where:

$X_m = \omega_s L_m$

R_{Lp}, X_{Lp} = The parallel equivalent load resistance and reactance connected across the generator's terminals.

In order to evaluate the performance of induction generators, one should take into account that the only power entering the circuit is the one from the prime mover, which the active power is given by equation (12.25). The reactive power should be balanced as in equation (12.26). The first term of equation (12.25) is the energy supplied to the generator from the mechanical shaft. The sign of the first term is inverted when $s > 1$ or $s < 0$, depicting the generating mode of the induction machine as a generator or brake, respectively [5, 22, 25]. Dividing equation (12.25) by I_2^2 after simplification one gets

$$\frac{R_2}{s} + R_1 + \frac{V_{ph}^2}{I_2^2}\left(\frac{1}{R_m} + \frac{1}{R_{Lp}}\right) = 0 \tag{12.27}$$

From Figure 12.4 we obtain

$$\frac{V_{ph}^2}{F^2 I_2^2} = \left(\frac{R_2}{Fs} + \frac{R_1}{F}\right)^2 + (X_1 + X_2)^2 \tag{12.28}$$

and with equation (12.27) the following holds

$$\left(\frac{R_2}{s} + R_1\right)R_{mL} + \left(\frac{R_2}{s} + R_1\right)^2 + F^2(X_1 + X_2)^2 = 0 \tag{12.29}$$

where :

$R_{mL} = R_m / R_{Lp}$

Equation (12.29) is of the second order, and the slip can be calculated with equation (12.30) (assuming constant parameters):

$$s = \frac{2R_2}{-2R_1 - R_{mL} \pm \sqrt{R_{mL}^2 - 4F^2 (X_1 + X_2)^2}} \tag{12.30}$$

The roots of the denominator in equation (12.30) will always be negative in order to satisfy the practical condition that if $X_1 + X_2$ is very small as usual, the load R_{mL} would have almost no influence on s, a constraint not always true. Therefore, it may be convenient to approximate equation (12.30) by the following expression:

$$s \cong -\frac{R_2}{R_1 + R_{mL}} = -\frac{R_2 (R_m + R_{Lp})}{R_1 R_m + R_1 R_{Lp} + R_m R_{Lp}} \tag{12.31}$$

An iterative method may be used when using equation (12.30) instead of equation (12.31), because the computation of F depends on ω_s, which, in turn, depends on s, which again is dependent on F. The efficiency can be estimated by

$$\eta = \frac{P_{out}}{P_{in}} = \frac{P_{in} - P_{losses}}{P_{in}} \tag{12.32}$$

The input mechanical power must subtract the copper losses $R_1 + R_2$ and the loss represented by the resistance R_m. Using equation (12.12) with equations (12.25) and (12.28) for the three-phase case without the portion corresponding to the load power ($R_{Lp} \to \infty$),

$$P_{out} = P_{in} - P_{losses} = 3I_2^2 R_2 \frac{1-s}{s} + 3I_2^2 (R_1 + R_2) + \frac{3V_{ph}^2}{R_m}$$

where:

$$P_{in} = 3I_2^2 R_2 \frac{1-s}{s}$$

Therefore, using these values in equation (12.32), after simplification one gets

$$\eta = \frac{I_2^2 R_2 \dfrac{1-s}{s} + I_2^2 (R_1 + R_2) + \dfrac{V_{ph}^2}{R_m}}{I_2^2 R_2 \dfrac{1-s}{s}} \tag{12.33}$$

Dividing the numerator and the denominator of equation (12.33) by I_2^2, and using once more equation (12.28), results in

$$\eta = \frac{\left(\dfrac{R_2}{s} + R_1\right) + \dfrac{1}{R_m}\left[\left(\dfrac{R_2}{s} + R_1\right)^2 + F^2\left(X_1 + X_2\right)^2\right]}{R_2 \dfrac{1-s}{s}} \quad (12.34)$$

12.8 Grid-Connected and Stand-Alone Operations

When an induction generator is connected directly with the utility grid, the machine must change its speed (increased up to or above the synchronous speed). The absorbed mechanical power at the synchronous rotation is minimally enough to withstand the mechanical friction and resistance of the air. In case the speed is increased slightly above the synchronous speed, a regenerative action happens, yet without supplying energy to the distribution network. This will happen only when the demagnetizing effect current of the rotor is balanced by a stator magnetic component capable of supplying just its own iron losses. Above this speed, the generator will supply power to the network.

When considering grid-connected mode, it should be observed that there is a rotation just above the synchronous speed in which the efficiency is very low. This effect is caused by the fixed losses related to the low level of the generated power and torque at such low speeds (see Figure 12.5) [5, 23–26].

Another important aspect to consider is the maximum shaft rotation. At such maximum value, disconnection from the distribution network should occur; otherwise a braking initiative should be taken for the turbine, probably put it under speed-controlled operation. The disconnection should be performed for electric safety of the generator in the case of a flaw in the control brake and for the safety of the local electric power company maintenance teams, when the generator should be disconnected from the distribution network.

Direct connection of induction generators to the distribution network is a quite simple process as long as the interconnection and protection guidelines are followed by the local utilities. Technically, the rotor is turned in the same direction as that in which the magnetic field is rotating as close as possible to the synchronous speed to avoid unnecessary speed and voltage perturbations. A similar circumstance happens with the connection of motors or transformers to the distribution network. There is a transient exchange of active power between generator and mechanical shaft plus a reactive power circulation between generator and the grid, with some flow of active power to the grid. The load can be any ordinary electrical load or a power inverter for interconnection with the distribution network or any ac load [24].

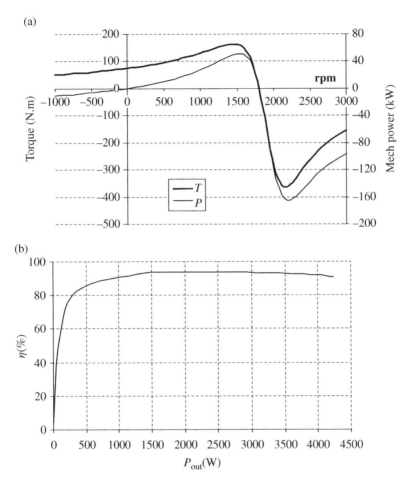

Figure 12.5 Typical performance of a four-pole 50-kW induction generator: (a) speed versus torque and output power; (b) generator efficiency characteristic.

The active power supplied to the electric load by an induction generator, similarly to what happens with synchronous generators, can be controlled by speed change, which is related to controlling the mechanical primary power. However, active power is converted only if reactive power is furnished for excitation, that is, providing a magnetizing current circulation through L_m. For the case of stand-alone operation, the magnetizing current is to be obtained from the self-excitation process, as discussed earlier in the chapter. The generator supplies a capacitive current, because the current I_1 is leading V_{ph}, as depicted in Figure 12.1. This is because the mechanical energy of rotation can influence only the active component of the current, with no effect on the reactive component. Therefore, the cage rotor illustrated in Figure 12.5

shows a typical performance of a four-pole 50-kW induction generator, depicting the speed versus torque and output power. In addition, the generator efficiency characteristic does not supply reactive current.

As a component of the magnetizing current is necessary in the main field direction, $I_{mx(B)}$, to maintain this flux, another reactive current to keep the dispersion flux also becomes essential. A leading current with respect to the stator current should supply both currents.

It is interesting to study when an induction generator is in parallel with a synchronous machine. The excitation depends on the relative speed between them, so the short-circuit current supplied depends on the voltage drop produced across the terminals of the synchronous generator. However, a very intense short-circuit transient current arises with extremely short duration. If the voltage across the terminals goes to zero, the steady-state short-circuit current is zero. A small current is supplied in the case of a partial short circuit because the maximum power the induction generator can supply with fixed slip factor and frequency is proportional to the square of the voltage across its terminals (see equation (12.19)). Such incapacity of sustaining short circuits reduces possible damages caused by electrical and mechanical stresses. As a result, induction generators have a lower short-circuit power and lower rating protection circuit needs, which makes lower cost and more improved on such features, when compared with the synchronous generators providing electrical power alone.

The induction generator dampens oscillations. All the load variation is followed by a speed variation and small phase displacement. The mechanical speed variations of the primary machine driving the generator are so small, with high inertia shafts, that they produce only minor variations of load.

Some steady-state conditions can be used for design and performance prediction induction generators in operation with wind, diesel, or hydroturbines. Aspects such as high slip factor, torque curve, winding sized to support higher saturation current, and increased number of poles can enhance the overall performance of the system.

12.9   Speed and Voltage Control

Alternative sources of energy require several levels of control, such as system- and equipment-based controls. Figure 12.6 shows a basic control diagram for rotating generators, some variables are used in accordance with the overall strategy, such as active and reactive power levels or shaft rotation. Such a system is usually a multiple-input/multiple-output control system where the output voltage is associated with the reactive power, frequency is associated with the mechanical output power, and variables are monitored continuously and fed back to follow reference signals [28–31].

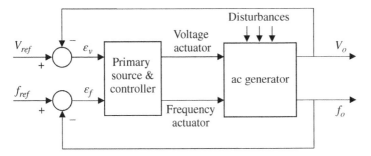

Figure 12.6 Basic control diagram for rotating generators.

When designing the control of an alternative source of energy for electricity generation, it is necessary to consider whether the source is of the rotating or static type. Typical rotating types are related to the wind energy and hydroelectric energy, whereas static ones are related to photovoltaic and fuel cells. For the rotating type, three options of generators can be used: dc generator, synchronous generator, or asynchronous generator.

In principle, all generators operate internally as ac, but their constructional features will make their output become dc (it is very rare to have dc generators nowadays) or ac. The ordinary type of dc generator is also known as a commutating machine, because an internal mechanism is used such that the machine output becomes a direct current while internally the flux alternates. The brushless dc generator, which is in fact an ac generator, is so-called for two reasons: (i) because the alternating current must be variable frequency and so derived from a permanent magnet or dc supply and (ii) its speed-torque characteristics are very similar to those of an ordinary dc generator (with brushes). The rotor consists of strong permanent magnets passing between the winding poles of the stator to induce an alternating current. A major advantage is that there is no need for induced current as in the induction generator, making dc generators more efficient and with slightly greater specific power.

A problem with the brushless dc generator is that the torque may be unsteady, causing some stress on the turbine and the mechanical drive train. This characteristic may be improved by distributing three or more harnessing coils around the stator, which, in turn, increases the frequency of the output voltage for the same rotation (increased number of poles). Such a feature is particularly interesting for wind energy because it may eliminate the gearbox. Dc generation has been replaced by ac generation because most modern ac power sources are lighter in construction, are mass produced, have very sophisticated power electronic controllers, and are very easy and flexible to operate. A complete discussion of other types of generators is not included here, but Chapter 13 gives an introduction to permanent magnet synchronous generators (PMSG) for renewable and alternative energy applications. The reader is encouraged to refer to an electric machines text.

Two main variables are important to control ac generators: the output voltage level and the power frequency. Synchronous generators have two control inputs (field current i_f and motor torque T_m) and four control outputs (output active power, output reactive power, and voltage magnitude, represented by P_o, Q_o, and $|V|$, respectively, and frequency f). All these variables are related to each other under a very complex interaction: torque variation causes changes in all four outputs, and there is no stronger relationship between T_m and Q_o. The degree of cross-coupling will depend on the structure and system size. Renewable energy-based systems are usually small, with reduced shaft inertia easily subject to oscillations, except for static sources [26–29].

One point in common between induction generators and other asynchronous sources of energy, such as photovoltaic and fuel cells, is the asynchrony with respect to the utility grid voltage. Therefore, a similar control philosophy may be extended to them. In the case of induction generators, the control variables are flux, torque, and impressed frequency voltage terminal. For photovoltaic and fuel cells, a dc link is a loop control associated with machine speed, and all the other variables (torque, flux, and frequency) correspond. Induction generators may control the input power by mechanical or electromechanical means, or electrically by load matching of the electrical terminals. PV arrays can have mechanical controls for positioning the array. Fuel cells may have hydrogen feed line control, and both can have electrical load matching through power electronic inverters.

12.9.1 Frequency, Speed, and Voltage Controls

Mechanical controls are applied in controlling the input shaft power of induction generators. For example, centrifugal weights might be used to close or open the admission of water for hydroturbines or for exerting pitch blade control in wind turbines. A gas turbine will have a throttle control of the gas–air mixture; a geothermal-based power plant will have a flow control of their fluid on the specific turbine that may impress mechanical power on an induction generator shaft. Some sophisticated mechanical controls include flywheels (to store transient energy), anemometers, or flow meters (to determine the wind intensity and water flow, respectively), in addition to wind tails and wind vanes (for detection of the wind direction), used for control and to limit the turbine rotor speed. Despite being robust, they are slow, bulky, and rough. However, they are still used; they are considered proven, reliable, and efficient solutions. Electromechanical controls are more accurate and relatively lighter and faster than mechanical controls. They use governors, actuators, servomechanisms, electric motors, solenoids, and relays to control the primary energy.

Electronic controls use sensors and power converters to regularize energy, speed, and power. They can be quite accurate in adjusting voltage, speed, and frequency. They are lightweight and are easily suitable for remote control and telemetric applications, as discussed in the following section.

Power electronics technology has enabled the use of alternative energy sources for many types of electrical machines: permanent magnet-based, reluctance, and, particularly, the induction generator. From the ordinary closure of ac and dc switches to the very fast and precise electronic switching techniques used in pulse width modulation (PWM) and pulse density modulation (PDM) control, it is possible to adjust the machine to obtain any desired operating control. To obtain the best features for the system required, it is important for the hardware designer to be aware of what can be done in terms of control.

A very important machine, typically used for high-power applications, is the doubly fed induction generator (DFIG), known in the past as a Scherbius variable-speed driver. The DFIG is a wound rotor machine with the rotor circuit connected to an external variable voltage and frequency source via slip rings, and the stator connected to the grid network as illustrated in Figure 12.7. It is also possible to alter the rotor reactance by effectively modulating some inductors in series with the original rotor reactance. Adjusting the frequency of the external rotor source of current controls the speed of the DFIG, which is usually limited to a 2 : 1 ratio [5, 29, 30].

Doubly fed induction machines were not popular in the past due to the maintenance required for the slip rings. However, recently, with the development of new materials, powerful digital controllers, and power electronics, the DFIG became an excellent solution in power generation for ratings of up to several hundred of kilowatts, up to 1 or 2 MW. Power converters (whose cost for large units becomes irrelevant compared with the entire system) usually create a need for a variable-frequency source for the rotor.

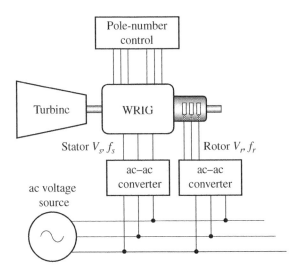

Figure 12.7 Experimental setup for overall stator and rotor parameter control.

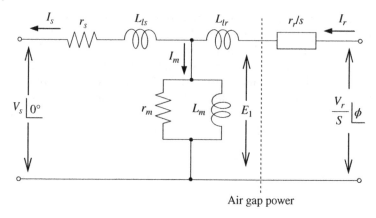

Figure 12.8 Equivalent model of a DFIG.

As said earlier, the control of induction generators can be exerted through either the stator or the rotor variables. The design of the stator will include variables such as the number of poles, the voltage, and the frequency, whereas the design of the rotor will include variables such as rotor resistance, leakage reactance, and speed of operation. The DFIG also has similar design parameters, and its controllable variables are current, voltage, frequency, and voltage phase shift with respect to the stator voltage angle. An experimental setup to implement and monitor changes in the stator and rotor characteristics is depicted in Figure 12.7. Obviously, in most applications, this setup can be simplified, for example, if there is no interest in the stator frequency or voltage changes, the stator power converter will not be required.

Figure 12.8 helps to understand the operation of a DFIG. This figure is considered a voltage source $|V_s| \underline{|0°}$ impressed at the stator side, while a voltage $|V_r/s| \underline{|\phi}$ is introduced across the rotor circuit to control either rotor voltage or frequency that would be exactly what a PWM control would do. This configuration introduces an apparent and effective change in the rotor reactance by manipulation of the rotor voltage or current source, and a family of performance curves can be derived from it.

From the equivalent model in Figure 12.8, the rotor current and torque may be obtained as

$$\bar{I}_r = \frac{V_s\underline{|0^o} - \dfrac{V_r}{s}\underline{|\varphi}}{\sqrt{\left(r_s + \dfrac{r_r}{s}\right)^2 + \omega_e^2\left(L_{ls} + L_{lr}\right)^2}\, \left|tan^{-1}\left(\dfrac{\omega_e\left(L_{ls} + L_{lr}\right)}{r_s + \dfrac{r_r}{s}}\right)\right.} \tag{12.35}$$

$$T_e = 3\left(\frac{p}{2}\right)\frac{r_r}{s\omega_e}\frac{\left(V_s^2 - 2\dfrac{V_sV_r}{s}\cos\varphi + \left(\dfrac{V_r}{s}\right)^2\right)}{\left(r_s + \dfrac{r_r}{s}\right)^2 + \omega_e^2\left(L_{ls} + L_{lr}\right)^2} \qquad (12.36)$$

As supported by the equivalent model, from an increase in stator voltage, increases in output torque and power are obtained. There is an almost direct proportion among these three variables and the increase in rotor speed, as shown in Figure 12.9a. Similar proportional results are obtained when there is an increase in the rotor external source voltage for the power injected through the rotor as shown in Figure 12.9b. Notice that in this torque versus speed characteristic, there is a shift to the left (lower rotor speeds) of the pullout torque caused by changes in the rotor parameters. The pullout torque is defined as the maximum torque to which generator is subject. This is a significant parameter because the maximum power transfer, from the induction generator to the load, happens whenever the source impedance is equal to the load impedance:

$$\frac{R_r}{s} = \sqrt{R_s^2 + \left(X_s + X_r\right)^2}$$

which when replaced in the torque equation (12.36) for $V_r = 0$ gives

$$T_e = \left(\frac{3p}{2\omega_e}\right)\frac{V_s^2\sqrt{R_s^2 + \left(X_s + X_r\right)^2}}{\left(r_s + \sqrt{R_s^2 + \left(X_s + X_r\right)^2}\right)^2 + \omega_e^2\left(L_{ls} + L_{lr}\right)^2} \qquad (12.37)$$

Further variations are obtained through higher rotor voltages. The shift to the left in the pullout torque becomes much more pronounced if the phase angle *f* between the stator voltage and rotor voltage is varied as shown in Figure 12.9c.

The slope of the torque curve close to the synchronous speed determines the variation range of the rotor speed during load operation. The steeper the torque is when crossing the synchronous speed point, the stiffer the system becomes and the narrower the speed regulation. This slope increases to the left as the rotor resistance increases in the torque characteristics, yet there is not much effect on the magnitude of the pullout torque, as demonstrated in Figure 12.9d. One problem here is that the heat so generated in the rotor needs to be dissipated to the outside, so limiting the slip factor in large machines. The stall control used in these large machines has been replaced by a fast pitch control to compensate for the rotor speed stiffness caused by the small slip factor.

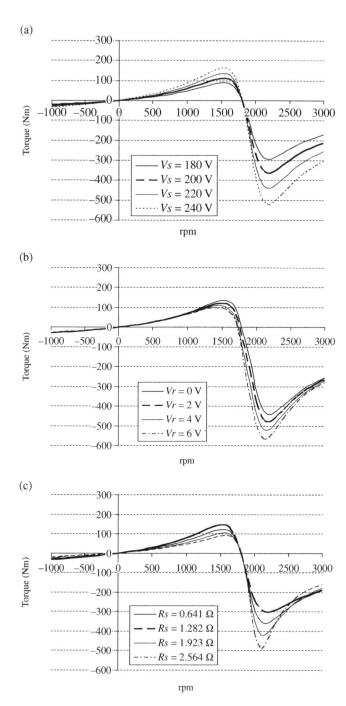

Figure 12.9 Control variables in the torque-speed characteristics of the induction generator: (a) stator voltage; (b) rotor voltage; (c) stator resistance; (d) rotor resistance; (e) stator inductance; (f) rotor inductance; (g) frequency of the stator voltage; (h) stator–rotor voltage phase shift; and (i) number of stator poles.

(d)

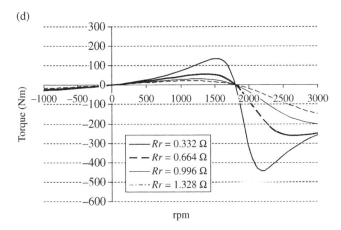

(e)

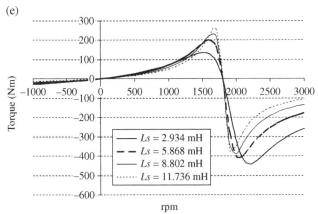

(f)

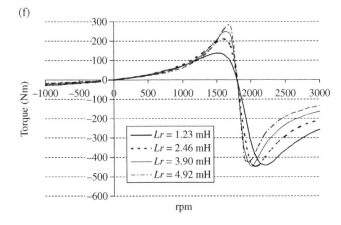

Figure 12.9 (Continued)

(g)

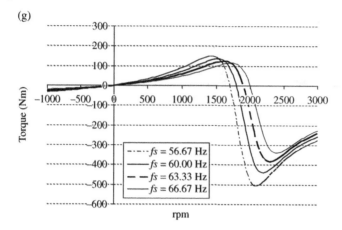

(h)

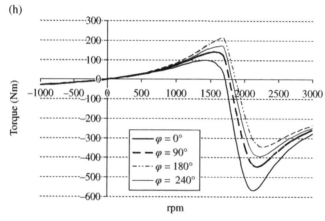

(i)

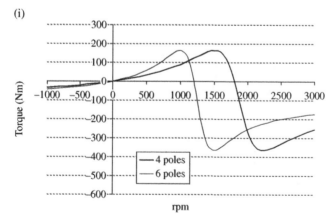

Figure 12.9 (Continued)

Rotor reactance can be changed through rotor design, rotor frequency control, or external insertion of modulated values of inductance. The quadratic effects of the rotor reactance on the torque versus speed characteristics are noticeable in Figure 12.9e. Different rotor angular frequencies can only be observed by the fact that the resistance does depend on the slip factor. It is interesting to point out here that these changes in X_r have very little effect on the rotor losses, which is not the case with the actual rotor resistance changes. It is also important to note that mismatching the rotor frequency too much may cause interaction between rotor frequency and stator frequency, which can lead to an undesirable resonant and/or antiresonant effect.

A very effective way of changing the synchronous speed of the induction generator is through stator frequency changes. These changes are proportionally related to the load frequency and the line crossing over the rotor speed, as illustrated in Figure 12.9f. One way to change the grid speed ranges of the induction generator is through machine pole changing. The phase circuitry of the stator is designed so that a different configuration of poles (pole number) can produce quite different synchronous speeds. None of the other variables is comparable in their effects on the pullout torque, as is a change in the pole number (see Figure 12.9g).

Notice that in all parts of Figure 12.9 except part (d), the changes proposed in the stator and rotor parameters have a pronounced influence not only on the crossover of the torque characteristic of the rotor speed axis but also on the magnitude of the corresponding pullout torques.

12.9.2 The Danish Concept: Two Generators on the Same Shaft

The Danish concept allows a directly grid-connected squirrel-cage induction generator solution where there is one large generator with another one smaller induction generator connected alternately in parallel (see Figure 12.10) [6, 11, 13]. Such solution became very successful in commercial wind turbines for its simple and inexpensive design. The large generator initially operates as a motor fed by the grid to put it at a speed slightly above the synchronous speed. If the wind speed is just above the cut-in speed of the turbine, the small generator is connected to the grid. As the wind speed increases, the small generator is switched off and the large generator is switched on. The wind speed difference between the two generators may be quite large to accommodate small and large rated electrical loads, from a range of two to three times in magnitude. As explained in Chapter 4, the best energy conversion is obtained for a wind turbine when their power coefficient C_p is kept at its maximum value.

For high wind intensities and small loads, a large electrical generator tends to overspeed, and vice versa, with small generators tending to under-speed. This is the main reason that small generators are convenient to use alternately in parallel with large generators, as in the Danish concept.

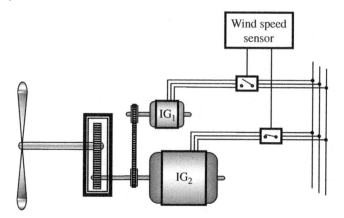

Figure 12.10 The Danish concept.

Although the Danish concept results in a bulkier configuration, it is advantageous for the relatively low cost of the control mechanism and better suitability. They fit well for pronounced natural variations in wind speed. The disadvantages refer to the heavy mechanical stress on the gearbox because of sudden interchanges of generators and the need for larger slip factor designs for the large generator that allow them to work smoothly. This concept is suitable only for relatively small wind power units; larger units may use other techniques, such as pole changing. Power electronics have become a very powerful tool to control torque, speed, and flux of the induction generator for high performance systems. In addition, simple solutions are possible, for example, changing the rotor resistance through an external control, as suggested in Figure 12.9d, as well as pole changing through a stator, or externally driven rotor circuit, as suggested in Figure 12.7.

12.9.3 Variable-Speed Grid Connection

For a variable-speed grid connection, the generator may have a slip factor control. This control may typically be exerted through changes in the rotor resistance R_2 to allow more room for variations in the rotor speed of large generators (see Figure 12.9d). Costs for the ac–ac power converters in these systems are relatively small, and they act to decouple, as much as possible, the rotor speed and the grid frequency (changes in the slip factor), keeping the wind turbine loads and power fluctuations within limits.

Very large turbines must use as much ordinary speed control (pitch control) as possible to avoid power dissipation in the rotor circuit of the turbine. The rotor resistance is increased only during transient peaks of load or wind intensity. In the usual operation, programming of the experimental circuit depicted in Figure 12.7 involves the stator and rotor converters, both supervised by a control system that monitors the wind speed. For low-power turbines, the decoupling between rotor speed and the ac system is usually made with cascaded back-to-back converters with an intermediate dc–dc converter in between them. This system somehow works like the Danish concept except by the fact that an electronic power circuit replaces the smaller generator. This type of power electronics is considered an over-synchronous static Kraemer system. For lower wind speeds, the generator operates as a squirrel-cage rotor, and the power electronics circuitry, which is equivalent to the small generator in Figure 12.7, is switched off. For the design of normal operating ranges of wind speed, the wind turbine operates at the maximum power coefficient C_p (optimum power), as detailed in Chapter 4. For very high wind speeds, there is speed limiting by pitch control until the unit is cut off.

12.9.4 Control by the Load versus Control by the Source

Electronic controls may exert their action by the load, so changes in the load may be used to control speed, frequency, and voltage. Figure 12.11 illustrates an equivalent circuit for such scheme. The load represented by R should be seen as regenerative, because the energy may be transferred from the generator to the public network or to a secondary load. Such a secondary load could be a battery charger, back pumping of water, irrigation, hydrogen production, heating, or fluid tank freezing. Degenerative dissipation can also be achieved by burning out the excess energy in normal power resistors, dissipating such loss to heat water, or to the wind or to the flow of running water or any fluid.

Notice that electronic control by load compromises efficiency with the power generated. For example, with wind and run-of-river turbines, the energy available in nature has no means of storage. In those cases, if the available energy is not used at the instant it is available in nature, it will not be converted, and that can be considered as a "wasted energy."

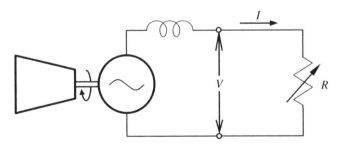

Figure 12.11 Principle of the electronic control by the load.

Figure 12.11 shows that the primary source may be a turbine, a waterwheel, an ICE, a gas turbine, a diesel set, or another primary source. The electrical load of the rotating generator may be a dc load if the output is rectified, a power converter, or, in per-phase terms, represented by R, RL, RLE, and RLI loads.

The literature discusses many approaches in varying the load through electronic control by load, from the simple step-by-step (discrete modulation) method to a continuous variation of load. For essentially additive loads such as resistors, it is possible to optimize their rated values to obtain a smoother variation of load with a minimum number of resistors by using the 2^n rule, depicted in Figure 12.12a for current control by discrete modulation of load. For additive/subtractive loads such as power transformers, optimization may be obtained even with a fewer elements by using the 3^n rule, depicted in Figure 12.12b for a series connection of secondary windings. Similar 2^n and 3^n rules can also be used in association with capacitors and inductors. When necessary, these approaches use electronic circuits with commutation under load.

Computers, electronic devices, and other sophisticated loads do not accept large fluctuations of voltage. For these loads, a continuously adjusted duty cycle is used for smooth control of turbine speed and load voltage for secondary loads. Figure 12.13 illustrates the smooth change of an effectively variable resistor used as a secondary load. It is common to use isolated gate bipolar

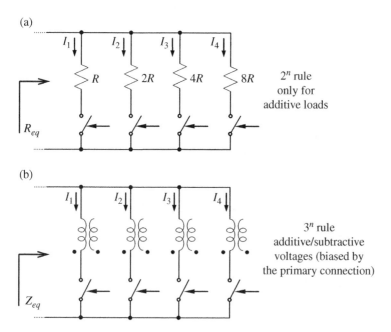

Figure 12.12 Discrete steps of load control: (a) current control by discrete modulation; (b) voltage control by discrete modulation.

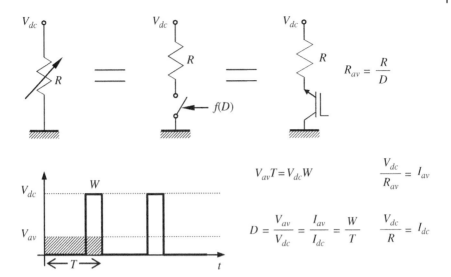

Figure 12.13 Electronically variable resistor.

transistors (IGBTs) for these purposes with switching frequencies of 20 kHz. The advantages of this type of control are the good speed and voltage regulation within certain ranges, use of a single power resistor with an IGBT and heat sink, ease in modular manufacturing, suitability for ac and dc loads through power electronic ac switches, and an absence of switching surges. Its disadvantages are the electromagnetic interference, power switching complications for power above 50 kW, use of filters, noise, the short life span without maintenance, moderated costs, and needs for qualified servicing.

An example of secondary loads using the additive $2''$ rule is the rural irrigation of plantations where water pumps of different sizes, such as 1, 2, 4, and 8 kW, could be used. Therefore, they could be programmed to be activated or deactivated whenever the turbine speed surpasses certain limits of control. Steps of 1-kW power will limit the speed within a reasonable range. An example of the $3''$ rule is the voltage control of large loads that are too sensitive to power changes.

12.10 Economics Considerations

To determine economics aspects, several parts of an alternative system must be accessed for a project evaluation. For example, wind energy will consist of a tower, a rotor (with blades, hub, and shaft), a generator, control equipment, and power conditioning and protection equipment. A gearbox is generally used to match the generator speed with the natural impressed velocity by the wind. Some sort of braking is used to protect the system against high winds and severe weather conditions. Combinations with other sources of generation, such as photovoltaic and sometimes diesel generators, form more reliable hybrid systems [26–28].

For a modern turbine design, the typical lifetime is assumed 20–25 years, during which capital costs will be amortized. Currently, a rule of thumb for costs is $1000/kW of capacity on average, with the marginal tower cost at $1600/m. Installation costs include foundations, transportation, road construction, utilities, telephones and other communications, substation, transformer, controls, and cabling. Depending on soil conditions and distance to power lines, a margin of safety must be used. Operation and maintenance costs generally increase throughout the life of a wind power project. Costs may range from about $0.008 to $0.014/kWh generated and increase at a rate of about 2.5% per year. The high end ($0.014/kWh) would include a major overhaul of equipment such as rotor blades and gearboxes, which are subject to a higher rate of wear and tear. Other operation costs include plant monitoring and periodic semester inspections.

The potential income from wind energy is based on annual energy production, which in turn is dependent on average annual speed. Wind speed varies globally, regionally, and locally following seasonal patterns. The duration and force of the wind are critical to the lucrative operation of wind turbines, to pay for investment on equipment, maintenance, and operations. Large-scale wind turbines require an annual average wind speed of 5.8 m/s (13 mph) at 10 m height; small-scale wind turbines require 4 m/s (9 mph) [28–31].

Energy production is also a function of turbine reliability and availability. Even though the induction generator is a small fraction of the investment required (an IG converts the mechanical energy from shaft to electricity), it makes important that the better efficiency of the induction generator will pay off for even fractions of improvement. Better electrical and magnetic design associated with insulation, thermal, and mechanical design will make the energy converter unit to extract power more reliably. Controllers capable of placing the induction generator at the best operating point, of programming the optimum magnetic flux level, or of producing required lagging or leading unit power-factor energy contribute to faster amortization and a more productive investment.

Hydropower converts the potential and kinetic energy of water conveyed through a pipeline or canal to the turbine into electrical energy. The energy in the water enters a high pressure, rotates the generator shaft, and leaves at a lower pressure. Power generation can be amplified with an increase of water elevation height, the flow of the river or stream, and the size of the watershed. Small hydropower generators are usually run-of-the-river systems that use a dam or weir for water diversion but not for reservoir storage. The amount of pressure at the turbine is determined by the actual flow of the stream or river and the head, or the height of the water above the turbine.

Controlling the flow of a river can have uses other than power generation. Those uses would include irrigation, flood control, municipal and industrial water supply, and recreation. These parameters affect the financial return from such projects and sometimes have long-term benefits that may be attractive locally. Hydropower is characterized by extremely high up-front costs, very low operating costs, and very long life cycles. Such capital cost scenarios make projects very sensitive to financial variables, construction timing, interests, and discount rates. Hydro projects usually have long lead times, typically taking 10 years between analysis and deployment. Such projects require resource assessment and environmental and social considerations (such as water use, displacement of homes, and distance to transmission lines).

Small power plants, of approximately 10–200 kW for rural uses of distributed energy, do not have the same requirements as those of large power plants (greater than 1 MW). However, small-scale applications for a target cost of $2000/kW will require a 35% investment in civil engineering, 40% for plant installations and commission, 7% for overall design and management, 8% for electrical engineering, and contingencies of approximately 10%. For either wind or hydropower energy sources, the induction generator will need a peak power tracking control, a dummy load or any type of storage system, and the pump-back schemes sometimes used in conjunction with water flow needs [16, 31].

References

1 S.J. Chapman, Electric Machinery Fundamentals, 3rd ed., McGraw-Hill, New York, 2011.
2 A.S. Langsdorf, Theory of Alternating Current Machines, McGraw-Hill, New York, 1977.
3 R.L. Lawrence, Principles of Alternating Current Machinery, McGraw-Hill, New York, 1953, p. 640.
4 M. Kostenko and L. Piotrovsky, Electrical Machines, Vol. II, Mir Publishers, Moscow, 1969.
5 M.G. Simões and F.A. Farret, Modeling and Analysis with Induction Generators, CRC Press, Boca Raton, FL, 2015.
6 S.S. Murthy, O.P. Malik, and A.K. Tandon, Analysis of self-excited induction generators, Proceedings of IEE, Vol. 129, No. 6, pp. 260–265, 1982.
7 IEEE Std. 1547, Standard for Interconnecting Distributed Resources with Electrical Power Systems, IEEE Press, Piscataway, NJ, 2014.
8 IEEE Std. (Draft) P1547.1, Standard Conformance Test Procedures for Interconnecting Distributed Energy Resources with Electric Power Systems, IEEE Press, Piscataway, NJ, 2013.
9 IEEE Std. (Draft) P1547.2, Application Guide for IEEE Standard 1547: Interconnecting Distributed Resources with Electric Power Systems, IEEE Press, Piscataway, NJ, 2008.

10 IEEE Std. (Draft) P1547.3, Guide for Monitoring, Information Exchange, and Control of Distributed Resources Interconnected with Electric Power Systems, IEEE Press, Piscataway, NJ, 2008.

11 IEEE Std. 112–1991, Standard Test Procedure for Polyphase Induction Motors and Generators, IEEE Press, Piscataway, NJ.

12 I.R. Smith and S. Sriharan, Transients in induction machines with terminal capacitors, Proceedings of IEE, Vol. 115, pp. 519–527, 1968.

13 S.S. Murthy, B. Sing, and A.K. Tandon, Dynamic models for the transient analysis of induction machines with asymmetrical winding connections, Electric Machines and Electromechanics, Vol. 6, No. 6, pp. 479–492, 1981.

14 D.B. Watson and I.P. Milner, Autonomous and parallel operation of self-excited induction generators, Electrical Engineering Education, Vol. 22, pp. 365–374, 1985.

15 C. Grantham, Determination of induction motor parameter variations from a frequency stand still test, Electrical Machine Power Systems, Vol. 10, pp. 239–248, 1985.

16 S.S. Murthy, H.S. Nagaraj, and A. Kuriyan, Design-based computational procedure for performance prediction and analysis of self-excited induction generators using motor design packages, Proceedings of IEE, Vol. 135, No. 1, pp. 8–16, 1988.

17 K.E. Hallenius, P. Vas, and J.E. Brown, The analysis of a saturated self-excited induction generator, IEEE Transactions on Energy Conversion, Vol. 6, No. 2, 1991.

18 N.N. Hancock, Matrix Analysis of Electric Machinery, Pergamon, Press, Oxford, 1964, p. 55.

19 M. Liwschitz-Gärik and C.C. Whipple, Alternating Current Machines, Compañia Editorial Continental, Col Barrio del Niño Jesús, Mexico, 1970.

20 B.C. Doxey, Theory and application of the capacitor-excited induction generator, Engineer, pp. 893–897, November 1963.

21 C.H. Lee and L. Wang, A novel analysis of parallel operated self-excited induction generators, IEEE Transactions on Energy Conversion, Vol. 13, No. 2, June 1998.

22 C. Grantham, Steady-state and transient analysis of self-excited induction generators, IEE Proceedings, Vol. 136, Pt. B, No. 2, pp. 61–68, 1989.

23 D. Seyoum, C. Grantham, and F. Rahman, The dynamics of an isolated self-excited induction generator driven by a wind turbine, Proceedings of the 27th Annual Conference of the IEEE Industrial Electronics Society, IECON'01, Denver, CO, December 2001, pp. 1364–1369.

24 S.P. Singh, B.M. Singh, and P. Jain, Performance characteristics and optimum utilization of a cage machine as capacitor excited induction generator, IEEE Transactions on Energy Conversion, Vol. 5, No. 4, pp. 679–684, 1990.

25 E. Muljadi, J. Sallan, M. Sanz, and C.P. Butterfield, Investigation of self-excited induction generators for wind turbine applications, IEEE, Industry Applications Conference, 1999. Thirty-Fourth IAS Annual Meeting. Conference Record of the 1999, pp. 509–515.

26 R.P. Tabosa, G.A. Soares, and R. Shindo, The High Efficiency Motor (Motor de Alto Rendimento), Technical Handbook of the PROCEL Program, Eletrobrás/Procel/CEPEL, Rio de Janeiro, Brazil, August 1998.

27 I. Barbi, Fundamental Theory of the Induction Motor (Teoria Fundamental do Motor de Indução), Federal University of Santa Catarina, Press-Eletrobrás, Florianópolis, Brazil, 1985.

28 G.W. Stagg and A.H. El-Abiad, Computer Methods in Power Systems Analysis, McGraw-Hill, New York, 1968.

29 O.I. Elgerd, Electric Energy Systems Theory: An Introduction, McGraw-Hill, TMH Edition, New York, New Delhi, 1971.

30 R. Gasch and J. Twele, Wind Power Plants: Fundamentals, Design, Construction and Operation, Solarpraxis, Berlin, 2002, pp. 299–313.

31 C. Ortega and A.A. Xavier del Toro, Novel direct torque control for induction motors using short voltage vectors of matrix converters. IEEE, Transactions on Industry Applications, Vol. 1, pp. 1353–1358, 2005.

13

Permanent Magnet Generators

13.1 Introduction

A permanent magnet synchronous generator (PMSG) is an alternating current (AC) machine not only used for high-performance and high-efficiency motor drives but also used as a generator. In order to achieve control in either motoring or generating mode, vector control techniques, also called field-oriented control (FOC), can be used for PMSM and PMSG, where the control algorithm decomposes the machine three-phase stator current into a magnetic field-generating part and a torque-generating part, that is, an equivalent DC machine model, which runs in real time. Both components can be controlled separately after decomposition, and the structure of the motor or generator controller (vector control controller) makes the system to have an equivalent DC motor–generator behavior and control structure, which makes easier the implementation of advanced control systems for torque and speed in four quadrants of operation.

The PMSG may have neither a brush nor any commutators, and the flux is supplied by permanent magnets (PM). Therefore, the maintenance work is minimized. As the field winding is replaced by a PM, the copper losses of the field winding are absent. The stator is made of laminated steel to reduce the eddy current losses.

Some features of the PMSG, such as reduced needs for maintenance and robustness for variable speed, made synchronous generators with PM excitation an attractive alternative for hydro and wind power generation. The cost of PM was low for some time, but that will not be sustained in long term, because rare-earth mining is possible just in a few regions of the planet. In comparison with the conventional electrical excitation, the PM excitation favors a reduced active weight and decreased copper losses, and the output energy must be somewhat higher. A number of studies have been conducted investigating different topologies of PMSG suited for direct-driven low-speed hydro and

Integration of Renewable Sources of Energy, Second Edition. Felix A. Farret and M. Godoy Simões.
© 2018 John Wiley & Sons, Inc. Published 2018 by John Wiley & Sons, Inc.

wind generators [1–3]. In general, they are suitable for many applications, such as small power plants, remote area off-grid power supply systems, and isolated loads.

The PM generator has variable-frequency AC impressed by power electronic circuits. The characteristic is very similar to a brushless DC (BLDC) generator, which has trapezoidal current waveforms, while PMSG has sinusoidal ones.

There are features in the PMSG construction optimized for application in wind turbines. They can be constructed with mechanical drive shaft for water or wind application. The magnetic field may create electrical currents through bars of metallic components, and high-end machines will have polymer or Teflon for limiting leakage current and weather oxidation. Mechanical design is optimized toward fewer components and need of less maintenance along their lifetime.

The PMSG can be implemented for run-of-river hydro and small wind turbines. A compact design and more energy-efficient machine allow fluid moving through the turbine blades. Hydro turbines have smaller and larger numbers of blades than wind turbines. Permanent neodymium–iron–boron (NdFeB) magnets integrated into smaller hydraulic turbine blades make them turn more efficiently and are less prone for damages when compared with other ordinary types of generators. The PMSG can be used for the generation of electrical energy from rivers, using variable-speed hydro turbines, or from sea waves by using tides and waves, as well as from other forms of water motion energy [4–6].

13.1.1 PMSG Radial Flux Machines

Radial flux machines are the conventional type of machines, where the flux swings along the radius of the yoke. The manufacturing technology is well established, which makes the production cost lower compared with the axial-type ones because they are extensively used in ship propulsion, robotics, traction and automation, and factory systems. Furthermore, they are very flexible for scaling, as the higher power ratings of the machine are achieved by increasing its length. In other words, completely new design and completely new geometry can be avoided. They have been extensively used in ship propulsion, robotics, traction, and wind systems. Figure 3.3 shows cross-sectional view in the radial and axial directions, respectively, for a typical radial flux PMSG.

13.1.2 Axial Flux Machines

Various axial flux topologies have been proposed in recent years. Those pros and cons are categorized. Generally, axial flux machines have a smaller length when compared with radial flux machines. Their main advantage is high torque density, so they are mostly recommended for traction applications with size constraints especially in axial direction. They have found application in gearless elevator systems, and they are rarely used in traction, servo applications, micro-generation, and propulsion systems. Figure 3.4 shows cross-sectional view in the radial and axial direction, respectively, for a typical axial flux PMSG.

One of the disadvantages of the axial flux machines is that they are not balanced in their single-rotor–single-stator edition version. Usually, for a better performance the rotor is sandwiched between two stators or vice versa. Unlike radial flux machines, the stator windings are located in the radial direction. A circumferentially laminated stator is required for reduction of iron losses, which complicates the manufacturing process. Scaling of axial flux machine is another drawback. Unlike radial flux machines, any increase in length is accompanied by increase in air gap diameter. Hence, to increase the power rating, a new design and a new geometry is needed. One other way to increase the power rating is by increasing the number of stators and rotors, making the machine more expensive and less standard for high acceptance. This, however, makes the machine costly. Therefore, it seems that small energy systems based on PM generators will be dominated by the traditional radial flux design [6–8].

There is a lot of confusion in terminology and in understanding how to use and operate machines based on PM. These are PMDC, PMAC, brushless AC, PMSM (permanent magnet synchronous motor), and BLDC. The PMDC motors typically employ PM mounted to the inside of the motor frame, rotating wound armature and commutator (brushes). PMAC, PMSM, and brushless AC are synonymous terms; they are PM machines that operate with a PWM AC drive or control with advanced control and real-time software. BLDC motors are very similar to PMAC machines in terms of construction, but they use DC (trapezoidal) drives rather than AC ("sine") drives, which are used to control PMAC motors [6–8].

13.1.3 Operating Principle of the PMSG

PM generators have a number of benefits compared with other types of generators since they do not require an external field circuit. Because no field windings are required, they do not have the field circuit copper losses, so they can be smaller, lower losses, and lighter. However, PM cannot produce very high flux density, as the ones impressed with field current. Therefore, a PMSG will have a lower induced torque per ampere of armature current only for high flux density (stronger magnets) or wound field structures. When subjected to high temperature during prolonged periods of overload, or mechanical shock or an oscillating electrical field, PMSGs can suffer from demagnetization. As a rule, the armature current in any machine produces an armature magnetic field of

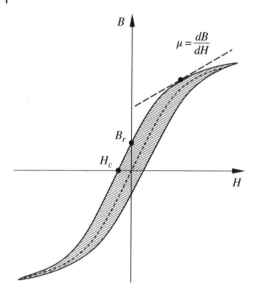

Figure 13.1 Typical magnetization curve of ferromagnetic materials.

its own subtracts magnetomotive force (mmf) from some portions of the pole faces and adds them to the mmf of the poles under other portions of the pole faces as an armature reaction effect, and proper design must avoid material saturation because of such composed effect [9–12].

A number of new magnetic materials have been developed mostly in China and some other Asian countries. They have desirable characteristics for making PM (Greenland also has rare earth, so Denmark can start to produce PM if they decide to do it). Figure 13.1 shows a typical plot of flux density B versus magnetizing intensity H for a typical ferromagnetic material where a residual flux B_r is a result of a strong external mmf applied to this material and then suddenly removed. A good material for the poles of a PMSG should have a residual flux density as large as possible while keeping a large coercive magnetizing intensity given by H_c and related to B by the derivative $\mu = dB/dH$ [8, 13, 14].

The major types of magnetic materials are the ceramic (ferrite) and the rare-earth magnetic. In Table 13.1 it is noticeable that rare-earth magnet combines both a high residual flux and a high coercive magnetizing intensity. Neodymium magnets are the strongest magnet material available, with strengths ranging from 30 megagauss-oersted (MGOe) to 52 MGOe for power production.

From the parameters given in Table 13.1, it is clear that the magnetization curves of some typical ceramic and rare-earth magnets make them comparable with the magnetization curve of a conventional ferromagnetic alloy (e.g., Alnico 5). Therefore, from this comparison it can be inferred that the best

Table 13.1 Characteristics of permanent magnets.

Type of PM	B_r (tesla)	H_c (A/m)	Performance	Cost	μ_r	ρ kg/m^3	Origin
Alnico 5 (Al, Ni, Co)	High (0.15)	Low	Robust	Relatively inexpensive	2.00	7300	China, India, Taiwan
Ferrite (Oe MagnetY30 BH)	Low (0.37)	Medium	Fragile	Moderately inexpensive	1.06	4700	India, China, United States, Japan, Taiwan
NdFeB (Vacodym 633 PT)	High (1.32)	High	Excellent performance	Moderately expensive	1.06	7700	China, Slovakia, Germany, United States
Rare-earth samarium–cobalt	High (1.30)	High	Small temperature coefficient	Very expensive	1.04	8300	India, China, Jamnagar

rare-earth magnets can produce the same residual flux as the best conventional ferromagnetic alloys while simultaneously being largely immune to demagnetization problems due to the armature reaction.

13.2 Permanent Magnets Used for PMSGs

PM are essential ingredients to the PMSG. In the last two decades, PM technology has been substantially improved. When a PM has been magnetized by an external magnetic field, it will remain magnetized even if the magnetic field intensity is gone. The magnetic flux density at this point is called the remanence, denoted B_r. The remanence is the maximum flux density that the magnet can produce by itself. On the other hand, if magnetic field intensity has increased in the opposite polarity, the flux density will eventually become zero. The magnetic field intensity at this point is called the coercivity, denoted H_c.

The absolute value of the product of the magnetic flux density, B_r, and the magnetic field intensity, H_c, at each point along the magnetization curve in the second quadrant region is called the energy product or the density of magnetic energy as it is illustrated in Figure 13.1. The higher the density of magnetic energy is, the smaller the machine dimensions and core losses are.

The operating point can be determined by the intersection between the permeance line and the hysteresis curve in the second quadrant region. The permeance line is determined by the characteristics of the machine structure: air gap length, magnetic path length, and number of coil turns. The large B_r produces a large flux in the machine, while the large H_c means that a very large current would be required for demagnetizing the poles [1, 8, 15].

Nowadays, many different types of PM materials are available. The types available include alnico, ferrite (ceramic), rare-earth samarium–cobalt, and NdFeB. The NdFeB magnets, which have been developed recently, have the highest energy product. Ferrite magnets are the most commonly used due to their cost effectiveness and widespread usage. Each magnet type has different properties, leading to different constraints and different levels of performance. The characteristics of the PM materials are compared in Table 13.1. NdFeB magnets have the highest remanence and coercivity, but the initial price is very high. Ferrite magnets are widely used because the material and production costs are both low, even though they have low remanence, coercivity, and power product [8–15].

13.3 Modeling a Permanent Magnet Synchronous Machine

There are several types of PM generators: the conventional PM synchronous machine, the conventional PM synchronous machine with flux concentration, the slotted axial flux PM machine, TORUS, the surface-mounted transverse-flux PM machine, and the flux-concentrating transverse-flux PM machine. In order to select the best type of PMSG, one has to consider four main characteristics among the various topological variants of the PM machines: (i) air gap orientation with respect to the rotational axis: radial or axial; (ii) stator core orientation with respect to the direction of the movement: longitudinal or transverse; (iii) PM orientation with respect to the air gap: surface mounted or flux concentrating; and (iv) copper housing: slotted or slot less [13, 14]. By all these alternatives, it is important to select the PMSG that better fits small wind turbines and small hydropower plants.

A PMSG is the same machine as the permanent magnet synchronous machine. In order to understand the basics, suppose that a wire is moving through a magnetic field. A voltage e will be induced on it. If a voltage is applied to the wire, a current i will circulate through it, and a force f will act on this wire. The magnetic circuit to allow this magnetic flux is such as the length of the wire that is perpendicular to the magnetic flux, and both are in a plane such as the velocity v of this wire that is normal to this plane, in effect, making each one of them perpendicular to each other.

The instantaneous induced electromotive force (emf) is expressed as equation (13.1):

$$e = B\ell v \qquad (13.1)$$

where:

$e =$ The voltage induced in the wire (V)
$B =$ The magnetic flux density (tesla)
$\ell =$ The conductor length in the magnetic field (m)
$v =$ The velocity of the wire passing through the magnetic field (m/s)

Equation (13.1) is Faraday's law, also known as the flux-cutting rule, that is, an emf is induced in a conductor if it "cuts" magnetic flux lines of density. The emf induced in the conductor will circulate a current through it if a closed circuit is allowed by an external connection. The direction of the current in the conductor will be such as to oppose the cause of it as stated by Lenz's law.

A current carrying conductor located in a magnetic field will experience a force given by equation (13.2). That is, whenever a change in flux linkages occurs, an emf is induced that tends to set up a current in such a direction as to produce a magnetic flux that opposes the cause of it. Therefore, if a current carrying conductor is placed in a magnetic field, it will produce another magnetic field. The interaction of them sets up a force when moving the conductor (generator mode) in the magnetic field, as illustrated in Figure 13.2 (Fleming's right-hand rule). For the case of a generator, the conductor must be moved against this counterforce or the opposing force, and a current is supplied to the conductor against the emf generated (known as the counter emf, counter

Figure 13.2 Right-hand rule applied to a wire moving in magnetic field.

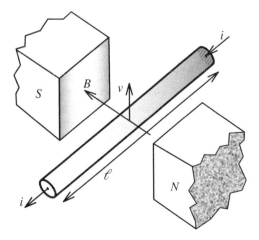

(a)

(b)

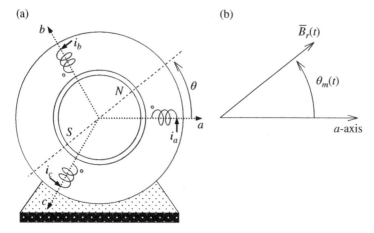

Figure 13.3 (a) Hypothetical three-phase concentrated winding permanent magnet-based synchronous motor/generator and (b) direction of the induced field.

electromotive voltage (cemf), or back emf). As a result, the same machine can be operated either as a motor or as a generator, depending on whether mechanical power or electrical power is supplied:

$$f = B\ell i \tag{13.2}$$

where:

f = The force in the wire (N)
B = The magnetic flux density (tesla)
ℓ = The conductor length in the magnetic field (m)
i = The current flowing in the wire due to an external circuit connection

In a hypothetical cylindrical structure such as illustrated in Figure 13.3, the rotor has a PM with two poles, north and south. The stator has concentrated windings. Then, the machine can be explained by a sequential turning on and turning off of those windings; in respect to the shaft position of the PM, more slots could be used for concentrated windings. This type of machine is common in electric home appliances and some industrial actuators. On the other hand, the use of distributed windings is common in high power applications including electric vehicle (EV), hybrid electric vehicle (HEV), traction drive, and wind turbine generators. In distributed winding, the air gap flux produces a sinusoidal distribution, and three-phase sinusoidal voltages and currents will control the three-phase windings.

Figure 13.4 shows the basic electronic principles in driving a PMSG. Although a synchronous large-power generator can be connected directly to the grid without any power electronic system, the small and medium PM synchronous

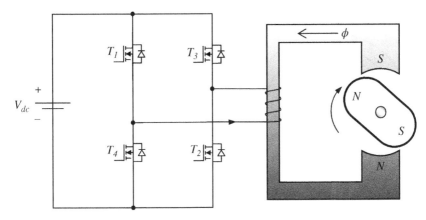

Switch states	Position	Switch states	Position	Motor torque	Generator torque
1–2	On	3–4	Off	Positive spin	Negative spin
1–2	Off	3–4	Off	Inertia	Primary drive
1–2	Off	3–4	On	Positive spin	Negative spin
1–2	Off	3–4	Off	Inertia	Primary drive

Figure 13.4 Operating principle of the PMSG.

motor or synchronous generator will always have a power electronics drive associate to its terminals [1, 16]. Figure 13.4 shows a simplified machine, just for understanding the power electronics principles. The main idea is to allow the electric current to be connected to the exterior of the power generator. The basic difference between a conventional DC generator and a PMSG like this one would be the commutator function placed on the outside of the magnetic field moving cyclically on the spindle and thus eliminating so the brushes. In this single-phase one-pole PMSG machine, the PM are assembled on the rotor. Depending on the relative position of the rotor in respect to the pole, the power electronic switches will select the winding to have current one way or another.

The PMSG model with a single winding as portrayed in Figure 13.4 is not usual because its torque is very unsteady, so a higher number of coils would be advisable. Once the rotor begins its movement by one of the poles, the next one will be just near enough to receive the next magnetic push and so on. With a higher number of poles, the rotor movement becomes smoothed out. More typically, the number of poles could be three or a multiple of three as shown in Figure 13. This generator is considered similar to a salient pole synchronous machine, because there is an interpole region where the leakage inductance will be minimum compared with the pole region. The stator windings can be constructed in a sinusoidal distribution and ideally shifted from each other by

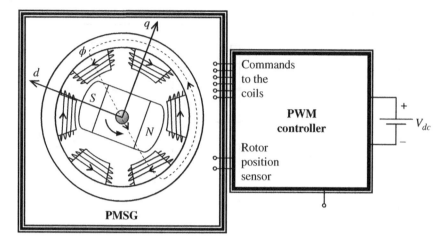

Figure 13.5 Internal structure of a multi-salient pole PMSG.

120° as the phases represented in Figure 13.5 by resistors and inductances. In the rotor magnetically coupled to the stator, there is an equivalent winding to the magnet that is represented also by a resistance and an inductance. The current going through the stator coils is assumed as the positive sense to describe the generator excitation.

13.3.1 Simplified Model of a PMSG

A sinusoidal pulse width modulation (SPWM) can be implemented with a three-phase bridge with six transistors and six diodes, as indicated in Figure 13.6. Such converter can work as a bidirectional inverter/rectifier; the DC-link is a voltage, so this is a type of a voltage source inverter (VSI). It can be used for current control of the three-phase machine, in either leading or lagging power factor, as long as the switches are commanded by an instantaneous d–q decoupled current control with outer loop for active and reactive power commands. In the given circuit, a sinusoidal current reference for the d–q controller makes the distribution of phase currents such to keep them in phase with the generator voltage (for unity power-factor operation). In the generation mode, voltage and current are out of phase, and depending on the power-factor requirement, it could be somewhat leading in phase. For variable input power from a wind turbine (provided by a rectifier), or from a PV or a fuel cell, the maximum power and efficiency will have to be programmed at the power electronics control. For several machines, maybe the PMSG is non-sinusoidal (it could be a trapezoidal emf waveform) [1–4, 17].

A mathematical representation of a PM generator can be made by using phases a, b, and c of the rotor represented by the direct d and quadrature q axis

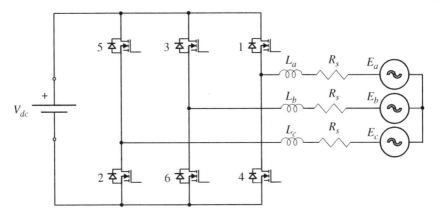

Figure 13.6 Three-phase converter connected to an equivalent PMSG model.

with the rotor magnetic axis coinciding with the quadrature axis q, as illustrated by Figure 13.7. The equivalent circuit of the PMSG generator is shown in Figure 13.7. The stator resistances are defined as R_s; the self-inductances for phases a, b, and c are defined respectively as L_a, L_b, and L_c; and the mutual inductances are defined as M_{ij} where i,j refer to the phase combination as presented in the matrix equation.

The analysis of a PMSG is based on the following assumptions: (i) the generator is operated within the rated condition, and the magnetic core of the machine is not saturated; (ii) all three phases have an identical and balanced induced emf shape; (iii) iron losses are negligible; and (iv) power semiconductor devices in the converter are ideal. Using the notation of Figure 13.7, we can write the following matrix equation:

$$\begin{bmatrix} v_a \\ v_b \\ v_c \end{bmatrix} = \begin{bmatrix} R_s & 0 & 0 \\ 0 & R_s & 0 \\ 0 & 0 & R_s \end{bmatrix} \begin{bmatrix} i_a \\ i_b \\ i_c \end{bmatrix} + \frac{d}{dt} \begin{bmatrix} \psi_a \\ \psi_b \\ \psi_c \end{bmatrix} \tag{13.3}$$

where:

$$\begin{bmatrix} \psi_a \\ \psi_b \\ \psi_c \end{bmatrix} = \begin{bmatrix} L_a & M_{ab} & M_{ac} \\ M_{ab} & L_b & M_{bc} \\ M_{ca} & M_{bc} & L_c \end{bmatrix} \begin{bmatrix} i_a \\ i_b \\ i_c \end{bmatrix} + \frac{d}{dt} \begin{bmatrix} \psi_{pma} \\ \psi_{pmb} \\ \psi_{pmc} \end{bmatrix}$$

ψ_{pma}, ψ_{pmb}, and ψ_{pmc} = The coupled magnetic fluxes through phase poles a, b, and c, respectively

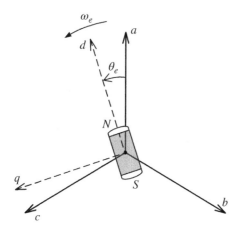

Figure 13.7 Coordinate transformation from *abc* to *dq* frame.

The rotor position is defined by an electrical angle θ_e with respect to the fixed frame a, b, and c as follows [7, 8]:

$$L_a = L_0 + L_m \cos(2\theta_e)$$

$$L_b = L_0 + L_m \cos\left(2\theta_e - \frac{2\pi}{3}\right)$$

$$L_c = L_0 + L_m \cos\left(2\theta_e + \frac{2\pi}{3}\right)$$

$$M_{ab} = -\frac{1}{2}L_0 + L_m \cos\left(2\theta_e + \frac{2\pi}{3}\right)$$

$$M_{bc} = -\frac{1}{2}L_0 + L_m \cos(2\theta_e) \tag{13.4}$$

$$M_{ca} = -\frac{1}{2}L_0 + L_m \cos\left(2\theta_e - \frac{2\pi}{3}\right)$$

In equation (13.4), the physical reactive parameters of the generator are represented by L_0 and L_m. The varying magnetic sinusoidal coupling between stator and rotor is expressed as a function of the maximum magnetic phase flux ψ_{pm} for phases a, b, and c as follows:

$$\psi_{pma} = \psi_{pm} \cos(\theta_e)$$

$$\psi_{pmb} = \psi_{pm} \cos\left(\theta_e - \frac{2\pi}{3}\right) \tag{13.5}$$

$$\psi_{pmc} = \psi_{pm} \cos\left(\theta_e + \frac{2\pi}{3}\right)$$

It is usual to transform the three-phase matrixes represented in equations (13.4) and (13.5) in a two-phase system with invariant inductances with respect to θ_e where the winding magnetic axis are denoted by a, b, and c; the direct and quadrature axes are denoted d and q, as depicted in Figure 13.7.

The *abc–dq* transformation matrix for a balanced system is

$$
T_{dq0} = \frac{2}{3}\begin{bmatrix} \cos(\theta_e) & \cos\left(\theta_e - \dfrac{2}{3}\right) & \cos\left(\theta_e + \dfrac{2}{3}\right) \\ -\sin(\theta_e) & -\sin\left(\theta_e - \dfrac{2}{3}\right) & -\sin\left(\theta_e + \dfrac{2}{3}\right) \\ \dfrac{1}{2} & \dfrac{1}{2} & \dfrac{1}{2} \end{bmatrix}
\tag{13.6}
$$

The voltage equation for the synchronous coordinates is

$$
\begin{bmatrix} v_d \\ v_q \end{bmatrix} = \begin{bmatrix} R_s & 0 \\ 0 & R_s \end{bmatrix}\begin{bmatrix} i_d \\ i_q \end{bmatrix} + \omega_e \begin{bmatrix} 0 & 1 \\ -1 & 0 \end{bmatrix}\begin{bmatrix} \psi_d \\ \psi_q \end{bmatrix} + \frac{d}{dt}\begin{bmatrix} \psi_d \\ \psi_q \end{bmatrix}
\tag{13.7}
$$

where:

$$
\begin{bmatrix} \psi_d \\ \psi_q \end{bmatrix} = \begin{bmatrix} L_d & 0 \\ 0 & L_q \end{bmatrix}\begin{bmatrix} i_d \\ i_q \end{bmatrix} + \begin{bmatrix} \psi_{pm} \\ 0 \end{bmatrix}
$$

$$
L_d = \frac{3}{2}\left(L_0 - L_m\right)
$$

$$
L_q = \frac{3}{2}\left(L_0 + L_m\right)
$$

The state equation is given by equation (13.8), and direct and quadrature linked fluxes can be calculated by equations (13.9) and (13.10):

$$
\frac{d}{dt}\begin{bmatrix} i_d \\ i_q \end{bmatrix} = \begin{bmatrix} -\dfrac{R_s}{L_d} & \dfrac{\omega_e L_q}{L_d} \\ -\dfrac{\omega_e L_d}{L_q} & -\dfrac{R_s}{L_q} \end{bmatrix}\begin{bmatrix} i_d \\ i_q \end{bmatrix} + \begin{bmatrix} \dfrac{1}{L_d} & 0 \\ 0 & \dfrac{1}{L_q} \end{bmatrix}\begin{bmatrix} v_d \\ v_q \end{bmatrix} + \omega_e\begin{bmatrix} 0 \\ -\dfrac{\psi_{pm}}{L_q} \end{bmatrix}
\tag{13.8}
$$

$$
\lambda_d = -L_d i_d - L_m i_d
\tag{13.9}
$$

$$
\lambda_q = -L_q i_q + L_m\left(-i_q + i_m'\right)
\tag{13.10}
$$

(a)   (b)

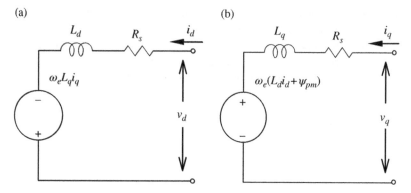

Figure 13.8 Equivalent circuits of the *d–q* axis: (a) direct axis and (b) quadrature axis.

From equations (13.7) and (13.8), the equivalent circuit in synchronous coordinates like the circuits shown in Figure 13.8 can be built.

The electrical angular velocity is related to the torque by the differential equation, where T_e and T_m are the machine electrical impressed generator and the mechanical prime moving torque, impressed by a turbine, where the electrical torque impressed by the generator is calculated by the cross product of the stator flux and the direct-quadrature current components, in accordance to equation (13.12):

$$T_e - T_m = \left(\frac{2}{P}\right)\frac{d\omega_r}{dt} \tag{13.11}$$

or

$$T_e = \left(\frac{3}{2}\right)\left(\frac{P}{2}\right)\left[\lambda_d i_q - \lambda_q i_d\right] \tag{13.12}$$

Using digital signal processing or some advanced RISC based microcontroller, it is possible to make a multiphase converter with position sensors for synchronizing the switching current with the rotor position. These sensors are either optical or Hall effect devices, using the rotor magnetism to sense the instantaneous position. It is expected that torque is reduced as the speed increases since the rotor generates it proportionally to a back emf across the closest winding to the pole. There will be a current reduction and a reduction of the mmf and, consequently, reduced torque. The maximum speed will be reached when the cemf is equal to the equivalent DC source voltage across the terminals of the machine. This feature makes the PMSG generator to be used for dynamic or regenerative breaking.

13.4   Core Types of a PMSG

The PMSG can be categorized in accordance to the way the PM are mounted on the rotor and the shape of the induced emf. The PM can either be surface mounted, interior mounted, or buried in the rotor, and the induced emf shape in the stator can either be sinusoidal or trapezoidal [9].

Most electrical machines can be manufactured in two different ways: with inner rotor (external stator) or outer rotor (internal stator). For a given torque, the air gap area remains the same in either case. However, from the design point of view, an outer rotor is preferable, as the machine's outer diameter becomes smaller. This results in a better utilization of the available volume. The windings are also easier to place in the slots of the internal stator. On the other hand, it requires a considerably more complex supporting structure like bearings and mounting in order to keep the air gap constant. The internal stator design might also have a poorer heat transfer capability as the main heat source—copper losses in the conductors—is located further away from the cooling surface in case of an air-cooled machine.

The heat transfer of the outer rotor can be improved by introducing an internal water cooling system. Depending on the type of excitation, a transverse-flux machine can be either electrically or magnetically excited. An electrically excited machine normally has a more massive rotor and larger weight as compared with the magnetically excited one. The recently reduced magnet prices are making PM machines more feasible than ever.

Depending on the way the PM are placed in the rotor, the PM machine may have several design possibilities. As stated earlier, the most common designs used in the conventional radial flux machine are the surface-mounted, interior-mounted, and buried designs. Selection of the proper design would affect the machine's performance, as well as its weight and the overall production cost. The output voltage and current of a three-rotor design with the direct d and quadrature q axes are represented in Figure 13.9.

In the rotor type of a surface-mounted PM, each PM is mounted on the round surface of the rotor. There is always a possibility for the attached PM to fly away during a high-speed operation or fast transient. Typically, for this type of generator, the inductance variation by rotor position is negligibly small. In this configuration, the magnets are polarized radially or sometimes circumferentially. The bandaging of such a machine is necessary in order to protect those magnets from the centrifugal forces. The reactances in the d- and q-axis are nearly same. The construction of such rotor is relatively simple, when compared with the other rotor designs.

The interior-mounted rotor has radially polarized magnets embedded in slots on the rotor surface. The q-axis synchronous reactance is larger than in the d-axis. The rotor is likely to be lighter in this case since the emf induced by the magnets is generally lower than in surface-mounted design due to the larger flux leakage. The use of the interior-mounted PM is not as common as the surface-mounted type, but the interior-mounted PM is a good candidate for high-speed operation, for example, in micro turbines powered by gas, or compressed air energy storage, because of its robustness and smaller diameter. It is noted that there is an inductance variation for this type of machine because the effective air gap varies by rotor position.

In the buried-magnet rotor, the magnets are circumferentially magnetized. The synchronous reactance in the q-axis is larger than in the d-axis. The thickness of the bridge between the magnets should be carefully chosen. In this configuration, a nonmagnetic shaft should preferably be used. The advantage of this rotor design is that the air gap flux density can be greater than the remanent flux density of the PM [10–12].

13.5 PSIM Simulation of the PMSG

The PM generator driven by a wind turbine was simulated in PSIM, as shown in Figure 13.9, using the data presented in Tables 13.2 and 13.3 to feed a three-phase RL load. Figure 13.10 illustrates the load voltage and current result of a noncontrolled system whose voltage level and rotation would strongly depend on the load. This is a typical situation when the load is used to electronically control voltage and rotation/frequency (electronic control by the load) in systems with a constant load [5, 6].

13.6 Advantages and Disadvantages of the PMSG

The PMSG has many advantages. Such machine is the most efficient of all electric machines since it has a magnetic source inside itself. The use of PM for the excitation consumes no extra electrical power. Therefore, copper loss of

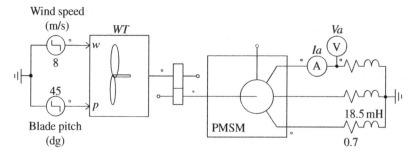

Figure 13.9 PMSG connected to the grid.

Table 13.2 Parameters of the PMSG and load.

Parameter	Magnitude
Rated voltage	60 V
Rated current	12 A
Rotor speed	100 rpm
p	4 poles
R_{load}	0.7 Ω
L_{load}	18.5 mH
L_d	0.027 H
L_q	0.067 H
R_{stator}	3.3 Ω
L_d	22.6 mH
L_q	67.0 mH
$V_{pk}/$krpm	98.67 V/rpm
Inertia (J)	1 kg*m*m
Shaft time constant	1 s

Table 13.3 Parameters of the wind turbine.

Parameter	Magnitude
Rated power	3 kW
Base wind speed	0.185 m/s
Base rotational speed	0.2 rpm
Initial rotational speed	5 rpm
Moment of inertia (J)	350 kg*m*m
Gear box	600 : 1

the exciter does not exist, and the absence of mechanical commutator and brushes or slip rings means lower mechanical friction losses. Another advantage is its compactness.

The recent introduction of high-energy density magnets (rare-earth magnets) has allowed the achievement of extremely high flux densities in the PMSG. Therefore, rotor winding is not required, in addition, making the generator smaller, lighter, and with a more rugged structure. As there is no current circulation in the rotor to create a magnetic field, the rotor of a PMSG does not

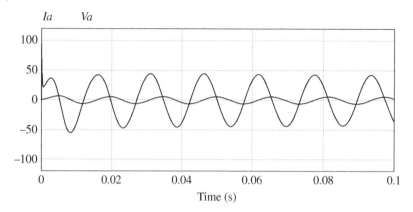

Figure 13.10 Voltage and current of a PMSG connected to an RL load.

contribute to temperature increase. The only heat production is on the stator, which is easier to cool down than the rotor because it is on the periphery of the generator and it is static.

The absence of brushes, mechanical commutators, and slip rings suppresses the need for associated regular maintenance and decreases the risk of failure in those elements. They have very long lasting winding insulation, bearing, and magnet life. Since no noise is associated with the mechanical contacts, the driving converter switching frequency could be above 20 kHz, producing only ultrasound, which is inaudible for humans.

When the PMSG is compared with respect to the conventional ones for low-speed water flows or low wind speed, the list of advantages increases: (i) No speed multipliers or gears since there may be multiple permanent or electromagnets in the rotor for more current production; (ii) few maintenance services because of simplified mechanical design; (iii) easy mechanical interface; (iv) cost optimization; (v) highest power-to-weight ratio in a direct drive; (vi) location of a moving magnetic field being generated in the center of the field; (vii) more precise operations since a microprocessor controls the generator/motor electrical output and current instead of mechanical brushes; (viii) higher efficiency for its brushless generation of electrical current and it is digitally controllable with flexible adjustment of the generator speed with less friction, fewer moving components, less heat, and reduced electrical noise; and (ix) since the permanent or electromagnets are located on the rotor, they are kept cooler and thus have a longer life.

The PMSG has some inherent disadvantages. Two of them are related to (i) the high cost of the PM and (ii) their scarce commercial availability. The cost of higher energy density magnets prohibits their use in applications where initial costs are a major concern. Another constraint is about the limited field-weakening operation for the PMSG machine. It is somewhat difficult to implement it because some flux must be aligned to counteract the flux of the

PM. An accidental speed increase might damage the power electronic components by voltages and currents above the converter rating, especially for vehicle applications. In addition, the surface-mounted PM generators cannot reach high speeds because of the limited mechanical strength of the assembly between the rotor yoke and the PM. Finally, the demagnetization of the PM is possible by the large opposing mmf, high temperatures, excessive vibrations, or harsh short-circuit conditions.

The critical demagnetization force is different for each magnet material (see Table 13.2) where variable B_r stands for relative flux density, μ_r is the corresponding relative magnetic permeability, and ρ is the volumetric density. In addition, extreme care must be taken to cool down the generator core, especially if it is compact. Furthermore, it is not possible to control the speed of a PMSG by only varying the field current or flux. The only methods of speed control available for a PMSG are armature voltage control and armature resistance control. For more information about PMSGs, see Refs. [1, 6, 10, 14].

References

1 S.J. Chapman, Electric Machinery Fundamentals, 3rd ed., McGraw-Hill, New York, 2011.
2 B.J. Chalmers, W. Wu, and E. Spooner, An axial-flux permanent-magnet generator for a gearless wind energy system, IEEE Trans. Energy Conversion, Vol. 14, No. 2, pp. 251–267, 1999.
3 P. Anpalahan, J. Soulard, and H.P. Nee, Design steps towards a high power factor transverse flux machine, Proceedings of the European Conference on Power Electronics and Applications, Graz, Austria, 27–29 August, 2001.
4 M.G. Simões, F.A. Farret, and F. Blaabjerg, Small wind energy systems, Electric Power Components and Systems, Vol. 43, pp. 1388–1405, 2015.
5 H.W. Lee, Advanced control for power density maximization of the brushless DC generator, PhD Dissertation submitted to the Office of Graduate Studies of Texas A&M University, December 2003.
6 I. Boldea, Variable Speed Synchronous Generators, Electrical Power Engineering Series, CRC Press, Boca Raton, FL, 2005.
7 E. Afjei, O. Hashemipour, M.A. Saati, and M.M. Nezamabadi, A new hybrid brushless DC motor/generator without permanent magnet, IJE Transactions B: Applications, Vol. 20, No. 1, pp. 77–86, 2007.
8 D. Svechkarenko, On analytical modeling and design of a novel transverse flux generator for offshore wind turbines, Licentiate Thesis, KTH Teknikringen, Stockholm, Sweden, 2007.
9 W. Zhao and T.A. Lipo, Byung-Il Kwon, comparative study on novel dual stator radial flux and axial flux permanent magnet motors with ferrite magnets for traction application, IEEE Transactions on Magnetics, Vol. 50, No. 11, 2014, DOI: 10.1109/TMAG.2014.2329506.

10 R.G. Kloeffler, R.M. Kerchner, and J.L. Brenneman, Direct Current Machinery, Rev. ed., The Macmillan Co., New York, 1948.

11 J. Pyrhönen, J. Nerg, P. Kurronen, J. Puranen, and M. HAAVISTO, Permanent magnet technology in wind power generators, IEEE 2010 XIX International Conference on Electrical Machines (ICEM), Rome, Italy, September 2010, pp. 1–6.

12 O. Danielsson, K. Thorburn, M. Eriksson, and M. Leijon, Permanent magnet fixation concepts for linear generator, 5th European Wave Energy Conference, Division for Electricity and Lightning Research, Department of Engineering Sciences, Uppsala University, Sweden, September 2003.

13 M.R.J. Dubois, Optimized permanent magnet generator topologies for direct-drive wind turbines, MSc Dissertation, Technische Universiteit Delft, Canada, January 2004.

14 Q. He and Q. Wang, Optimal design of low-speed permanent magnet generator for wind turbine application, 2012 Asia-Pacific Power and Energy Engineering Conference, 2012, pp. 1–3, DOI:10.1109/APPEEC.2012.6307131.

15 C. Heck, Magnetic Materials and Their Applications, Butterworth & Co., London, 1974.

16 L.N. Elevich, Application note on 3-phase BLDC motor control with hall sensors using 56800/E digital signal controllers, Freescale Semiconductor Inc, AN1916, Rev. 2.0, November 2005.

17 D.M. Saban, C. Bailey, K. Brun, and D. Gonzalez-Lopez, Beyond IEEE Std 115 & API 546: test procedures for high-speed, multi-megawatt permanent-magnet synchronous machines, 2009 Proceedings of Industry Applications Society 56th Annual Petroleum and Chemical Industry Conference, 2009, pp. 1–9, DOI: 10.1109/PCICON.2009.5297158.

14

Storage Systems

14.1 Introduction

Electrical energy storage is considered a critical technology for the successful implementation of renewable and alternative energy systems. At the beginning of the twentieth century, electrochemical batteries were used to power telephones and telegraphs, and enormous and heavy flywheels were common in rotating generators to smooth out load oscillations. Electric vehicles were initially more common than gasoline-powered vehicles. Batteries were important in such transportation applications because they could store energy and give reasonable transportation autonomy. Water dams have been used to store potential energy.

Energy storage systems play an important function in the current state of the art in unifying, distributing, and augmenting the capabilities of alternative and renewable energy-distributed generating systems. In these systems, the energy has to be captured to their maximum power point from solar and wind power, river streams, ocean waves and tide, and the energy storage will play a fundamental role to keep steadily such energy supply, by saving the surplus or compensating the lack of power. The centralized electrical generation system of utilities, with long-distance transmission, and their distribution has been considered large enough in order to not be affected by residential and commercial load changes, particularly low instantaneous load power variation. Figure 14.1 illustrates a typical household peak-day load curve and an individual load. However, stand-alone microgrid and distributed generation schemes are exposed to such fluctuations of individual loads. Because of being smaller in size, they do not undergo any averaging effect seen by large power utility generators.

For diesel generator (DG)-based industrial applications, the base load can be considerably higher, and it is possible that a generator will operate cost-effectively. However, small systems do not have this capacity. Energy storage systems capable of smoothing out load fluctuations and making up for seasonal

Integration of Renewable Sources of Energy, Second Edition. Felix A. Farret and M. Godoy Simões.
© 2018 John Wiley & Sons, Inc. Published 2018 by John Wiley & Sons, Inc.

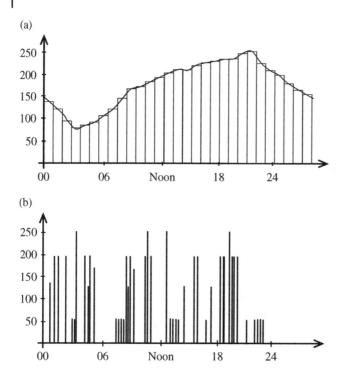

Figure 14.1 Residential load: (a) area average and (b) one household.

variations in renewable sources plus reacting to fast transient power quality may contribute to efficient energy management policies and faster economic investments in new projects.

Energy storage technologies are classified according to their energy, time, and transient response required for their operation [1, 2]. It is convenient to define storage capacity in terms of the time that the nominal energy capacity can cover the load at rated power. Storage capacity can be categorized in terms of energy density requirements (for medium- and long-term needs) or in terms of power density requirements (for short- and very short-term needs). Table 14.1 shows how storage objectives determine storage features depicting the functionality of storage systems in terms of time response.

Energy storage enhances DG in three ways: (i) it stabilizes and permits DG to run at a constant and stable output despite load fluctuations and required maintenance services, (ii) it provides energy to ride-through instantaneous lacks of primary energy (such as those of the sun, wind, and hydropower sources), and (iii) it permits DG to operate seamlessly as a dispatchable unit. Energy storage may be designed for rapid damping of peak surges in electricity demand to counter momentary power disturbances, to provide a few seconds

Table 14.1 Functionality of storage system in terms of time response.

Storage capacity	Energy storage features
Transient (microseconds)	Compensates for voltage sags
	Rides through disturbances (backup systems)
	Regenerates electrical motors
	Improves harmonic distortion and power quality
Very short term (cycles of the grid frequency)	Covers load during startup and synchronization of backup generator
	Compensates transient response of renewable-based electronic converters
	Increases system reliability during fault management
	Keeps computer and telecommunication systems alive for safe electronic data backup
Short term (minutes)	Covers load during short-term load peaks
	Smoothens renewable energy deficits for online capture of wind or solar power
	Decreases needs of startup backup generator
	Improves maintenance needs of fossil fuel-based generators
	Allows ride-through of critical medical, safety, and financial procedures
Medium term (a few hours)	Stores renewable energy surplus to be used at a later time
	Compensates for load-leveling policies
	Allows stored energy to be negotiated on net-metering basis
	Integrates surplus energy with thermal systems
Long term (several hours to a couple of days)	Stores renewable energy for compensation of weather-based changes
	Provides reduction in fuel consumption and decreases waste of renewable energy
	Possibly eliminates fossil fuel-based generator backup
	Requires civil constructions for hydro and air systems
	Produces hydrogen from renewable sources
Planning (weeks to months)	Includes large power storage systems, such as pumped hydro and compressed air systems
	Uses fossil fuel storage to offset economic fluctuations
	Stores hydrogen from biomass or renewable-based systems

of ride-through while backup generators start in response to a power failure, or to reserve energy for future demand. These characteristics contribute to different strategies.

For example, long-term storage systems installed on the customer side of a meter save demand charges on the customer utility bill. Charge and discharge of these storage systems during off-peak periods according to a dispatch strategy minimizes the peak load, which is billed monthly. Such a system can also provide energy savings and may improve the power factor. Similarly to such a customer ownership scenario, peak shaving can be provided by storage technologies owned and operated by the utility. With a transportable storage system, the deferral of transmission and distribution can be provided at multiple sites over the system life. Short-term storage technologies can provide enhanced reliability at a customer site by providing ride-through during momentary utility outages or transition to standby generation.

Storage can also supplement a fuel cell or a DG, providing load following the service and enhancing the generator efficiency (full load versus part load). A storage energy facility can operate in parallel with a generating unit in order to meet temporary peaks of demand higher than the rated generating capacity. A storage system can also support voltage-through mechanisms of reactive power control in a microgrid system.

Small-scale storage can provide spinning reserve so that output can be ramped down in a controlled manner without disrupting the grid. On a larger scale, storage systems can provide load-leveling or commodity storage capability with controlled dispatchability, in which stored energy is held back for release when prices on the spot market reach a threshold.

Figure 14.2 shows the maturity and the emerging storage technologies organized in categories of ride-through, power quality, and energy management. Applications include lead-acid batteries, advanced batteries, low- and high-energy flywheels, ultracapacitors, superconducting magnetic energy storage (SMES) systems, heating systems, pumped hydro, geothermal underground, and compressed air energy storage (CAES). Hydrogen storage is not included in this section, but it can be understood in the chapter about fuel cells. Feasibility, economical evaluation studies, and details of fuel cell technology must be fully covered when discussing this topic.

14.2 Energy Storage Parameters

Energy storage for DG can be compared with the parameters that define their performance criteria: capacity, specific energy, energy density, specific power, efficiency, recharge rate, self-discharge, lifetime, capital cost, and operating cost.

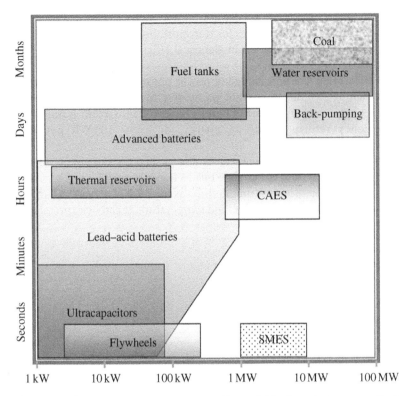

Figure 14.2 Energy management, power quality, and ride-through storage applications.

Capacity: the available energy storage capability. The SI unit of capacity is the joule, but this is a very small unit. Usually, the watt-hour (i.e., the energy equivalent of working at a power of 1 W for 1 hour or 3600 J) is used instead.

Specific energy: the electrical energy stored per mass, in units of Wh/kg. Specific energy is ordinarily used when the energy capacity of a battery needed in a certain system is known; it is then divided by the specific energy to give an approximation of battery mass.

Specific power: the amount of power obtained per kilogram of a storage system in W/kg. It is a very sensitive parameter because several storage systems cannot operate at this maximum power for long, so they may affect lifetime or operate very inefficiently.

Energy density: a measure of energy stored per volume in Wh/m^3. It can be used to give an approximation of the energy storage volume for a given application.

Physical efficiency: how much power is stored in a given volume and mass. It is usually considered in batteries for transportation applications, in which industry may accept degradation in electrical efficiency in return for good physical efficiency.

Electrical efficiency: the percentage of power put into a unit that is available to be withdrawn; an important parameter for DG applications. It is measured by the energy capable of being converted into work. A unit with 90% efficiency returns 9 kWh of energy for every 10 kWh put into storage.

Recharge rate: the rate at which power can be pushed for storage. A storage system might take 10 hours to deplete but 14 hours to refill.

Specific gravity: a dimensionless unit defined as the ratio of density of a material to the density of water at a specified temperature. It can be expressed as

$$SG = \frac{\rho}{\rho_{H_2O}}$$

where:

SG = The specific gravity

ρ = The density of the electrolyte

ρ_{H_2O} = The density of water at 4 °C (because the density of water at this temperature is at its highest)

Self-discharge: the indicator of the duration of batteries take to discharge when unused. This is usually due to current leakage and heat dissipation.

Lifetime: the service life of a unit, which varies with technology and intensity of use. Batteries are noted for having short lifetimes in applications in which they are repeatedly charged and discharged completely.

Capital cost: the initial cost in dollars per kilowatt for design, specification, civil works, and installation.

Operating and maintenance costs: including the costs in dollars per kilowatt-hour for periodic inspection, fueling, maintenance, parts (bearings and seals) replacement, recalibration, and so on.

Table 14.2 shows nominal parameters for lead–acid batteries. They are a common storage solution, but their lifetime is a concern. Therefore, continual over- or undercharging is not recommended, and a charger system should control the operating cycle. Tables 14.3 and 14.4 compare the parameters of large-scale energy storage. The parameters in the tables are guidelines only and are useful for broad comparative purposes.

Table 14.5 compares cost projections of storage systems. In the following sections, several storage systems are discussed to support the design engineer's understanding of the fundamentals of each technology [2].

Table 14.2 Typical lead–acid battery parameters.

Specific energy	20–35 Wh/kg
Energy density	50–90 Wh/L
Specific power	About 250 W/kg
Nominal cell voltage	2.0 V
Electrical efficiency	About 80%, depending on recharge rate and temperature
Recharge rate	About 8 hours (possible to quick recharge 90%)
Self-discharge	1–2% per day
Lifetime	About 800 cycles, depending on the depth of cycle

Table 14.3 Typical battery parameters.

Battery type	Specific energy $(Wh{\cdot}kg^{-1})$	Energy density $(Wh{\cdot}L^{-1})$	Specific power $(W{\cdot}kg^{-1})$	Nominal cell voltage
Lead–acid	30	75	250	2.0 V
Nickel cadmium	50	80	150	1.2 V
Nickel metal hydride	65	150	200	1.2 V
Zebra	100	150	150	High voltage, integrated stack
Lithium ion	90	150	300	3.6 V
Zinc–air	230	270	105	1.65 V

14.3 Lead–Acid Batteries

The lead–acid battery is an electrochemical device invented by Planté in 1859. Two electrodes react through a sulfuric acid electrolyte. During discharge, both electrodes are converted to lead sulfate, as described by the following charge–discharge reaction [3]:

$$\overset{cathode}{Pb} + 2H_2SO_4 + \overset{anode}{PbO_2} \underset{charge}{\overset{discharge}{\rightleftarrows}} \overset{cathode}{PbSO_4} + 2H_2O + \overset{anode}{PbSO_4}$$

When the battery is charged, the anode is restored to lead dioxide and the cathode to metallic lead. However, irreversible changes in the electrodes limit the number of cycles, and failure may occur after a couple of thousand cycles, depending on battery design and depth of discharge.

Table 14.4 Comparison of large-scale energy storage systems.

Criteria	Ultracapacitor	Flywheels (low speed)	Flywheels (high speed)	Pumped hydro	CAES	SMES
Capital cost/MWh	$25,000,000	$300,000	$25,000,000	$7,000	$2,000	$10,000
Weight/MWh	10,000 kg	7,500 kg	3,000 kg	3,000 kg	2.5 kg	10 kg
Efficiency	95%	90%	95%	80%	85%	95%
Operating cost/MWh	$5	$3	$4	$4	$3	$1
Capacity	0.5 kWh	50 kWh	750 kWh	22,000 MWh	2,400 MWh	0.8 kWh
Lifetime	40 years	20 years	20 years	75 years	30 years	40 years

Table 14.5 Cost projection of energy storage systems.

System	Typical size range (MW)	$/kW	$/kWh
Lead–acid batteries	0.5–100	100–200	150–300
Advanced batteries	0.5–50	200–400	
Ultracapacitors	1–10	300	3600
Flywheels	1–10	200–500	100–800
SMES	10–1000	300–1000	300–3000
CAES	50–1000	500–1,000	10–15
Pumped hydropower	100–1000	600–1000	10–15

Many rechargeable batteries are suitable for DG applications, but comprehensive descriptions are outside the scope of this book. The lead–acid battery is still the most common because of its widespread use and its relatively economic power density, and it will probably remain as the most used devices for the next few years. Several improvements underway by hybrid cars and electric car manufacturers may allow their utilization in DG. It is expected that other technologies may surpass the lead–acid battery on the basis of energy density and lifetime. For example, nickel–cadmium batteries are common in applications that require sealed batteries capable of operating in any position with higher energy density. Other advanced batteries, such as nickel–metal hydride (MH) and several lithium technologies, may become cost-effective in the future for residential and commercial applications.

14.3.1 Constructional Features

The basic building block of a lead–acid battery is a 2-V cell. Each battery unit will supply power in case of outages or low production from renewable energy sources. The batteries are wired together in series to produce 12-, 24-, or 48-V strings or other voltages. But very high-voltage strings are not recommended because of lower reliability. These strings are then connected together in parallel, allowing higher current capacity in order to make up a battery bank. The battery bank supplies dc power to an inverter, which produces ac power that can be used to run domestic, residential, or industrial loads and home appliances in general. The battery bank voltage and current rating are determined by the inverter input requirements, the type of battery, and the energy storage required. The flooded lead–acid battery is used in automobiles, forklifts, and uninterruptible power supply systems. Flooded-cell batteries have two sets of lead plates that are coated with chemicals and immersed in a liquid electrolyte. As the battery is used, the water in the electrolyte evaporates and needs to be replenished with distilled water.

During the mid-1970s, maintenance-free lead–acid batteries were developed by transforming liquid electrolyte into moistened separators and sealing the enclosure. Safety valves were added to allow gas venting during charge and discharge. Two lead–acid systems emerged the small sealed lead–acid (SLA) battery, also known under the brand name gel cell, and the large valve-regulated lead–acid (VRLA) battery.

VRLA batteries do not require adding water to keep their electrolyte functioning or mixing the electrolyte in order to prevent stratification. The oxygen recombination and the valves of VRLAs prevent the venting of hydrogen and oxygen gases and the input of air. However, the battery subsystem may need to be replaced more frequently than a flooded lead–acid battery, which can increase the leveled cost of the system. The major advantages of VRLA batteries over flooded lead–acid cells are (i) the dramatic reduction in maintenance, (ii) the reduced footprint and weight because of the sealed construction, and (iii) the immobilized electrolyte, which allows battery cells to be packaged more tightly.

VRLA batteries are less robust, more expensive, and have a shorter lifetime than flooded lead–acid batteries. VRLA batteries are perceived as maintenance-free and safe, and they became popular for standby power supplies in telecommunication applications and for uninterruptible power supplies, making unnecessary special rooms set aside for batteries.

Unlike flooded lead–acid batteries, both SLA and VRLA batteries are designed with a low overvoltage potential to prevent them from reaching their gas-generating potential during charge. Excess charging would cause gassing and water depletion. Consequently, these batteries can never be charged to their full potential. They are not really the best choice for DG and renewable energy applications. Table 14.6 lists the advantages and limitations of lead–acid batteries.

14.3.2 Battery Charge–Discharge Cycles

The batteries used to start automobiles are known as shallow-cycle batteries because they are designed to supply a large amount of current for a short time and to withstand mild overcharge without losing electrolyte. Unfortunately, they cannot tolerate being deeply discharged. If they are repeatedly discharged more than 20%, their lifetime will be severely affected. Lead–acid batteries for automotive use are not designed for deep discharge and should always be kept at maximum charge using constant voltage at 13.8 V (for six-element car batteries). These batteries are not a good choice for DG systems.

Deep-cycle batteries are designed to be discharged repeatedly by as much as 80% of their capacity, so they are a good choice for DG systems. In addition to being designed to withstand deep cycling, these batteries have a longer life if the cycles are shallower. Of course they have a bigger size when compared with shallow-cycle units.

Table 14.6 Characteristics of lead-acid batteries.

Advantages	Limitations
Inexpensive and simple to manufacture	Low energy density; poor weight-to-energy ratio limits use to stationary and wheeled applications
Mature, reliable, and well-understood technology	Cannot be stored in discharged condition; cell voltage should never drop below 2.10 V
When used correctly, durable and provides dependable service	Allows only a limited number of full discharge cycles; well suited for standby applications that require only occasional deep discharges
Self-discharge is among the lowest of rechargeable battery systems	Lead content and acid electrolyte make environmentally unfriendly
Low maintenance requirements, no memory; no electrolyte to fill on sealed version	Transportation restrictions on flooded lead–acid; environmental concerns regarding spillage; thermal runaway can occur with improper charging
Capable of high discharge rates	

Flooded-cell batteries must be replenished with distilled water. In gel cell batteries, the electrolyte is suspended in a gelatin-like material so it will not spill even if the battery is operating by its side. Gel cells are often called recombinant batteries because oxygen gas given off at the positive plate is recombined with hydrogen given off at the negative plate to keep the electrolyte moist. This means that distilled water does not need to be added. Absorbed glass mat batteries have a highly porous microfiber sponge-like glass mat between the plates to absorb the electrolyte and have no free liquid. Similar to gel cells, they use seal and recombinant gas effects so they do not lose liquid during use.

A lead–acid battery consists of a lead cathode and a lead oxide (PbO_2) anode immersed in a sulfuric acid solution. The discharging reaction at the anode consists of the exchange of oxygen ions from the anode with sulfate ions of the electrolyte. At the cathode, the discharge involves sulfate ions from the electrolyte combining with lead ions to form lead sulfate. Two electrons must enter the anode terminal, and two electrons must leave the cathode terminal via the external circuit for every two-sulfate ions that leave the electrolyte. This corresponds to the current supplied by the battery to the external circuit. Removal of sulfate ions from the solution reduces the acidity of the electrolyte. When an external voltage, greater than the voltage produced by the reactions at the anode and cathode, is applied across the battery terminals, the current will flow into the anode rather than out, thus charging the battery. The chemical processes can be reversed, and sulfate ions are liberated to the solution, which increases the concentration of sulfuric acid in the electrolyte.

Finding the ideal charge voltage limit is critical. A high voltage limit (more than 2.4 V per cell) produces good battery performance but shortens service life because of grid corrosion on the positive plate. Such corrosion is permanent. A low voltage (<2.4 V per cell) is safe if charged at a higher temperature, but it is subject to sulfation on the negative plate. If excessive lead sulfate builds up on the electrodes, their effective surface areas are reduced, which affects performance. It is important to completely avoid discharging a lead–acid battery. Conversely, if excessive charging is imposed and there is no more sulfate at the cathode to maintain continuity of the charging current, hydrogen is liberated. This is a potential safety hazard. However, an occasional charging to the gassing stage with a slight degree of bubbling is sometimes conducted to clean up the electrodes and provide a mixing action on the electrolyte. The electrolyte freezing point depends on the state of charge. Therefore, fully charged batteries may operate at low temperatures, while batteries in a lesser degree of states of charge will need to operate in warmer environments.

A charge controller is essential to maintain batteries under optimized levels. The charge controller shuts down the load when a prescribed level of charge is reached. The controller can also shut down the connection to the main source (e.g., a photovoltaic (PV) array) when the battery is fully charged. When a charge controller manages a bank of batteries, it performs an equalization of charges electronically. Equalization is a charge about 10% more than the normal float or trickle charge, ensuring that the cells are equally charged and, in flooded batteries, that the electrolyte is fully mixed by the gas bubbles. Gelled and sealed batteries should be equalized at a much lower rate than flooded batteries. Usually, the final charge cycle on a three-stage charger is sufficient to equalize the cells.

Depending on the depth of discharge and operating temperature, an SLA battery provides 200–300 discharge–charge cycles for the entire lifetime. The primary reasons for such relatively short life is the corrosion of the positive electrode, depletion of the active material, and expansion of the positive plates. These changes are most prevalent at higher operating temperatures. Keeping batteries in temperature-controlled rooms will increase their life, at increased installation costs, and temperature control expenses, of course. Cycling does not prevent or reverse the trend.

14.3.3 Operating Limits and Parameters

The optimum operating temperature for a lead–acid battery is 25 °C (77 °F). The Arrhenius law describes how the rate of a chemical reaction doubles for every 10 °C rise in temperature. In this case, it applies to the rate at which the slow deterioration of the active chemicals increases. It is possible to assume that every 8 °C (15 °F) rise in temperature will cut the battery life in half, just to be on the safe side. A VRLA battery would last for 10 years at 25 °C (77 °F).

It will only be good for 5 years if operated at 33 °C (95 °F). Theoretically, the same battery would endure a little more than 1 year at a desert temperature of 42 °C (107 °F). Among modern rechargeable batteries, the lead–acid battery family has the lowest energy density, making it unsuitable for handheld devices that demand compact size. In addition, performance at low temperatures is poor.

The capacity of a battery is referred to as C. Thus, if a load is connected to a battery such that it will discharge in n hours, the discharge rate is C/n. Figure 14.3 shows how the terminal voltage depends on the charge or discharge rate and the state of charge for a typical lead–acid battery.

It is extremely difficult to accurately measure the state of charge of a lead–acid battery and predict the remaining capacity. In a battery, the rate at which an input or output current is drawn affects the overall energy available from the battery. For example, a 100-Ah battery at a 20-hour rate means that over 20 hours, 100 Ah is available (i.e., the user can expect to draw up to 5 A per hour for up to 20 hours). If a quick discharge is imposed, the effective ampere-hours will be fewer.

There are some commercial state-of-charge meters, called E-meters. They sample the rate of discharge every few minutes, recalculate the time remaining

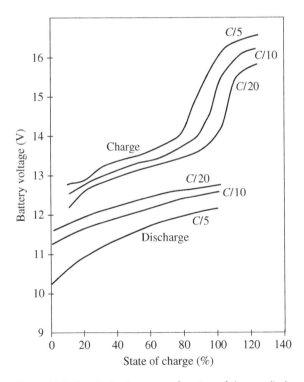

Figure 14.3 Terminal voltages as a function of charge–discharge rate and state of charge.

before the battery is discharged, and update and display a fuel gauge bar graph. The E-meter also measures kilowatt-hours and historical battery information such as the number of cycles, deepest discharge, and average depth of discharge. When an E-meter is not available, a measurement of the open-circuit voltage is a fair approximation of battery discharge.

The SLA battery is rated at a 5-hour discharge or $C/5$. Some batteries are rated at a slow 20-hour discharge. Longer discharge times produce higher capacity readings. The lead–acid battery performs well on high load currents. Note that if the battery is charged at a $C/5$ rate, full charge will be reached at a terminal voltage of 16 V. However, if the battery is charged at $C/20$, the battery will reach full charge at a terminal voltage of 14.1 V. When the charging current is zeroed, the terminal voltage will drop below 13 V. As an example, a six-cell battery has the following characteristic voltage set points:

- Quiescent (open-circuit) voltage: 12.6 V
- Unloading end voltage: 11.8 V
- Charge: 13.2–14.4 V
- Gassing voltage: 14.4 V
- Recommended floating voltage for charge preservation: 13.2 V

After full charge, the terminal voltage drops quickly to 13.2 V and then slowly to 12.6 V.

14.3.4 Maintenance of Lead–Acid Batteries

A routine checkup of battery banks should be done every month in order to inspect the water level and corrosion on the terminals. At least two or three times a year, the state of charge should also be evaluated. As the battery discharges, more water is produced. A lower specific gravity is found in discharged batteries. State of charge, or the depth of discharge, can be determined by measuring the voltage or specific gravity of the acid with a hydrometer. The measurement of these parameters alone will not support the battery conditions assessment; only a sustained load test can do that.

Voltage on a fully charged battery is typically 2.12–2.15 V per cell or 12.7 V for a 12-V battery. At 50%, the reading will be 2.03 volts per cell (VpC), and at 0% it will be 1.75 VpC or less. Specific gravity will be about 1.265 for a fully charged cell and 1.13 or less for a totally discharged cell, but these numbers can vary. The user should measure them when the batteries are brand new, by fully charging them and leaving to settle down for a while and then taking a reference measurement. Of course, hydrometer readings are not possible for sealed batteries, and only voltage readings are used for evaluation of the depth of discharge. Maintenance personnel should only use distilled water in order to avoid adding minerals that can reduce the batteries' effectiveness. The batteries should be brought to a full charge before adding water because the electrolyte expands as the state of charge increases.

14.3.5 Sizing Lead–Acid Batteries for DG Applications

Sizing lead–acid batteries for DG applications must be considered if cycling of the battery is required and if the battery bank will be used for (i) transient compensation of source response and high bandwidth response and (ii) autonomous operation for source outage, where battery storage will be used in conjunction with diesel engines, microturbines, or fuel cells and typically connected to the grid. For such autonomous operation, the energy is stored from renewable sources and used for a long period of time when the main source is not available. It is a longer-term use with a lot of idling time, whereas for transient compensation, usually a hybrid solution such as battery–supercapacitor, battery–fuel cell, or battery–flywheel will be used for the rapid and more frequent changes. Therefore, sizing batteries for a DG system for autonomous operation will depend on the prescribed autonomy. If the system is grid-connected, battery size can be reduced for users that do not typically sell power to the grid. It must be increased if users sell power through net metering. The sizing for autonomous operation is typically a procedure utilized by small residential system designers, so it is described next. It is possible to oversize a battery, allowing it for transient compensation, by making an assumption of the typical transient response of the energy source (say, a fuel cell or a microturbine).

The following procedure for sizing batteries for DG applications was adapted from Refs. [3–8]. The dc link where the battery charger is connected defines the string connection of battery cells. The dc-link voltage will usually fluctuate with the intermittence of the main source of energy (solar, wind, hydro, or other), unless it is controlled by a specific converter connected to it, with closed-loop control of the dc-link voltage. Wind and hydro resources can be connected to the dc link with a rectifier or directly to the ac grid. If only a PV array is connected to the dc link, it will define the maximum dc-link voltage.

Batteries are specified with a nominal cell voltage equal to their open-circuit voltage (V_{cn}) for a 100% state of charge and that is the recommended float voltage of the cell. Let us define the cell voltage at the completion of discharge by their final volts per cell (FV_{pC}) parameter, which should be selected as high as possible, typically within 80 to 90% of V_{cn}:

$$FV_{pC} = 0.8V_{cn} \tag{14.1}$$

Let V_{bat} represent the terminal voltage of the battery (consisting of a series string of N_c cells) after discharge. Therefore, the output voltage V_{bat} must be equal to the minimum dc-link bus voltage, which is related to the minimum voltage at which the inverter connected to the dc link will operate. Then the number of cells can be selected by the following ratio, taking the next higher integer value:

$$N_c = \frac{V_{bat}}{FV_{pC}} \tag{14.2}$$

The decision to whether the battery should be sized for transient compensation or autonomous operation must be made at this point. When selecting a battery to compensate the main source (e.g., fuel cell or microturbine) for changes in the load demand, the typical transient response of the fuel cell or microturbine source must be known empirically and defined by the mathematical function $g(t)$, which starts at zero and ends at the unity value at a time $T_{r,max}$.

This function represents the source response to a step change in power, and either a ramp or an exponential function could be used to approximate it. Then the maximum power drawn from the battery to compensate for this transient is

$$P_{b,max}(t) = P_{fl} g(t) \tag{14.3}$$

where:

P_{fl} = The full-load power (kilowatts)

The maximum energy storage required in the battery can be determined by

$$E_{b,max} = \int_0^{T_{r,max}} P_{b,max}(t) dt \tag{14.4}$$

The integral can be approximated by

$$E_{b,max} = K_1 P_{fl} T_{r,max} \tag{14.5}$$

where:

$K_1 = \int_0^{T_{r,max}} g(t) dt$ = Evaluated to be 0.5 and 0.2 for ramp function and exponential function, respectively

The full-load power P_{fl} can be obtained by

$$P_{fl} = \frac{kVA_{fl} \cdot PF}{\eta} 10^5 \tag{14.6}$$

where:

kVA_{fl} = The full-load kilovolt-amperes of the inverter
PF = The power factor of the load connected to the inverter
η = The inverter efficiency

Combining equations (14.5) and (14.6), the battery energy is given by

$$E_{b,max} = \frac{K_1 \cdot kVA_{fl} \cdot PF \cdot T_{r,max}}{\eta} 10^5 \tag{14.7}$$

Equation (14.8) gives the ampere-hour per cell, which is the C rate of the battery:

$$Ah = \frac{E_{b,max}}{N_c \cdot FVpC \cdot 3600} = \frac{K_1 \cdot kVA_{fl} \cdot PF \cdot T_{r,max}}{\eta \cdot N_c \cdot FVpC \cdot 3600} 10^5 \qquad (14.8)$$

Considering d as the multiplier that determines the discharge rate of the battery, then

$$d = \frac{3600}{T_{r,max}} \qquad (14.9)$$

and the discharge rate of the battery is expressed as dC. The average current drawn from the battery during discharge is d Ah for a duration of $T_{r,max}$.

Batteries should not be discharged excessively. A deep-cycle lead–acid battery (the main battery option) will last longest if it is discharged only at 50%. By multiplying the total ampere-hours from equation (14.8) by 2, the optimal battery capacity can be determined.

After this procedure, it becomes straightforward to size batteries for autonomous operation. The designer needs to find the "daily energy use" as $pkVA_{fl}$ and to decide the number of days to have power in storage when the sun is not shining adequately for PV-based systems or when any other sources are not capable of supplying power. Three to five days is recommended. If 3 days is used, $T_{r,max} = 72$ hours and K_1 is assumed to be united. For PV-based systems installed in a latitude higher than 40, 5 days of storage is recommended. The flowchart in Figure 14.4 demonstrates this procedure.

14.4 Ultracapacitors (Supercapacitors)

Regular capacitors store energy in their dielectric material calculated by $CV^2/2$, where C is the capacitance (farads) and V (volts) is the voltage across its terminals. The maximum voltage of a regular capacitor is dependent on the breakdown characteristics of the dielectric material. The charge Q (coulombs) stored in the capacitor is given by $Q = CV$. The capacitance of the dielectric capacitor depends on the dielectric constant (ε) and the thickness (d) of the dielectric material plus its geometric area:

$$C = \varepsilon \frac{A}{d} \qquad (14.10)$$

During the past few years, electric double-layer capacitors (EDLCs) with very large capacitance values have been developed. Those capacitors are frequently called supercapacitors, ultracapacitors, or electrochemical capacitors [9–11]. The term ultracapacitor is used in this book because the power

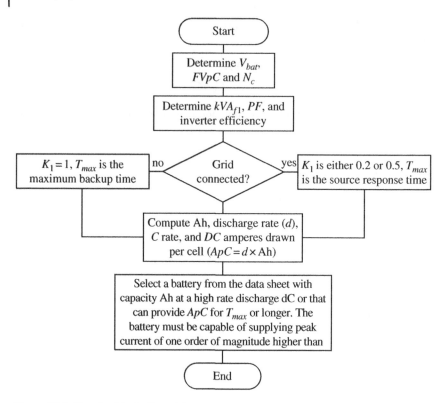

Figure 14.4 Flowchart for battery sizing.

industry seems to use it more frequently. An ultracapacitor is an electrical energy storage device that is constructed much like a battery because it has two electrodes immersed in an electrolyte with a separator between them. The electrodes are fabricated from porous high surface-area material that has pores of diameter in the nanometer range. Charge is stored in the micropores at or near the interface between the solid electrode material and the electrolyte. The charge and energy stored are given by the same expressions as those for an ordinary capacitor, but the capacitance depends on complex phenomena that occur in the micropores of the electrode.

14.4.1 Double-Layer Effect

Figure 14.5 shows the construction details of a double-layer ultracapacitor. The capacitor contains two particulate-carbon electrodes formed on conductive polymer films. An ion-conductive membrane separates the two electrodes, and a potassium hydroxide electrolyte permeates the capacitor. The micropores in the carbon particles result in an enormous surface area and yield extremely high-capacitance values, which conventional capacitors cannot attain. Energy

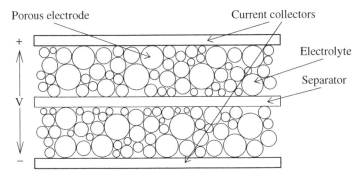

Figure 14.5 Construction details of a double-layer ultracapacitor.

is stored in the double-layer capacitor as charge separation in the double layer formed at the interface between the solid electrode material surface and the liquid electrolyte in the micropores of the electrodes. The ions displaced in forming the double layers in the pores are transferred between the electrodes by diffusion through the electrolyte. The separator prevents electrical contact between the two electrodes. It is very thin, with high electrical resistance, but ion permeable, which allows ionic charge transfer. Polymer or paper separators can be used with organic electrolytes and ceramic, or glass fiber separators are often used with aqueous electrolytes.

When stored in batteries as potentially available chemical energy, electrical energy requires faradic oxidation and reduction of the electroactive reagents to release charges. These charges can perform electrical work when they flow between two electrodes of different potentials. A dielectric material sandwiched between the two electrodes stores electrostatic energy by accumulation of charges, but it has a very limited storage capability compared with that of batteries. The energy and charge stored in the electrochemical capacitor are given by the same equations as for ordinary capacitors. However, the capacitance is dependent primarily on the characteristics of the electrode material (i.e., surface-area and pore size distribution). The specific capacitance of an electrode material depends on the effective dielectric constant of the electrolyte and the thickness of the double layer formed at the interface. These very complex phenomena are not fully understood. The thickness of the double layer is very small (a fraction of a nanometer in liquid electrolytes), which results in a high value for the specific capacitance.

The performance of electrochemical capacitors depends on the specific capacitance (F/g or F/cm^3) of the electrode material and the ionic conductivity of the electrolyte used in the device. The specific capacitance of a particular electrode material depends on whether the material is used in the positive or negative electrode of the device and whether the electrolyte is aqueous or organic. Most carbon materials exhibit higher specific capacitance, in the ranges of 75–175 F/g for aqueous electrolytes and 40–100 F/g for organic electrolytes, which allow much larger ions. Although a lower specific capacitance is achieved with organic electrolytes, they have the advantage of higher operating voltage. For aqueous electrolytes, cell voltage is about 1 V; for organic electrolytes, cell voltage is 3–3.5 V [9–11].

The following example illustrates the principles of high-capacitance devices. Suppose that two carbon rods are separated from each other and immersed in a thin sulfuric acid solution, which is subjected to a voltage slowly increased from zero. Under such conditions, almost nothing happens up to 1 V. Then, at about 1.2 V, gas bubbles appear on the surface of both electrodes because of the electrical decomposition of water. Just before decomposition occurs, where there is no current flow yet, an electric double layer (EDL) occurs at the boundary of electrode and electrolyte (i.e., electrons are charged across the double layer and form a capacitor). Gassing comes up at voltages above 1 V. This indicates that the capacitor is breaking down by overvoltage, which causes decomposition of the electrolyte. The electrical double layer works as an insulator only below the decomposing voltage because water-soluble or aqueous electrolytes constrain the application at a withstanding operating voltage of about 1 V. Organic electrolytes are used to increase the operating voltage (currently at 2.5 V) and provide overall stability against electrolyte decomposition. In organic electrolytes, there is a salt in the solvent that determines the peak capacity; aqueous electrolytes do not have this limitation.

14.4.2 High-Energy Ultracapacitors

The high-energy content of ultracapacitors originates in the activated carbon electrode material, which has an extremely high specific surface area and an extremely short distance (on the order of a few nanometers) between the opposite charges of the capacitor. Because the dielectric is extremely thin, consisting only of the phase boundary between electrode and electrolyte, capacitances approximately a few thousand farads are possible in a very small volume. Researchers have been exploring the possibility of using carbon nanotubes for ultracapacitor electrodes because they have nanopores (about 0.8 nm in diameter), which could, in theory, store much more charge if the nanotubes could be properly assembled into macroscale units.

When such a device is called an electrochemical capacitor, it is because reduction–oxidation (redox) chemical reactions take place. In a true capacitor, no chemical reactions are involved. An ultracapacitor uses the effect of an EDL on both positive and negative electrodes. Therefore, it is often called symmetric EDLC.

To increase energy density, a combination of a battery and a capacitor has been designed. In it, one electrode is for a battery, and the other electrode is for a capacitor. Such a hybrid configuration is called asymmetric EDLC. Asymmetric capacitors tend to be capable of higher capacitance because the battery electrode, physically smaller than the carbon electrode, allows more space. Most EDLCs use aqueous electrolytes, which are limited by the gassing voltage of the water. Most symmetrical capacitors use organic electrolytes to achieve voltages rated at 2.5–2.7 V.

The leading manufacturers of ultracapacitors are Maxwell Technologies in the United States, NESS Capacitor Co. in South Korea, Saft in France, Panasonic and Okamura in Japan, and EPCOS in Germany. These companies manufacture carbon–carbon or symmetric ultracapacitors. Some companies in Russia are developing aqueous technology. ESMA produces asymmetrical devices, and ECOND and ELIT produce symmetrical units.

14.4.3 Applications of Ultracapacitors

The energy and power density of ultracapacitors is for applications on the range of the ones of batteries and conventional capacitors, as depicted in Figure 14.6; they have more energy than a capacitor but less than a battery and they have more power than a battery but less than a capacitor. Unlike batteries, the ultracapacitor voltage varies linearly with the state of charge, as depicted in Figure 14.7. The voltage range between full charge and end of charge is higher for ultracapacitors than for batteries. Because of such drastic variation with state of charge, series connection is often required for high-voltage applications, and power electronic circuits must be integrated with the stack to control charge, discharge, and voltage equalization, as portrayed in Figure 14.8.

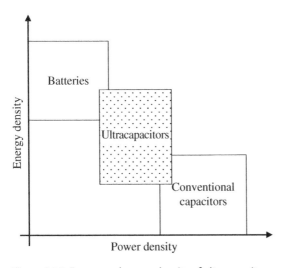

Figure 14.6 Energy and power density of ultracapacitors.

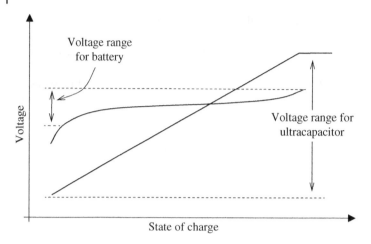

Figure 14.7 Comparison of voltage profile for ultracapacitor with battery.

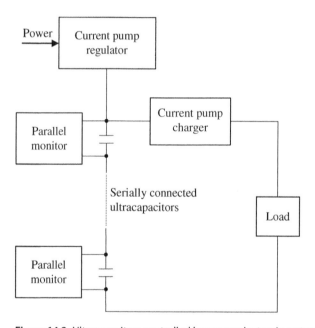

Figure 14.8 Ultracapacitors controlled by power electronic systems.

A current pump charger is a switching voltage regulator that maintains a constant load current under varying terminal voltages. A current pump is inherently suited to the charging requirements of nickel–cadmium and nickel–MH batteries or constant current requirements of Peltier coolers and, in this case, is used for an ultracapacitor system.

Ultracapacitors have the following typical uses:

- Pulse power, in which a load receives short, high-current pulses
- Holdup or bridge power to a device or piece of equipment for seconds, minutes, or days when the main power or battery fails or when the battery is swapped out

Ultracapacitors can be used for several applications:

- As efficient battery packs, achieved by paralleling a low-impedance, high-power aerogel capacitor and a high-impedance, high-energy battery to create a low-impedance, high-power, and high-energy hybrid battery capacitor
- In wireless communications, in which they are used for pulse power during transmission in GSM cell phones, 1.5- and 2-way pagers, and other data communication devices
- In mobile computing, in which they are used for holdup and pulse power in portable data terminals, personal digital assistants, and all other portable devices using microprocessors
- For industrial applications, such as solenoid and valve actuation, electronic door locks, and uninterruptible power supplies

Ultracapacitors have several advantages over batteries. They tend to have a longer cycle life due to the absence of chemical reactions, yielding a stable electrode matrix and no wear-out. Ultracapacitors typically work more than 100,000 cycles with energy efficiency greater than 90%. They are currently able to bridge power for seconds in the hundreds of kilowatts of power range, and their use for residential peak shaving is under consideration (see the hybrid PV system depicted in Figure 14.9). The use of ultracapacitors for multimegawatt and transmission and distribution applications that require several hours of energy storage for peak shaving and load leveling is not yet feasible [8–10].

Ultracapacitors have been considered for use in association with uninterruptible power supplies and fuel cells because they respond faster than batteries and have a higher energy density. Ultracapacitors could provide a longer period of autonomy for an uninterruptible power supply and cope with heavier transients than fuel cells are able to handle because of their slow response to current surges. Ultracapacitors can also be used in a programmable array, where their association will search a dc voltage across a PV in order to optimize the PV power transfer in high-power applications, making the use of boost converters unnecessary [8, 9].

14.5 Flywheels

In the past, it was very common to add inertia to a motor–generator with flywheels. That was the primary method of limiting power interruptions in critical loads. Flywheels were first used in steam engines. In the 1970s, when the

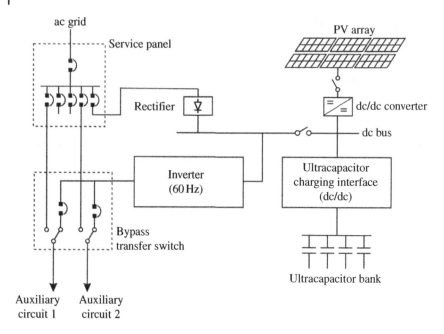

Figure 14.9 Hybrid system composed of ultracapacitor storage and photovoltaic array.

technology was used for centrifugal enrichment of uranium, it was rediscovered as a potential new energy storage system. Of course, a lot has changed, and heavy flywheels may not be cost-effective nowadays when compared with power electronic circuits that might do the same for a fraction of a cost and less bulky and greasy. However, flywheels can still be used and can be modernized, rotating in a higher speed with a specialized machine for conversion. They have very simple physics; they store energy in a spinning mass. Through a coupled electric machine, the system can store power by speeding up the shaft and retrieve power by slowing it down. The stored energy depends on the moment of inertia of the rotor, I, which depends on the distribution of mass density around the rotating axis, $p(x)$, and geometrical radius of the rotor, r [12, 13]. The energy will depend on the moment of inertia and the square of the angular speed of the flywheel shaft:

$$E = \frac{1}{2}I\omega^2 \qquad (14.11)$$

$$I = \int p(x)r^2 dx \qquad (14.12)$$

14.5.1 Advanced Performance of Flywheels

Although it appears that only high angular velocity and large masses around the axis of rotation would allow high-energy storage, it is important to evaluate how to obtain the highest energy density given a prescribed material strength. Therefore, the tensile strength (i.e., the highest stress not leading to a permanent deformation of the material) should be used to define the optimal flywheel shape. The components in x, y, and z directions of the force per unit volume are related to the stress tensor t_{ij} by equation (14.13), and the tensile strength s_i, in a direction specified by i, is given by equation (14.14):

$$f_i = \sum_j \frac{\partial \tau_{ij}}{\partial x_j} \tag{14.13}$$

$$\sigma_i = max\left(\sum_j \tau_{ij} n_j\right) \tag{14.14}$$

where:

n_j = A unit vector normal to the cut in the material and the maximum sustainable stress in the direction i is to be sought by varying the angle of slicing

The value of the shape factor or form factor is a measure of the efficiency with which the flywheel geometry uses the material strength, and a general expression for any flywheel made from material of uniform mass density is

$$E_{m,max} = K \frac{\sigma_{max}}{\rho} \tag{14.15}$$

where:

$E_{m,max}$ = The maximum kinetic energy per unit mass

If the material is isotropic, the maximum tensile strength is independent of directions and may be denoted by σ_{max}. By calculating the internal forces in the flywheel and considering a given flywheel shape symmetrical around the axis of rotation, a resulting shape factor K may be calculated as indicated in Table 14.7. Typical materials for flywheel rotors of energy storage are listed in Table 14.8.

Expression (14.15) can be interpreted as being a given maximum acceptable stress σ_{max} for a given maximum energy storage density (i.e., it is not dependent on rotational speed, and it is maximized for light materials and high stress-based design). For a given geometry, the limiting energy density (energy per unit mass) of a flywheel is proportional to the ratio of material strength to weight density (specific strength).

14.5.2 Applications of Flywheels

Flywheels have a fast actuation when compared with electrochemical energy storage. Therefore, flywheel systems have regained consideration as a means of supporting a critical load during power grid interruptions. Advances in power

Table 14.7 Flywheel shape factor *K*.

Flywheel type	Shape	K
Constant section		1.000
Constant stress disc		0.931
Approx. constant section		0.834
Conic disk		0.806
Flat unpierced disk		0.606
Thin rim		0.500
Rod or circular brush		0.333
Flat pierced disc		0.305

Table 14.8 Flywheel rotor materials for energy storage.

Material	Density (kg/m³)	Design stress (MN/m²)	Specific energy (kWh/kg)
Aluminum alloy	2700		
Birch plywood	700	30	
Composite carbon fiber—40% epoxy	1550	750	0.052
E-glass fiber—40% epoxy	1900	250	0.014
Kevlar fiber—40% epoxy	1400	1000	0.076
Maraging steel (Ni–Fe alloy)	8000	900	0.024
Titanium alloy	4500	650	0.031
"Super paper"	1100		
S-glass fiber/epoxy	1900	350	0.020

electronics and digital controller fields initially provided an alternative to flywheels and have now actually enabled better flywheel designs in order to deliver a cost-effective alternative to the power quality market. In typical systems, a motor inputs electrical energy to the flywheel and a generator is coupled on the same shaft and outputs the energy through an electronic converter. It is also

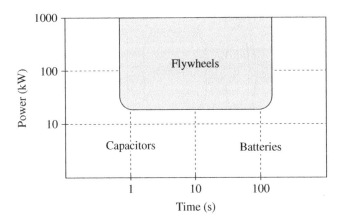

Figure 14.10 Power and time boundaries for flywheels, capacitors, and batteries.

possible to design a bidirectional power electronic system with one machine that is capable of motoring and regenerating operations.

Figure 14.10 shows that flywheels can be used in applications for capacitors and batteries, for power greater than 80 kW in a time span of events happening from 1 to 100 seconds. Of course, flywheels cannot store energy for a long time, but their stored energy can be released in a short time. This makes them suitable for uninterruptible power supply and electric vehicle applications [12, 13].

14.5.3 Design Strategies

The energy storage capability of flywheels can be improved by using the right material and increasing the moment of inertia or spinning it at higher rotational velocities. Some designs use hollow cylinders for the rotor, which allows the mass to be concentrated at the outer radius of the flywheel and improves storage capability with a smaller weight increase in the storage system. Flywheel energy storage systems have capacities from 0.5 to 500 kWh. The flywheel itself loses less than 0.1% of the stored energy per hour if magnetic bearings are used.

Two strategies have been used in the development of flywheels for power applications. One option is to increase the inertia by using a steel mass with a large radius and rotational velocities up to a few thousand revolutions per minute. A standard motor and power electronic drive can be used as the power conversion interface for this type of flywheel. This design results in relatively large and heavy flywheel systems. The effective increase in run time for such systems rarely exceeds 1 second at rated load, which corresponds to a delivery of less than 5% of the additional stored energy in the wheel. Delivery of more energy would result in reduced rotational speed and hence reduced electrical frequency, which is usually unacceptable. Rotational energy losses also limit

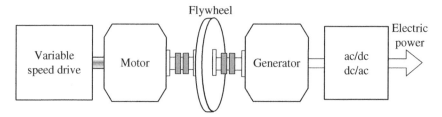

Figure 14.11 Standard flywheel system with motoring and generating sets.

the long-term storage ability of this type of flywheel. To decouple the decrease in rotational speed from the stable electrical frequency needed by the grid, a system with variable speed controlled by a power electronic drive with rectification and inversion electronics is usually implemented, as indicated in Figure 14.11. For such standard systems, the rotational speed is limited to approximately 10,000 rpm. Induction machines operating in field-weakening mode can be employed.

The second design strategy is to produce flywheels with a lightweight rotor that turns at high rotational velocities of up to 100,000 rpm. This approach results in compact, lightweight energy storage devices. Modular designs are possible, with a large number of small flywheels possible as an alternative to a few large flywheels. However, rotational losses because of air drag and bearing losses result in significant self-discharge, which affects long-term energy storage. Recent advances in composite materials technology may allow nearly an order of magnitude advantage in the specific strength of composites when compared with even the best engineering metals. The result of this continuous research in composites has supported the design of flywheels that operate at rotational speeds in excess of 100,000 rpm and with tip speeds in excess of 1000 m/s.

The ultrahigh rotational speeds that are required to store significant kinetic energy in flywheel systems virtually rule out the use of conventional mechanical bearings. High-velocity flywheels have to be operated in vacuum vessels in order to eliminate air resistance, and their magnetic bearings must withstand the difference in pressure of the vessel and the outside environment. There is some R&D on superconducting magnetic bearings for extremely high-speed flywheels and magnetic forces to levitate a rotor and eliminate the frictional losses inherent in rolling element and fluid film bearings. Induction machines are not used for this application because the vacuum chamber will preclude the machine rotor to dissipate power (there is no air for heating exchange). Rare-earth permanent-magnet machines are used. Because the specific strength of these magnets is typically just fractions of that of the composite flywheel, they must spin at much lower tip speeds; in other words, they must be placed very near the hub of the flywheel, which compromises the power density of the generator.

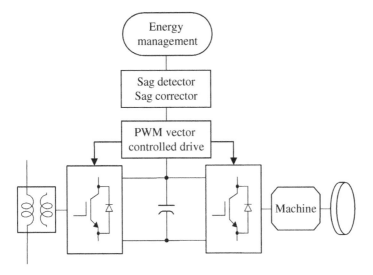

Figure 14.12 Flywheel systems for voltage restoration.

Figure 14.12 shows the use of a high-speed flywheel for a dynamic voltage restorer in which an energy control system keeps the balance in the shaft, while a sag detector injects the correctional voltage in series with the power system.

Advanced flywheel storage systems promise compact size, no emissions, lightweight, and low maintenance. In addition, they offer long life cycles, do not suffer from multiple discharges, and have minimal sensitivity to operating temperature. However, a mass rotating at very high speeds has inherent safety complexities that might be considered in application. When the tensile strength of a flywheel is exceeded, the flywheel will shatter and release all its stored energy at once. It is usual for flywheel systems to be contained in vessels filled with red hot sand as a safety precaution. Fortunately, composite materials tend to disintegrate quickly once broken (rather than producing large chunks of high-velocity fragments of the vessel) [12, 13].

14.6 Superconducting Magnetic Storage System

SMES is the stored energy in the magnetic field created by the flow of direct current in a superconducting coil as illustrated in Figure 14.13 with the persistence resistor with approximately $100\,\Omega$. High-temperature superconductors (high-Tc or HTSs) are materials that have a superconducting transition temperature above $30\,K$ ($-243.2\,°C$). The superconducting coil is currently cryogenically cooled to such temperature below its superconducting critical temperature. The energy content in an electromagnetic field is determined by

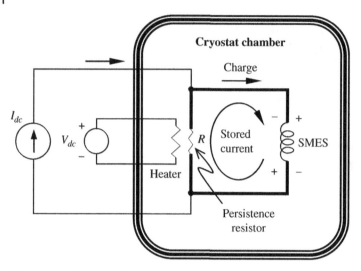

Figure 14.13 Superconducting magnetic storage system.

the current through the N turns of the magnet coil with inductance L. The product NI is called the magnetomotive force. By assuming the electromotive force in the coil as in equation (14.16), the energy is given by equation (14.17). The stored energy is given by integrating the magnetic field strength H over the entire volume in which the induction B is significant. By assuming a linear relationship between H and B, the volume energy density can be obtained from those equations as shown in equation (14.18). The energy density achieved in a magnetic field can be estimated by equations (14.17) and (14.18):

$$e = -N\frac{d\varphi}{dt} \tag{14.16}$$

$$E = \int_o^\varphi Ni(t)\,d\varphi = \int_0^B \ell HAdB = \int_{volume}\int_0^B HdB \tag{14.17}$$

$$E_{volume} = \frac{B^2}{2\mu} \tag{14.18}$$

As an example, typical ferromagnetic materials at $B = 2T$ are approximately $2 \cdot 10^6\,J/m^3$, which is an order of magnitude greater than the electrostatic field but still small compared with electrochemical batteries. An inductor stores energy proportional to the inductance value and to the square of the current, as indicated by equation (14.19). Therefore, it seems plausible that high values of energy density can only be possible by using such media as air or vacuum with very high current values, but electrical resistance of the coil is a limiting factor [14–16]:

$$E(t) = \frac{1}{2} Li(t)^2 \tag{14.19}$$

A superconductor is a material whose resistance is zero when it is cooled to a very low temperature, known as a cryogenic temperature. In an SMES system, an inductor wound with superconducting wire, such as niobium–titanium, is used to create a dc magnetic field [14–16]. The current-carrying capacity of the wire is dependent on temperature and the local magnetic field. A cryogenic system keeps the operating temperature such that the wire becomes a superconductor.

The critical temperature is the point at which the electrical resistance drops dramatically. For all superconductors, this used to be nearly 4 K using liquid helium. After the 1980s, new superconductors made of copper oxide ceramic became available. They only have to be chilled to around 100 K, using liquid nitrogen or special refrigerators. These newer materials are classified as HTSs. Low-temperature superconductors must be cooled to around 269 °C.

Some research has been conducted in the use of the cable-in-conduit conductor concept. In the design, a superconducting cable is placed inside a conduit (or jacket) filled with helium. The conductor is not only the main electrical path but also the helium containment element. This combination of functions allows for more flexibility with the potential for simplification and lower cost.

14.6.1 SMES System Capabilities

SMES systems are typically able to store up to about 10 MW. Higher capacities are possible for shorter time spans. Because a coil of around 150–500 m radius would be able to support a load of 5000 MWh, at 1000 MW it is expected that SMES can potentially store up to 2000 MW.

One problem associated with SMES systems is the requirement of compensation for a stray field. The field falls at the inverse cube of the distance from the center of the coil, and 1 km of radius would still keep a few milliteslas. Because the Earth's field is about one-20th of a millitesla, a guard coil with a magnetic moment opposing the SMES systems must be added around an outer circle to avoid interference on aircraft, power transmission lines, communications, people, and bird navigation. A special requirement for SMES systems is a protection system capable of detecting and avoiding quenches (i.e., loss of superconductivity because of thermodynamic critical coupling between superconductors and cooling tubes), which leads to ohmic heat release that would cause irreversible damage to the superconducting coil.

Figure 14.13 shows the basic idea of controlling the charge and discharge of a superconducting inductor for single- and three-phase systems. The inductor

(a)

(b)

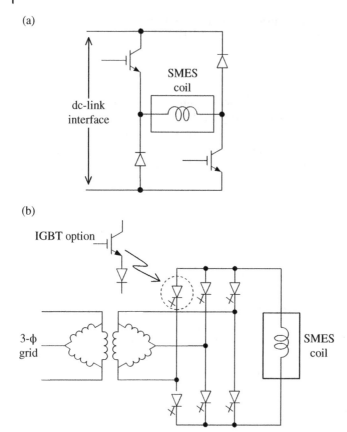

Figure 14.14 Power electronics topology for superconducting inductor: (a) single-phase and (b) three-phase.

is kept with current flowing in the same direction. As energy is stored, a positive voltage is applied across the superconducting coil, and, as energy is released, the voltage is reversed. For average zero voltage, the system is put in persistent-current mode, so that the only loss is the energy required to run the cryogenic refrigeration and activate the power electronic circuitry. The topology presented in Figure 14.14 requires that the power flow be bidirectional with the grid.

14.6.2 Developments in SMES Systems

The ACCEL Instruments GmbH team in Germany has designed a 2-MJ SMES system to ensure the power quality of a laboratory plant of Dortmunder Elektrizitäts und Wasserwerke. Figure 14.15 shows a SMES system connected

Figure 14.15 SMES integrated to a dc-link system.

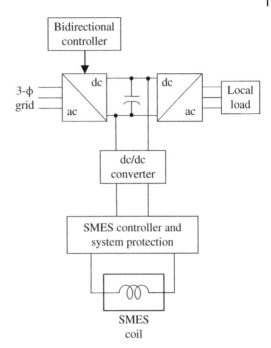

by a dc-link converter to the electrical grid of the plant. The system is designed for a carryover time of 8 s at an average power of 200 kW. The SMES system is wound of NbTi-mixed matrix superconductor cooled with helium, with a two-stage cryocooler. The SMES system has a sophisticated quench protection system, and the interface to the dc link makes the system more flexible than a direct grid connection.

Technical improvements and a better knowledge of how to control cryogenic systems have allowed SMES to penetrate the power system market in the past few years. American Superconductor has a commercial system called D-SMES, which is a shunt-connected flexible ac transmission device designed to increase grid stability, improve power transfer, and increase reliability, as illustrated in Figure 14.16. The D-SMES injects real power as well as dynamic reactive power quickly to compensate for disturbances on the utility grid. This system injects in the plant load a correcting voltage in series with the utility grid for power quality improvement.

When selecting an SMES system as a storage solution, several aspects must be considered. Because of its fast response to power demand and high-efficiency features, it has the capability of providing frequency support (spinning reserve) during loss of generation, transient and dynamic stability by damping transmission-line oscillations, and dynamic voltage support.

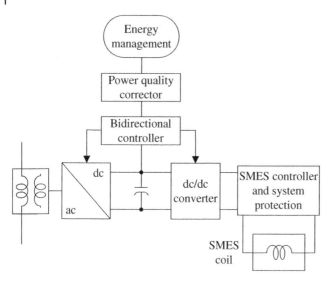

Figure 14.16 SMES integrated to a capacitor bank for voltage restoration.

14.7 Pumped Hydroelectric Storage

Water plays an important role in our lives, and the overall global climate is intertwined with the water cycle. Water is naturally lifted to the atmosphere and released to flow down the land. This can be used to generate power.

Pumped hydroelectric energy storage is the oldest kind of large-scale energy storage. It has been in use since the beginning of the twentieth century. In fact, until 1970, it was the only commercially available option for large energy storage. Pumped hydroelectric stations are in active operation, and new ones are still being built:

$$E_{initial}^{potential} = mgh = m' \cdot \frac{1}{2}v^2 + (m-m')\frac{P}{\rho} = E^{kinetic} + enthalpy \tag{14.20}$$

$$E_{pumping} = \frac{\rho g h V}{\eta_p} \tag{14.21}$$

$$E_{generator} = \rho g h V \eta_g \tag{14.22}$$

As indicated in Figure 14.17, pumped storage projects differ from conventional hydroelectric projects in that they pump water from a lower reservoir to an upper reservoir when demand for electricity is low. The initial potential energy associated with the head is transformed into kinetic energy. One part of

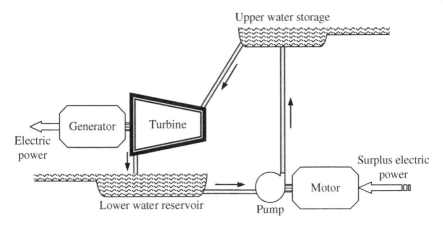

Figure 14.17 Pumped hydroelectric storage system.

this energy is associated with the mass m' moving with velocity v. The other is the pressure part, with the enthalpy given by pressure P over the density of water multiplied by the remaining mass $(m - m')$.

There are frictional losses, turbulence, and viscous drag, and the turbine has an intrinsic efficiency. For the final conversion of hydropower to electricity, the generator efficiency must also be considered. Therefore, the overall efficiency of pumped hydro systems must consider the ratio of the energy supplied to the consumer and the energy consumed while pumping. The energy used for pumping a volume V of water up to a height h with pumping efficiency η_p is given by equation (14.21), and the energy supplied to the grid while generating with efficiency η_g is given by equation (14.22).

During peak demand, water is released from the upper reservoir. It drops downward through high-pressure pipelines. It passes through turbines and ultimately pools in the lower reservoir. The turbines drive power generators that create electricity. Therefore, when releasing energy during peak demand, a pumped hydroelectric storage system works similarly to traditional hydroelectric systems. When production exceeds demand, water is pumped up and stored in the upper reservoir, usually with an early morning surplus.

14.7.1 Storage Capabilities of Pumped Systems

The amount of electricity that can be stored depends on two factors: the net vertical distance through which the water falls, called the head, h; and the flow rate, Q (or the volume of water per second from the reservoir and vice versa). A common false assumption is that head and flow are substitutable, but in fact simply increasing the volume per second cannot compensate for the lack of head. This is because high-head plants can quickly be adjusted to meet the electrical demand surge. For a lower head, the diameter of the pipe would have to be enormous to produce the same power except for the run-of-river plants. This would not be feasible economically, which is why most plants are of the high-head variety.

In the early days, pumped hydroelectric plants used a separate motor and dynamo, mainly because of the low efficiency of dual generators. This increased cost because separate pipes had to be built. A majority of modern plants use a generator that can be run backward as an electric motor. The efficiency of the generator has increased, and it is now possible to retrieve more than 80% of the input electrical energy. This leads to significant cost savings. In addition, modern generators are of suspended vertical construction. This allows better access to the thrust bearing above the rotor for inspection and repair.

Of all large-scale energy storage methods, pumped hydroelectric storage is the most effective. It can store the largest capacity of electricity (more than 2000 MW), and the period of storage is among the longest. A typical plant can store its energy for more than half a year. Because of their rapid response speed, pumped hydroelectric storage systems are particularly useful as backup. An example is the large man-made cavern under the hills of North Wales. Its six huge pump turbines can each deliver 317 MW, producing together up to 1800 MW from the working volume of 6 million cubic meters of water and a head of 600 m. These pump turbines are second in size to the 337-MW devices installed at Tianhuangping in China. The North Whales storage system benefits from more than 50 m extra head compared with the Chinese storage plant.

The value of the hydroelectric storage system is enhanced by the speed of response of the generators. Any of the turbines can be brought to full power in just 10 seconds if they are spinning initially in air. Even starting from a complete standstill takes only 1 minute. Because of the rapid response speed, pumped hydroelectric storage systems are particularly useful as backup in case of sudden changes in demand.

Partly because of their large-scale and relatively simple design, pumped hydroelectric storage systems are among the cheapest by operating costs per unit of energy. The cost of storing energy can be an order of magnitude lower in pumped hydroelectric systems than in, for example, superconducting magnetic storage systems.

Unlike hydroelectric dams, pumped hydroelectric systems have little effect on the landscape. They produce no pollution or waste. However, pumped hydroelectric storage systems also have drawbacks. Probably the most fundamental is their dependence on unique geological formations. There must be two large-volume reservoirs and a sufficient head for building to be feasible. This is uncommon and often forces location in remote places, such as in the mountains, where construction is difficult and the power grid is not present or is too distant.

14.8   Compressed Air Energy Storage

CAES is a mature storage technology for high-power, long-term load-leveling applications. Compressed air is the medium that allows elastic energy to be embedded in the gas. The ideal gas law relates the pressure P, the volume V, and the temperature T of a gas as indicated by

$$PV = nRT \tag{14.23}$$

where:

$n =$ the number of moles and R is the universal gas constant

It is very difficult to calculate the energy density for any unusual shape volume. Therefore, if a piston without friction is assumed under an isobaric process, a simplified volumetric energy density can be calculate by

$$E_{volume} = \frac{1}{V_0} nRT \int_{V_0}^{V} \frac{dV}{V} = \left(\frac{1}{V_0}\right) P_0 V_0 \; ln\left(\frac{V_0}{V}\right) = P_0 \; ln\left(\frac{V_0}{V}\right) \tag{14.24}$$

where:

$P_0 =$ The pressure
V_0 and $V =$ The initial volumes

By assuming a starting volume of $1\,m^3$ and a pressure of $2 \times 10^5\,Pa$, if the gas is compressed to $0.4\,m^3$ at constant temperature, the amount of stored energy is $1.8 \times 10^5\,J$, which is a much higher energy density than that of magnetic or electric fields.

In the past few years, research has been conducted to improve the efficiency of turbines and heat transfer mechanisms used to pump and retrieve compressed air. In CAES, air is compressed and stored under pressure. Release of the pressurized air is subsequently used to generate electricity, most efficiently in conjunction with a gas turbine. Figure 14.18 shows a typical CAES system, in which air is used to drive the compressor of a gas turbine. This makes up 50–60% of the total energy consumed by the gas turbine system.

The most important part of a CAES plant is the storage facility for compressed air. This is usually a man-made rock cavern, salt cavern, or porous rock created by water-bearing aquifers or because of oil and gas extraction. Aquifers, in particular, can be attractive as a storage medium because the compressed air will displace water, setting up a constant-pressure storage system. The pressure in alternative systems varies when adding or releasing air.

The largest CAES plant was built at Huntorf, Germany. This plant rated 290 MW and operated for 10 years with 90% availability and 99% reliability. Despite the decommissioning of this plant, CAES technology was highly promoted, and the Alabama Electric Cooperative built a 110-MW commercial

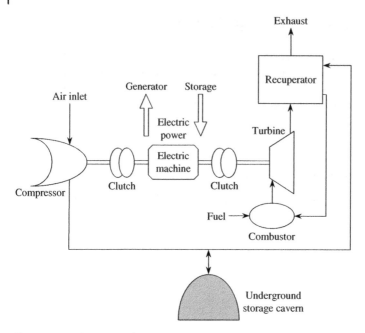

Figure 14.18 Compressed air energy storage system.

project in McIntosh, Alabama. The plant started to operate in May 1991 and has since supplied power during peak demand periods. It is the first in the world to use a fuel-efficient recuperator, which reduces fuel consumption by 25%. Off-peak electricity is used to compress air. The electricity consumed during compression is 0.82 kWh of peak load generation. The plant provides enough electricity to supply the demand of 11,000 homes for 26 hours. The storage is a salt cavern, 1500 ft underground, capable of holding 10 million cubic feet of air. At full charge, air pressure is 1100 pounds per square inch (psi). At full discharge, cavern air pressure is 650 psi. The air flows through the CAES plant generator at a rate of 340 pounds per second, which is as fast as the rate of a wide-body jet engine. Fuel consumption during generation is 4600 Btu higher heating value per kilowatt-hour of electricity. Italy tested a 25-MW installation, and a 1050-MW project has apparently been proposed in the Donbass region on the Russia–Ukraine border.

The annual investment costs for a CAES system are estimated at between $90 and $120/kW (uniform cash flow), depending on air storage. With a 9% discount rate and a 10-year life cycle, these correspond to necessary initial investments between $580 and $770/kW. The development of a large-scale CAES is limited by the availability of suitable sites. Therefore, current research is focused on the development of systems with man-made storage tanks.

Although the high air pressure drives turbines quite often, the air is mixed with natural gas. They are burned together in the same way as in a conventional turbine plant. This method is actually more efficient because the compressed air loses less energy.

Many geological formations can be used in this scheme. In general, rock caverns are about 60% more expensive to mine for CAES purposes than salt caverns. Aquifer storage is the least expensive method and is therefore used in most current locations. The other approach is called compressed air storage (CAS) in vessels. In a CAS system, air is stored in fabricated high-pressure tanks. However, the technology is not advanced enough to allow the manufacture of high-pressure tanks at a feasible cost.

The components of a basic CAES installation include the motor–generator, the air compressor, the turbine train, the air compressor that may require two or more stages, the intercoolers and aftercoolers, the recuperator, the turbine train high- and low-pressure turbines and equipment control, and the compressor and auxiliaries. The motor–generator employs clutches to provide for alternative engagement to the compressor or turbine trains. The intercoolers and aftercoolers serve to achieve economy of compression and reduce moisture content. The high- and low-pressure turbines and equipment control serve for operating the combustion turbine. The compressor and auxiliaries are used to regulate and control changeover from generation mode to storage mode.

Recently, research has been conducted to make small-scale CAS systems commercially available. In such cases, because air pressure varies widely, a mechanism for peak power tracking and identifying the best operating point is usually employed.

14.9 Heat Storage

The amount of energy stored by heating a material from temperature T_0 to T_1 without changing its phase or chemical composition at constant pressure is

$$E = m \int_{T_0}^{T_1} c_P \, dT \qquad (14.25)$$

where:

m = The mass heated
c_P = The specific heat capacity at constant pressure

Materials suitable for heat storage have a large heat capacity, are chemically stable, and are used to add heat to and to subtract heat from. Water, soil, solid metals, and rock beds are often used for heat storage. Other configurations are also common, such as a solar pond, where a man-made salty lake collects and stores solar energy. Being heavier than freshwater, salt water does not rise or mix by natural convection, creating a large temperature gradient within the pond. Freshwater forms a thin insulating surface layer at the top, and underneath it is the salt water, which becomes hotter with depth, close to 90 °C at the bottom. Then heated brine is draw from the bottom and piped into a heat exchanger, where the heat converts a liquid refrigerant into a pressurized vapor that can spin a turbine, eventually generating electricity. An Israeli project covering 62 acres near the Dead Sea produces up to 5 MW of electricity during peak demand periods, and a 6000-m^2 solar pond in Bhuj (India) is capable of delivering 80,000 L of hot water daily at 70 °C for a dairy plant.

The fundamentals of heat transfer calculations of thermally isolated reservoirs are discussed in Section 5.3 and a solar tower power station in Section 5.4. Central receivers or power tower systems use thousands of individual tracking mirrors called heliostats to reflect solar energy onto a receiver located at the top of a tall tower. This energy heats molten salt flowing through the receiver, and the salt's heat energy is then used to generate electricity in a steam generator. The molten salt retains heat, so it can be stored for hours or even days before being used to generate electricity. The liquid salt at 290 °C is pumped from a cold storage tank through the receiver, where it is heated to 570 °C and then to a hot tank for storage. When power is needed, hot salt is pumped to a steam-generating system that produces superheated steam to power a turbine and generator. From the steam generator the salt is returned to the cold tank for reuse.

Fuel cells are also conveniently hybridized with heat storage. Lower temperatures can fulfill space heating, and by capturing unused low-temperature heat energy rejected from the electric production process, fuel energy is used more efficiently. Combining heat and power production reduces the net fuel demands for energy generation by supplying otherwise unused heat for residential, commercial, and industrial consumers that have thermal needs. Sections 7.7 and 7.8 cover the integration of fuel cell heat into electric power systems [17–19].

14.10 Hydrogen Storage

Hydrogen can be stored in various forms. The most typical is to store hydrogen as compressed gas, but it can also be stored as a liquid or as an intermediate compound:

Hydrogen storage in gas form: In this natural state, hydrogen has low density (large volumes), leading to difficulties in its storage process. One solution is to store the hydrogen gas by compressing it in high-pressure cylinders. The hydrogen mass becomes 6–10% volume of the total storage at pressures of 35–70 MPa. Although commercially available, this form of storage is expensive with high-energy consumption, taking about 30% of the hydrogen energy, troubles with mechanical fracture and insecurity, and the need of an appreciable compressor structure [1, 2].

Hydrogen storage in liquid form: It is the more compact storage means. The hydrogen is stored in cryogenic tanks, typically in isolated cylinders, with a 20% hydrogen mass at pressures of 0.1 MPa at −253 °C. The drawbacks of this method are the problems associated with the loss of energy during the H_2 liquefaction and evaporation stages [20, 21].

Hydrogen storage in intermediate compounds: In this method of storage, the hydrogen is adsorbed in solid state of MH. Under these circumstances, the hydrogen takes a potential maximum of about 7% of mass with 90 kg/m^3, at pressures of 0.1–6.0 MPa. The drawbacks of this type of storage are the distinct adsorption and desorption temperatures. The main advantages are the lower working pressure, representing higher security of operation, and small recharging times, besides occupying lower volumes. The formation/dissociation processes in MH storage tanks require a thermal exchanging system. During formation of the hydrides (exothermic reaction), the tank needs cooling, whereas during dissociation (endothermic reaction), the tank requires heat to recover the stored hydrogen [20, 21].

14.11 Energy Storage as an Economic Resource

Any renewable energy project must be assessed on key socioeconomic variables. After deciding on the alternative energy source and corresponding topology, whether grid-connected or stand-alone, the storage system is the next most important technical decision that might affect the overall impact. Issues such as employment, added value, and imports must be considered. For example, in the cases of pumped hydro, CAES, and SMES systems, the neighbor community will be strongly affected, and qualified operating personnel will have to be hired. Biomass will create important jobs in developing countries, decrease pollution in nearby rivers and the atmosphere, and retain a portion of society in rural areas. Effects on the imports of a country are also relevant because normally, building renewable energy plants requires imports from third-world countries.

Energy commodities are used to provide energy services such as lighting, space heating, water heating, motive power, and electronic activity. Energy economics studies human use of energy resources and energy commodities and the consequences of that use. Energy resources such as crude oil, natural gas, coal, biomass, uranium, wind, sunlight, and geothermal deposits can be harvested to produce energy commodities. Figure 14.19 is a block diagram of an intelligent energy management system that functions to optimize generation in accordance with forecasting of the renewable energy sources and

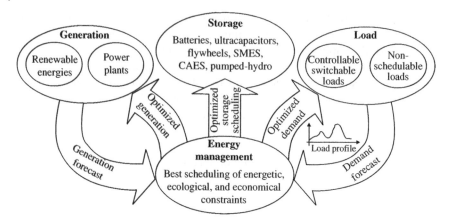

Figure 14.19 Intelligent energy management.

optimizing demand of controllable and switchable loads in accordance with demand forecasts. The best scheduling of energy, ecological, and economic constraints will define the optimized storage scheduling [17–19].

The goal of a module is to optimize energy flows among generation, supply, and points of use to minimize the operating cost or to maximize profit on systems with DG. The output of the module is a set of total energy flows for a given period, issued as a set of vectors that represent recommended energy flows from source to destination during each hour. The optimization module minimizes the cost function for the system. The fixed costs are not considered because they are not influenced by the power dispatch [1, 2, 9, 10, 22]. The goal function is made of:

1) The cost of generation by nonrenewable sources and storages
2) The cost associated with the grid
3) The penalty cost

The function of optimization is to minimize the operating cost of a household with DG while maximizing the energy output. As such, it seeks to apply electricity supplied from the generation system or the grid to the load or the grid in the most cost-effective manner. The system can use available storage to offset expensive energy purchases with cheaper energy from an earlier period or to store energy for an anticipated price spike. To capture the dynamics of real-time pricing properly, the system must optimize over a long-enough span of time to ensure reasonable coverage of the real-time energy costs.

The costs of generation include all costs that depend on the energy generated by the renewable sources and storage. The cost of buying or selling energy to a utility is considered while defining the goal function. Violating boundary conditions such as underutilizing a unit or failing to maintain contracts with consumers or a utility causes penalty costs. The penalty cost term in the cost function greatly helps in convergence. Allowing penalty costs for violation of "soft" boundary conditions helps the optimization algorithms find global minimum more rapidly.

Cost of generation is estimated by

$$C_G = \sum_{k=1}^{N}\left(\sum_{i=1}^{M} G_{i,k} \cdot CG\$_{i,k}\right) \tag{14.26}$$

Cost associated with grid:

$$C_{grid} = \sum_{k=1}^{N}\left(GP_k \cdot CGP\$_k - GS_k \cdot CGS\$_k\right) \tag{14.27}$$

Penalty costs:

$$C_P = \sum_{k=1}^{N}\left(\sum_{i=1}^{M}\left|G_{i,k}^{MAX} - G_{i,k}\right| \cdot CP\$G_{i,k}\right) \tag{14.28}$$

Total costs:

$$C = C_G + C_{grid} + C_P \tag{14.29}$$

where:

C_G = Total cost of generation by renewable sources
$G_{i,k}$ = Amount of energy generated by source i during period k
$CG\$_{i,k}$ = Unit cost of energy in $ generated by source i during period k
C_{grid} = Total cost associated with grid
GP_k = Amount of energy purchased from grid during period k
$CGP\$_k$ = Unit cost of energy in $ purchased from grid during period k
GS_k = Amount of energy sold to grid during period k
$CGS\$_k$ = Unit cost of energy in $ sold to grid during period k
C_P = Total penalty costs
$G_{i,k}^{max}$ = Maximum amount of energy
$CP\$G_{i,k}$ = Unit cost of the unused capacity of the source i during period k
M = Total number of renewable sources
N = Total number of periods for which optimization is performed.

This minimization process is constrained by the functional and physical limits of the problem. First, enough energy must be supplied from generation of the grid to meet the home load. The storage capacity is constrained by a maximum storage size. The supply of energy from generation is limited to what is available at the time, especially in the case of renewable sources. The transfer of energy to and from the grid is limited by the rating of the electrical service to the household. These constraints can be expressed by the following equations:

$$S_{k+1} = S_k + \sum_{i=1}^{M} G_{i,k} + GP_k - GS_k - L_k \tag{14.30}$$

$$S_N = S_0 \tag{14.31}$$

$$0 \le L_k \le L^{MAX} \tag{14.32}$$

$$0 \le S_k \le S^{MAX} \tag{14.33}$$

$$0 \le G_{i,k} \le G_{i,k}^{MAX} \tag{14.34}$$

where:

S_k, S_{k+1} = Storage levels in periods k and $k+1$
L_k = Load connected to the microgrid during period k
L^{MAX} = Maximum possible load
S^{MAX} = Maximum amount of storage

The structure of the constraint equations is such that all energy is routed through storage on its way to a destination. The aim is to include the energy balance, but there are many different ways that could be expressed. Given the complexity of the problem, however, this expression is more intuitively obvious than other methods. Either way, the solution provided by equations (14.25) through (14.34) would be adequate for the situation. It should also be noted that equation (14.31) sets a constraint that storage at the end of N periods should be equal to the starting storage state. This constraint is necessary to provide a bound to the entire sequence of storage states. Where the other constraints provide limits on magnitude of energy in any given period, the constraint in equation (14.30) provides a limit to the entire problem over N periods. The constraint applies only to storage because the energy balance equation is expressed in terms of the state of storage. The cost (objective) function and constraint equations provide the basis for finding an optimal solution. The constraint equations confine the operating region of the system to an N-dimensional hyperspace. Within the hyperspace, every point has an associated cost in accordance with equation (14.29). The goal of the optimization program is to find the single point of operation within that hyperspace that provides the lowest operating cost.

The goal function, *C*, which is to be minimized, emerges from the addition of these three components, and it is highly nonlinear, discontinuous, and large in dimensions. The inclusion of storage devices makes the goal function time dependent.

The on–off solution, as provided by the mixed integer linear programming, is a popular choice in such an optimization process in a grid-connected network, where it is required to either operate the power plants at the maximum power or switch them off. This cannot be applied to the microgrid where sometimes the sources, and usually the storage system, are to be operated in partial load [8, 23, 24].

Energy supply is an intricate task that depends on import dependencies, seasonal and random differences in energy supply and use, and daily fluctuations in consumption. A very sophisticated management of energy resources and conversion, or energy distribution and resource intermittence, is required. As a result, storage plays a critical role, and energy storage will balance the daily fluctuations and seasonal differences of energy resource availability resulting from physical, economical, or geopolitical constraints.

References

1 J. Jensen and B. Sorensen, Fundamentals of Energy Storage, John Wiley & Sons, Inc., New York, 1984.

2 S.M. Schoenung and W.V. Hassenzahl, Long vs. Short-Term Energy Storage Technologies Analysis: A Life-Cycle Cost Study for the DOE Energy Storage Systems Program, Report SAND2003-2783, Sandia National Laboratories, Albuquerque, NM, and Advanced Energy Analysis, Piedmont, CA, 2003.

3 Z.M. Salameh, M.A. Casacca, and W.A. Lynch, A mathematical model for lead-acid batteries, IEEE Transactions on Energy Conversion, Vol. 7, pp. 93–98, March 1992.

4 R.A. Jackey, A Simple, Effective Lead-Acid Battery Modeling Process for Electrical System Component Selection, The MathWorks Inc., Novi, MI, 2007.

5 N. Moubayed, J. Kouta, A. EI-AIi, H. Dernayka, and R. Outbib, Parameter identification of the lead-acid battery model, Government College of Technology, 2008.

6 M.S. Illindala and G. Venkataramanan, Battery energy storage for stand-alone microsource distributed generation systems, presented at the 6th IASTED International Conference on Power and Energy Systems, Marina Del Rey, CA, May 13–15, 2002.

7 S. Barsali and M. Ceraolo, Dynamical models of lead-acid batteries: implementation issues, IEEE Transactions on Energy Conversion, Vol. 15, No. 4, pp 16–23, 2002.

8 Y. Kim and N. Chang, Maximum power transfer tracking for a photovoltaic-supercapacitor energy system, International Symposium on Low Power Electronics and Design, ISLPED'10, Austin, TX, August 18–20, 2010.

9 M. Okamura and H. Nakamura, Energy capacitor system, parts 1 and 2: capacitors and their control, presented at the 11th International Seminar on Double Layer Capacitors and Similar Energy Storage Devices, December 3–5, 2001.

10 A. Rufer and P. Barrade, Key developments for supercapacitive energy storage: power electronic converters, systems and control, Proceedings of the International Conference on Power Electronics, Intelligent Motion and Power Quality PCIM 2001, Nurenberg, Germany, June 19–21, 2001, pp. 17–24.

11 P. Barrade, S. Pitte, and A. Rufer, Energy storage system using a series connection of supercapacitors with an active device for equalizing the voltages, presented at the International Power Electronics Conference, IPEC 2000, Tokyo, Japan, April 3–7, 2000.

12 M.E. Amiryar and K.R. Pullen, A Review of Flywheel Energy Storage System Technologies and Their Applications, Academic Editor: F. Blaabjerg, Applied Sciences, Basel, Switzerland, Vol 7, p. 286–292, March 2017.

13 B.B. Plater and J.A. Andrews, Advances in flywheel energy-storage systems, http://www.powerpulse.net/powerpulse/archive/aa_031901c1.stm, accessed March 7, 2017.

14 P.D. Baumaan, Energy conservation and environmental benefits that may be realized from superconducting magnetic energy storage, IEEE Transactions on Energy Conversion, Vol. 7, pp. 253–259, 1992.

15 V. Karasik, K. Dixon, C. Weber, B. Batchelder, G. Campbell, and P.F. Ribeiro, SMES for power utility applications: a review of technical and cost considerations, IEEE Transactions on Applied Superconductivity, Vol. 9, pp. 541–546, 1999.

16 A.B. Arsoy, W.Z. Wang, L.Y. Liu, and P.F. Ribeiro, Transient modeling and simulation of a SMES coil and the power electronics interface, IEEE Transactions on Applied Superconductivity, Vol. 9, No. 4, pp. 4715–4724, 1999.

17 D. Gallo, C. Landi, M. Luiso, and R. Morello, Optimization of experimental model parameter identification for energy storage systems, Energies, Vol. 6, pp. 4572–4590, 2013, DOI: 10.3390/en6094572.

18 G. Winkler, C. Meisenback, M. Hable, and P. Meier, Intelligent energy management of electrical power systems with distributed feeding on the basis of forecasts of demand and generation, presented at CIRED 2001, June 18–21, 2001.

19 O. Tremblay and L.A. Dessaint, Experimental validation of a battery dynamic model for EV applications, World Electric Vehicle Journal, Vol. 3, 2009.

20 F. Gonzatti, V.N. Kuhn, F.Z. Ferrigolo, M. Maicon, and F.A. Farret, Theoretical and practical analysis of the fuel cell integration of an energy storage plant using hydrogen, 11th IEEE/IAS International Conference on Industry Applications INDUSCON 2014, Juiz de Fora, MG, Brazil, 2014.

21 F. Gonzatti, F.Z. Ferrigolo, V.N. Kuhn, D. Franchi, and F.A. Farret, Automation of thermal exchanges of metal hydrides cylinders integrated with energy storage, 11th IEEE/IAS International Conference on Industry Applications INDUSCON 2014, Juiz de Fora, MG, Brazil, 2014.

22 G. Venkataramanan, M.S. Illindala, C. Houle, and R.H. Lasseter, Hardware Development of a Laboratory-Scale Microgrid Phase 1 Single Inverter in Island Mode Operation, NREL Report SR-560-32527, National Renewable Energy Laboratory, Golden CO, November 2002.

23 R.A. Huggins, Energy Storage, Springer, New York, 2010.

24 Y. Brunet (ed.), Energy Storage, John Wiley & Sons, Inc., Hoboken, NJ, 2011.

15

Integration of Alternative Sources of Energy

15.1 Introduction

Interconnection of alternative sources of energy into a network for DG requires small-scale power generation technologies located close to the loads served. The move toward on-site distributed power generation has been accelerated because of deregulation and restructuring of the utility industry and the feasibility of alternative energy sources. DG technologies can improve power quality, boost system reliability, reduce energy costs, and defray utility capital investment. Four major issues related to DG are covered in this and the next three chapters: hardware and control, effect on the grid, interconnection standards, and economic evaluation.

Alternative sources of energy are either fossil fuel based (systems that use diesel, gas, or hydrogen from reforming hydrocarbons) or renewable based (systems that use solar, wind, hydro, tidal, or geothermal energy or hydrogen produced from renewable sources). For an energy source to be considered renewable, the recovery cycle of the source must be considered. For example, a sugarcane plantation used for the production of alcohol has a recovery cycle of one year. In comparison, fossil fuels can take thousands of years to recover.

The integration of renewable sources of energy poses a challenge because their output is intermittent and variable and must be stored for use when there is demand. If only one renewable energy source is considered, the electric power system is simple. The source can be connected to a storage system to deliver electricity for stand-alone use or interconnected with the grid, as discussed in Chapter 11. In the grid-connected application, the grid acts as energy storage. However, if multiple renewable energy sources are used, the electric power system can be rather complex and a microgrid will be formed.

The concept of a microgrid was developed recently. It is defined as a cluster of loads and DG units that operate to improve the reliability and quality of the power system in a controlled manner. For customers, microgrids can provide power and heat in a reliable way. For the distribution network, microgrids

Integration of Renewable Sources of Energy, Second Edition. Felix A. Farret and M. Godoy Simões.
© 2018 John Wiley & Sons, Inc. Published 2018 by John Wiley & Sons, Inc.

are dispatchable cells that can respond quickly autonomously or from signals from the power system operator. Information technology achievements, along with new DG systems with intelligent controllers, will allow system operators and microgrid operators to interact in an optimal manner [1, 2].

In this chapter, we first present the concept of integrating one renewable source of energy into the grid and provide a fundamental understanding of interactive inverters. The integration of multiple sources into a microgrid is then covered, and three electrical buses for integration (direct current (dc), alternating current (ac), and high-frequency ac (HFAC)) are discussed.

Power electronics is the enabling technology that allows the conversion of energy and injection of power from renewable energy sources to the grid. Therefore, it is suggested that the reader review some background information on power electronics from the technical literature to complement this material.

15.2 Principles of Power Interconnection

When applying electrical energy conversion technology to renewable energy systems, two classes of electrical systems must be considered: stationary and rotatory. The stationary type usually provides dc. Photovoltaic (PV) arrays and fuel cells are the main renewable energy sources in this group. The rotatory type usually provides ac. Induction, synchronous, and permanent magnet generators are the main drivers for hydropower, wind, and gas turbine energy sources. Dc machines are the rotatory type but are not usually employed because of their high cost, bulky size, and maintenance needs. All these systems are asynchronous with respect to grid frequency and need a phase lock loop (PLL) type of control to match these frequencies [3, 4].

15.2.1 Converting Technologies

Figure 15.1 illustrates the main electrical conversion technologies required for injection of renewable energy power. If only PV and fuel cell systems are used, a dc-link bus might be used to aggregate them. Ac power can be integrated through dc–ac conversion (inverter) systems. If only hydro or wind power is used, a variable-frequency ac voltage control must be aggregated into an ac link through an ac–ac conversion system. Ac–ac conversion systems can be put together by several approaches discussed later in the chapter.

Of course, alternative energy sources such as diesel and gas can also be integrated with renewable sources. They have a consistent and constant fuel supply, and the decision to schedule their operation is based more on straightforward economic reasons than any other. Diesel generators are commercially available with synchronous generators that supply 60 Hz, and direct interconnection

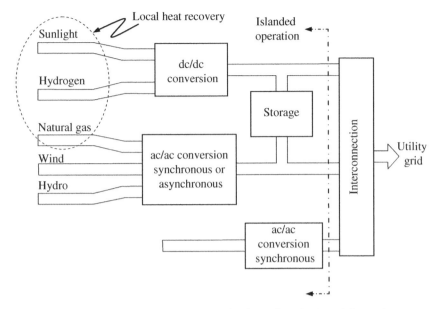

Figure 15.1 Alternative energy conversion technologies for injection of alternative energy power to the grid.

with the grid is typically easier to implement. Gas microturbines are most frequently implemented with a power electronic inverter-based technology. When integrating and mixing all those sources, a microgrid has to be based on a dc or ac link. However, it seems promising to use an HFAC link, for reasons discussed later in this chapter.

Figure 15.1 illustrates the use of some alternative energy conversion technologies for injection of alternative energy power to the grid, but it does not include how energy storage is integrated. As discussed in Chapter 11, the ultimate design of a microgrid must incorporate energy storage with seamless control and integration of source, storage, and demand and allow for system expansion or interconnection.

The stationary group of renewable energy sources does not absorb power; the energy flows uniquely outward to the load. On the other hand, the rotatory group of renewable energy sources require bidirectional power flow, either to rotate in motoring mode (for pumping, braking, or start-up) or to absorb reactive power (for induction generators and transformers). The storage systems can also be stationary (e.g., batteries, super capacitors, and magnetic) or rotatory (generators, flywheels, and pumping hydro). Therefore, a generalized concept of power electronics is broadly needed for renewable energy systems to embrace such categories, as portrayed in Figure 15.2.

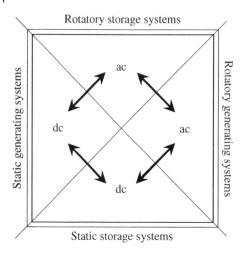

Figure 15.2 Ac and dc conversion needs for renewable energy systems.

Static systems appear as a dc input converted to a dc or ac output and unidirectional power flow (no moving parts, low time constant, silent, no vibrations, minimally polluting, and low power up to now). The rotating systems appear as an ac input converted to a dc or ac output and bidirectional power flow (angular speed, noise, mechanical inertia, weight and volume, and inrush). Storage systems are necessarily bidirectional systems with requirements of ac and dc conversion (depending on the application). The voltage generated by variable-speed wind power generators, PV generators, and fuel cells cannot be directly connected with the grid. Therefore, power electronics technology plays a vital role in matching the characteristics of the dispersed generation units and the requirements of grid connection, including frequency, voltage, active and reactive power controls, and harmonic minimization [5–8].

15.2.2 Power Converters for Power Injection into the Grid

A power electronic converter uses a matrix of power semiconductor switches to enable it to convert sources of stationary electric power into rotating electrical power, and vice versa, at high efficiency. The standard definition of a converter assumes one input source connected to an output source or a load, as indicated in Figure 15.3.

The inherent regenerative capabilities of renewable energy systems require fewer restrictions for input–output or supply–load when connecting two sources. The links between a power converter and the outside world can be perceived as a possibility of power reversal, where the sink can be considered as a negative source, in accordance with Figure 15.3. Therefore, it is a better fit to define voltage source (VS) and current source (CS) when describing the immediate connection of converters to entry and exit ports.

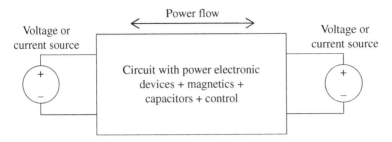

Figure 15.3 Connection of two dc sources.

The VS is driven to maintain a prescribed voltage across its terminals, irrespective of the magnitude or polarity of the current flowing through the source. The prescribed voltage may be constant dc, sinusoidal ac, or a pulse train. The CS maintains a prescribed current flowing through its terminals, irrespective of the magnitude or polarity of the voltage applied across these terminals. Again, the prescribed current may be constant dc, sinusoidal ac, or a pulse train.

Current source inverters (CSIs) have several advantages over variable-voltage inverters because they can operate under a wide input voltage range and a boost converter stage may not be required. They provide protection against short circuits in the output stage and can easily handle temporary overrating situations such as wind gusts or faulty turbines. In addition, CSIs have relatively simple control circuits and good efficiency. As disadvantages, CSIs produce torque pulsations at low speed, cannot handle undersized motors, and are large and heavy. The phase-controlled bridge rectifier CSI is less noisy than its chopper-controlled counterpart, losses are smaller, and it does not need high-speed switching devices, but it cannot operate efficiently from direct dc voltage. The chopper-controlled CSI can operate from batteries and produces more noise because of its need for high-speed switching devices. With the advent of fast, high-power controlled devices, voltage source inverters (VSI) have become popular, but much R&D is needed before they are applied to very high-power systems.

Figure 15.3 shows that a power electronic circuit with magnetic devices and capacitors can control the power flow between two VS or CS. Usually, those power-converting circuits are based on the two conventional topologies: VS and CS. A VS converter is typically a dc–dc converter—such as buck, boost, and SEPIC [6]—or a three-phase inverter [7]. A CS converter is typically used for very-high-power GTO-based applications with an inductor imposing a constant current in a dc link [8]. For most small- and medium-power applications, VSI are used, with some inherent constraints. For example, a VSI might be a buck converter that cannot produce an output voltage greater than the source voltage. This limitation often requires an additional dc–dc boost

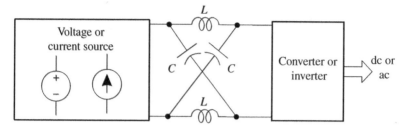

Figure 15.4 Z-source converter system.

converter for stationary applications such as batteries, PV, and fuel cells, which makes the power conversion have lower efficiency, larger size, and higher cost.

A new type of power conversion circuit, known as an impedance-source (Z-source) power converter, is designed to overcome the characteristics associated with the conventional V-source and I-source converters [9, 10]. A Z-source inverter can produce any desired voltage regardless of its input voltage, which greatly reduces system complexity, cost, size, and power loss.

Figure 15.4 shows the concept of a Z-source converter. In it, there is a unique LC network in the dc link, making it possible for buck and boost operation. Even though such topology has gained attention in the past few years (with some interesting features such as application to almost all the conversion circuits: dc–ac inversion, ac–ac conversion, and dc–dc conversion), in this chapter we concentrate on the analysis of dc-link VS conversion because of a vast and solid application to the field of renewable energy-based conversion systems. In addition, a Z-source converter has more passive-type components in the link, and an economical breakthrough analysis must be carried out for high-power applications.

Figure 15.5 shows a renewable energy conversion that consists of a standard VSI. L_s represents the inductance between the inverter ac voltage V_{RI} and the ac system voltage V_s, and I_s is the current injected into the grid. The output waveform of a VSI can be decomposed into a fundamental component plus harmonics, but for the analysis of power transfer, the resistance and harmonics can be neglected.

15.2.3 Power Flow

Decoupling active and reactive power flow is a common practice in load flow studies. The variation of active power is strongly coupled with power angle and frequency variation, whereas reactive power is strongly coupled with voltage amplitude. The single-phase equivalent circuit of Figure 15.6 can evaluate the fundamentals of power flow control. Considering for simplicity that $V_{s1} = V_{r1}$, one gets the phasor diagram of Figure 15.7a and b, and when the output voltage V_{C1} leads by 90° with respect to V_{s1}, current I_1 is in phase with V_{s1}, which causes

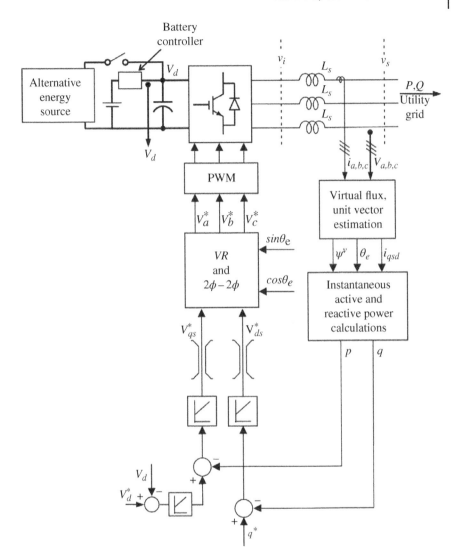

Figure 15.5 Renewable energy system injecting power into the grid.

active power flow from the source to the load. When V_{C1} is in phase with V_{s1}, current I_1 will lag by 90° with respect to V_{s1}, which results in reactive power flow. The fundamental voltage and current components can be represented by the phasor diagrams in Figure 15.8a and b, which correspond, respectively, to the fundamental voltage control mode and current control mode.

The following analysis considers the VSI connected with the grid, as indicated by Figure 15.5. In this figure, v_s is the grid voltage. The variables v_s and i_s

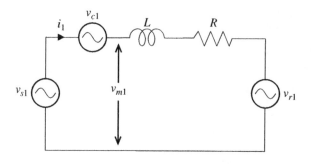

Figure 15.6 Equivalent circuit of power flow controller (From Ref [11].) © MyScienceWork - 2016.

(a)

(b)

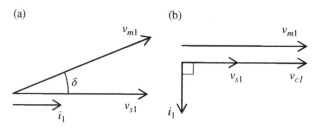

Figure 15.7 Simplified phasor diagrams for power flow: (a) active power control and (b) reactive power control.

are, respectively, the voltage and current impressed by the inverter, δ is the angle between voltage phasors, and ϕ is the angle between v_s and i_s (i.e., $cos(\phi)$ is the power factor seen by the utility). For the sake of simplicity, all voltages and currents are considered sinusoidal, and the discussion applies only to the fundamental of any of the distorted variables. If a more detailed analysis is required, the harmonic distortion must be included in the analysis. Such a steady-state phasor-based analysis is not possible.

The phasor diagram of Figure 15.8a is similar to that of typical synchronous generators. The real and active power injected into the grid by a VSI can be expressed in per-unit values as

$$P_s = P_i = \frac{V_s V_i}{X_s} sin(\delta) \tag{15.1}$$

$$Q_s = \frac{V_s V_l}{X_s} cos(\delta) - \frac{V_s^2}{X_s} \tag{15.2}$$

where:

$$X_s = 2\pi f_s L_s$$

$f_s =$ The grid frequency

(a)   (b)

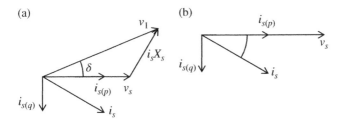

Figure 15.8 Phasor diagrams of grid-connected voltage source inverter: (a) voltage-controlled VSI and (b) current-controlled VSI.

Equation (15.1) shows that for a given grid voltage V_s and inductance L_s, the real power is P_s. Equation (15.2) shows that reactive power Q_s can be regulated by the magnitude of V_i and power angle δ. In accordance with equation (15.1), the reference for real power P^* and reactive power Q^* can be obtained by setting the following reference power angle δ^* and inverter ac terminal voltage V_i^*:

$$\delta^* = tan^{-1}\left(\frac{P^*}{Q^* + \dfrac{V_s^2}{X_s}}\right) \tag{15.3}$$

$$V_i^* = \frac{P^* X_s}{V_s sin\left(\delta^*\right)} \tag{15.4}$$

A VSI can also operate in closed-loop current-controlled mode [12]. The power control of a current-controlled VSI can be explained with the phasor diagram in Figure 15.8b. The power injected into the grid can be expressed in per-unit values as

$$P_s = V_s I_s cos\left(\varphi\right) \tag{15.5}$$

$$Q_s = V_s I_s sin\left(\varphi\right) \tag{15.6}$$

The reference real and active power can be controlled by regulating the current magnitude and the angle:

$$\varphi^* = tan^{-1}\left(\frac{P^*}{Q^*}\right) \tag{15.7}$$

$$I_s^* = \frac{P^*}{V_s cos\left(\varphi^*\right)} \tag{15.8}$$

15.3   Instantaneous Active and Reactive Power Control Approach

If the exact amount of power generated by the renewable energy source is transferred to the grid, as illustrated in Figure 15.9, the power absorbed by the load and the power injected into the grid are generated by the same source. However, if the ac load is absorbing active power P_L, and the reactive power Q_L is not supplied by the VSI, the power factor seen by the grid may fall out of the prescribed limits allowed by the utility and possibly force over-charges and penalties to the customer. To solve this problem, the VSI should supply active power and at the same time compensate reactive power to restrain the power factor within the limitations. To have a continuous con-trolled active and reactive power as demanded by a variable load and source conditions, the theory developed by Akagi [13] and applied to renewable energy power injection by Watanabe [14, 15] helps the design of a power electronic control system.

For a three-phase balanced system, equation (15.9) states that p and q (i.e., the instantaneous real power (W) and instantaneous imaginary power (VA)) are given by the average (i.e., overbar) and oscillatory (i.e., overcaret) components. The oscillatory components are present because of harmonics in the system. The average $\bar{p}$ and $\bar{q}$ are scalar values with same magnitudes as of the phasor quantities P and Q. The voltages and currents in equation (15.9) are obtained

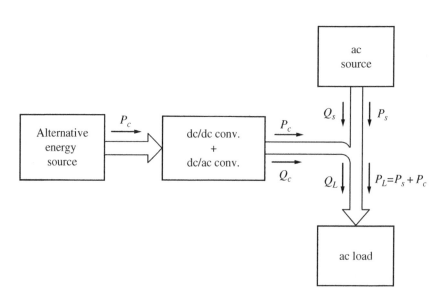

Figure 15.9 Active and reactive power flows from the energy system into the grid.

from the power invariance of the Clarke transformation, whose terms are indicated by equations (15.10) and (15.11):

$$
\begin{bmatrix} p \\ q \end{bmatrix} = \begin{bmatrix} \bar{p} \\ \bar{q} \end{bmatrix} + \begin{bmatrix} \tilde{p} \\ \tilde{q} \end{bmatrix} = \begin{bmatrix} v_\alpha & v_\beta \\ -v_\beta & v_\alpha \end{bmatrix} \cdot \begin{bmatrix} i_\alpha \\ i_\beta \end{bmatrix}
\tag{15.9}
$$

$$
\begin{bmatrix} v_0 \\ v_\alpha \\ v_\beta \end{bmatrix} = \sqrt{\frac{2}{3}} \begin{bmatrix} \dfrac{1}{2} & 1 & 0 \\ \dfrac{1}{2} & -\dfrac{1}{2} & \dfrac{\sqrt{3}}{2} \\ \dfrac{1}{2} & -\dfrac{1}{2} & -\dfrac{\sqrt{3}}{2} \end{bmatrix} \cdot \begin{bmatrix} v_a \\ v_b \\ v_c \end{bmatrix}
\tag{15.10}
$$

$$
\begin{bmatrix} v_a \\ v_b \\ v_c \end{bmatrix} = \sqrt{\frac{2}{3}} \begin{bmatrix} \dfrac{1}{2} & \dfrac{1}{2} & \dfrac{1}{2} \\ 1 & -\dfrac{1}{2} & -\dfrac{1}{2} \\ 0 & \dfrac{\sqrt{3}}{2} & -\dfrac{\sqrt{3}}{2} \end{bmatrix} \cdot \begin{bmatrix} v_0 \\ v_\alpha \\ v_\beta \end{bmatrix}
\tag{15.11}
$$

Equation (15.9) gives the definition of instantaneous real and imaginary power assuming measured voltages and currents. Therefore, to control the VSI connected to the renewable energy source, the current references in the $\alpha - \beta$ frame must be imposed as defined by equation (15.12) for given set-point values of p_c and q_c. Inversion of the voltage matrix in equation (15.9) yields

$$
\begin{bmatrix} i_\alpha^* \\ i_\beta^* \end{bmatrix} = \frac{1}{\left(v_m\right)^2} \cdot \begin{bmatrix} v_\alpha & -v_\beta \\ v_\beta & v_\alpha \end{bmatrix} \cdot \begin{bmatrix} p_c \\ q_c \end{bmatrix}
\tag{15.12}
$$

where:

$$
\left(v_m\right)^2 = \left(v_\alpha\right)^2 + \left(v_\beta\right)^2
$$

The pulse width modulation scheme of the VSI can produce i_α^* and i_β^* and then generate the switching pulses for the transistor by using a hysteresis current controller [14, 15] and an extra synchronous reference frame $d - q$ transformation to impress the voltage command with sinusoidal pulse width modulation through proportional–integral (PI) controllers or space vector modulation [16]. Details of such implementations are outside the scope of this book. The reader can learn more from the available technical literature on power electronics and drives.

The renewable energy source should be injected into the ac system by the pulse width modulation VSI through appropriate control of the reference currents. Neglecting the harmonics $p_c = p^*$, the real power set point p^* must be impressed by feedforward path (very hard for variable renewable energy source input) or a controller tied to the dc-link capacitor, which is responsible for maintaining the power balance. Of course, a hybrid system with storage would be able to compensate for any power mismatch. This scheme considers the ac grid to be the storage under power import/export net metering capabilities. The energy stored E_c in the dc-link capacitor C is given by

$$E_c = \int_{-\infty}^{t} \left(v_d i_s - p_c \right) d\lambda = \frac{1}{2} C v_d^2 \tag{15.13}$$

where:

v_d = The instantaneous DC capacitor voltage (V)
i_s = The source current supplied by the renewable energy source
p_c = The instantaneous power at the pulse width modulation VSI terminals (W)
λ = A dummy variable for time integration

Instantaneous active power control can be generated by a feedback loop, in which the capacitor voltage is used to indicate power mismatch. The small signal transfer function dv/dp indicated in equation (15.14) considers, for a constant input current, how the dc-link voltage deviation responds for small variations of the real power flowing through the dc bus:

$$\frac{\Delta \tilde{v}_d(s)}{\Delta \tilde{p}_c(s)} = \frac{1}{s C V_d} \tag{15.14}$$

where:

V_d = The quiescent voltage at the dc-link capacitor

The stationary renewable energy source is typically associated with a dc–dc converter. For example, PV arrays are connected through a maximum peak power tracking converter, and fuel cells are connected through a boost converter with optimized balance of plant control. Therefore, it is possible to assume that the i_s input source current can be modeled for derivation of the control law as a CS. Therefore, dc-link voltage rising from a reference value indicates a surplus of generated input power, and decreasing voltage indicates a lack of input power. The PI controller depicted in Figure 15.10 shows that the measured dc-link voltage v_d is compared with the set point v_d and the error is fed back through a PI controller that automatically produces the tracking real power $p^* = p_c$. By properly fine-tuning the PI gains, the intrinsic low-pass filter behavior will eliminate the switching harmonics, and the dc-link capacitor voltage will have a time response compatible with the power balance.

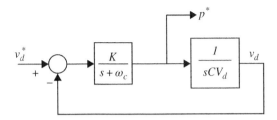

Figure 15.10 Real power tracking by dc-link voltage control.

The VSI must also program the amount of reactive power by impressing a q set point in accordance with

$$q^* = q_{load} - \left(p_{load} - p^*\right) \cdot tan\left(\varphi^*\right) \tag{15.15}$$

in which $cos\,(\phi^*)$ is a prescribed power factor as seen by the utility grid. The pulse width modulation VSI must be rated in accordance with the apparent power required by the total power flow. Equation (15.16) can be used to specify the capacity of such converters:

$$S_{converter} = \sqrt{\left[q_{load} - \left(p_{load} - p^*\right)tan\left(\varphi^*\right)\right]^2 + \left(p^*\right)^2} \tag{15.16}$$

15.4 Integration of Multiple Renewable Energy Sources

The integration of multiple renewable energy sources can be considered a facet of DG. As introduced in Chapter 1, DG systems consist of small generators, typically 1 kW to 10 MW, scattered throughout the system to provide electrical and, sometimes, heat energy close to consumers. When interconnected with distribution systems, these small modular generation technologies can form a new type of power system, the microgrid. Microgrids are broadly illustrated in Figure 15.11.

The microgrid concept assumes a cluster of loads and micro sources operating as a single controllable system that can provide power and heat to the local area. The electrical connection of sources and loads can be done through a dc link, an ac link, or an HFAC link. Converters are usually connected in parallel, although series arrangements of sources and loads are possible to allow better use of high voltages and currents.

An example of series connection is a group of small run-of-river hydroelectric systems that drive induction generators connected. All configurations require a controlled voltage at the load link bus that is capable of supplying power to all the loads with reasonable power quality. On the other hand, a

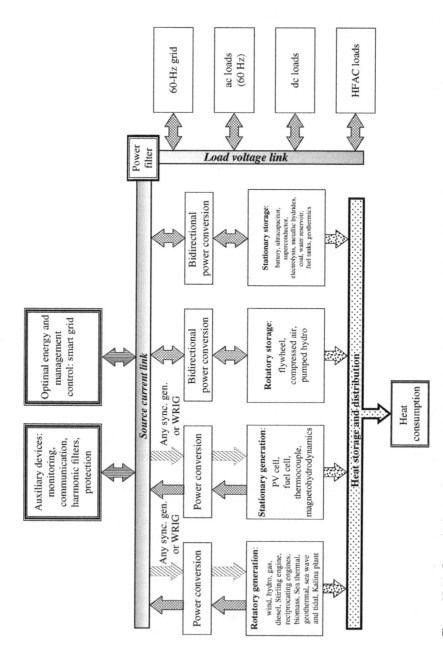

Figure 15.11 General schemes for integration of electrical sources, loads, and thermal dissipation.

controlled CS is more convenient to adjust the voltage of every electrical source to the common voltage of the source bus, according to the strategy of power control. In the configurations depicted in Figure 15.12, thyristors or GTOs are still recommended for their robustness, guaranteed reliability, handling of higher power levels, low cost, easy control, and high efficiency. The major problem with these components is that they cannot operate easily with frequencies much higher than 1 kHz. Furthermore, GTOs demand a high control current to interrupt their load currents. Even so, for very high-power applications, they are absolute in the industry, and they will be preferred in the following discussions.

15.4.1 DC-Link Integration

The simplest and oldest type of electrical energy integration is through a dc link (see Figure 12.12a). An example is the straightforward connection of a DC source to a battery and load scheme. Most of the first cars used this type of integration. With the advent of power diode rectifiers and controlled rectifiers, dc-link integration has widened its capabilities with ac–dc connections, controlled links, voltage matching levels, and dc transmission and distribution. The advantages of dc-link integration include the following:

1) Synchronism is not required.
2) There are lower distribution and transmission losses than with ac link.
3) It has high reliability because of parallel sources.
4) Although the terminal needs of a dc link are more complex, the dc transmission infrastructure per kilometer is simpler and cheaper than in ac links.
5) Long-distance transmission (for high-voltage links) is possible, which enables integration of offshore wind turbines and other energies with inland networks.
6) The converters required are easily available.
7) Single-wired connections allow balanced terminal ac systems.

The disadvantages are:

1) The need for careful compatibility of voltage levels to avoid current recirculation between the input sources
2) The need for robust forced commutation capabilities in circuits at high power levels
3) The possibility of current recirculation between sources
4) Corrosion concerns with the electrodes
5) A large number of components and controls
6) More complex galvanic isolation
7) Higher costs of terminal equipment
8) Difficulties with multiterminal or multi-voltage-level operation for transmission and distribution

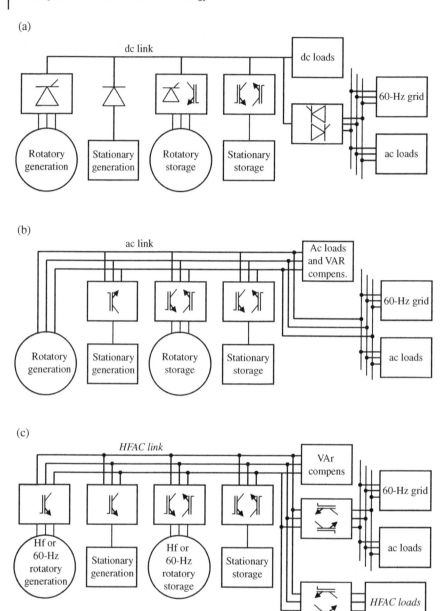

Figure 15.12 Varieties of energy integration: (a) dc-link integration, (b) ac-link integration, and (c) HFAC-link integration.

In the rotatory generation with synchronous or induction generators shown in Figure 15.12a, two separate converters would be necessary to feed the dc link. For stationary generation, it is used only in a diode scheme to prevent back-feeding to the dc source. In some cases, such as for fuel cell stacks, a dc–dc converter with filtering features is required for voltage matching and smoothing out undesirable ripples. The dc–ac converter connecting the dc link to the 60-Hz grid is necessary if the grid is used to store energy; otherwise, it could be an ordinary transistor or thyristor inverter. Control is exerted individually on each source because current contribution to the dc link can occur just by raising the output voltage above other sources.

The same solution can be used to store energy in a battery, making sure that the battery can supply energy when the bus voltage is lower or receive energy when the bus voltage is higher. A typical case of a current control source is that of small wind power plants for a boat or remote load. Other alternatives have to be adapted according to the needs of the generating system.

15.4.2 AC-Link Integration

Another possibility for integrating renewable energy sources is the ac-link bus operating at either 50 or 60 Hz, as indicated in Figure 15.13. This bus can be the public grid or a local grid for islanded operation. In this case, the interconnection of a local bus to the public bus must follow the standards and requirements for interconnection (Institute of Electrical and Electronics Engineers standards and others; see Chapter 14).

Positive features of this configuration include:

- Utility regulation and maintenance of the operational voltage, which makes it easier to inject power into the grid
- The possibility, in some cases, of eliminating the electronic converters (e.g., by using synchronous generators or induction generators that establish their own operating point)
- Easy multi-voltage and multiterminal matchings
- Easy galvanic isolation
- Well-established scale economy for consumers and existing utilities

Negative features include:

1) The need for rigorous synchronism, voltage-level matching, and correct phase sequence between sources during interconnection as well as during operation
2) Leakage inductances and capacitances in addition to the skin and proximity effects that cause losses in long distributions
3) Electromagnetic compatibility concerns
4) The possibility of current recirculation between sources
5) The need for power factor and harmonic distortion correction
6) Reduced limits for transmission and distribution

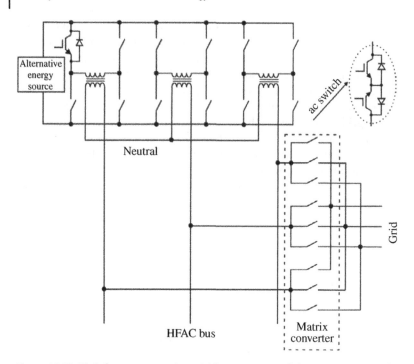

Figure 15.13 High-frequency ac microgrid for integration of alternative sources of energy.

The rotatory generation in Figure 15.12b is feeding the ac link to the grid directly. The stationary generation will need a dc–ac inverter or, in some cases, a dc–dc booster and then a dc–ac converter to allow minimum levels of efficient conversion. No dc loads are assumed in this case. Volt-amperes reactive compensation and harmonic compensation are easily implemented in this scheme using the storage facilities and the p–q theory concepts, as explained in Section 15.3.

15.4.3 HFAC-Link Integration

An HFAC microgrid system is a power electronics solution that is promising as an interface for the utility grid and stand-alone operation. It provides embedded fault protection, small dimensions, and a configuration to serve various power quality functions. Frequencies in industrial use are 400 Hz, as is the case in spaceships, boats, buses, planes, submarines, and other loads of the kind. Researchers are investigating increased levels of these frequencies for high-power electronic circuits and devices.

Power electronic-based systems connected with the grid can be controlled and monitored remotely to allow real-time optimization of power generation and power flow, which allows aggregation of distributed power generation resources into a "virtual utility." Several HFAC power distribution systems were implemented for aerospace applications, and NASA evaluated this scheme for space station applications [12, 17]. It has been proposed as a workable power distribution system for hybrid electric vehicles as well [18] using traditional ac–ac conversion through semiconductor switches. This is performed in two stages, from ac–dc and then from dc–ac (dc-link converters), or directly by cycloconverters. A microgrid is illustrated in Figure 15.12c, and details are provided in Figure 15.13.

Historically, cycloconverters have been used in high-power applications driving induction and synchronous motors, using phase-controlled thyristors, because of their easy zero-current commutation features [13]. The cycloconverter output waveforms have complex harmonics. Higher-order harmonics are usually filtered by machine leakage inductance. In a cycloconverter, there are no storage devices; therefore, instantaneous output power equals input power minus losses. To maintain this balance on the input side with sinusoidal voltages, the input current is expected to have complex harmonic patterns.

The matrix converter is another technology under development. First proposed in the early 1980s, it consists of a matrix of ac switches that connects the three input phases to the three output phases directly [14]. This circuit allows any of the input phases to be connected to any output phase at any instant, depending on the control strategy. Power semiconductor manufacturers are expending efforts to come up with modern, fast-power semiconductor devices (ac switches) capable of symmetric voltage blocking, gate controlled or commanded by either zero-current or zero-voltage conditions. It is expected that these devices will accelerate the widespread use of matrix converters to convert HFAC into utility voltage and frequency.

Series resonant converters, using zero-voltage or zero-current switching, can be used with each of the sources to generate the HFAC link [17]. In this way, the overall losses in the converters can be reduced. An HFAC microgrid system has the following inherent advantages:

1) The harmonics are of higher orders and are easily filtered out.
2) Fluorescent lighting will experience improvement because, with higher frequency, the luminous efficiency is improved, flicker is reduced, and dimming is accomplished directly. The ballast inductance is reduced proportionally to the frequency, with a corresponding reduction in size and weight.
3) High-frequency induction motors can be used for compressors, high-pressure pumps, high-speed applications, and turbines. Ac frequency changers based on matrix converters can be used to soft start high-frequency induction motors. A safe operating area is not a restriction for soft switching, and therefore modern power electronic devices will be advantageous.
4) Harmonic ripple current in electric machines will decrease, improving efficiency.

5) High-frequency power transformers, harmonic filters for batteries, and other passive circuit components become smaller.
6) Auxiliary power supply units are easily available by tapping the ac link. They would be smaller with higher efficiencies.
7) Batteries have been the traditional energy storage source, but in HFAC microgrids, dynamic storage is an alternative.

The disadvantages are:

1) The high cost of transformers
2) The large number of devices (because of the use of bipolar ac switches)
3) Very complex control
4) The dependence on future advances of power electronic components
5) Concerns about electromagnetic compatibility
6) Extremely reduced limits and technological problems for transmission and distribution at high frequencies

When an HFAC microgrid is adopted, all the active and reactive power flow inside the grid must be considered, and a high-level hierarchical control, such as the one depicted in Figure 15.14, must be adopted to control the voltage and reactive power exchange with the interacting grid. For each energy source,

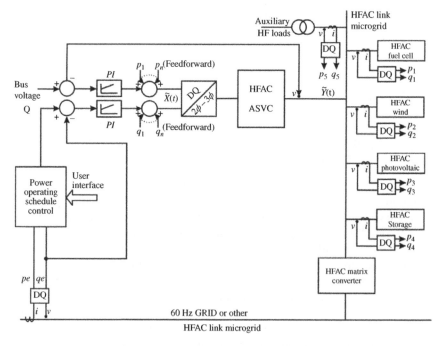

Figure 15.14 Active and reactive power flow control of an HFAC microgrid.

HFAC loads and the 60-Hz grid or other HFAC microgrids will have a power electronic controlled unit. At the coupling point, the active and reactive power injection or demand can be calculated by Akagi's *p–q* theory or by measuring the voltage angles and computing the equations for incoming bus *i*:

$$P_i = \sum_{\substack{j=1 \\ j \neq i}}^{n} V_i V_j B_{ij} sin\left(\delta_i - \delta_j\right) \tag{15.17}$$

$$Q_i = \sum_{j=1}^{n} V_i V_j B_{ij} cos\left(\delta_i - \delta_j\right) \tag{15.18}$$

where:

V_i and V_j = Voltage magnitudes at bus *i* and *j*
B_{ij} = Line susceptance (inverse of reactance) between bus *i* and *j*
δ_i and δ_j = Voltage angles at bus *i* and *j*

15.5 Islanding and Interconnection Control

An important issue related to microgrids is islanding. This condition occurs when the microgrid continues to energize a section of the main grid after that section has been isolated from the main utility. Unintentional islanding is a concern for the utility because sources connected to the system but not controlled by the utility pose a possibility of harm to utility personnel and damage from uncontrolled voltage and frequency excursions. In addition, utility equipment such as surge arresters may be damaged by overvoltages during a shift-neutral reference or resonance. Furthermore, special synchronization and voltage-level matching measures will have to be taken for reconnection of utility power. Islanding can lead to asynchronous reclosure, which can damage equipment. It is therefore important that microgrid systems incorporate methods to prevent unintentional islanding. However, intentional islanding may occur when a secure aggregation of loads and sources capable of operating in parallel with, or to, end users allows power flow only within the microgrid or safely exports it to the utility grid [19–21].

Islanding phenomenon can be categorized into intentional and unintentional islanding. However, most of the works in islanding detection (ID) area focus on unintentional ID. Technically, the introduced methodologies and algorithms in ID can be classified under two major classes: (i) local and (ii) remote approaches. Usually, remote methods are founded over a communication scheme with the main grid (by monitoring the status of all breakers in the network), whereas local techniques are developed based on the available local information and observations. Although remote techniques are highly reliable, implementing such approaches can be expensive due to required extra infrastructure for direct communication between the DGs and utility through an interstitial technology such as fiber-optic or wireless communication networks. Moreover, the practical implementation of these schemes can be inflexible and complex because of the high penetration and wide area positioning of DGs in large-scale complex systems. However, there are some communication methods that are working by transporting a carrier signal through the power transmission lines [22]; however, this may result in a power system signal distortion. Therefore, for simplicity and applicability, a cost-effective local technique is preferred in general. Remote methods are more reliable but are neither more cost-effective nor simpler to implement than local methods. As a result, they are widely used in the case of highly sensitive load in microgrids such as data centers or banking system operational rooms.

The local methods are either passive or active. The core concept of local ID techniques remains the same as that of some system parameters that are supposed to be steady during the grid-tied operation but significantly change during the transition state from the grid-connected to island mode operation.

Passive ID relies on changes to electrical parameters (such as frequency, voltage, and power signals) to determine whether islanding had occurred. These local type methods were the first to be developed. Actually, passive ID are relying on the parameter thresholds. Their advantages include easy implementation (controller not required), no degradation of the inverter power quality, and inexpensiveness. Their primary drawbacks are a relatively large non-detection zone (NDZ) [22] and their ineffectiveness in multi-inverter systems. Conventional passive techniques utilize the measurement of electrical quantities, such as voltage, current, or frequency, to estimate the system island state using undervoltage protection/overvoltage protection (UVP/OVP), underfrequency protection/overfrequency protection (UFP/OFP), and vector shift (VS) functions. These are protection relays used by power systems utilities, usually installed at the distribution feeder.

The current technology aims to overcome the limitations within passive methods by introducing the active ID frameworks. The fundamental idea besides the active ID is to inject a small disturbance (or signature signal) at the inverter output and track its feedback behavior in order to detect anomalies and faults. The main advantage of such an approach is the relatively smaller NDZ than that in passive methods [22]. However, this reliability improvement will be reached on the cost of increasing the possibility of deteriorating output power quality, destabilizing the inverter, and complexity of the system due to the need (usually) for additional controllers.

Due to enormous variety of grid fault types and definitions, each of the methods discussed in the previous paragraphs has found its own application. Both techniques are used to monitor whether the grid voltage/frequency exceeds the limits imposed by the relevant standard. However, the passive ID strategies are more preferable specially to ensure the power quality perspective. Beyond passive approaches, the major drawback related to the UVP/OVP and UFP/OFP is the challenging situation that may happen within the cases where the load and generation in an islanded microgrid is almost equal; as a result the risk of the large NDZ will increase [23].

In order to mitigate the large NDZ issues, some improvements have been proposed in the literature. A method is proposed in Ref. [24] where the risk of NDZs of UVP/OVP and UFP/OFP can be decreased within a comparison between the P-V and P-Q characteristics of inverters equipped with a constant current controller. Authors in Ref. [25] reported another phase difference-based monitoring scheme that tracks the phase differences between the inverter terminal voltage and output current. The major advantageous of this approach is the easiness of implementation. In practice, this method only needs the modification of the PLL within the inverters for utility synchronization. In Ref. [26] a total harmonic distortion (THD-based) tracking approach at the point of common coupling (PCC) has been investigated, where the inverter disconnects the microgrid system from the power grid whenever the THD value exceeds a predefined threshold.

Computational intelligence and data mining has found extensive interests in the ID especially within the passive ID frameworks. The basic idea within this new strategy is to perform a near real-time estimation of the off-grid operation without need of any extra information from utility component and without causing any issue for power quality. Besides the fast and real-time estimation, these techniques are providing high computational efficiency with good reliability and accuracy. To this end, signal processing has been vastly integrated through pattern recognition approaches. In Ref. [27] authors are implementing the fast Fourier transform (FFT) in addition to the immunological principle in order to respond to inverter islanding [27]. However, due to raising of nonstationary behavior within the corresponding signal behavior during islanding period, this approach may be unrealistic in practical cases.

Wavelet signal decompositions are beyond the most well-known nonstationary signal analysis techniques. As a result, wavelet-based feature extraction methods have been merged into famous artificial neural network (ANN) classifier in order to implement a robust ID, as described in Ref. [28]. Due to high efficiency in their operational time and computational complexity, discrete wavelet transform (DWT) has been applied on the current signal seen at DG terminal in order to extract some signature features to identify the islanding occurrence.

Other islanding classification techniques, such as decision tree, radial basis function (RBF), and probabilistic neural network (PNN), were also used as classifiers in different schemes [23, 26]. In another try, authors in Ref. [29] developed a new technique called phase space method, which has been vastly utilized in various classification and detection algorithms. This algorithm is

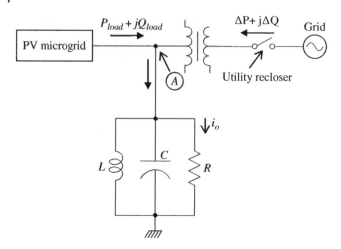

Figure 15.15 Microgrid power flow to the utility grid.

based on the mathematical method, which reconstructs the data of a time series in a higher-dimensional space. It has been shown in the literature that due to time efficiency, this new approach can be a promising alternative in the case of real-time applications. The artificial intelligence-based frameworks are going to be the most exciting future topic in ID, which is still under investigation and needs a comprehensive evaluation to be aggregated within the current industrial relay and protection technologies.

Consider the configuration shown in Figure 15.15. The microgrid is connected to a feeder line with a local load, which is, in turn, connected to the utility grid through a transformer and some sort of switch (a recloser, breaker, or fuse). If the switch is opened under certain conditions, it is possible for the microgrid to continue energizing the isolated section of the grid and to supply power to the local load. This is an unintentional island, and the isolated section of the utility being powered by the PV system is referred to as an island of supply or, simply, an island. Because of this, grid-connected alternative energy systems are required to have protective relays to sense conditions of over- and undervoltage/frequency and to signal the unit to be disconnected from the microgrid utility in the event that the magnitude or frequency goes beyond limits.

Now consider Figure 15.16. When the separation device is closed, real and reactive power P_{mic} and Q_{mic} flow from the microgrid to node a. Power P_{load} and Q_{load} flow from a to the load, and the utility provides

$$\Delta P = P_{load} - P_{mic} \tag{15.19}$$

$$\Delta Q = Q_{load} - Q_{mic} \tag{15.20}$$

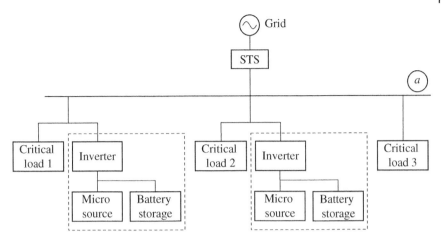

Figure 15.16 Microgrid interaction with grid.

The microgrid and any local load are considered a lumped load by the utility. The frequency dependence on the load can be approximated by equation (15.21), where D is typically within 4% of the base frequency. The utility generators will have a frequency variation in accordance with equation (15.22). In it, the constant R is defined by utility operational policies, and Δf is the frequency regulation constant or frequency droop, normally given in MW/Hz:

$$\Delta P_{load} = D\Delta f \tag{15.21}$$

$$\Delta P_{grid} = -R\Delta f \tag{15.22}$$

Assuming that the load may have a leading or lagging power factor, an RLC circuit, as shown in Figure 15.16, can represent such an operation. The following equations relate the active and reactive power at the load with the voltage at node a:

$$P_{load} = \frac{V_a^2}{R_{load}} \tag{15.23}$$

$$Q_{load} = \left[\frac{V_a^2}{\omega L} - \frac{V_a^2}{\frac{1}{\omega C}} \right] \tag{15.24}$$

When the separation device opens, the grid active power change ΔP and reactive power change ΔQ will go to zero. The behavior of the isolated system will depend on ΔP and ΔQ at the instant before the separation device opens to form the island, denoted by ΔP_- and ΔQ_-.

The overvoltage relay (OVR), undervoltage relay (UVR), overfrequency relay (OFR), or underfrequency relay (UFR) can prevent unintentional islanding in four cases:

1) $\Delta P_- > 0$, in which the microgrid system produces less real power than is required by the local load ($P_{load} > P_{mic}$). Therefore, when the switch opens and P becomes zero, P_{load} will decrease, and voltage V_a at node a will decrease because R_{load} can be assumed to be constant over this short time span. This decrease can be detected by the UVR, and islanding is prevented.

2) $\Delta P_- > 0$, in which power flows into the utility ($P_{load} < P_{mic}$). When ΔP becomes zero, P_{load} will decrease, and V_a will increase. This condition can be detected by the OVR, and again islanding is prevented.

3) $\Delta Q_- > 0$, which corresponds to a lagging power factor load or a load whose reactive component is inductive. After the separation device opens, $\Delta Q = 0$. For a microgrid unity power factor operation at the PCC, $Q_{mic} = 0$ and $Q_{load} = 0$. In accordance with equation (15.24), the right-hand term has to become zero, meaning that the inductive term must drop and the capacitive term must increase. That occurs for the frequency ω increasing, which can be detected by the OFR.

4) $\Delta Q_- > 0$, which corresponds to a leading power factor load or one that is primarily capacitive. When ΔQ becomes zero, the inductive and capacitive terms of equation (15.24) must balance, requiring the frequency ω to decrease, which can be detected by the UFR.

There are other possible cases when ΔP_- or ΔQ_- is zero. One is a case in which the microgrid power production is matched to the load power requirement and the load displacement factor is unity. In this case, when the switch is opened, no change occurs in the isolated system, and the OVR/UVR and OFR/UFR will not detect any voltage or frequency deviation. The magnitude and frequency of the utility voltage can be expected to deviate slightly from nominal values, and therefore the thresholds for the four relays cannot be set arbitrarily small, or the microgrid system will be subject to nuisance trips. This limitation leads to the formation of an NDZ. The probability of ΔP_- or ΔQ_- falling into the NDZ of the OVR/UVR and OFR/UFR is significant. It is therefore important that microgrid systems incorporate methods to prevent unintentional islanding in the case in which the microgrid is injecting a lagging or leading power factor or when $\Delta P_- \approx 0$ or $\Delta Q_- \approx 0$.

When the separation device opens to isolate a utility fault, the DG units indicated in Figure 15.16 must immediately share the increased power demand in a predetermined manner among themselves to continue supplying power adequately to all critical loads within the microgrid. This sharing of power can be achieved with no physical communication links between the DG systems by

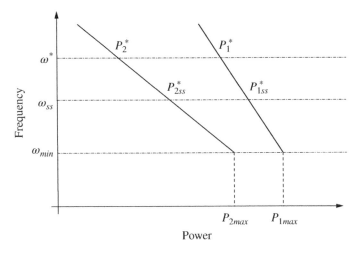

Figure 15.17 P-ω droop characteristics.

introducing artificial real power versus supply frequency and reactive power versus voltage droop characteristics into the DG controllers.

As an illustration, consider the droop characteristics of the two DG systems represented in Figure 15.17. The droop characteristics should be coordinated to make each DG system supply real and reactive power in proportion to its power ratings. It is expressed mathematically as

$$\omega_j(t) = \omega^* - R\left[P_j^* - P_j(t)\right] \tag{15.25}$$

$$R = \frac{\omega^* - \omega_{min}}{P_j^* - P_{j,max}} \tag{15.26}$$

where:

$P_j(t)$ = The actual real power output of the DG system j (in accordance with Figure 3.16, $j = 1$ or 2)

$P_{j,max}$ and ω_{min} = The maximum real power output of the DG system and the minimum allowable operating frequency

P_j^* and ω^* are, respectively, the dispatched real power and operating frequency of the DG system j when in the grid-connected mode, and $R < 0$ is the slope of the droop characteristic.

A block diagram of equations (15.25) and (15.26) is shown in Figure 15.19. Similarly, the magnitude set point of each DG output voltage can be tuned according to a specified $Q–V$ droop characteristic to control the flow of reactive power within the microgrid. Mathematically, those characteristics can be expressed by equations (15.27) and (15.28), with a block realization function indicated in Figure 15.20:

$$V_j(t) = V^* + \varepsilon\left[Q_j^* - Q_j(t)\right] \tag{15.27}$$

$$\varepsilon = \frac{V^* - V_{min}}{Q_{j,max} - Q_j^*} \tag{15.28}$$

When the utility grid returns to normal operating conditions, the microgrid has to resynchronize with it for reconnection. Synchronization can be achieved by aligning the voltage phasors at the microgrid and utility ends of the separation device. This can be implemented conveniently by adding two separate synchronization compensators to the external real and reactive power control loops, as shown in the dashed frames in Figures 15.18 and 15.19. Inputs to these synchronization compensators are the magnitude and phase errors of the two voltage phasors at both ends of the separation device. Their outputs are fed to the real and reactive power loops to make the voltage phasor at the microgrid end closely to track the phasor at the utility end (in both magnitude and frequency). Once synchronized and upon closing the separation device,

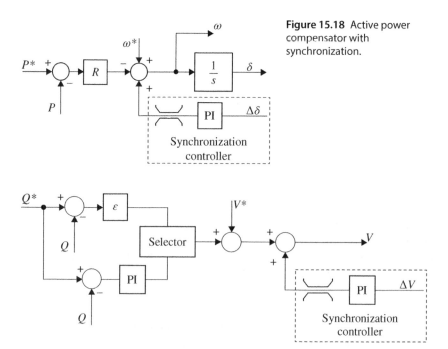

Figure 15.18 Active power compensator with synchronization.

Figure 15.19 Reactive power compensator with synchronization.

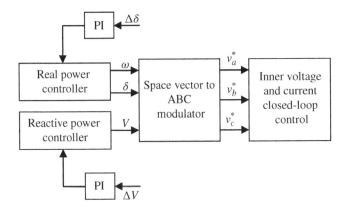

Figure 15.20 Unified power and reactive microgrid power controller with synchronization.

the synchronization compensators must be deactivated by setting their outputs to zero so they will not interfere with proper operation of the real and reactive power control loops in grid-connected mode. The droop control requires that each DG output voltage be different from that of the utility grid to allow for proper reactive power control in the grid-connected mode.

The inverter at the microgrid side can incorporate the active and reactive power controller, as depicted in Figure 15.20. When connected to the utility grid, only the PI compensator is selected for reactive power control, which forces the DG reactive power output to track its desired value to a zero steady-state error. For this PI compensator, low gains are used to give a sufficiently long response time, which, in turn, allows the dynamics of the inner voltage and current loops and external power loops to be decoupled. When a utility fault occurs and the microgrid "islands," the control switches over to select the droop characteristics for reactive power sharing between the DG systems, which ensure a smooth transition from grid-connected to islanding mode. The active and reactive powers transmitted across a lossless line are approximated by equations (15.29) and (15.30):

$$P = \frac{EV}{X} sin\varphi \tag{15.29}$$

$$Q = \frac{EV cos\varphi - V^2}{X} \tag{15.30}$$

where:

E = The impressing generator or DG voltage, with phase shift ϕ
V = The PCC bus voltage considered to have phase shift 0°
X = The reactance connecting the DG voltage to the bus

Since the power angle ϕ is typically small, we can have the further simplification (assuming that $sin\varphi \approx \varphi$ and $cos\varphi \approx 1$, indicated by the equations (15.31) and (15.32):

$$\phi \approx \frac{PX}{EV} \tag{15.31}$$

$$E - V \approx \frac{QX}{E} \tag{15.32}$$

From those previous equations, it can be derived that the active power is predominately dependent on the phase shift and the reactive power depends on the bus voltage. Therefore, simple linear relationships can be derived for relating the voltage angular frequency and amplitude on load, with coefficients (called droop coefficients) for frequency and amplitude that will provide a linear variation of the angular frequency falling with active power decrease, or a bus voltage rising with a reactive power increase. When frequency falls, the output power of DG device increases; multiple parallel units with the same droop characteristic can respond to the fall in frequency by increasing their output active power simultaneously. The increase in output active power will counteract the reduction in frequency, and the system will settle at output active power and frequency at a certain steady point. The droop characteristic allows multiple units to share load without the units trying to overcome in controlling the load on its own. The same rationale is also applicable for the bus voltage droop control.

15.6   DG PLL with Clarke and Park Transformations

Figure 15.21 represents the classical *abc* system taking into account that $V_{line} = \sqrt{3}V_a$, $V_a = V_m/\sqrt{2}$, and $V_{line} = \sqrt{3/2}\,V_m$. Therefore $V_m = \sqrt{2/3}\,V_{line}$:

$$\begin{bmatrix} v_a \\ v_b \\ v_c \end{bmatrix} = V_m \begin{bmatrix} sin(\omega t) \\ sin\left(\omega t - \dfrac{2\pi}{3}\right) \\ sin\left(\omega t + \dfrac{2\pi}{3}\right) \end{bmatrix} = \sqrt{\frac{2}{3}}V_{line} \begin{bmatrix} sin(\omega t) \\ sin\left(\omega t - \dfrac{2\pi}{3}\right) \\ sin\left(\omega t + \dfrac{2\pi}{3}\right) \end{bmatrix} \tag{15.33}$$

15.6.1   Clarke Transformation for AC-Link Integration

The Clarke transformation is the translation from the three-phase reference frame *abc* to the two-axis orthogonal stationary reference frame $\alpha\beta$ as shown in Figure 15.22 considering an unbalance voltage V_0 represented only for

Figure 15.21 Classical *abc* system.

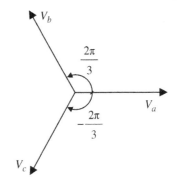

Figure 15.22 Clarke transformation.

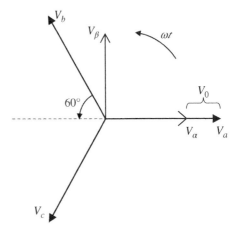

phase *a*, that is, to lock the coordinate α to one of the *abc* phases, usually to phase *a*. Coordinate β is shifted by 90° from coordinate α. Voltage vectors of phases *abc* are projected on this $\alpha\beta$ frame expressed as [30]

$$V_a = V_\alpha + V_0$$
$$V_b = -V_\alpha cos60° + V_0 + V_\beta cos30°$$
$$V_c = -V_\alpha cos60° + V_0 - V_\beta cos30°$$

or

$$V_a = V_\alpha + V_0 \tag{15.34}$$

$$V_b = -\frac{1}{2}V_\alpha + V_0 + \frac{\sqrt{3}}{2}V_\beta \tag{15.35}$$

$$V_c = -\frac{1}{2}V_\alpha + V_0 - \frac{\sqrt{3}}{2}V_\beta \tag{15.36}$$

$$V_a + V_b + V_c = 3V_0 \tag{15.37}$$

In matrix form, equations (15.34)–(15.37) become

$$
\begin{vmatrix} V_a \\ V_b \\ V_c \end{vmatrix} = \begin{bmatrix} 1 & 0 & 1 \\ -\dfrac{1}{2} & \dfrac{\sqrt{3}}{2} & 1 \\ -\dfrac{1}{2} & -\dfrac{\sqrt{3}}{2} & 1 \end{bmatrix} \begin{vmatrix} V_\alpha \\ V_\beta \\ V_0 \end{vmatrix}
$$

Simultaneous solution of equations (15.34) and (15.37) gives

$$
V_\alpha = \frac{2}{3}\left(V_a - \frac{V_b + V_c}{2} \right) = \frac{2}{3}V_a - \frac{1}{3}V_b - \frac{1}{3}V_c
$$

$$
V_\beta = \frac{V_b - V_c}{\sqrt{3}} = 0V_a + \frac{1}{\sqrt{3}}V_b - \frac{1}{\sqrt{3}}V_c
$$

$$
V_0 = \frac{V_a + V_b + V_c}{3}
$$

In matrix form for instantaneous values,

$$
\begin{bmatrix} v_\alpha \\ v_\beta \\ v_0 \end{bmatrix} = \sqrt{\frac{2}{3}}V_m \begin{bmatrix} 1 & -\dfrac{1}{2} & -\dfrac{1}{2} \\ 0 & \dfrac{\sqrt{3}}{2} & -\dfrac{\sqrt{3}}{2} \\ \dfrac{1}{2} & \dfrac{1}{2} & \dfrac{1}{2} \end{bmatrix} \begin{bmatrix} sin(\omega t) \\ sin\left(\omega t - \dfrac{2\pi}{3} \right) \\ sin\left(\omega t - \dfrac{4\pi}{3} \right) \end{bmatrix}
\tag{15.38}
$$

15.6.2 Blondel or Park Transformation for AC-Link Integration

The Blondel or Park transformation represents a three-phase *abc* axis system in a rotating reference frame with two *dq* axis (direct and quadrature) [31–33]. That is, coordinate *d* rotates by some angle θ with respect to one of the *abc* phases, usually phase *a*. Coordinate *q* is shifted by 90° from coordinate *d*. In Figure 15.23, phases *abc* are projected on this *dq* frame as

$$
\begin{bmatrix} V_d \\ V_q \\ V_0 \end{bmatrix} = \begin{bmatrix} cos\theta & sin\theta & 0 \\ -sin\theta & cos\theta & 0 \\ 0 & 0 & 1 \end{bmatrix} \begin{bmatrix} V_\alpha \\ V_\beta \\ V_0 \end{bmatrix}
\tag{15.39}
$$

Figure 15.23 Park transformation.

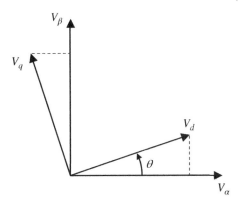

or its inverse

$$
\begin{bmatrix} V_\alpha \\ V_\beta \\ V_0 \end{bmatrix} = \begin{bmatrix} cos\theta & -sin\theta & 0 \\ sin\theta & cos\theta & 0 \\ 0 & 0 & 1 \end{bmatrix} \begin{bmatrix} V_d \\ V_q \\ V_0 \end{bmatrix}
$$

From equation (15.39), when V_a is superimposed with V_a and $V_a + V_b + V_c$ is zero, V_a, V_b, and V_c can be transformed to only V_α and V_β. In a and b line values,

$$
\begin{bmatrix} V_d \\ V_q \\ V_0 \end{bmatrix} = \sqrt{\frac{2}{3}} V_{line} \begin{bmatrix} cos\theta & sin\theta & 0 \\ -sin\theta & cos\theta & 0 \\ 0 & 0 & 1 \end{bmatrix} \begin{bmatrix} cos\omega t \\ sin\omega t \\ 0 \end{bmatrix}
$$

$$
= \sqrt{\frac{2}{3}} V_{line} \begin{bmatrix} cos\theta \cdot cos\omega t + sin\theta \cdot sin\omega t \\ -sin\theta \cdot cos\omega t + cos\theta \cdot sin\omega t \\ 0 \end{bmatrix}
$$

(15.40)

$$
\begin{bmatrix} V_d \\ V_q \\ V_0 \end{bmatrix} = \sqrt{\frac{2}{3}} V_{line} \begin{bmatrix} cos(\omega t - \theta) \\ sin(\omega t - \theta) \\ 0 \end{bmatrix}
$$

(15.41)

When PLL angle θ is close to the actual voltage vector angle $\omega' t$, the difference $(\omega' t - \theta)$ is small or close to zero and then $sin(\omega' t - \theta) \cong (\omega' t - \theta)$. Therefore, it can be said for a balanced three-phase system when PLL is locked, the q-axis component in the rotating reference frame reduces to zero, and

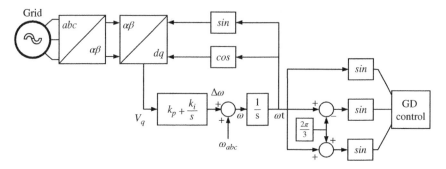

Figure 15.24 Block diagram of a PLL scheme to synchronize a GD to the grid.

when it is not locked or has small error, the q-axis component is linearly proportional to the error $V_q \cong \sqrt{2/3}\, V_{line}\,(\omega't - \theta)$.

The difference between the voltage transformations presented here is that coordinate α in the Clarke transformation is locked on phase a, while coordinate d of the Park transformation can be adjusted by an angle θ with respect to phase a. The resulting voltage difference between these voltages (rotating and stationary voltage frames) constantly projected on the abc voltage frame makes a dc voltage since the two ac-transformed voltages are at the same speed ωt but shifted by θ from each other. When $\theta = 0$, both of them have the same mathematical representation. In other words, it is possible to adjust θ to cause some effect on the coordinate frame $\alpha\beta$ and then reflect it onto the abc phases, for example, to lock the frequency of two distinct alternate voltages to each other or to lock a GD frequency to the grid frequency through a power converter (see Figure 15.24). A second example of application could be the selection of an angle θ to reduce reactive power of the abc system. A third example could be harmonic minimization in power systems by locking a distorted wave shape to a local voltage-controlled oscillator free from any distortion.

15.7 DG Control and Power Injection

Combining loads with sources, allowing for intentional islanding, and using available waste heat, one can implement a microgrid. Figure 15.25 shows how the traditional role of central generation, transmission, and distribution is transformed by aggregation of distributed resources (DR), which results in microgrid architecture. There is a single point of connection to the utility called the PCC. In the microgrid, some feeders can have sensitive loads that require local generation. Static switches that can separate them in less than a cycle provide intentional islanding from the grid.

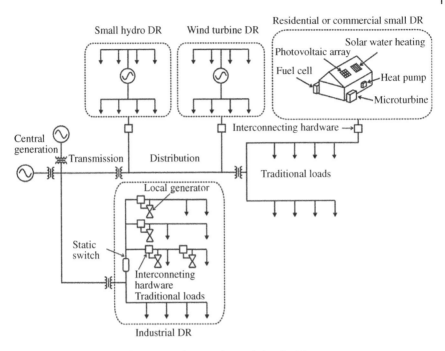

Figure 15.25 Distributed resources shaping the traditional grid.

When a microgrid is connected, power from local generation can be directed to the feeder with noncritical loads or be sold to the utility if agreed or allowed by net metering. In addition to allowing better efficiency, waste recovery, and tailored reliability, a microgrid is designed for the requirements of end users, a stark difference from the central generation paradigm. Key to this characteristic is the reliance on the flexibility of advanced power electronics that control the interface between DR and their surrounding ac system.

In Figure 15.25, the expression "distributed resource" is used for generating units in a microgrid. In addition, to the power electronics control, a microgrid requires operational control to ensure economic commitment and dispatch within environmental and other constraints. In a DG power system, there are needs of coordination of control layers. A hierarchical system consists of several decision-making components and has an overall goal, which is distributed among its individual components. The levels of the hierarchy exchange information (usually vertically) among themselves in an iterative mode, and as the level increases, the time horizon increases (i.e., lower-level components or modules are usually faster than their higher-level counterparts). In a multilayer hierarchical structure, the first layer acts as the regulation or direct control layer. It is followed by optimization, adaptation, and self-organization functions.

A multiechelon structure (Figure 15.26) consists of a number of subsystems situated in levels such that each one can coordinate lower-level units while it is itself coordinated by a higher-level unit. Among the particular tasks of higher-level echelons is conflict resolution in achieving specified objectives. The most time-demanding level of control, the first layer (actuator), is related to the switching of high-power transistors in the power electronic converters. This task requires high-speed, real-time control with sampling rates about microseconds. Vector control, space vector, pulse width modulation, current regulation, suppression of harmonics, and power electronic device protection are within this layer.

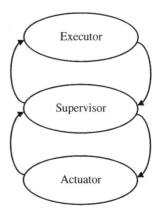

Figure 15.26 Hierarchical control of DG systems.

A second level of control (supervisor) is required to manage the system; to generate power set points to control the power flow among the energy source, energy storage, and load; to control dc or ac bus voltage; and to monitor faulty signals. The particular signals to be controlled depend on the specific DG technology, but sampling rates about milliseconds are typically required.

The third level of control (executor) is related to communications to external equipment and the outside world, which provides a variety of remote monitoring and control. The executor level is responsible for implementation of control schemes to produce as much energy from the system as possible to recover the installation cost. Policies related to the costs of fuel, maintenance, and negotiation with neighbor sites are implemented in this level. Typically, sampling times are approximately minutes, and hours are required to implement such policies.

A crucial and usually overlooked aspect of DG system control is the operation of these three levels under a real-time structure. The concepts of software design and hardware integration are very complex and require close interaction and cooperation of engineers with very diverse areas of expertise. If well done, a high-performance and highly integrated system will result.

Conventional power systems with multiple generators use a load-sharing technique (droop scheme) in which the generators share the system load by drooping the frequency of each generator with the real power P delivered by the generator. This allows each generator to share changes in total load in a manner determined by its frequency droop characteristic. This simple idea essentially uses the system frequency as a communication link between the generator control systems. Similarly, a droop in the voltage amplitude with reactive power Q is used to ensure reactive power sharing. This load-sharing technique is based on the power flow theory in an ac system. This theory states that flow of active power P and reactive power Q between two sources can be

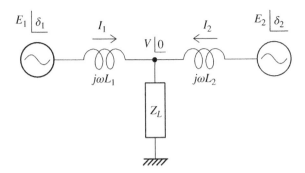

Figure 15.27 Parallel connection of two power converters with a common load.

controlled by adjusting the power angle and voltage magnitude of each system. This means that the active power flow P is controlled predominantly by the power angle, while the reactive power Q is controlled predominantly by the voltage magnitude.

For simplified analysis, Figure 15.27 indicates variables for load sharing of two power converters arranged in parallel configuration. In it, two inverters are represented by two VS connected to a load through line impedances represented by pure inductances L_1 and L_2. The complex power at the load because of the i-th inverter is given by

$$S_i = P_i + jQ_i = V \cdot I_i^* \tag{15.42}$$

where:

$$i = 1,2$$

$I_i^* =$ The complex conjugate of the inverter current I and is given by

$$I_i^* = \left[\frac{E_i cos\delta_i + jE_i sin\delta_i - V}{j\omega L_i} \right]^* \tag{15.43}$$

$$S_i = V \left[\frac{E_i cos\delta_i + jE_i sin\delta_i - V}{j\omega L_i} \right]^* \tag{15.44}$$

Such equations allow the calculation of active and reactive power flowing from the ith inverter as

$$P_i = \frac{VE_i}{\omega L_i} sin\delta_i \tag{15.45}$$

$$Q_i = \frac{VE_i cos\delta_i - V^2}{\omega L_i} \tag{15.46}$$

Equations (15.42)–(15.46) show that if δ_1 and δ_2 are small enough, the real power flow is influenced primarily by the power angles δ_1 and δ_2, whereas the reactive power flow depends predominantly on the inverter output voltages E_1 and E_2. This means that to a certain extent, real and reactive power flow can be controlled independently. Because controlling the frequencies dynamically controls the power angles, the real power flow control can be achieved equivalently by controlling the frequencies of the voltages generated by the inverters. Therefore, as mentioned previously, the power angle and the inverter output voltage magnitude are critical variables that can control the real and reactive power flow directly for proper load sharing of power converters connected in parallel.

Figure 15.28 is a simplified diagram that represents a DG control model with frequency and voltage command. The inverters respond instantaneously to decoupled current references (p and q) to impress current into the grid with a prescribed amplitude and angle. The same approach can be applied to parallel operation of distributed energy systems in a stand-alone ac power supply application.

In general, there is a large distance between inverter output and load bus, so each DR is required to operate independently using only locally measurable voltage and current information. There is also a long distance between DR units, and proper load sharing between each unit must be guaranteed despite impedance mismatches. Voltage and current measurement error mismatches and interconnection tie-line impedance can heavily affect the performance of load sharing.

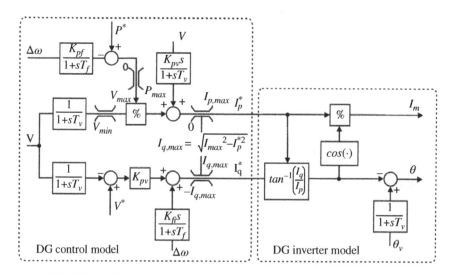

Figure 15.28 DG control interacting with inverter under frequency and voltage command.

Conceptually, the isolated microgrid is like a scaled-down version of a large-scale utility grid, and most of the technical requirements are the same. To supply reliable good-quality power, the microgrid must have mechanisms to regulate voltage and frequency in response to changes in consumer loads and disturbances.

The grid-connected microgrid should be designed and operated such that it presents the appearance of a single predictable and orderly load or generator to the grid at the point of interconnection. This arrangement provides several advantages. For example, the DG owners may be able to rate and operate their generation more economically by being able to export (and import) power to the microgrid. Load consumers may be able to have continued service (possibly at a reduced level) when connection to the host utility is lost. The host utility may be able to depend on the microgrid to serve load consumers in such a way that the substation and bulk power infrastructure need not be rated (or expanded) to meet the entire load [22, 34, 35]. The microgrid could be controlled in such a fashion as to be an active asset to bulk system reliability (e.g., by providing spinning reserve or black-start services).

Another form of simple and effective control, perhaps suitable up to 100 kW, is heuristic hill-climbing control. It is based on the idea that all conversion systems for alternative sources of energy, either rotatory or stationary, have a similar power versus current characteristic, presented in Figure 15.29. Therefore, it is possible to adapt them to operate with hill-climbing control or with fuzzy control for maximum power point tracking [22, 35].

The most obvious applications of hill-climbing control are wind, solar, and hydroelectric power plants. This method is not suitable for fuel cells because

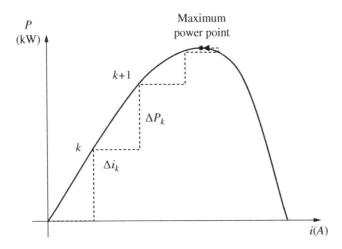

Figure 15.29 Typical shape of the power characteristic of electrical generators.

their maximum power does not coincide with their best efficiency ranges. In this case, the coincidence of maximum power and efficiency ranges is an important feature for hydrogen consumption and longer life span of cells, as discussed in Chapter 7. Chapter 13 deals with DG and its basic concepts, applications, and optimization. It assumes that the means of injection of energy into the grid discussed in this chapter serve as the connection between the renewable/alternative source of energy and the grid.

References

1 G. Venkataramanan and M. Illindala, Microgrids and sensitive loads, Proceedings of the IEEE Power Engineering Society Winter Meeting, Vol. 1, January 27–31, 2002, pp. 315–322.

2 R. Dugan and S. Price, Issues for distributed generation in the U.S., Proceedings of the IEEE Power Engineering Society Winter Meeting, Vol. 1, January 27–31, 2002, pp. 121–126.

3 M. Malinowski, M.P. Kazmierkowski, S. Hansen, F. Bllabjerb, and G.D. Marques, Virtual flux-based direct power control of three-phase PWM rectifiers, IEEE Transactions on Industry Applications, Vol. 37, No.4, pp. 1019–1027, 2001.

4 J.L. Duarte, A. Van Zwam, C. Wijnands, and A. Vandenput, Reference frames fit for controlling PWM rectifiers, IEEE Transactions on Industrial Electronics, Vol. 46, No. 3, pp. 628–630, 1999.

5 H. Stemmler and P. Guggenbach, Configurations of high-power voltage source inverter drives, Proceedings of the 5th European Conference on Power Electronics and Applications, Brighton, East Sussex, England, Vol. 5, 1993, pp. 7–14.

6 N. Mohan, T.M. Undeland, W.P. Robbins, Power Electronics: Converters, Application and Design, John Wiley & Sons, Inc., New York, 2003.

7 B.K. Bose, Modern Power Electronics and AC Drives, Prentice Hall, Upper Saddle River, NJ, 2001.

8 S. Nonaka and Y. Neba, A PWM GTO current source converter-inverter system with sinusoidal inputs and outputs, IEEE Transactions on Industry Applications, Vol. 25, No. 1, pp. 76–85, 1989.

9 F.Z. Peng, X. Yuan, X. Fang, and Z. Qian, Z-source inverter, IEEE Transactions on Industry Applications, Vol. 38, pp. 504–510, 2003.

10 P.C. Loh, D.M. Vilathgamuwa, Y.S. Lai, G.T. Chua, and Y. Li, Pulse-width modulation of Z-source inverters, Conference Record of the IEEE 39th Industry Applications Annual Meeting Conference, Vol. 1, October 3–7, 2004, pp. 155–162.

11 H. Fujita, Y. Watanabe, and H. Akagi, Control and analysis of a unified power flow controller, IEEE Transactions on Power Electronics, Vol. 14, No. 6, pp. 1021–1027, 1999.

12 M.A. Rahman, R.S. Radwan, A.M. Osheiba, and A.E. Lashine, Analysis of current controllers for a voltage source inverter, IEEE Transactions on Industrial Electronics, Vol. 44, No. 4, pp. 477–485, 1997.

13 H. Akagi, Y. Kanazawa, and A. Nabae, Instantaneous reactive power compensators comprising switching devices without energy storage components, IEEE Transactions on Industrial Applications, Vol. 20, No. 3, pp. 625–630, 1984.

14 E.H. Watanabe, R.M. Stephan, and M. Aredes, New concepts of instantaneous active power in electrical systems with generic loads, IEEE Transactions on Power Delivery, Vol. 8, No. 2, pp. 697–703, 1993.

15 P.G. Barbosa, L.G.B. Rolim, E. H. Watanabe, and R. Hanitsch, Control strategy for grid connected dc-ac converters with load power factor correction, IEE Proceedings: Generation, Transmission and Distribution, Vol. 145, No. 5, pp. 487–491, 1998.

16 V. Vlatkovic, Alternative energy: state of the art and implications on power electronics, Proceedings of the IEEE Applied Power Electronics Conference and Exposition, APEC'04, Vol. 1, 2004, pp. 45–50.

17 Z. Chen and E. Spooner, Voltage source inverters for high-power variable-voltage dc power sources, IEE Proceedings: Generation, Transmission and Distribution, Vol. 148, No. 5, pp. 439–447, 2001.

18 M. Mohr, B. Bierhoft, and F.W. Fuchs, Dimensioning of a current source inverter for the feed-in of electrical energy from fuel cells to the mains, Paper 41, presented at the Nordic Workshop on Power and Industrial Electronics, NORPIE 2004, Trondheim, Norway, June 14–16, 2004.

19 M.E. Ropp, M. Begovic, and A. Rohatgi, Prevention of islanding in grid-connected photovoltaic systems, Progress in Photovoltaics: Research and Applications, Vol. 7, pp. 39–50, 1999.

20 V. John, Y. Zhihong, and A. Kolwalkar, Investigation of anti-islanding protection of power converter based distributed generators using frequency domain analysis, IEEE Transactions on Power Electronics, Vol. 19, No. 5, pp. 1177–1183, 2004.

21 Y. Zhihong, A. Kolwalkar, and Y. Zhang, Evaluation of anti-islanding schemes based on nondetection zone concept, IEEE Transactions on Power Electronics, Vol. 19, No. 5, pp. 1171–1176, 2004.

22 K. Jia, T. Bi, B. Liu, D. Thomas, and A. Goodman, Advanced islanding detection utilized in distribution systems with DFIG, International Journal of Electrical Power & Energy Systems, Vol. 63, pp. 113–123, 2014.

23 A. Khamis, H. Shareef, A. Mohamed, and E. Bizkevelci, Islanding detection in a distributed generation integrated power system using phase space technique and probabilistic neural network, Neurocomputing, Vol. 148, pp. 587–599, 2015.

24 W.-J. Chiang, H.-L. Jou, J.-C. Wu, K.-D. Wu, and Y.-T. Feng, Active islanding detection method for the grid-connected photovoltaic generation system, Electric Power Systems Research, Vol. 80, pp. 372–9, 2010.

25 H. Zeineldin and J. Kirtley, A simple technique for islanding detection with negligible non-detection zone, Proceedings of the '09 IEEE Power & Energy Society General Meeting, PES, 2009, pp. 1–6.

26 W.Y. Teoh and C.W. Tan, An overview of islanding detection methods in photo-voltaic systems, World Academy of Science, Engineering and Technology, Vol. 58, pp. 674–682, 2011.

27 S. Syamsuddin, N.A. Rahim, and S.J. Krismadinata, Implementation of TMS320F2812 in islanding detection for Photovoltaic Grid Connected Inverter, International Conference for Technical Postgraduates (TECHPOS), 2009, pp. 1–5.

28 A. Timbus, A. Oudalov, and C.N.M. Ho, Islanding detection in smart grids, Energy Conversion Congress and Exposition (ECCE), Atlanta, GA, September 2010, pp. 3631–3637.

29 H.H. Zeineldin, E.F. El-Saadany, and M.M.A. Salama, Impact of DG interface control on islanding detection and nondetection zones, IEEE Transaction on Power Delivery, Vol. 21, No. 3, pp. 1515–1523, 2006.

30 W.C. Duesterhoeft, M.W. Schulz, Jr., and E. Clarke, Determination of instantaneous currents and voltages by means of alpha, beta, and zero components, AIEE Transactions, Vol. 70, pp. 1248–1255, 1951.

31 O. Elgerd, Electric Energy Systems Theory: An Introduction, Tata McGraw Hill, New York, 1971, p. 88.

32 R.H. Park, Two-reaction theory of synchronous machines generalized method of analysis-part I, Transactions of the American Institute of Electrical Engineers-Vol. 48, No. 9, pp. 716–727, 1929.

33 Texas Instruments, Software phase locked loop design using C2000 microcontrollers for three-phase grid connected applications, Application Report SPRABT4A, November 2013.

34 Y. Li, M. Vilathgamuwa, and P.C. Loh, Design, analysis, and real time testing of a controller for a multibus microgrid system, IEEE Transactions on Power Electronics, Vol. 19, No. 5, pp. 1195–1204, 2004.

35 M.G. Simões and F.A. Farret, Modeling and Analysis with Induction Generators, CRC Press, Boca Raton, FL, 2015.

16

Distributed Generation

16.1 Introduction

Economic and industrial growth in the twentieth century allowed electricity to be generated centrally and transported over long distances. Economies of scale in electricity generation led to an increase in power output and massive power systems. A balance of demand and supply was possible by the average combination of large, instantaneously varying loads. Security of supply increased because other power plants compensated the failure of one power plant in an interconnected system. For the purposes of this book, the electric power system grid is composed of the generation system, the transmission system, the distribution system, and the loads. Although the electric power market had experienced steady growth for several decades, changes in fuel economy, congestion, and required investments in transmission, distribution, and generation were required to meet an ever-increasing demand. This demand eventually reached a critical level that threatened system integrity, environment, reliability, and efficiency.

Deregulation of the power industry made the transmission network accessible in a nondiscriminatory manner. Distributed generation (DG) made remote central power plants a feasible and cost-effective strategy for power generation. This type of clean, full-time on-site power generation is based on technologies such as turbines and engines powered by natural gas, biogas, propane, wind, and small-scale hydropower as well as hydrogen-powered fuel cells and photovoltaic panels.

DG has the potential of being less costly, more efficient, and more reliable. Most existing power plants, central or distributed, deliver electricity to user sites at an overall fuel-to-electricity efficiency in the range 28–32%. This represents a loss of around 70% of the primary energy provided to the generator.

To reduce energy loss, it is necessary to increase the fuel-to-electricity efficiency of the generation plant or to use the waste heat. The use of waste heat in DG close to the user increases further the overall efficiency for space

Integration of Renewable Sources of Energy, Second Edition. Felix A. Farret and M. Godoy Simões.
© 2018 John Wiley & Sons, Inc. Published 2018 by John Wiley & Sons, Inc.

heating or industrial processes. The ability to avoid transmission losses and make effective use of waste heat makes on-site cogeneration or combined heat and power (CHP) systems 70–80% efficient.

Industrial, commercial, and residential DG systems can be tailored for efficiency improvement by using, for example, heat exchangers, solar-heated geothermal, absorption chillers, or desiccant dehumidification to reach overall fuel-to-useful energy efficiencies of more than 80%. For example, Capstone manufactures a 60-kW microturbine that uses waste heat to heat water. This system has an energy efficiency of fuel-to-useful energy that approaches 90%. The use of waste heat through cogeneration or combined cooling, heating, and power implies an integrated energy system that delivers both electricity and useful heat from an energy source.

Unlike electricity, heat cannot be transported over long distances easily or economically, so CHP systems typically provide heat for local use. Because electricity is transported more readily than heat, the generation of nearby heat to the load usually makes more sense than the generation of heat close to the generator site. Figure 16.1 illustrates total energy efficiency as a function of loading ratio. Three thermal recovery efficiencies are shown on the plot. Two systems assume separate generation of electricity and heat; the third is a CHP system. The assumed thermal generation efficiency for the non-CHP examples is 85%; the electrical efficiencies are 60% and 30%, respectively. If a loading ratio of 1 is assumed, the overall efficiencies of the separate systems are 70%

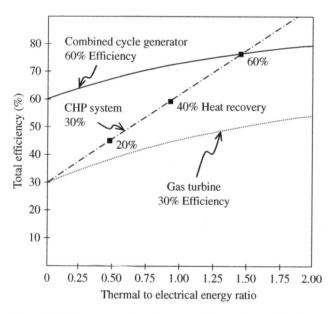

Figure 16.1 Comparison of total energy efficiency for combined power cycles.

and 44%. For the case in which the waste heat is near the heat load, it can be used instead of fuel to provide the required heat.

Typical thermal recovery efficiencies range from 20 to 80%, according to the size of the station and fuel as can be observed in Figure 16.2. Best estimates of the realizable average lifetime maximum fuel efficiency for various technologies of generation as a function of unit size show a definite economy of scale in all cases. The maximum ratio of heat to electricity is limited. For example, if the electrical efficiency is 30%, 70% of the fuel will result in waste heat. If this waste heat can be converted into useful heat, assuming a thermal recovery efficiency of 40%, the total energy efficiency is 58%, and the ratio of thermal energy to electrical energy is 1. This is the maximum loading and maximum total efficiency for this system. If the system is not loaded to this level, the total efficiency drops linearly. As observed in [1–4], CHP systems can greatly improve total energy efficiency through loading levels and thermal recovery efficiencies, as depicted in Figure 16.1.

Although there is still a considerable economy of scale in cost, the advantage that a large generator had over a small one constantly shrank down during the last half of the twentieth century as technological advance tended to improve small generator performance more than that of the larger units. Coupled with the increasing proportion of costs and losses that must be devoted to transmission and distribution, this trend was more than enough to make DG competitive against larger generators in many situations [5–9].

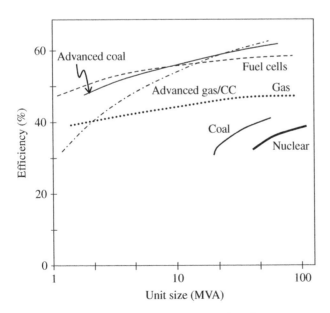

Figure 16.2 Comparison of energy efficiency for different unit sizes.

16.2 The Purpose of Distributed Generation

DG is the application of small generators from 15 to 10,000 kW scattered throughout a power system and either interacting with the grid or providing power to isolated sites. Dispersed generation is sometimes used as an interchangeable term, but it should be used for very small generation units, in the range 1–100 kW, sized to serve individual households or small businesses. DG technologies may be renewable (e.g., photovoltaic, thermal, wind, geothermal, and ocean-source systems) or nonrenewable (e.g., internal combustion engines, combined cycle engines, combustion turbines, microturbines, and fuel cells).

Household and rural users are concerned with the deployment of DG because of the overwhelming investments required to connect to a distant grid. For these users, DG is more economical than the central station system plus associated transmission and distribution expansion. Because of cost and reliability, industrial and commercial institutions may decide to install DG as a match with the electric utility system. This can happen when the particular application is of very high reliability and high cost or very low reliability and low cost, as discussed in [10]. The trend of DG to win at either end of the cost-reliability spectrum is depicted in Figure 16.3.

Reliability concerns (e.g., in medical, defense, and financial institutions) require uninterruptible power supply systems for conditioning input voltage. A backup generation system based on a rotating machine can be used as an

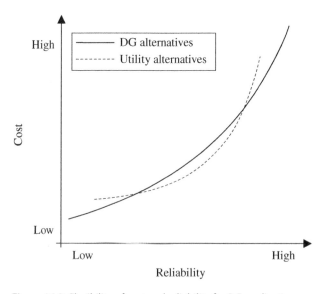

Figure 16.3 Flexibility of cost and reliability for DG applications.

alternative to a UPS system to supply power to sensitive loads during temporary interruptions in grid power. In this case, a full-fledged DG approach can be taken.

The benefits associated with DG projects are described below.

16.2.1 Modularity

Central station power relies on the economy of scale to increase efficiency and reduce the cost of power generation (i.e., with the approach of building larger generators to amortize costs over a larger consumer base). Distributed power relies on the economy of production to reduce the cost of power generation. The capital investment made in a manufacturing plant produces thousands of low-cost units, which can penetrate markets that large-scale generation cannot reach. This model naturally leads to modular unit sizing. Modular system design increases the ability of a system designer to respond quickly and locally to near-term load growth without risking large sums of capital on capacity that may not be used in the future.

16.2.2 Efficiency

Central station generation has developed energy-efficient technologies such as natural gas combined cycle systems. However, these technologies require continuous, reliable, and inexpensive access to a natural gas fuel source and are unsuited for small-scale deployment. DG technologies generally use renewable resources with high-energy efficiencies. Renewable energy technologies use fundamentally local energy supplies, and some nonrenewable DG technologies use easily transported fuels other than natural gas. Other resources—such as energy storage, cogeneration, and demand-control devices—help improve the efficiency of whatever source of generation is being used. Increased energy efficiency decreases both energy costs and greenhouse gas emissions per unit of generation [1–4].

16.2.3 Low or No Emissions

DG that uses renewable resources is inherently emission-free. However, some advanced distributed power technologies can also reduce emissions of conventional fossil fuels such as oil, natural gas, biogas, and propane. They accomplish this through increased efficiency and alternative energy conversion processes, such as found in a fuel cell, CO sequestration reform, and production of gas.

16.2.4 Security

This includes system reliability and power quality indices like those listed in Table 16.1. Distributed power provides inherent redundancy. When one on-site generator fails, the spare capacity in the remaining system resources

Table 16.1 Reliability and power quality indices.

Measurement	Formula
System average interruption frequency	$SAIFI = \dfrac{\sum N_{sustained}}{N_{served}}$
System average interruption duration	$SAIDI = \dfrac{\sum \left(N_{sustained} D_{sustained} \right)}{N_{served}}$
Customer average interruption frequency	$CAIFI = \dfrac{\sum N_{sustained}}{\sum N_{affected}}$
Customer average interruption duration	$CAIDI = \dfrac{\sum \left(N_{sustained} D_{sustained} \right)}{N_{sustained}}$
Average system interruption frequency	$ASIFI = \dfrac{\sum kVA_{sustained}}{kVA_{served}}$
Average system interruption duration	$ASIDI = \dfrac{\sum \left(kVA_{sustained} D_{sustained} \right)}{N_{served}}$
Momentary average interruption	$MAIFI = \dfrac{\sum N_{momentary}}{N_{served}}$
Average system RMS variation frequency	$SARFI_x = \dfrac{\sum N_{sag}}{N_{served}}$

can provide instantaneous reserve power (typically known as spinning reserve). Even if the primary generator fails, critical loads can be supported from on-site generators or overall system capacity. Distributed resources support power quality by preventing system-wise problems and mitigating line problems before a consumer load detects them.

16.2.5 Load Management

This implies modifying the load profile by peak-load clipping, valley filling, load shifting, reducing voltage (brownout), reducing load, and load building. Conservation of energy involves reducing the entire energy load. Energy efficiency measures fall under this category. Conservation can also be interpreted as reducing waste and reusing waste for production of energy.

Demand-side management (DSM) means modifying energy use to maximize energy efficiency, as discussed in Section 16.4. In contrast to supply-side strategies, which increase or redistribute supplies (by building new power plants or changing system reconfigurations), the goal of DSM is to smooth out the peaks and valleys in electric (or gas) demand. It makes the most efficient use of energy resources and defers the need to develop new power plants. This may entail shifting energy use to off-peak hours, reducing overall energy requirements, or increasing demand for energy during off-peak hours. DSM strategies can be classified as peak-clipping or valley-filling strategies. In peak-clipping strategy, a controller seeks to reduce energy consumption at the time of peak load. In programs to reduce peak load, the utility or consumer generally exerts control over appliances such as air conditioners or water heaters.

DG systems such as photovoltaic and solar thermal power supplies can play a role in clipping peak demand if it coincides with their output. In valley-filling strategies, the goal is to build up off-peak loads to smooth out the load and improve the economic efficiency of the utility. An example of valley filling is charging electric vehicles or electrolyzing water to produce hydrogen and oxygen at night, when the utility is not required to generate as much power as during the day. Large battery storage can also be operated at night, and the energy stored can be used for peak clipping during the day. Economic assessment, including charging and discharging losses, must be accounted for to validate the feasibility of this option.

Another possible strategy is load shifting, in which thermal energy storage enables a consumer to use electricity to make ice or chilled water at night, when overall electricity consumption is low. The ice or chilled water is then used to cool buildings during the day, when overall electricity consumption is high. Water heating is another application that is widely used during off-peak hours.

Figure 16.4 illustrates a reason that utilities invest in small DG modules, which can be added closely in step with demand. The marked areas show construction and financing times required by central units. In addition to avoiding overshooting, small units have short lead times and reduce the risk of buying

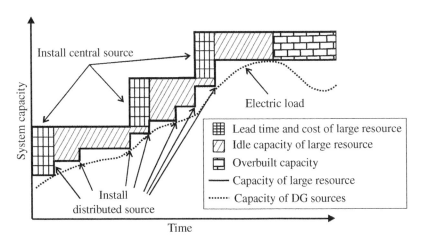

Figure 16.4 Addition of central or distributed sources to match demand.

technology that will become obsolete. Distributed resources also exploit agile manufacturing techniques with a standardized workforce, so they have fewer turnovers, less retraining, and better management than that for a very long project.

16.3 Sizing and Siting of Distributed Generation

In several US states and in other countries, consumers can install small utility grid-connected renewable energy systems, such as solar or wind systems, to reduce their electricity bills using a protocol called net metering. Under net metering, electricity produced by renewable energy systems can flow into the utility grid and spin the electricity meter backward when the production of energy is more than local consumption. Other than the renewable energy, a system does not need any special equipment.

Even in the absence of net metering, consumers can use the electricity they produce to offset their electricity demand instantaneously. However, if the consumer produces excess electricity (beyond what is needed to meet his own needs at the time), the utility purchases that electricity at the wholesale or "avoided cost" price, which is lower than the retail price. Net metering simplifies this arrangement by allowing the consumer to use excess electricity to offset electricity used at other times during the billing period.

There are three reasons to support net metering procedures. First, as consumers that are more residential install renewable energy systems, they are getting used to standardized protocol for connecting their systems with the electricity grid and ensuring safety and power quality. Second, most residential consumers are not at home to use electricity during the day, and net metering allows them to receive full value for the electricity they produce then without expensive battery storage systems. Third, net metering provides a simple, inexpensive, and easily administered mechanism for encouraging the use of renewable energy systems, which provides important local, national, and global benefits such as economic development, electrical energy availability, and a cleaner environment. As a result, microgrid architectures can be alternative source-based, in which local generators supply electricity for industrial and commercial needs, or renewable-based, in which residential consumers install dispersed generation under net metering policies. For utility companies, a major economical drawback of net metering is accounting in tariffs for the costs of installation and maintenance of the transmission and distribution systems used by energy-producing consumers [5–8].

The addition of DG to the energy matrix has significant effects on the power quality of the system. The power quality issues related to DG include sustained interruptions, voltage regulation, overall stability, and harmonics. To size DG on preexisting grids, the voltage profile and losses can be analyzed. Voltage regulation plays a vital role in determining how much DG can be accommodated on a distribution feeder while keeping the voltage level of the system within specified limits. Studying the effects of DG on voltage level and system losses, the optimal size and location of DG can be identified. Under this approach, a model of the system with all the operating scenarios during open and closed states of the circuit breakers must be formed. DG injection is assumed in different scenarios, and power flow analysis is conducted for the scenarios. The result of the power flow study gives the voltage level of the buses and losses in the system. Voltage level and voltage regulation at the consumer end are then tabulated in the order of their degree of impact to find voltage regulation and system losses for critical cases. Then DG optimal size and location are identified [9, 10].

16.4 Demand-Side Management

Conventional operation of grid power systems depended on changing generation to match demand. As suggested in Figure 16.5, DG can be applied for contingency capacity support at only one feeder (higher priority) or both feeders. Various methods can then be used to change load according to the needs of generation (e.g., DSM).

The most obvious example is a low-price electricity tariff to maintain load for nuclear power and large coal-burning stations at night. There are several options for implementing nighttime tariff measures [11–14]:

- Consumer meters are switched to different energy flow registers by a local clock or a radio signal (such as in the United Kingdom, often named "Economy 7"), which allows all load to change tariff. Consumers' bills show energy consumed at each tariff, and they are charged accordingly.

(a)

(b)

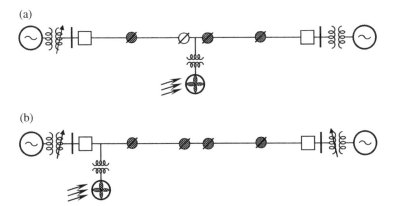

Figure 16.5 DG applied for contingency capacity support: (a) support for both feeders and (b) support for only one feeder.

- Specific loads (e.g., storage heaters and water-tank heaters) are enabled by time clocks or ripple control (a signal "down the wire"), as is common in Australia for water heaters and in the United Kingdom for space heaters. These loads are wired on a separate circuit with a separate meter register.

The technology for remote switching (ripple control) is widely available (e.g., from Landis + Gyr and Siemens). There are many ways to communicate such control, including through long-wave radios, mobile telephone networks, and signals sent on power lines. Modern digital electronics offer accurate, low-cost communication for tariff information or direct control. In addition, modern communication methods allow remote metering to measure and monitor electricity consumption at short intervals (e.g., minutes). Analysis of such information allows recommendations for DSM and reduced consumption. The communication technologies for remote metering are similar to those needed for remote switching.

The range of options for DSM, whereby consumers reduce electricity costs by responding over short periods to options offered by suppliers, is large. Most attention is given to peak shaving, but a range of techniques, tariffs, and other methods (e.g., on-site generation, interruptible loads, real-time pricing, and peak pricing) are lumped together as demand response.

Storage systems for electricity regeneration absorb electrical power from the grid, store it as potential energy, and then release it to regenerate electrical power back into the grid. The operation is usually a commercial enterprise. Examples are large-scale pumped-hydro reservoirs, compressed gas accumulators, and flywheel motor/generators. Such systems are called dedicated storage. They are judged by their efficiency and the cost of absorbing electrical energy and regenerating it back into the grid when required.

16.5 Optimal Location of Distributed Energy Sources

Advances in generation technology and new directions in electricity industry regulation should cause a significant increase in DG use all over the world (Table 16.2). It is predicted that small sources of energy may account for up to 25% of all new generation units online since 2010. It is critical that distribution system effects be assessed as accurately as possible to avoid degradation in efficiency, power quality, reliability, and control.

The energy-saving policy used in the past years has resulted in a significant growth reduction of electric loads. To continue this progress, a reformulation of the traditional strategies of centralization and increases of generating unit sizes, very typical during the 1970s and 1980s, is needed. Of course, the traditional way demands massive investments in large power plants with relatively

Table 16.2 Factors influencing realized load reductions.

Aspect	Examples of effect on savings
Lighting quality	Lower lumens or light quality from a compact fluorescent bulb may result in the owner turning on more light fixtures than were used previously
Comfort level	A more efficient device lowers the customer's bill, but the lower bill may encourage the customer to set the thermostat to a more comfortable setting and thus consumes some of the expected energy savings
Output level	An industrial customer's use may vary with plant production levels. Changes due to plant production have to be separated from energy saving because of DSM measures
Weather	Moderate weather may result in greater-than-expected energy savings. Similarly, extreme weather conditions could reduce expected savings by triggering more energy-consuming alternatives
System balance	Benefits from DSM measures often depend on decisions made regarding other aspects of a user facility. Power consumption decreases with the square of voltage reduction, but it may be compensated by an increase in bulb wattage to improve illumination levels. Better housing insulation may be compensated by more efficient energy supply

Source: California Energy Commission.

long return periods and an increase in the length of networks, which, in some ways, causes negative effects on the environment and electrical losses. The appearance of new and efficient generation technologies increases the cost of electric power transmission and distribution. Restructuring of the energy sector, together with the previous considerations, will lead to the growth of energy consumption in the near future, which will be satisfied primarily through generation resources (including alternative sources) with relatively low output powers located along distribution systems [10–14].

Most alternative source generating units belong to independent producers, who began to have relatively free access to the power system after the energy sector's deregulation. However, this causes some conflicts of interest. The energy supply companies do not have an interest in reducing their energy revenues. In addition, if installed in a random way, distributed energy resources may hinder the operational control of distribution systems. It is therefore necessary to emphasize that DG located at strategic points of distribution systems allows energy distributors to reduce investments in development (e.g., in the reinforcement of distribution lines or additional control equipment) and cuts operational costs (through reductions of power and energy losses and increases in the reliability of supply and quality of marketable energy).

In the following sections, we analyze important aspects of DG optimal location for power distribution companies.

16.5.1 DG Influence on Power and Energy Losses

Power losses in distribution systems can be reduced by altering load flows through certain network sections, located between substations and the DG installation point. For a quantitative estimation of loss alterations, a hypothetical feeder with uniformly distributed load is analyzed. Suppose that a linear distribution of load along the distribution line such that $J_0 = I_0/\ell$ is the feeder load per unit of length of a unitary line resistance R_0. In any line section of the feeder, as depicted in Figure 16.6a, with a current J_0 through it, the three-phase power losses of this configuration, during any random period of time and at a point of distance x from the beginning of the section, are defined by

$$\Delta P = \int_0^\ell 3\left(J_0 x\right)^2 R_0 dx = 3J_0^2 R_0 \int_0^\ell x^2 dx = J_0^2 R_0 \ell^3 \tag{16.1}$$

where:

$J_0 = $ The linear density of phase current in A/km
$x = $ The distance from the line end in kilometers

(a)

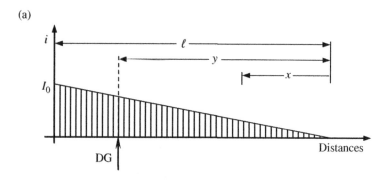

(b)

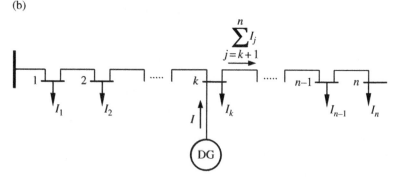

Figure 16.6 Examples of distributed loads and concentrated DG: (a) a feeder with uniformly distributed loads and DG integration and (b) a feeder with discrete loads.

Assume that the DG power is equivalent to a load on the distribution system and that it is installed a distance y from the network line end. In this case, power losses are calculated by

$$\Delta P = \int_0^y 3(J_0 x)^2 R_0 dx + \int_y^\ell 3(J_0 x - I)^2 R_0 dx$$

$$= J_0^2 \ell^3 R_0 - 3J_0 I \ell^2 R_0 + 3J_0 I y^2 R_0 + 3I^2 \ell R_0 - 3I^2 y R_0$$

(16.2)

where:

I = The phase current in amperes
ℓ = The length of the line in kilometers

The coordinates of the DG connecting point may then be defined to guarantee a minimum of power losses:

$$\frac{\partial(\Delta P)}{\partial y} = 6J_0 I y R_0 - 3I^2 R_0 = 0$$

By this condition, it follows that

$$y = \frac{I^2 R_0}{2J_0 I R_0} = \frac{I}{2J_0}$$

(16.3)

Equation (16.3) means that in the case of a uniformly distributed load along the line, its middle point would be the best location for the alternative source. That is, when half the current injected by the renewable source fully compensates or minimizes the current going to each half of the feeder, the power losses will be a minimum because the load current will be a minimum. The example analyzed here does not correspond to a real load distribution in distribution networks with discrete locations of random intensity and uneven dispersal. Nevertheless, the results can be used to construct rules that define the best location of a DG for power loss minimization. For example, point k in Figure 16.6b, which may be a considerable point for connecting DG, is defined for minimum power losses as

$$\left| \frac{I - I_k}{2} - \sum_{i=k+1}^n I_i \right| \to min$$

(16.4)

where:

I = The DG injection current into the network
I_k = The load node under consideration

The DG location for power loss minimization becomes appropriate when the resources are used for a limited time (e.g., only during the load peak period).

Considering the daily alterations of node loads, their heterogeneity, and that DG can generate energy with constant power during the entire day, the condition presented in equation (16.4), in general, does not guarantee minimum energy losses in the same feeder for a long period (e.g., for an entire day). Naturally, considering node load alterations and a DG constant output, the solution for optimal location, based on equation (16.4), may be different for distinct periods. The difference between the connecting point defined by the analysis of the maximum demand state and the point that guarantees minimum energy losses depends on the consumer load curve.

Returning to the analysis of a feeder with uniformly distributed load, the coordinates of the DG point of connection can be defined for energy loss minimization, assuming that the generator operates at constant load during the entire day. In this case, energy losses can be defined as

$$
\begin{aligned}
\Delta W &= \int_0^y 3\sum_{t=1}^T \left(J_{0t}x\right)^2 R_0 dx + \int_y^\ell 3\sum_{t=1}^T \left(J_{0t}x - I\right)^2 R_0 dx \\
&= R_0 y^3 \sum_{t=1}^T J_{0t}^2 + R_0 \ell^3 \sum_{t=1}^T J_{0t}^2 - R_0 y^3 \sum_{t=1}^T J_{0t}^2 - 3R_0 \ell^2 \sum_{t=1}^T J_{0t}I + \cdots \\
&\quad + 3R_0 y^2 \sum_{t=1}^T J_{0t}I + 3R_0 \ell \sum_{t=1}^T I^2 - 3R_0 y \sum_{t=1}^T I^2 \\
&= R_0 \ell^3 \sum_{t=1}^T J_{0t}^2 - 3R_0 \ell^2 \sum_{t=1}^T J_{0t}I + 3R_0 y^2 \sum_{t=1}^T J_{0t}I + 3R_0 \ell \sum_{t=1}^T I^2 - 3R_0 y \sum_{t=1}^T I^2
\end{aligned}
$$

$$(16.5)$$

where:

T = The time duration of DG injection

Similar to equation (16.3), the condition of a DG connection is defined to guarantee minimum daily energy losses in a feeder with uniformly distributed loads:

$$
\frac{\partial(\Delta W)}{\partial y} = 6R_0 y + I\sum_{t=1}^T J_{0t} - 3R_0 TI^2 = 0 \tag{16.6}
$$

From equation (16.6), it follows that

$$
y = \frac{TI}{2\sum_{t=1}^T J_{0t}} \tag{16.7}
$$

Assuming now that the voltage across the FD and the power factor across the terminals of the distribution transformers do not change during the period T (e.g., one day), equation (16.7) can be transformed into the following form:

$$y = \frac{\sqrt{3} \, V_n \, \cos \phi \, TI}{2\sqrt{3} \, V_n \, \cos \phi \sum\limits_{t=1}^{T} J_{0t}} = \frac{W_d}{2W_0} \tag{16.8}$$

where:

$W_d =$ The DG-generated energy during period T
$W_0 =$ The energy consumed in the distribution system

So the optimal point for a DG connection depends on the amount of energy it generates and on the energy consumed by the loads connected across the FD. To define the coordinates of the best connecting point, initially it is necessary to find the energy flow using consumption data (for the entire period T) of the distribution transformer loads in the FD analyzed. In this way, under the point of view of minimization of energy losses, it is reasonable to connect the DG across the terminal of the distribution transformer k for which the following condition is satisfied:

$$\left| \frac{W_d - W_k}{2} - \sum_{i=k+1}^{n} W_i \right| \rightarrow min \tag{16.9}$$

where:

$W_k =$ The energy consumption of the distribution transformer k during the period T

$\sum_{i=k+1}^{n} W_i =$ The summation of the energy consumption during period T of all distribution transformers located after point k, for which installation of DG is foreseen

When the load elements of a load curve are arranged in the order of descending magnitudes, the curve thus obtained is called a load duration curve. The load duration curve is obtained from the same data as the load curve, but the ordinates are arranged in the order of descending magnitudes. In other words, the maximum load is represented to the left, and decreasing loads are represented to the right in the descending order. Hence, the area under the load duration curve and the area under the load curve are equal. As an example, assume that the FD contains 15 uniformly distributed distribution transformers with identical daily load duration curves like the ones represented in Figure 16.7.

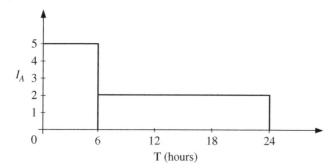

Figure 16.7 Load duration curve of distribution transformers.

Table 16.3 Power and energy losses related to various DG locations.

	DG installation point						
See Figure 16.6	9	10	11	12	13	14	Equation
P_{max} (kW)	3.56	3.3	3.11	3.0	2.96	2.99	(16.4)
W (kWh)	25.07	23.63	23.18	23.72	25.25	27.77	(16.9)

The DG used in this example has the following parameters: $S_n = 600\,kVA$, $I = 25\,A$, $cos\,\varphi = 1$, and $V_n = 13.8\,kV$. The calculation presented in Table 16.3 is only for adjacent points 9–14. The minimum $P_{max}(kW)$ of equation (16.4) and minimum $W(kWh)$ of equation (16.9) show that for power loss minimization within the load peak period, it is reasonable to locate the source at node 13.

Considering that daily consumption in the network nodes is 1450 kWh and supposing that the DG source is working 24 hours a day at constant load (which corresponds to 14,324 kWh), the optimal point of DG location, taking into account the minimization of energy losses, would be node 11. This solution is based on equation (16.7) and verified with the data of Table 16.3.

16.5.2 Estimation of DG Influence on Power Losses of Sub-transmission Systems

Load alteration or redistribution in distribution networks causes alterations in the operational state of sub-transmission systems. Particularly, the installation of DG on distribution systems not only causes reduction of loads in the same networks and substation system (SS) transformers but also alters the load flow (and consequently, power losses) in the networks of sub-transmission systems. For this reason, when selecting the optimal location of DG, it is desirable to estimate its influence on the sub-transmission system losses. According to Ref. [12], by the classical definition of complex power, power losses in the system are defined by

$$\Delta P + j\Delta Q = 3[V]^t [I]^*$$ (16.10)

where:

* = The conjugated symbol
t = Transposed matrix

Considering that $[V] = [Z][I]$ and that matrix $[Z]$ is symmetrical, the previous expression can be transformed into equation (16.10):

$$\Delta P + j\Delta Q = 3[I]^t [Z][I]^*$$ (16.11)

Assuming that $[Z] = [R] + j[X]$ and $[I] = [I_p] + j[I_q]$, equation (16.11) becomes

$$\Delta P + j\Delta Q = 3\left(\left[I_p\right] + j\left[I_q\right]\right)^t \left([R] + j[X]\right)\left(\left[I_p\right] - j\left[I_q\right]\right)$$ (16.12)

where the term of interest is

$$\Delta P = 3\left(\left[I_p\right]^t [R]\left[I_p\right] + \left[I_p\right]^t [X]\left[I_q\right] + \left[I_q\right]^t [R]\left[I_q\right] - \left[I_q\right]^t [X]\left[I_p\right]\right)$$
$$= 3\left(\left[I_p\right]^t [R]\left[I_p\right] + \left[I_q\right]^t [R]\left[I_q\right]\right)$$ (16.13)

where:

p and q = Active and reactive components, respectively

Again, assuming that I is the vector of the initial loads of each transformer in the SS and I' is the load vector considering the introduction of the DG in the distribution network of one of the SS, the reduction in power losses in the sub-transmission system lines will be given by

$$\delta(\Delta P) = 3\left\{\left[I_p\right]^t [R]\left[I_p\right] + \left[I_q\right]^t [R]\left[I_q\right] - \left[I_p'\right]^t [R]\left[I_p'\right] - \left[I_q'\right]^t [R]\left[I_q'\right]\right\}$$ (16.14)

or, after some simplifications:

$$\delta(\Delta P) = 3\left\{[I]^t [R][I] - [I']^t [R][I']\right\}$$ (16.15)

For instance, supposing that

$$
I = \begin{bmatrix} I_1 \\ I_2 \\ I_3 \end{bmatrix} \qquad I' = \begin{bmatrix} I_1 \\ I_2 \\ I_3 - \Delta I \end{bmatrix} \tag{16.16}
$$

where:

ΔI = The load reduction in one of the system transformers related to the DG implantation on the distribution network

In this case, we have

$$
\begin{aligned}
\delta(\Delta P)_3 &= 3 \left\{ \begin{bmatrix} I_1 I_2 I_3 \end{bmatrix} \begin{bmatrix} R_{11} & R_{12} & R_{13} \\ R_{21} & R_{22} & R_{23} \\ R_{31} & R_{32} & R_{33} \end{bmatrix} \begin{bmatrix} I_1 \\ I_2 \\ I_3 \end{bmatrix} \right\} \\
&\quad -3 \left\{ \begin{bmatrix} I_1 I_2 I_3 - \Delta I \end{bmatrix} \begin{bmatrix} R_{11} & R_{12} & R_{13} \\ R_{21} & R_{22} & R_{23} \\ R_{31} & R_{32} & R_{33} \end{bmatrix} - \begin{bmatrix} I_1 \\ I_2 \\ I_3 - \Delta I \end{bmatrix} \right\} \\
&= 3 \left\{ \Delta I^2 (R_{31} + R_{32} + R_{33}) - \Delta I \big[I_1 (R_{31} + R_{11} + R_{32} + R_{12} + R_{33} + R_{13}) \right. \\
&\quad \left. + I_2 (R_{31} + R_{21} + R_{32} + R_{22} + R_{33} + R_{23}) + I_3 (2R_{31} + 2R_{32} + 2R_{33}) \big] \right\}
\end{aligned}
\tag{16.17}
$$

where:

R_{ij} for $i, j = 1, 2, 3$ = Line resistances of phases 1, 2, and 3 when $i = j$ or the resistance between lines i and j when $i \neq j$

With these results, it is possible to build an equation to calculate the alterations of power losses in the lines of the sub-transmission system regarding the use of a DG source located in the distribution system belonging to the arbitrary SS m as

$$
\delta(\Delta P)_m = 3 \left\{ \Delta I_m^2 \sum_{\ell=1}^{n} R_{m\ell} - \Delta I_m \left[\sum_{\substack{i=1 \\ i \neq m}}^{n} I_i \left(\sum_{j=1}^{n} R_{ij} + \sum_{j=1}^{n} R_{mj} \right) + 2 I_m \sum_{\ell=1}^{n} R_{m\ell} \right] \right\}
\tag{16.18}
$$

where:

n = The total number of SS in the network

However, in most cases, power and distribution systems are operated separately. This causes difficulties in obtaining the necessary information related to sub-transmission systems during the strategic or operational planning of the distribution systems. At the same time, to solve these problems, it is often necessary to perform multiple experimental calculations. For this reason, it is reasonable to create a simplified equivalent of the sub-transmission system that can be used by the distribution department to assist its own interests. Evidently, this equivalence cannot have a universal character, but it should be suitable for the solution of an exact functional problem. As a rule, it is necessary to build a mathematical model (based on equation (16.11)) to estimate power loss alterations in the sub-transmission lines and choose the optimal DG location. To realize this task, a mathematical tool known as *experimental design* can be used.

16.5.3 Equivalent of Sub-transmission Systems Using Experimental Design

The choice of experimental design [13] as an instrument for the construction of functional models of sub-transmission systems is based on three considerations. First, this approach demands a minimum number of experimental calculations to construct a multifactor model. Second, unlike multiple regression analysis, this method produces a well-formalized statistical analysis that includes the significant estimation of each factor together with the adequacy of the model. Finally, in case the simplest (linear) models are not appropriate, experimental design allows the construction of more complex (nonlinear) models using the results of preliminary stages. Of course, these circumstances simplify the construction of complex system models. Experimental design is based on the study of changes in the response function due to factor alterations [9–12].

In general, two factor levels are considered, which correspond to their maximum and minimum values (x_p^+, x_p^-). In the case of considering all possible combinations of factor levels, it is necessary to realize $N = 2^k$ tests (where k is the combinational number of factors) to allow construction of a linear model. In general, the normalized values of the given factors are used as

$$\tilde{x}_p = \frac{x_p - x_p^0}{\Delta x_p}, \quad p = 1,\ldots,k$$

$$x_p^0 = \frac{x_p^- + x_p^+}{2}, \quad \Delta x_p = \frac{\left(x_p^+ - x_p^-\right)}{2}, \quad p = 1,\ldots,k \tag{16.19}$$

Factors x_p^0 are the common factors and x_p are the differential factors. This simplifies model construction and statistical analysis. Such an approach permits one to build a model as follows:

$$y = b_0 + \sum_{p=1}^{k} b_p \tilde{x}_p + \sum_{\substack{p=1 \\ p<q}}^{k} b_{pq} \tilde{x}_p \tilde{x}_q + \sum_{\substack{p=1 \\ p<q<r}}^{k} b_{pqr} \tilde{x}_p \tilde{x}_q \tilde{x}_r + \cdots \qquad (16.20)$$

where the coefficients of equation (16.20) are defined as

$$\begin{aligned}
\tilde{b}_0 &= \frac{1}{N} \sum_{n=1}^{N} \tilde{x}_{np} y_n, \\
\tilde{b}_{pq} &= \frac{1}{N} \sum_{n=1}^{N} \tilde{x}_{np} \tilde{x}_{nq} y_n, && p,q,r = 0,1,\ldots,s \\
\tilde{b}_{pqr} &= \frac{1}{N} \sum_{n=1}^{N} \tilde{x}_{np} \tilde{x}_{nq} \tilde{x}_{nr} y_n, && p \neq q \neq r
\end{aligned} \qquad (16.21)$$

Obviously, the increase in factors quickly increases the number of tests necessary for the construction of the model. At the same time, any changes inside the model increase in the number of members that reflect interactions among the factors. According to early experiences in the process of following statistical analysis, such members become insignificant most of the time. This serves as a base for the use of fractional experimental design. In this case, the interactions among factors that can be insignificant are used for presentation of new factors.

As an example, a simplified sub-transmission system is presented in Figure 16.8. Analysis of equation (16.18) allows one to define the following factors that can be included in the model for the estimation of the alteration of power losses in relation to the system presented in Figure 16.8. Assuming that one neglects interactions among the significance factors to construct the model, it is possible to use fractional experimental design 2, shown in Table 16.4.[7-14] To adjust the factor values in each test, it is first necessary to define the possible alterations of SS loads (considering their daily variations). The model has the following form:

$$\delta(\Delta P) = b_0 + b_1 \tilde{x}_1 + b_2 \tilde{x}_2 + b_3 \tilde{x}_3 + b_4 \tilde{x}_4 + b_5 \tilde{x}_5 + b_6 \tilde{x}_6 + b_7 \tilde{x}_7 \qquad (16.22)$$

It is important to emphasize that computer experiments do not permit the use of formal statistical analysis. In this case, to estimate the significance factors and verify model adequacy, some adaptation, as proposed in Ref. [14], is used:

$$\Delta I_3 = x_1, \quad \Delta I_3 I_3 = x_2, \quad \Delta I_3 I_1 = x_3, \quad \Delta I_3 I_2 = x_4,$$
$$\Delta I_3 I_4 = x_5, \quad \Delta I_3 I_5 = x_6, \quad \Delta I_3 I_6 = x_7$$

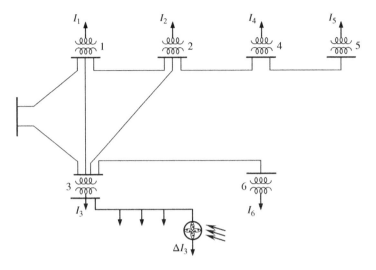

Figure 16.8 Sub-transmission system.

Table 16.4 Matrix of the experimental design 2^{7-14}.

N	x_0	x_1	x_2	x_3	$x_6=x_1x_2$	$x_7=x_1x_3$	$x_5=x_2x_3$	$x_4=x_1x_2x_3$	Y
1	+1	−1	−1	−1	+1	+1	+1	−1	y_1
2	+1	+1	−1	−1	−1	−1	+1	+1	y_2
3	+1	−1	+1	−1	−1	+1	−1	+1	y_3
4	+1	+1	+1	−1	+1	−1	−1	−1	y_4
5	+1	−1	−1	+1	+1	−1	−1	+1	y_5
6	+1	+1	−1	+1	−1	+1	−1	−1	y_6
7	+1	−1	+1	+1	−1	−1	+1	−1	y_7
8	+1	+1	+1	+1	+1	+1	+1	+1	y_8

16.6 Algorithm of Multicriterial Analysis

Reduction of power and energy losses is just one positive factor associated with the appropriate location of DG in the distribution system. In most cases, this indicator cannot serve as the only criterion for decision-making. The DG influences several operational characteristics of an electric network, including the voltage and reactive power operating modes, the loading level of the elements, and the system reliability. Each of these can be defined quantitatively.

Energy quality—in particular, voltage levels—can be estimated by the energy consumed out of the standard voltage deviations. The quantitative characteristics of reliability can be presented through an integral indicator. Evidently, calculations of each of these characteristics and the analysis of their relationships with DG demand special research and development of appropriate methods.

In the same way, building DG projects is very complex. In the process of decision-making, it is impossible to neglect cogeneration efficiency, necessary infrastructure, and safety of operation. In the preliminary analysis, there is no possibility of accomplishing a detailed project for each alternative point of installation. Expert estimates can be used, in particular, under a linguistic form. Therefore, during the process of decision-making, several quantitative and qualitative criteria should be taken into account.

This book uses an approach based on an algorithm of Bellman–Zadeh [13–17]:

$$\tilde{A}_j = \left\{ X, \mu_{A_j}(X) \right\}, \quad X \in D_x, j = 1, \ldots, n$$

where:

$\mu_{A_j}(X) =$ Membership function of $\tilde{A}_j$

Initially, with this algorithm, all objective functions $F_j(X), X \in D_x, j = 1, \ldots, n$ are represented by fuzzy objective functions. The primary problem with the Bellman–Zadeh approach is the formation of membership functions, which should be concave and reflect the proximity level of each objective function with respect to its own optimal solution. Experience with this approach shows the efficiency of the use of membership functions determined according to equations (16.23) and (16.24) [13–17]:

$$\mu_{A_j}(x) = \left[\frac{F_j(x) - \min_{x \in D_x} F_j(x)}{\max_{x \in D_x} F_j(x) - \min_{x \in D_x} F_j(x)} \right] \tag{16.23}$$

for objective functions that should be maximized and

$$\mu_{A_j}(x) = \left[\frac{\max_{x \in D_x} F_j(x) - F_j(x)}{\max_{x \in D_x} F_j(x) - \min_{x \in D_x} F_j(x)} \right] \tag{16.24}$$

for objective functions that should be minimized.

As presented in Refs. [15–19], equation (16.25) defines a fuzzy solution $\tilde{D}$ of the initial problem:

$$D = \bigcap_{j=1}^{n} \tilde{A}_j \tag{16.25}$$

In this case, the membership function of the fuzzy solution is calculated by

$$\mu_D(x) = \overset{n}{\underset{j=1}{\wedge}} \mu_{A_j}(x) = \underset{j=1,\ldots n}{min} \mu_{A_j}(x), \quad x \in D_x \tag{16.26}$$

According to the proposed algorithm, the optimal solution is the one that presents a maximum value of the membership function:

$$\underset{x \in D_x}{max} \mu_D(x) = \underset{x \in D_x}{max} \underset{j=1,\ldots,n}{min} \mu_{A_j}(x) \tag{16.27}$$

A general multicriterial approach to resolve the problem of DG optimal location in distribution systems shows that the optimal location of sources of DG allows a general increase in the efficiency of the generating units, including capability and energy quality. Special attention has to be given to the reduction of power and energy losses. The model of a sub-transmission system presented in this section reflects loss reduction on DG sources located in distribution systems. Other legal and technical restrictions related to integration and to interconnection of DG are dealt with in Chapters 12 and 15, respectively.

16.6.1 Voltage Quality in DG Systems

The maximum limit of DG installation on a feeder will depend on the thermal limits of the conductors, circuit breakers, fuses, switches, and any effects on the voltage quality. A utility is a power quality concern related to the voltage quality and reliability of the system. Therefore, the power quality of a power system with DG can be evaluated in steady state by their voltage profile and line losses. Exceeding the feeder capacity will make system losses increase. The economic penalty is not sufficient to drive investment decision and higher losses alone to not affect system reliability. Higher currents may lead to thermal loss of life in transformers and other equipment, which may lead to service interruptions, which is an indirect effect on system reliability. If the limit of DG installation is reached, excessive voltage drop may occur, but there is a time limitation (if short it is called voltage sag), and voltage regulators (VSs) (autotransformers with automatic control) can correct such a voltage drop. Figure 16.9 shows a simple equivalent circuit for the evaluation of voltage effect of a load connected to an ideal source. The voltage variation is approximated, because for small phase shift, it is easier to calculate by an algebraic sum by a difference ε [20–23].

Voltage supplied to each customer is an important measure of service quality. The satisfactory voltage level is required to operate appliances, lights, equipment, domestic, residential, commercial, and industrial loads. Injecting power from a DG device into the power system will offset load current, thus reducing the voltage drop on the utility lines. The DG device may inject leading reactive power (capacitive) in the power system or draw lagging reactive power

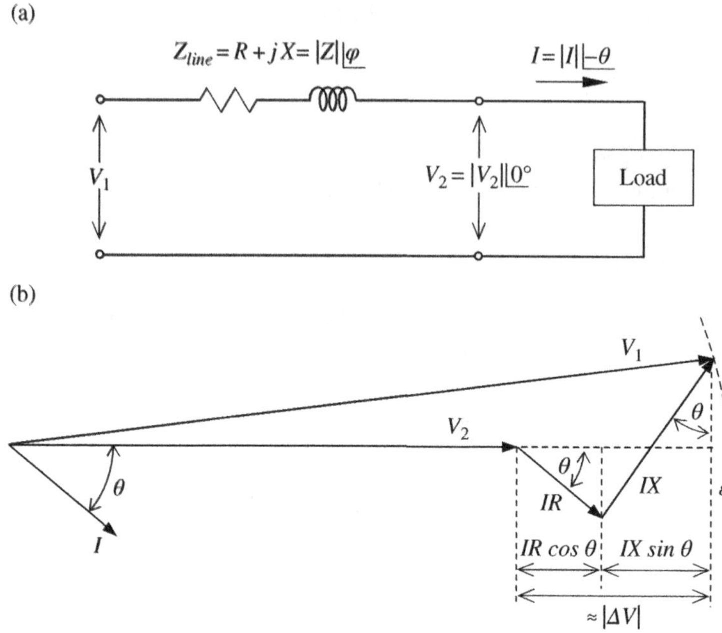

(a)

(b)

Figure 16.9 Evaluation of the voltage effect of a load connected to an ideal *source*: (a) per-phase equivalent circuit and (b) phasor diagram.

(inductive) from the power system, thus, affecting the voltage drop, and very probably some reactive power requirements. Using phasor analysis, we can estimate the magnitude of the approximate voltage drop as

$$|\Delta V| \approx I(R\cos\theta + X\sin\theta) \tag{16.28}$$

The voltage drop estimation can also be based on P and Q for two points, as indicated in Figure 16.10. When $|\Delta V| = |V_A| - |V_B|$ is small, one approximation is to consider that the voltage at the denominator could be either $|V_A|$ or $|V_B|$ and calculate the voltage drop at one end of the line compared to the other, for a given active and reactive power transfer over the line. This equation is very useful for studying the connection of DG devices to distribution feeders. This formula is simple to be used, and it incorporates both real and reactive power together:

$$V_{AB} = \frac{S \cdot Z}{V_A} = \frac{(P - jQ)(R + jX)}{V_A} = \frac{PR + QX}{V_A} + j\frac{PX - QR}{V_A}$$

or

$$\Delta V \approx \frac{PR + QX}{V_A}$$

(a)

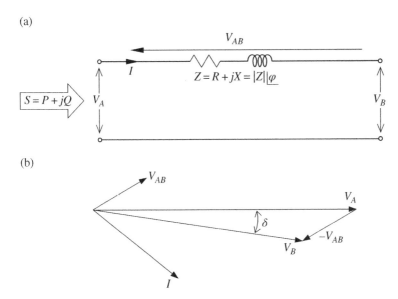

(b)

Figure 16.10 Equivalent circuit for the evaluation of voltage effect of a load connected to an ideal source: (a) per-phase equivalent circuit; (b) phasor diagram.

A second approximation is applicable mostly for transmission networks or higher layers of distribution or sub-transmission, where the geometry and size of the conductors make the impedance Z to be dominated by the reactance X. These relationships provide an important insight into the flow of complex power in transmission networks where $X \gg R$. Therefore, for X/R ratio high, the following are very useful rules of thumb:

- When $\Delta V \propto Q$, these network voltages are determined largely by reactive power flow. It can be affirmed that voltage magnitude differences between the ends of a transmission line is primary driver of reactive power flow.
- When $\delta \propto P$, the phase angles are determined largely by the reactive power flow. It can be affirmed that angular phase differences between the voltages at adjacent nodes are a primary driver of active power flow.

The voltage drop on the feeder is a consequence of current flow and the impedance of the feeder conductor, transformer, and load. In order to improve voltage quality, the following actions are possible: (i) increasing the feeder conductor size, (ii) transferring of loads to other feeders, (iii) increasing primary voltage levels, (iv) changing the feeder sections from single phase to three-phase, and (v) installing new substations and primary feeders. However, there are very simple electrical methods to control the feeder voltage: (i) VRs, (ii) on-load tap-changing transformers, and (iii) capacitor banks (fixed or switched) [24–27].

(a)

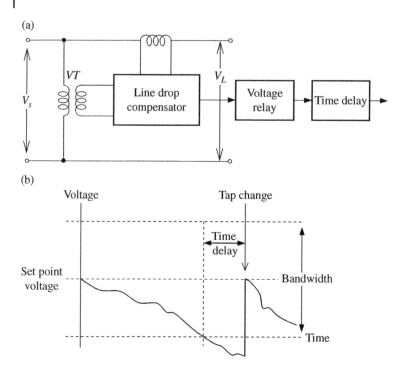

(b)

Figure 16.11 Voltage regulation circuit for correction of feeder voltage profile: (a) per-phase equivalent circuit; (b) bandwidth and voltage profile.

(a)

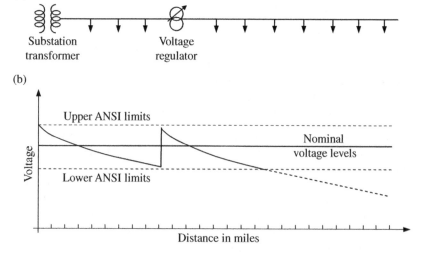

(b)

Figure 16.12 Voltage regulation operation in one particular point of a feeder: (a) addition of a voltage regulator; (b) effects on the voltage regulation.

(a)

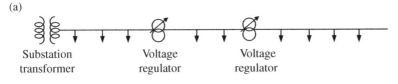

(b)

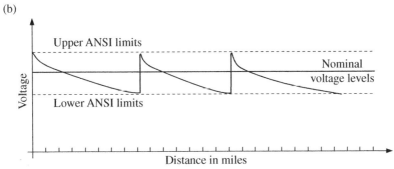

Figure 16.13 Voltage regulation operation with two VR on a feeder : (a) addition of two voltage regulators; (b) effects on the voltage regulation.

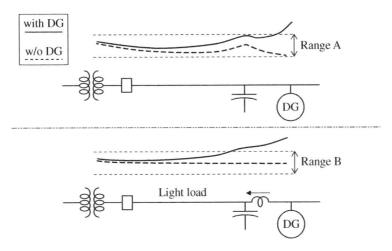

Figure 16.14 Effect of voltage drop with/without capacitors with nondirectional current-based capacitive control.

A VR is shown in Figure 16.11a with the voltage effects of a tap change. The set point voltage is the desired output of the regulator. The bandwidth in Figure 16.11b is related to when the difference exceeds one-half of the bandwidth will a tap change start. There is a time delay, that is, a waiting time from when the voltage goes out of band until the controller changes a tap.

The effect of voltage drops and addition of one VR along the distribution line can be visualized in Figure 16.12. By its turn, Figure 16.13 shows the effect of voltage drop and addition of two VRs. Figure 16.14 shows the effect of voltage drop with/without capacitors and one DG [28–30].

References

1 P.G. Barbosa, L.G.B. Rolim, E.H. Watanabe, and R. Hanitsch, Control strategy for grid connected dc-ac converters with load power factor correction, IEE Proceedings: Generation, Transmission and Distribution, Vol. 145, No. 5, pp. 487–491, 1998.

2 G. Perpermans, J. Driesen, D. Haeseldonckx, W. D'haeseleer, and R. Belmans, Distributed generation: definition, benefits and issues, Working Paper Series 2003–8, K.U. Leuven Energy Institute, Heverlee, Belgium, August 2003.

3 M.N. Marwali and A. Keyhani, Control of distributed generation systems, part I: Voltages and currents control, IEEE Transactions on Power Electronics, Vol. 19, No. 6, pp. 1541–1550, 2004.

4 M.N. Marwali, J.W. Jung, and A. Keyhani, Control of distributed generation systems, part II: Load sharing control, IEEE Transactions on Power Electronics, Vol. 19, No. 6, pp. 1551–1561, 2004.

5 R.H. Lasseter and P. Piagi, Micro-grids: a conceptual solution, Proceedings of the 35th Annual IEEE Power Electronics Specialists Conference, Aachen, Germany, June 20–25, 2004, pp. 4285–4290.

6 H. Zang, M. Chnadorkar, and G. Venkataramanan, Development of static switchgear for utility interconnection in a microgrid, Proceedings of the Conference on Power and Energy Systems, PES 2003, Palm Springs, CA, February 24–26, 2003, pp. 235–240.

7 S.W. Park, I.Y. Chung, J.H. Choi, S.I. Moon, and J.E. Kim, Control schemes of the inverter interfaced multi-functional dispersed generation, Proceedings of the IEEE Power Engineering Society General Meeting, Vol. 3, pp. 1924–1929, July 13–17, 2003, Toronto, Canada.

8 F. Blaabjerg, Z. Chen, and S.B. Kjaer, Power electronics as efficient interface in dispersed power generation systems, IEEE Transactions on Power Electronics, Vol. 19, No. 5, pp. 1184–1194, 2004.

9 Z. Chen and E. Spooner, Voltage source inverters for high-power variable-voltage dc power sources, IEE Proceedings: Generation, Transmission and Distribution, Vol. 148, No. 5, pp. 439–447, 2001.

10 H.L. Willis and W.G. Scott, Distributed Power Generation: Planning and Evaluation, Marcel Dekker, New York, 2000.

11 L.N. Canha, V.A. Popov, A.R. Abaide, F.A. Farret, A.L. König, D.P. Bernardon, and L. Comassetto, Multicriterial analysis for optimal location of distributed energy sources considering the power system reaction, presented at the 9th

CIGRE' Symposium of Specialists in Electric Operational and Expansion Planning, Rio de Janeiro, Brazil, May 2004.

12 O.E. Elgerd, Electric Energy Systems: An Introduction, McGraw-Hill, New Delhi, 1975.

13 B. Barros-Neto, I.S. Scarmino, and R.E. Bruns, Planning and Optimization of Experiments (Planejamento e Otimização de Experimentos), Editora Unicamp, Campinas, Brazil, 1995.

14 F.G. Guseinov and S.M. Mamediarov, Experimental Design in Problems of Electrical Engineering (in Russian), Energoatomizdat, Moscow, 1988.

15 R. Bellman and L. Zadeh, Decision making in a fuzzy environment, Management Sciences, Vol. 17, No. 4, pp. 141–164, 1970.

16 D.P. Bernardon, M. Sperandio, V.J. Garcia, J. Russi, L.N. Canha, A.R. Abaide, and E.F.B. Daza, Methodology for allocation of remotely controlled switches in distribution networks based on a fuzzy multi-criteria decision making algorithm, Electric Power Systems Research (Print), Vol. 81, pp. 414–420, 2011.

17 D.P. Bernardon, V.J. Garcia, A.S.Q. Ferreira, and L.N. Canha, Multicriteria distribution network reconfiguration considering subtransmission analysis, IEEE Transactions on Power Delivery, Vol. 25, pp. 2684–2691, 2010.

18 A. Barin, L.F. Pozzatti, L.N. Canha, R.Q. Machado, A.R. Abaide, and G. Arend, Multi-objective analysis of impacts of distributed generation placement on the operational characteristics of networks for distribution system planning, International Journal of Electrical Power & Energy Systems, Vol. 32, pp. 1157–1164, 2010.

19 D.P. Bernardon, V.J. Garcia, A.S.Q. Ferreira, and L.N. Canha, Electric distribution network reconfiguration based on a fuzzy multi-criteria decision making algorithm, Electric Power Systems Research (Print), Vol. 79, pp. 1400–1407, 2009.

20 N. Angela, M. Liserre, R.A. Mastromauro, and A.D. Aquila, A survey of control issues in PMSG-based small wind-turbine systems, IEEE Transactions on Industrial Informatics, Vol. 9, No. 3, pp. 1211–1221, 2013.

21 L. Barote, C. Marinescu, and M.N. Cirstea, Control structure for single-phase stand-alone wind-based energy sources, IEEE Transactions on Industrial Electronics, Vol. 60, No. 2, pp. 764–772, 2013.

22 R. Carnieletto, D. Brandao, S. Suryanarayanan, F. Farret, and M.G. Simoes, Smart grid initiative, IEEE Industry Applications Magazine, Vol. 17, No. 5, pp. 27–35, 2010.

23 S. Chakraborty, M.G. Simões, and W.E. Kramer, Power Electronics for Renewable and Distributed Energy Systems, Springer, London, 2013.

24 S. Chung, Phase-locked loop for grid-connected three-phase power conversion systems, IEE Proceedings—Electric Power Applications, Vol. 147, No. 3, pp. 213–219, 2000.

25 F. Harirchi, M.G. Simões, A. Al-Durra, and S. Muyeen, Short transient recovery of low voltage-grid-tied DC distributed generation, Energy Conversion Congress & Expo (ECCE), Montreal, Canada, September 20–24, 2015.

26 F. Harirchi, M.G. Simões, M. Babakmehr, A. Al-Durra, and S Muyeen, Designing smart inverter with unified controller and smooth transition between grid-connected and islanding modes for microgrid application, IEEE Industry Applications Society Annual Meeting (IAS), Dallas, TX, 2015.

27 S.-H. Hu, C.-Y. Kuo, T.-L. Lee, and J.M. Guerrero, Droop-controlled inverters with seamless transition between islanding and grid connected operations, 2011 IEEE Energy Conversion Congress and Exposition (ECCE), 2011, pp. 2196–2201.

28 B. Kroposki, C. Pink, R. DeBlasio, H. Thomas, M. Simoes, and P. Sen, Benefits of power electronic interfaces for distributed energy systems, IEEE Power Engineering Society General Meeting, Montreal, QC, Canada, 2006, pp. 1–8.

29 J. Lagorse, M.G. Simoes, and A. Miraoui, A multiagent fuzzy-logic-based energy management of hybrid systems, IEEE Transactions on Industrial Applications, Vol. 45, No. 6, pp. 2123–2129, 2009.

30 C.D. Lute, M.G. Simoes, D.I. Brandao, A.A. Durra, and S. Muyeen, Development of a four pulse floating interleaved boost converter for photovoltaic systems, IEEE Energy Conversion Congress and Exposition (ECCE), Pittsburgh, PA, 2014, pp. 1895–1902.

17

Interconnection of Alternative Energy Sources with the Grid

Benjamin Kroposki[1], Thomas Basso[1], Richard Deblasio[1], and N. Richard Friedman[2]

[1] Power System Engineering Center, National Renewable Energy Laboratory (NREL), Golden, CO, USA
[2] Resource Dynamics Corporation, Vienna, Virginia, USA

17.1   Introduction

Historically, utility electric power systems (EPSs) were not designed to accommodate active generation and storage at the distribution level. In addition, a multitude of utility companies employed numerous EPS architectures based on differing designs and choices of equipment. As a result, major factors must be addressed when interconnecting alternative energy sources (AESs) with the utility grid. The interconnection system consists of the hardware and software that make up the physical link between AESs and the area EPS (usually the local electric grid). The interconnection system is the means by which an AES unit connects electrically with the outside EPS. It can also provide monitoring, control, metering, and dispatch of the AES unit.

Figure 17.1 shows the major functional components required for the interconnection of an AES with the utility grid (area EPS). The interconnection system is composed functionally of the components within the dashed lines. These functions are not necessarily independent or discrete objects as shown in the figure. For example, some of the functions may be colocated in the AES. Similarly, some of the interconnection functions shown as discrete objects may be combined in equipment. In Figure 17.1 the boxes shown within the interconnection system are generally associated with power equipment functions, and the ellipses are associated with monitoring, information exchange, communications, and command and control functions. Therefore, there may not be independent, discrete demarcations among these power and communication functions in the interconnection system [1].

Integration of Renewable Sources of Energy, Second Edition. Felix A. Farret and M. Godoy Simões.
© 2018 John Wiley & Sons, Inc. Published 2018 by John Wiley & Sons, Inc.

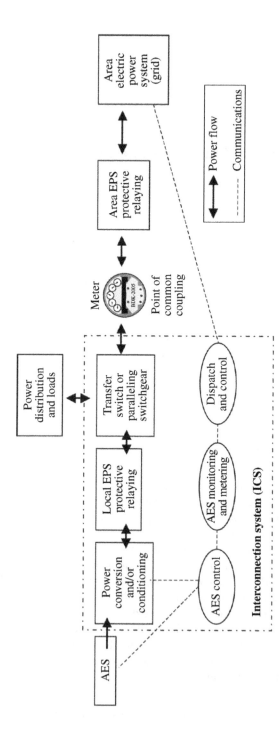

Figure 17.1 Interconnection system functional schematic.

Functions that may be included in whole or in part in an interconnection system include:

1) Power conversion and conditioning
   a) Power conversion. If necessary, power conversion functions change one type of electricity into another to make it EPS-compatible. For example, photovoltaics (PVs), fuel cells, and battery storage produce direct-current (dc) power, whereas microturbines produce high-frequency alternating current (ac).
   b) Power conditioning. This function provides the basic power quality to supply clean ac power to the load.
2) Protection functions monitor the EPS point of common coupling (PCC) and the input and output power of the AES and disconnect from the EPS when operating conditions exceed interconnection requirements. Examples of these functions are over- and undervoltage and frequency protective settings and anti-islanding schemes.
3) Autonomous and semiautonomous functions and operations.
   a) AES and load controls. These control the status and operation of the AES and local loads. Status can include on/off and power-level commands. These functions can also control hardware to disconnect from the EPS.
   b) Ancillary services. These services include voltage support, regulation, operating reserve, and backup supply.
   c) Communications. Communications allow the AES and local loads to interact and operate as part of a larger network of power systems or microgrids.
   d) Metering. The metering function allows billing for AES energy production and local loads.

AES interconnection technology development is at a crossroads. Electromechanical discrete relays, which dominated utility interconnection, protection, and coordination for years, are being supplanted by digital equipment, frequently with multifunction capability. Utilities are gravitating toward programmable digital relays. The rise of inverter technology as an alternative to rotating power conversion technology (i.e., induction and synchronous generators) has opened the door to integrated, inverter-based protective relaying.

This trend has created a major hurdle to streamlined interconnection because utility engineers have only recently begun to reach a comfort level with digital circuitry. In addition, digital circuit designs are often proprietary, which makes the approval process challenging for utilities and the limited group of third-party certification organizations.

17.2   Interconnection Technologies

The prime mover or engine most often differentiates one AES from another. It is typically the most expensive subsystem of an AES technology. Prime movers may be powered by traditional fuel or renewable energy, and they can change energy from one form to another. For example, prime movers may use direct physical energy to rotate a shaft (e.g., in a hydro-turbine or wind energy device), or they may use chemical energy conversion to change thermodynamic energy to physical energy (e.g., by burning diesel fuel, natural gas, or propane to move a piston or turbine to rotate a shaft).

The electric generator converts prime mover energy to electrical energy. Historically, most electrical generation has been accomplished via rotating machines. The prime mover drives an electric generator that is synchronous or induction (asynchronous). The generator has a stator that consists of a set of ac windings and a rotor with windings that are either ac or dc. The stator is stationary, and the prime mover rotates the rotor. The means of power conversion is through interaction of the magnetic fields of the stator and rotor circuits.

Synchronous generators are used when power production from the prime mover is relatively constant (e.g., for internal combustion engines). Most large utility generators are synchronous. Induction generators, on the other hand, are well suited for rotating systems in which prime mover power is not constant (e.g., for wind turbines and small hydro-turbines). However, they can also be used with engines and combustion turbines. Synchronous and induction generator electrical output then undergoes another power conversion to achieve the sinusoidal power-quality level compatible with utility grid interconnection.

Inverters are electronic systems that convert power statically. Inverters are based on power semiconductor devices, microprocessor or digital signal processor technologies, and control and communications algorithms (see Chapter 12). Generally, inverters are used with prime movers that provide dc electricity (such as PV or fuel cell units) or microturbines, which are small, high-speed, rotating combustion turbines coupled directly to a synchronous-type electric generator. In the case of the microturbine, the generator output voltage waveform has very high frequency, well beyond the utility grid's 50 or 60 Hz. The high-frequency waveform is rectified to dc electricity, and the dc electricity is then synthesized to a sinusoidal waveform suitable for connection with the grid.

17.2.1   Synchronous Interconnection

Most generators in service today are synchronous. A synchronous generator is an ac machine in which the rotational speed of normal operation is constant and in synchronism with the frequency of the area EPS with which it is connected.

In synchronous generators, a separate motor–generator set, a directly coupled self-excited dc generator, or a brushless exciter that does not require an outside electrical source supplies field excitation. Therefore, this type of generator can run in stand-alone mode or interconnected with the area EPS. When interconnected, the generator output is exactly in step with area EPS voltage and frequency. Note that separately excited synchronous generators can supply sustained fault current under nearly all operating conditions.

Synchronous generators require more complex control than grid-connected induction generators to synchronize with the area EPS and control field excitation. They also require special protective equipment to isolate them from the area EPS under fault conditions. An advantage is that this type of machine can provide power during area EPS outages. In addition, it permits AES owners to control the power factor at their facilities by adjusting the dc field current.

17.2.2 Induction Interconnection

Induction generators are asynchronous machines that require an external source to provide the magnetizing (reactive) current necessary to establish the magnetic field across the air gap between the generator rotor and stator. Without such a source, induction generators cannot supply electric power and must always operate in parallel with an area EPS, a synchronous machine, or a capacitor that can supply the reactive requirements.

In certain instances, an induction generator may continue to generate electric power after the area EPS source is removed. This phenomenon, known as self-excitation, can occur when there is sufficient capacitance in parallel with the induction generator to provide excitation and the load connected has certain resistive characteristics (see Chapter 10). This external capacitance may be part of the AES system or may consist of power factor correction capacitors located on the area EPS circuit with which the AES is directly connected.

Induction generators operate at a rotational speed determined by the prime mover, slightly higher than that required for exact synchronism. Below synchronous speed, these machines operate as induction motors and thus become a load on the area EPS. Some advantages of induction generators are as follows:

Grid-connected induction generators need only basic control systems because their operation is relatively simple.

They do not require special procedures to synchronize with the area EPS because this occurs essentially automatically.

They will normally cease to operate when an area EPS outage occurs.

A disadvantage of induction generators is their response when connected with the area EPS at speeds significantly below synchronous speed. Depending on the machine class, potentially damaging inrush currents and associated torques can result. Regardless of load, induction generators draw reactive power from the area EPS and may adversely affect the voltage regulation on the circuit with which it is connected. Induction generators then consume vars from the system. It is important to consider the addition of capacitors to improve the power factor and reduce reactive power draw.

17.2.3 Inverter Interconnection

Some AESs produce electric power with voltages not in synchronism with those of the area EPS with which they are to be connected. An electric power converter provides an interface between a non-synchronous AES and an area EPS so that the two may be properly interconnected.

The two categories of nonsynchronous AES output voltage are:

1) dc voltages generated by dc generators, fuel cells, PV devices, storage batteries, or ac generators through rectifiers
2) ac voltages generated by synchronous generators running at nonsynchronous speed or by asynchronous generators

The two categories of electric power converters that can connect AESs with area EPSs are:

1) Dc-to-ac power converters (inverters). In this case, the input voltage to the device is generally a nonregulated dc voltage. The output of the device is at the frequency and voltage magnitude specified by the local utility or grid operator. This is the dominant means of small and renewable energy interconnection.
2) Ac-to-dc electric power converters. In this case, the input frequency, voltage magnitude to the device, or both do not meet area EPS requirements. The output of the converter device is at the appropriate frequency and voltage magnitude as specified by the area EPS in cases in which dc power can be used. This approach is not widely used.

The profusion of data centers and other customers using dc power supplies (such as the power supplied by electronic ballasts) has opened the door to either a dc or dc-to-ac converter designed to deliver the dc output of small AES units directly to the application.

Static power converters are built with diodes, transistors, and thyristors and have ratings compatible with AES applications. These solid-state devices are configured into rectifiers (to convert ac voltage into dc voltage), inverters (to convert dc voltage into ac voltage), and cycloconverters (to convert ac voltage at one frequency into ac voltage at another frequency). Some types require the area EPS to operate; others may continue to function normally after area EPS failure. The major advantages of solid-state converters are their higher efficiency and potentially higher reliability than rotating machine converters. In addition, this technology offers increased flexibility with the incorporation of protective relaying, coordination, and communications options.

17.3 Standards and Codes for Interconnection

Recent advancements in AESs have allowed their increased use and integration into the EPS. When interconnected with the EPS at or near load centers, these technologies can provide increased efficiency, availability, and reliability; better power quality; and a variety of economic and power system benefits.

The interconnection of AESs with the distribution system is regulated by codes and standards to address performance, safety, and power quality issues [2–5]. Three organizations are major players in the interconnection codes and standards arena: (i) the Institute of Electrical and Electronics Engineers (IEEE), (ii) the National Fire Protection Association (NFPA), and (iii) Underwriters Laboratories (UL).

17.3.1 IEEE 1547

IEEE develops voluntary consensus standards for electrical and electronic equipment. These standards are developed with the involvement of equipment manufacturers, users, utilities, and general interest groups. Most state public utility commission guidelines and utility interconnection requirements reference IEEE standards. In addition, most UL standards that relate to interconnection ensure that the equipment is manufactured to comply with IEEE standards.

IEEE approved IEEE 1547, Standard for Interconnecting Distributed Resources with EPSs, which established technical requirements for using electricity from dispersed sources, including AESs. The 1547 standard focuses on the technical specifications for, and testing of, the interconnection itself. It provides requirements relevant to the performance, operation, testing, safety, and maintenance of the interconnection. It covers general requirements, response to abnormal conditions, power quality, islanding, and test specifications and requirements for design, production, installation evaluation, commissioning, and periodic tests. These requirements are needed universally for the interconnection of AESs (including synchronous generators, induction generators, and power inverters or converters) and are sufficient for most installations. The requirements are applicable to all AES technologies with an aggregate capacity of 10 MVA or less at the PCC that are interconnected with EPSs at typical primary or secondary distribution voltages. The installation of AESs on radial primary and secondary distribution systems is the emphasis of the 1547 standard, but installation on primary and secondary network distribution systems is also considered [6, 7].

IEEE Standards Coordinating Committee (SCC) 21 is developing a series of 1547 standards. IEEE P1547.1 will provide detailed test procedures for proving or validating that interconnection specifications and equipment conform to the functional and test requirements of IEEE 1547. IEEE P1547.2 will provide a technical background and application details to make IEEE 1547 easier to use and will characterize distributed resource technologies and associated interconnection issues. IEEE P1547.3 will aid interoperability by offering guidelines for monitoring, information exchange, and control among fuel cells, PV, wind turbines, and other distributed generators interconnected with an EPS. IEEE P1547.4 will address engineering aspects of how local facilities can function as "electrical islands" to provide power when utility power is not available [8, 9].

17.3.2 National Electrical Code

NFPA has been a worldwide leader in the provision of fire, electrical, and life safety to the public since 1896. The mission of this nonprofit organization is to reduce the worldwide burden of fire and other hazards on quality of life by providing and advocating scientifically based consensus codes and standards, research, training, and education. The NFPA publishes the National Electrical Code (NFPA-70), which covers electrical equipment wiring and safety on the customer side of the PCC. The NFPA also publishes other standards relating to AES interconnection.

17.3.2.1 NFPA 70: National Electrical Code

The National Electrical Code covers electric conductors and equipment installed within or on public and private buildings or other structures, including mobile homes and recreational vehicles, floating buildings, and premises such as yards, carnivals, parking and other lots, and industrial substations; conductors that connect the installations to a supply of electricity and other outside conductors and equipment on the premises; optical fiber cable; and buildings used by the electric utility, such as office buildings, warehouses, garages, machine shops, and recreational buildings that are not an integral part of a generating plant, substation, or control center. Some National Electrical Code articles related to interconnection are described in the succeeding text.

Article 230: Services. This article includes provisions and requirements for electric service—including emergency, backup, and parallel power—to a building.
Article 690: Solar Photovoltaic Systems. This mentions interconnection to the grid but focuses on descriptions of components and proper system wiring.
Article 692: Fuel Cells. This article covers stationary fuel cells for power production.
Article 700: Emergency Systems. This includes provisions that apply to emergency power systems and information about interconnection (such as references to transfer switches).

Article 701: Legally Required Standby Systems. This includes provisions that apply to standby power systems. It also has information about interconnection (such as references to transfer switches, UPSs, and generators).

Article 702: Optional Standby Systems. This article includes provisions that apply to standby systems that are not required legally. It has some information about interconnection (such as references to transfer switches, grounding, and circuit wiring).

Article 705: Interconnected Electrical Power Production Systems. This article broadly covers the interconnection of AESs (other than PV systems and fuel cells).

17.3.2.2 NFPA 853: Standard for the Installation of Stationary Fuel Cell Power Plants

This applies to the design and installation of the following stationary fuel cell power plant applications: (i) a singular prepackaged, self-contained power plant unit; (ii) a combination of prepackaged, self-contained units; and (iii) power plant units composed of two or more factory-matched modular components intended to be assembled in the field.

17.3.3 UL Standards

UL is an independent, not-for-profit product safety testing and certification organization. UL has tested products for public safety for more than a century and is the leader in US electrical product safety and certification. UL has a number of certifications that apply to AES interconnection equipment.

17.3.3.1 UL 1741: Inverters, Converters, and Controllers for Use in Independent Power Systems

These requirements cover inverters, converters, charge controllers, and output controllers intended for use in stand-alone (not grid-connected) or utility-interactive (grid-connected) power systems. Utility-interactive inverters and converters are intended to be installed in parallel with an area EPS. An electric utility supplies common loads. This standard is harmonized with IEEE 1547 interconnection requirements and IEEE P1547.1 test procedures.

17.3.3.2 UL 1008: Transfer Switch Equipment

These requirements cover automatic, no automatic (manual), and bypass/isolation transfer switches intended to provide for lighting and power in ordinary locations. They include:

1) Automatic transfer switches and bypass/isolation switches for use in emergency systems in accordance with Articles 517 (Health Care Facilities), 700 (Emergency Systems), 701 (Legally Required Standby Systems) and 702 (Optional Standby Systems of the National Electrical Code) and the NFPA Standard for Health Care Facilities (ANSI/NFPA 99).

2) Transfer switches for use in optional standby systems in accordance with Article 702 of the National Electrical Code.
3) Transfer switches for use in legally required standby systems in accordance with Article 701 of the National Electrical Code.
4) Automatic transfer switches and bypass/isolation switches for use in accordance with the NFPA Standard for Centrifugal Fire Pumps, ANSI/NFPA 20.
5) No automatic transfer switches for use in accordance with Articles 517 (Health Care Facilities) and 702 (Optional Standby Systems) of the National Electrical Code and the NFPA Standard for Health Care Facilities.

An automatic transfer switch for use in a legally required standby system is identical to that used for an emergency system. These requirements cover transfer switch equipment rated at 6000 A or less and 600 V or less. These requirements also cover transfer switches with their associated control devices, including voltage-sensing relays, frequency-sensing relays, and time-delay relays.

An automatic transfer switch, as covered by these requirements, is a device that automatically transfers a common load from a normal supply to an alternative supply in the event of failure of the normal supply. It also automatically returns the load to the normal supply when it is restored. An automatic transfer switch is allowed to be provided with a logic control circuit that inhibits automatic operation of the device from either a normal to an alternative supply or from an alternative to a normal supply when the switch reverts to automatic operation upon loss of power to the load. Automatic transfer switches may be open or closed transition transfer.

As covered by these requirements, a no automatic transfer switch is a device operated manually by a physical action or electrically by a remote control for transferring a common load between normal and alternative supplies. A transfer switch may incorporate overcurrent protection for the main power circuits. UL 1008 requirements cover completely enclosed transfer switches and open types intended for mounting in other equipment, such as switchboards.

Transfer switches are rated in amperes and are generally considered to be acceptable for total system transfer, which includes control of motors, electric-discharge lamps, electric-heating loads, and tungsten–filament lamp loads. A transfer switch intended for a total system transfer is considered to be acceptable for the control of tungsten–filament lamp loads not exceeding 30% of the switch ampere rating unless the switch has been investigated for a higher percentage of lamp loads and marked accordingly. A transfer switch may be limited to use with one or more specific types of load if investigated accordingly and marked appropriately.

These requirements also cover bypass/isolation switches that can be used to manually select an available power source to feed load circuits and provide for total isolation of an automatic transfer switch. These switches may be completely enclosed, enclosed with a transfer switch, or of the open type, which is intended for mounting in other equipment.

A product that contains features, characteristics, components, materials, or systems new or different from those covered by the requirements in this standard and that involves a risk of fire or electric shock or injury to persons is to be evaluated using appropriate additional components and end-product requirements to maintain the level of safety as anticipated originally by the intent of the standard. A product whose features, characteristics, components, materials, or systems conflict with specific requirements or provisions of this standard does not comply with this standard. Revision of requirements can be proposed and adopted in conformance with the methods employed for development, revision, and implementation of the standard.

17.3.3.3 UL 2200: Standard for Safety for Stationary Engine Generator Assemblies

These requirements cover stationary engine generator assemblies rated 600 V or less that are intended for installation and use in ordinary locations in accordance with the National Electrical Code; the Standard for the Installation and Use of Stationary Combustion Engines and Gas Turbines (NFPA 37), the Standard for Health Care Facilities (NFPA 99), and the Standard for Emergency and Standby Power Systems (NFPA 110). These requirements do not cover generators for use in hazardous (classified) locations. That equipment is covered by the Standard for Electric Motors and Generators for Use in Hazardous (Classified) Locations (UL 674). In turn, these requirements do not cover UPS equipment, which is covered by the Standard for Uninterruptible Power Supply Equipment (UL 1778). These requirements do not cover generators for marine use, which are covered by the Standard for Marine Electric Motors and Generators (UL 1112).

17.4 Interconnection Considerations

When interconnecting an AES with an area EPS, several issues should be examined to ensure that the area EPS and AES integrate in a safe and reliable manner. These issues are discussed in the following sections.

17.4.1 Voltage Regulation

Voltage regulation describes the process and equipment used by an area EPS operator to maintain approximately constant voltage to users despite normal variations in voltage caused by changing loads. Voltage regulation and voltage stability are important factors that affect the operation of a power distribution system. If a system is not well regulated or stable, machines that receive power from the system will not operate efficiently.

AES interconnection equipment should not degrade the voltage provided to the customers of an area EPS to service voltages outside the limits of ANSI C84.1, Range A. Apart from the effect on the voltage of the area EPS because of the real power generation of the AES, the AES should not attempt to oppose or regulate changes in the prevailing voltage level of the area EPS at the PCC. An exception is that AES generators can use automatic voltage regulation when it is accomplished without detriment to either the area EPS or local EPS.

Voltage regulation is based on radial power flows from the substation to the loads. AESs introduce "meshed" power flows that may interfere with the effectiveness of standard voltage regulation practice. The effect of AESs on area EPS voltage regulation can cause changes in power system voltage by (i) the generator offsetting the load current or (ii) the AESs attempting to regulate voltage. Most types of AES generators and utility-interactive inverters should strive to maintain an approximately constant power factor at any voltage within their rating; accordingly, the primary effect of AESs on voltage regulation is the result of an AES offsetting the load current.

17.4.2 Integration with Area EPS Grounding

A grounding system consists of all interconnected grounding connections in a specific power system and is defined by its isolation or lack of isolation from adjacent grounding systems. The isolation is provided by transformer primary and secondary windings that are coupled only by magnetic means.

The interconnection of an AES with an area EPS needs to be coordinated with the neutral grounding method in use on the area EPS. Use of an AES that does not appear as an effectively grounded source connected to such systems may lead to overvoltage during line-to-ground faults on the area EPS. This condition is especially dangerous if a generation island develops and continues to serve a group of customers on a faulted distribution system. Customers on the unfaulted phases could, in the worst case, see their voltage increase to 173% of the prefault voltage level for an indefinite period. At this high level, utility and customer equipment would almost certainly be damaged. Saturation of distribution transformers will help slightly limit this voltage rise. Nonetheless, the voltage can still become quite high (150% or higher).

17.4.3 Synchronization

To synchronize an AES with an area EPS, the output of the AES and the input of the area EPS must have the same voltage magnitude, frequency, phase rotation, and phase angle. Synchronization is the act of checking that these four variables are within an acceptable range (or acceptable ranges). For synchronism to occur, the output variables of the AES must match the input variables of the area EPS. With polyphase machines, the direction of phase rotation must also be the same. This is typically checked at the time of installation. The phases should be connected to the switches such that the phase rotation will always be correct. Phase rotation is not usually checked again unless wiring changes have been made on the generator or inverter or the area EPS.

17.4.4 Isolation

When required by area EPS operating practices, a readily accessible, lockable, visible-break isolation should be located between the area EPS and the AES. Strategically located disconnect switches are an integral part of any EPS. These switches provide visible isolation points to allow for safe work practices. The National Electrical Code dictates the requirements for disconnect devices, which allow for safe operation and maintenance of the EPSs within public or private buildings and structures. This requirement deals specifically with disconnect switches required to ensure safe work practices on the area EPS and not addressed by the National Electrical Code.

All electric utilities have established practices and procedures similar to the National Electrical Code to ensure safe operation of the EPS under normal and abnormal conditions. Several of these procedures identify methods to ensure that the electrical system has been configured properly to provide safe working conditions for area EPS line and service personnel. Although these procedures may vary somewhat between utilities, the underlying intent of the procedures is to establish safe work area clearances to allow area EPS line and service personnel to operate safely in proximity to the EPS. To achieve this, electric utilities have developed procedures that require visible isolation, protective grounding, and jurisdictional tagging of the portion of the EPS where clearance is to be gained. These procedures, in unison with other safety procedures and sound judgment based on knowledge and experience, have resulted in an essentially hazard-free work environment for area EPS personnel.

In an AES installation, some equipment and fuses or breakers may be energized from two or more directions. Thus, disconnect switches should be strategically installed to permit disconnection from all sources. Typically, the load-side contacts (switch blades) of a disconnect switch are deenergized when the switch is open. However, this is not necessarily the case when an AES is connected with the area EPS, so a safety label should be placed on the switch to warn that the load-side contacts may still be energized when the switch is in the open position. Also, a means should be provided for fuse replacement (in fused switches) without exposing workers to energized parts.

17.4.5 Response to Voltage Disturbance

The protection functions of the interconnection system should measure the effective or fundamental frequency value of each phase-to-neutral or,

Table 17.1 Interconnection system response to abnormal voltages per IEEE 1547.

Voltage range (% of base voltage*)	Clearing time** (s)
$V < 50$	0.16
$50 \leq V < 88$	2.0
$110 < V < 120$	1.0
$V \leq 120\,V$	0.16

* Base voltages are the nominal system voltages stated in ANSI C84.1, Table 1.
** AES $\leq$ 30 kW, maximum clearing times; AES > 30 kW, default clearing times.

alternatively, each phase-to-phase voltage. When any of the measured voltages is in any voltage range noted in Table 17.1, the AES should cease to energize the area EPS within the clearing time indicated. Clearing time is the time between the start of the abnormal condition and the AES ceasing to energize the area EPS. For AESs less than or equal to 30 kW in peak capacity, the voltage set points and clearing times should be either fixed or field-adjustable. For AESs more than 30 kW, the voltage set points should be field-adjustable.

The voltages should be measured at the point of AES connection when any of the following conditions exist:

1) The aggregate capacity of AES systems connected to a single PCC is less than or equal to 30 kW.
2) The interconnection equipment is certified to pass a no islanding test.
3) The aggregate AES capacity is less than 50% of the total local EPS minimum electrical demand and export of real or reactive power to the area EPS is not permitted.

The purpose of the time delay is to ride through short-term disturbances to avoid excessive nuisance tripping. For systems less than 30 kW in peak capacity, the set points are to be protected against unauthorized adjustment. Adjustment by a qualified person (or automatic adjustment for prevailing conditions) is desirable to allow compensation for voltage differences between the inverter and the PCC.

17.4.6 Response to Frequency Disturbance

Under- and over-frequency protective functions are among the most important means of preventing an AES island. It is desirable for these protections to operate promptly, but nuisance trips need to be avoided. At the point of generation, the frequency in a typical area EPS is very stable. However, voltage phase-angle swings can occur in transmission and distribution lines because of sudden changes in feeder loading and load current. Over a short

Table 17.2 Interconnection system response to abnormal frequencies (60-Hz base) per IEEE 1547.

AES size	Frequency range (Hz)	Clearing time* (s)
≤30 kW	>60.5	0.16
	<59.3	0.16
>30 kW	>60.5	0.16
	<{59.8–57.0} (adjustable set point)	Adjustable 0.16–300
	<57.0	0.16

* AES ≤ 30 kW, maximum clearing times; AES > 30 kW, default clearing times.

enough time, these voltage swings can cause nuisance trips of under- or over-frequency protective functions.

Frequency excursions typically occur on the area EPS during distribution system operations. Maintaining stable area EPS operations depends on the AES clearing off line when area EPS voltage or frequency is out of agreed-upon operating ranges. Table 17.2 gives the interconnection system response times per IEEE 1547. AES units of less than 30 kW potentially have less effect on system operations and typically can disconnect from the area EPS well within 10 cycles. AES units larger than 30 kW can have an effect on distribution system security. The IEEE 1547 requirement takes this into account by allowing the area EPS operator to specify the frequency setting and time delay for underfrequency trips down to 57 Hz.

Area EPS security depends on the system's ability to withstand the outage of certain lines or equipment without being forced into a system emergency. Security also depends on the proper matching of system load and generation. When generation is matched inadequately with system load, the area EPS frequency will decline. When this happens, the system operator seeks to match load quickly with available generation. Under-frequency relays are installed on the distribution system to shed load automatically to stabilize operations. This is the purpose of allowing the area EPS to determine the setting of the AES underfrequency trip relay.

Some underfrequency relays are sensitive to the rate of area EPS frequency decay and provide information to the system operator to assist in the timing of load shedding. Similar problems on the area EPS can occur when generation exceeds available load. An example is when a large block load is suddenly lost or when the tie lines exporting power relay are quickly closed. Over-frequency is much less of a problem to system operations than underfrequency.

In large power systems, frequency changes are rare. However, with installed AESs, some frequency change is unavoidable when blocks of load are switched. If a modern synchronous governor or static transfer switch is used on a distribution system feeder, these disturbances should be under 5% frequency change and less than 5 seconds in duration, even for full-load switching.

Both the frequency and voltage trip pickup settings for induction generators and static power converters may be relaxed at the discretion of the area EPS if it appears that the AES will experience too many nuisance trips. Synchronous generator trip settings can also be relaxed—but not too much because of the increased threat of islanding.

The frequency trip points should be adjustable in increments with a setting resolution of 0.5 Hz or better. Internal microprocessor protection functions in static power converter units may be substituted for external relays if they provide suitable accuracy. External test ports for periodic utility testing of the trip pickup settings should be included in the interconnection package.

17.4.7 Disconnection for Faults

Short-circuit currents on distribution circuits in the United States are from more than 20,000 A to values less than 1 A for high-impedance single-phase-to-ground faults. The maximum fault can be controlled by system design. Area EPSs are designed not to exceed the rating of distribution line equipment. Maximum faults are limited by restricting substation transformer size, impedance, or both by installing bus or circuit reactors or inserting reactance or resistance in the transformer neutral. Minimum fault magnitude is largely dependent on fault resistance that cannot be controlled. These faults are the most dangerous and difficult to detect.

Clearing times for short circuits on distribution circuits vary widely and depend on magnitude and the type of protective equipment installed. In general, large current faults will clear in 0.1 second or faster. Low-current faults frequently clear in 5–10 seconds or longer, and some very low-level but potentially dangerous ground faults may not clear at all except by manual disconnection of the circuit.

The AES system should be designed with adequate protection and control equipment and include an interrupting device that will disconnect the generator if the area EPS that connects to the AES system or the AES system itself experiences a fault. The AES system should have, as a minimum, an interrupting device that is:

1) Of sufficient capacity to interrupt maximum available fault current at its location
2) Sized to meet all applicable ANSI and IEEE standards
3) Installed to meet all local, state, and federal codes

A failure of the AES system protection and control equipment, including loss of control power, should open the disconnecting device automatically and thus disconnect the AES system from the area EPS.

17.4.8 Loss of Synchronism

A synchronous generator typically employs a three-phase stator winding that when connected to the area EPS three-phase source creates a rotating magnetic field inside the stator and cuts through the rotor. The rotor is excited with a dc that creates a fixed field. If spun around at the speed of the stator field, the rotor will "lock" its fixed field into synchronism with the rotating stator field. Force (torque) applied to the rotor in this synchronous state will cause power to be generated as long as the force is not so great that the rotor pulls out of step with the stator field.

An island is formed when a relay-initiated trip causes a section of the area EPS that contains AESs to separate from the main section of the area EPS. The main section of the area EPS and the island then operate out of synchronism. If an isolation is reclosed between the main section of the area EPS and the island, a voltage and current transient will occur while the island is brought into synchronism with the remainder of the area EPS. The severity of this transient will depend on the voltage phase-angle separation magnitude across the isolation when the reclosing event occurs.

17.4.9 Feeder Reclosing Coordination

Experience has shown that 70–95% of line faults are temporary if the faulted circuit is quickly disconnected from the system. Most line faults are caused by lightning or contact with tree limbs. If the resulting arc at the fault does not continue long enough to damage conductors or insulators, the line can be returned to service quickly.

Modern distribution feeders reclose (reenergize the feeder) automatically after a trip resulting from a feeder fault. This trip–reclose sequence may be initiated by reclosing relays that control the corresponding feeder breaker at the substation or by pole-mounted reclosers or sectionalizers located on the feeder away from the substation. Pole-mounted reclosers or sectionalizers are strategically placed to limit the customers affected per given feeder fault. Automatic reclosing allows immediate testing of a previously faulted portion of the feeder and makes it possible to restore service if the fault is no longer present. Depending on the fault magnitude, the first reclosing try can occur very fast, sometimes within 0.2 second. This short interval assumes settings of an instantaneous trip followed by an instantaneous reclosing.

It is common practice for utilities to attempt to reclose their circuit breakers automatically following a relay-initiated trip. The time delay between tripping and the initial reclose attempt can range from 0.2 second (12 cycles) to 15 seconds (or more). For radial feeders, this initial attempt is usually followed by two more time-delayed attempts, normally with 30- to 90-second intervals. If none of the reclose attempts is successful, the feeder will lock out. The reclose attempts are normally performed without any synchronism-check supervision because the feeders are radial in design, with the area EPS being the only source of power.

In the case of an area EPS protection function initiating a trip of an area EPS protective device in reaction to a fault, the AES must be designed to coordinate with the area EPS reclosing practices of the protective device. The response of the AES must be coordinated with the reclosing strategy of the isolations within the area EPS. Coordination is required to prevent possible damage to area EPS equipment and equipment connected to the area EPS. The AES and the area EPS reclosing strategy will be coordinated if one or more of the following conditions are met for all reclosing events:

1) The AES is designed to cease energizing the area EPS before the reclosing event.
2) The reclosing device is designed to delay the reclosing event until after the AES has ceased to energize the area EPS.
3) The AES and reclosing device are controlled to ensure that the voltage phase-angle separation magnitude across the isolation is less than one-fourth of a cycle when the reclosing event occurs.
4) The AES capacity is less than 33% of the minimum load on the feeder.

17.4.10 Dc Injection

Dc injection produces a dc offset in the basic power system waveform. This offset increases the peak voltage of one-half of the power system waveform (and decreases the peak voltage in the other half of the waveform). The increased half-cycle voltage has the potential to increase saturation of magnetic components, such as cores of distribution transformers. This saturation, in turn, causes increased power system distortion.

Dc injection is an issue because of the economics of magnetic component design. These economics dictate using the smallest amount of magnetic core material possible to accomplish the needed task. This results in the magnetic circuit of the component operating near that part of the B–H curve where the curve begins to become very nonlinear.

There is a concern that transformerless inverters may inject sufficient current into distribution circuits to cause distribution transformer saturation. Distribution transformers range from 25 kVA to more than 100 kVA. A 25-kVA transformer would typically supply power for four to six houses. A 100-kVA unit typically supplies power for 14–18 houses. These numbers vary depending on the amount of electric heating used but average about 5 kVA per residence.

17.4.11 Voltage Flicker

Determining the risk of flicker problems because of basic generator starting conditions or output fluctuations is straightforward using the flicker curve

Table 17.3 Maximum harmonic current distortion in percent of current (I) per IEEE 1547[a]

Individual harmonic order (odd harmonics)	<11	$11 \leq h < 17$	$17 \leq h < 23$	$23 \leq h < 35$	$35 \leq h$	TDD
Percent (%)[b]	4.0	2.0	1.5	0.6	0.3	5.0

a I is the greater of the local EPS maximum load current integrated demand (15 or 30 minutes) without the AES unit or the AES unit rated current capacity (transformed to the PCC when a transformer exists between the AES unit and the PCC).
b Even harmonics are limited to 25% of the odd harmonic limits noted.

approach. This is particularly true if the rate of these fluctuations is well defined, the fluctuations are step changes, and there are no complex dynamic interactions of equipment. The dynamic behavior of machines and their interactions with upstream voltage regulators and generators can complicate matters considerably. For example, it is possible for output fluctuations of an AES (even smoother ones from solar or wind systems) to cause hunting of an upstream regulator, and, although the AES fluctuations alone may not create visible flicker, the hunting regulator may create visible flicker. Thus, flicker can involve factors beyond starting and stopping generation machines or their basic fluctuations. Dealing with these interactions requires an analysis that is far beyond the ordinary voltage drop calculation performed for generator starting. Identifying and solving these types of flicker problems when they arise can be difficult, and the engineer must have a keen understanding of the interactions between the AES and the area EPS. In short, the AES should not create objectionable flicker for other customers on the area EPS.

17.4.12 Harmonics

When an AES is serving balanced linear loads, harmonic current injection into the area EPS at the PCC should not exceed the limits stated in Table 17.3. The harmonic current injections should be exclusive of any harmonic currents because of the harmonic voltage distortion present in the area EPS without the AES connected.

Harmonic distortion is a form of electrical noise. Harmonics are electrical signals at multiple frequencies of the power line frequency. Many electronic devices, including personal computers, adjustable speed drives, and other types of equipment that use just part of the sine wave by drawing current in short pulses, cause harmonics.

Switched power supplies dominate equipment with this operating characteristic. These power supplies are an economical way to provide voltage to the equipment being served and are not affected by minor voltage changes in the power system. Switched power supplies feed a capacitor that supplies the voltage to the electronic circuitry. Because the load is a capacitor as seen from the power system, the current to the power supply is discontinuous. That is,

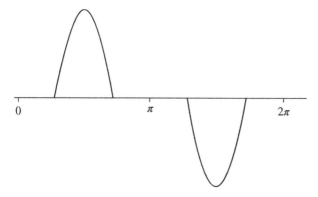

Figure 17.2 Current wave of switched-mode power supply.

current flows for only part of the half-cycle. Figure 17.2 shows the current waveform of such a power supply.

Linear loads, those that draw current in direct proportion to the voltage applied, do not generate large levels of harmonics. The nonlinear load of a switched power supply superimposes signals at multiples of the fundamental power frequency in the power sine wave and creates harmonics. The uses of nonlinear loads connected to area EPSs include static power converters, arc discharge devices, saturated magnetic devices, and, to a lesser degree, rotating machines. Static power converters of electric power are the largest nonlinear loads. Harmonic currents cause transformers to overheat, which, in turn, overheats neutral conductors. This overheating may cause erroneous tripping of circuit breakers and other equipment malfunctions. The voltage distortion created by nonlinear loads may create voltage distortion beyond the premise's wiring system, through the area EPS system to another user.

The type and severity of harmonic contributions from an AES unit depend on the power converter technology, its filtering, and its interconnection configuration. There has been particular concern about the possible harmonic current contributions inverters may make to the area EPS. Fortunately, these concerns are in part because of older, SCR-type power inverters that are line-commutated and produce high levels of harmonic current. Most new inverter designs are based on solid-state technology that uses pulse-width modulation to generate the injected ac and are capable of generating very clean outputs. When powered by small synchronous generators with high-impedance, nonlinear loads can result in voltage distortion.

In general, harmonic contributions from AES units are less of an issue than problems associated with other equipment on the distribution system. In some cases, equipment at the AES site may need to be derated because of added heating caused by harmonics elsewhere on the system. Filters and other mitigation approaches are sometimes required.

17.4.13 Unintentional Islanding Protection

Islanding occurs when an AES (or a group of AESs) continues to energize a portion of the area EPS that has been separated from the rest of the area EPS. This separation could be because of the operation of an upstream breaker, fuse, or automatic sectionalizing switch. Manual switching or "open" upstream conductors could also lead to islanding, which can occur only if the AES continues to serve a load in the islanded section.

In most cases, it is not desirable for an AES to island any part of the area EPS unplanned. This can lead to safety and power quality problems that affect the area EPS and local loads. During utility repair operations, AES islanding can expose utility workers to circuits that otherwise would be de-energized (and that the workers believe to be de-energized). This situation can pose a threat to the public as well. Service restoration can also be delayed as line crews seek to ensure that AES islanding is not a problem. IEEE 1547 requires that the AES cease to energize the area EPS within 2 seconds of the formation of an island. Several approaches can prevent unintentional islands from occurring.

The installed AES capacity can be limited to less than one-third of the minimum customer load. By limiting the installed AES capacity, the system protective functions for over- and undervoltage and frequency will be able to respond correctly to an area EPS outage. Reverse power protection can also be employed for unintentional islanding. This is done by placing a reverse power–current relay at the PCC. If the area EPS experiences an outage, any power that the AES tries to feed back onto the area EPS will cause the relay to trip.

AESs can be certified as no islanding by UL. This means that the interconnection system can pass the no islanding test specified in UL 1741 and IEEE 1547.1. This test sets up the worst-case balanced island conditions and confirms that the AES will disconnect in the required 2 seconds. Inverter-based systems typically have some type of active anti-islanding method with which they try to force the voltage or frequency outside normal trip limits. When the area EPS is connected, the inverter cannot move these parameters. When the area EPS experiences an outage, however, the inverter's anti-islanding algorithm forces the voltage or frequency outside the trip limits and disconnects from the area EPS. Direct transfer trip is often used for large AES systems. This involves a dedicated communications line that will trip the AES system when the area EPS has experienced an outage.

17.5 Interconnection Examples for Alternative Energy Sources

This section gives examples of the interconnection systems used for two types of AESs. Figure 17.3 shows a typical radial EPS. In a radial system, power flows from large central-station generators through transmission lines to the

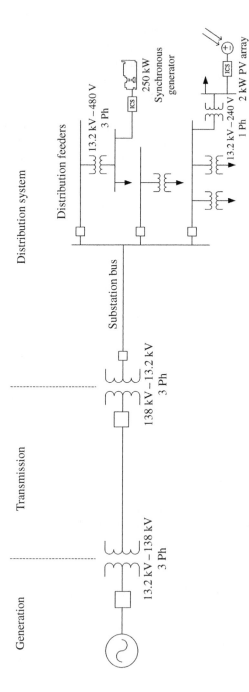

Figure 17.3 Typical EPS with AES and interconnection systems installed at the distribution system level.

distribution system, where customer loads are located. New AES systems are typically installed at the customer load site. This figure shows a synchronous generator and PV system installed at the distribution system level.

17.5.1 Synchronous Generator for Peak Demand Reduction

A typical application for a synchronous generator AES is peak shaving. In this application, the generator is installed at the customer site. Electric utilities typically sell power to commercial customers with a usage and demand charge. The usage charge is based on the energy (in kilowatt-hours) used, and the demand charge is based on the peak load (in kilowatts), typically over a 15-minute period. Figure 17.4 shows how running a 250-kW generator during times of peak load can reduce demand. Figure 17.5 shows the interconnection system functions that allow the customer or the area EPS to run the unit on demand. Currently, most utility dispatchable systems use a local utility remote terminal unit to communicate with the AES. The AES can be operated to assume all customer load or customer load and export excess power back to the utility. This type of interconnection system contains the paralleling switch as well as the generator and local EPS protective relays.

17.5.2 Small Grid-Connected PV System

Another common application is a small PV array that uses inverter-based interconnection systems. These systems are typically less than 10 kW. A functional diagram is given in Figure 17.6. Some state public utility commissions

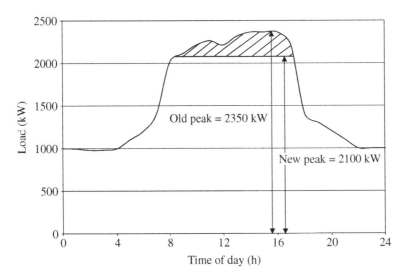

Figure 17.4 Reduction in peak load based on running a 200-kW generator.

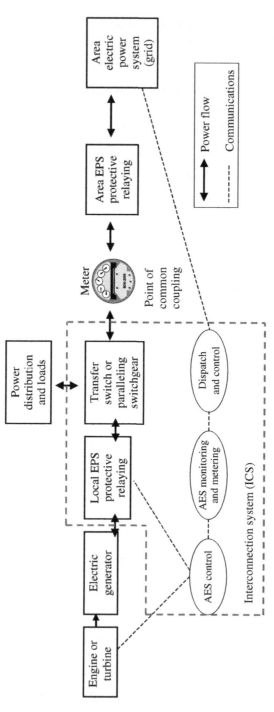

Figure 17.5 Synchronous generator interconnection system.

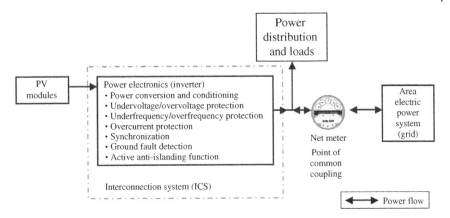

Figure 17.6 Small PV system with net metering.

require utilities to allow net metering systems. In a net-metered system, the customer is allowed to put power onto the grid when his system is producing energy. The meter keeps track of the net import of energy. The customer is typically allowed to have a monthly or yearly net energy balance of zero. Any energy in excess of the total customer load is fed back to the utility for free. In the PV system, the interconnection system consists of an inverter to convert dc power to ac power and all the necessary protective functions for compliance with interconnection standards.

References

1 B. Kroposki, T. Basso, and R. DeBlasio, Interconnection testing of distributed resources, presented at the 2004 PES General Meeting, Denver, CO, June 2004.
2 T. Basso and R. DeBlasio, IEEE 1547 series of standards: interconnection issues, IEEE Transactions on Power Electronics, Vol. 19, No. 5, pp. 1159–1162, 2004.
3 IEEE Std. 1547, Standard for Interconnecting Distributed Resources with Electric Power Systems, IEEE Press, Piscataway, NJ, July 2003.
4 IEEE P1547.1, Draft Standard for Conformance Test Procedures for Interconnecting Distributed Resources with Electric Power Systems, IEEE Press, Piscataway, NJ, February 2005.
5 IEEE P1547.2, Draft Application Guide for IEEE 1547 Standard for Interconnecting Distributed Resources with Electric Power Systems, IEEE Press, Piscataway, NJ, July 2004.
6 IEEE P1547.3, Draft Guide for Monitoring, Information Exchange, and Control of Distributed Resources Interconnected with Electric Power Systems, IEEE Press, Piscataway, NJ, 2008.

7 IEEE P1547.4, Draft Guide for Design, Operation, and Integration of Distributed Resource Island Systems with Electric Power Systems, IEEE Press, Piscataway, NJ, 2011.

8 Electrical Generating Systems Association, On-Site Power Generation: A Reference Book, 5th ed., EGSA, Boca Raton, FL, 2017.

9 N.R. Friedman, Distributed Energy Resources Interconnection Systems: Technology Review and Research Needs, NREL/SR-560-32459, National Renewable Energy Laboratory, Golden, CO, September 2002.

18

Micropower System Modeling with HOMER

Tom Lambert[1], Paul Gilman[2], and Peter Lilienthal[2]

[1] *Mistaya Engineering Inc., Calgary, Alberta, Canada*
[2] *Power System Engineering Center, National Renewable Energy Laboratory (NREL), Golden, CO, USA*

18.1 Introduction

The HOMER Micropower Optimization Model is a computer model developed by the US National Renewable Energy Laboratory (NREL) to assist in the design of micropower systems and to facilitate the comparison of power generation technologies across a wide range of applications. HOMER models a power system's physical behavior and its life-cycle cost, which is the total cost of installing and operating the system over its life span. HOMER allows the modeler to compare many different design options based on their technical and economic merits. It also assists in understanding and quantifying the effects of uncertainty or changes in the inputs.

A micropower system is a system that generates electricity, and possibly heat, to serve a nearby load. Such a system may employ any combination of electrical generation and storage technologies and may be grid connected or autonomous, meaning separate from any transmission grid. Some examples of micropower systems are a solar-battery system serving a remote load, a wind–diesel system serving an isolated village, and a grid-connected natural gas microturbine providing electricity and heat to a factory. Power plants that supply electricity to a high-voltage transmission system do not qualify as micropower systems because they are not dedicated to a particular load. HOMER can model grid-connected and off-grid micropower systems serving electric and thermal loads and comprising any combination of photovoltaic (PV) modules, wind turbines, small hydro, biomass power, reciprocating engine generators, microturbines, fuel cells, batteries, and hydrogen storage.

Integration of Renewable Sources of Energy, Second Edition. Felix A. Farret and M. Godoy Simões.

The analysis and design of micropower systems can be challenging due to the large number of design options and the uncertainty in key parameters, such as load size and future fuel price. Renewable power sources add further complexity because their power output may be intermittent, seasonal, and not dispatchable, and the availability of renewable resources may be uncertain. We designed HOMER to overcome these challenges.

HOMER performs three principal tasks: simulation, optimization, and sensitivity analysis. In the simulation process, HOMER models the performance of a particular micropower system configuration each hour of the year to determine its technical feasibility and life-cycle cost. In the optimization process, HOMER simulates many different system configurations in search of the one that satisfies the technical constraints at the lowest life-cycle cost. In the sensitivity analysis process, HOMER performs multiple optimizations under a range of input assumptions to gauge the effects of uncertainty or changes in the model inputs. Optimization determines the optimal value of the variables over which the system designer has control such as the mix of components that make up the system and the size or quantity of each. Sensitivity analysis helps assess the effects of uncertainty or changes in the variables over which the designer has no control, such as the average wind speed or the future fuel price.

Figure 18.1 illustrates the relationship among simulation, optimization, and sensitivity analysis. The optimization oval encloses the simulation oval to represent the fact that a single optimization consists of multiple simulations. Similarly, the sensitivity analysis oval encompasses the optimization oval because a single sensitivity analysis consists of multiple optimizations.

In order to limit the complexity of input variables and parameters, and also to permit fast enough computation to make optimization and sensitivity analysis practical, HOMER's simulation logic is less detailed than that of several other time-series simulation models for micropower systems, such as Hybrid2 [1], PV-DesignPro [2], and PV*SOL [3]. On the other hand, HOMER is more

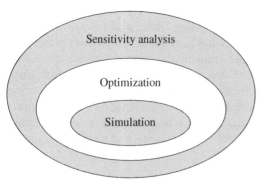

Figure 18.1 Conceptual relationship between simulation, optimization, and sensitivity analysis.

detailed than statistical models such as RETScreen [4], which do not perform time-series simulations. Of all these models, HOMER is the most flexible in terms of the diversity of systems it can simulate.

In this chapter, we summarize the capabilities of HOMER and discuss the benefits it can provide to the micropower system modeler. In Sections 18.2 through 18.4, we describe the structure, purpose, and capabilities of HOMER and introduce the model. In Sections 18.5 and 18.6, we discuss in greater detail the technical and economic aspects of the simulation process. A glossary defines many of the terms used in the chapter.

18.2 Simulation

HOMER's fundamental capability is to simulate the long-term operation of a micropower system. Its higher-level capabilities, optimization and sensitivity analysis, rely on this simulation capability. The simulation process determines how a particular system configuration, a combination of system components of specific sizes, and an operating strategy that defines how those components work together would behave in a given setting over a long period.

HOMER can simulate a wide variety of micropower system configurations, comprising any combination of a PV array, one or more wind turbines, a run-of-river hydro turbine, and up to three generators, a battery bank, an ac–dc converter, an electrolyzer, and a hydrogen storage tank. The system can be grid connected or autonomous and can serve ac and dc electric loads and a thermal load. Figure 18.2 shows schematic diagrams of some examples of the types of micropower systems that HOMER can simulate.

Systems that contain a battery bank and one or more generators require a dispatch strategy, which is a set of rules governing how the system charges the battery bank. HOMER can model two different dispatch strategies: load following and cycle charging. Under the load-following strategy, renewable power sources charge the battery, but the generators do not. Under the cycle-charging strategy, whenever the generators operate, they produce more power than required to serve the load with surplus electricity going to charge the battery bank.

The simulation process serves two purposes. First, it determines whether the system is feasible. HOMER considers the system to be feasible if it can adequately serve the electric and thermal loads and satisfy any other constraints imposed by the user. Second, it estimates the life-cycle cost of the system, which is the total cost of installing and operating the system over its lifetime. The life-cycle cost is a convenient metric for comparing the economics of various system configurations. Such comparisons are the basis of HOMER's optimization process, described in Section 18.3.

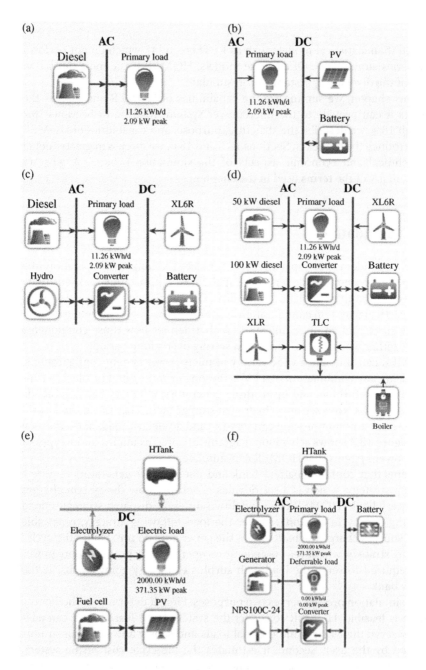

Figure 18.2 Schematic diagrams of some micropower system types that HOMER models: (a) a diesel system, (b) a PV–battery system, (c) a hybrid hydro–wind–diesel system, (d) a wind–diesel system, (e) a PV–hydrogen system, (f) a wind-powered system, (g) a grid-connected PV system, (h) a grid-connected combined heat and power (CHP) system, and (i) a grid-connected CHP system.

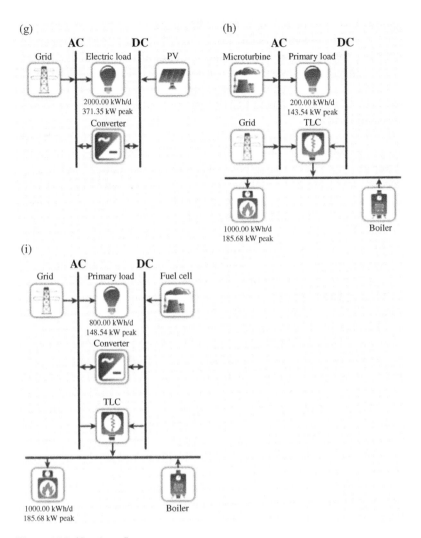

(g)

(h)

(i)

Figure 18.2 (Continued)

Some schematic diagrams of micropower system types that HOMER models are presented in Figure 18.2: (a) a diesel system serving an ac electric load; (b) a PV–battery system serving a dc electric load; (c) a hybrid hydro–wind–diesel system with battery backup and an ac–dc converter; (d) a wind–diesel system serving electric and thermal loads with two generators, a battery bank, a boiler, and a dump load that helps supply the thermal load by passing excess wind turbine power through a resistive heater; (e) a PV–hydrogen system in which an electrolyzer converts excess PV power into hydrogen, which a hydrogen tank stores for use in a fuel cell during times of insufficient PV power; (f) a wind-powered system using both batteries and hydrogen for backup, where the hydrogen fuels an internal combustion engine generator; (g) a grid-connected PV system; (h) a grid-connected combined heat and power (CHP) system in which a microturbine produces both electricity and heat; and (i) a grid-connected CHP system in which a fuel cell provides electricity and heat.

HOMER models a particular system configuration by performing an hourly time-series simulation of its operation over 1 year. HOMER steps through the year 1 hour at a time, calculating the available renewable power, comparing it to the electric load, and deciding what to do with surplus renewable power in times of excess or how best to generate (or purchase from the grid) additional power in times of deficit. When it has completed one-year worth of calculations, HOMER determines whether the system satisfies the constraints imposed by the user on such quantities as the fraction of the total electrical demand served, the proportion of power generated by renewable sources, or the emissions of certain pollutants. HOMER also computes the quantities required to calculate the system's life-cycle cost, such as the annual fuel consumption, annual generator operating hours, expected battery life, or the quantity of power purchased annually from the grid.

The variable used by HOMER to represent the life-cycle cost of the system is the *total net present cost* (NPC). This single value includes all costs and revenues that occur within the project lifetime, with future cash flows discounted to the present. The total NPC includes the initial capital cost of the system components, the cost of any component replacements that occur within the project lifetime, the cost of maintenance and fuel, and the cost of purchasing power from the grid. Any revenue from the sale of power to the grid reduces the total NPC. In Section 18.6 we describe in greater detail how HOMER calculates the total NPC.

For many types of micropower systems, particularly those involving intermittent renewable power sources, a one-hour time step is necessary to model the behavior of the system with acceptable accuracy. In a wind–diesel–battery system, for example, it is not enough to know the monthly average (or even daily average) wind power output, since the timing and the variability of that power output are as important as its average quantity. To predict accurately the diesel fuel consumption, diesel operating hours, the flow of energy through the battery, and the amount of surplus electrical production, it is necessary to know how closely the wind power output correlates to the electric load, and whether the wind power tends to come in long gusts followed by long lulls, or tends to fluctuate more rapidly. HOMER's one-hour time step is sufficiently small to capture the most important statistical aspects of the load and the intermittent renewable resources, but not as small as to slow computation to

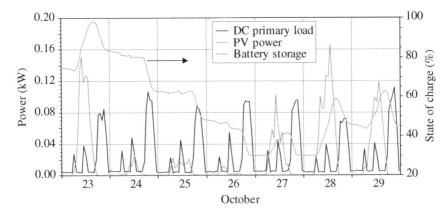

Figure 18.3 Sample hourly simulation results.

the extent that optimization and sensitivity analysis become impractical. Note that HOMER does not model electrical transients or other dynamic effects, which would require much smaller time steps.

Figure 18.3 shows a portion of the hourly simulation results that HOMER produced when modeling a PV–battery system similar to the one shown in Figure 18.2b. In such a system, the battery bank absorbs energy when the PV power output exceeds the load and discharges energy when the load exceeds the PV power output. The graph shows how the amount of energy stored in the battery bank drops during three consecutive days of poor sunshine, October 24–26. The depletion of the battery meant that system could not supply the entire load on October 26 and 27. HOMER records such energy shortfalls and at the end of the simulation determines whether the system supplied enough of the total load to be considered feasible according to user-specified constraints. HOMER also uses the simulation results to calculate the battery throughput (the amount of energy that cycled through the battery over the year), which it uses to calculate the lifetime of the battery. The lifetime of the battery affects the total NPC of the system.

HOMER simulates how the system operates over one year and assumes key results for that period (such as fuel consumption, battery throughput, and surplus power production) are representative of every other year in the project lifetime. It does not consider changes over time, such as load growth or the deterioration of battery performance with aging. The modeler can, however, analyze many of these effects using sensitivity analysis, described in Section 18.4. In Sections 18.5 and 18.6, we discuss in greater detail the technical and economic aspects of HOMER's simulation process.

18.3 Optimization

The simulation process approaches a particular system configuration, while the optimization process determines the best possible system configuration. In HOMER, the best possible, or optimal, system configuration is the one that satisfies the user-specified constraints at the lowest total NPC. Finding the optimal system configuration may involve deciding on the mix of components that the system should contain, the size or quantity of each component, and the dispatch strategy the system should use. In the optimization process, HOMER simulates many different system configurations, discards the infeasible ones (those that do not satisfy the user-specified constraints), ranks the feasible ones according to total NPC, and presents the feasible one with the lowest total NPC as the optimal system configuration.

The goal of the optimization process is to determine the optimal value of each decision variable that interests the modeler. A decision variable is a variable over which the system designer has control and for which HOMER can consider multiple possible values in its optimization process. Possible decision variables in HOMER include:

Size of the PV array
Number of wind turbines
Presence of the hydro system (HOMER can consider only one size of hydro system; the decision is therefore whether or not the power system should include the hydro system)
Size of each generator
Number of batteries
Size of the ac–dc converter
Size of the electrolyzer
Size of the hydrogen storage case
Dispatch strategy (the set of rules governing how the system operates)

Optimization can help the modeler find the optimal system configuration out of many possibilities. Consider, for example, the task of retrofitting an existing diesel power system with wind turbines and batteries. In analyzing the options for redesigning the system, the modeler may want to consider the arrangement of components shown in Figure 18.4, but would not know in advance what number of wind turbines, what number of batteries, and what size of converter minimize the life-cycle cost. These three variables would therefore be decision variables in this analysis. The dispatch strategy could also be a decision variable, but for simplicity this discussion will exclude the dispatch strategy. In Section 18.5.4 we discuss dispatch strategy in greater detail.

HOMER allows the designer to enter multiple values for each decision variable. Using a table like the one shown in Figure 18.5, the user enters any number of values for each decision variable. The spacing between values does

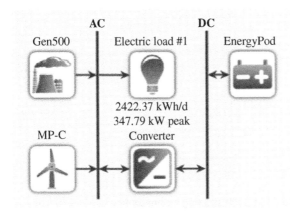

Figure 18.4 Wind–diesel system.

Converter Capacity (kW)	EnergyPod Strings (#)	Gen500 Capacity (kW)	MP-C Quantity (#)	
200	3	500	0	
150	2		1	
100	1		2	
50	0			

Figure 18.5 Search space comprising 48 system configurations (5 1 7 4 ¼ 140).

not have to be regular. In this example, the modeler chose to simulate five quantities of wind turbines, ranging from zero to four; the one existing generator size; seven quantities of batteries, ranging from 0 to 128; and four converter sizes, ranging from 0 to 120 kW. This table shows the search space, which is the set of all possible system configurations over which HOMER can search for the optimal system configuration. This search space includes 140 distinct system configurations because the possible values of the decision variables comprise 140 different combinations: five quantities of wind turbines multiplied by seven quantities of batteries multiplied by four sizes of converter.

In the optimization process, HOMER simulates every system configuration in the search space and displays the feasible ones in a table, sorted by total NPC. Figure 18.5 shows the results of the sample wind–diesel retrofit analysis.

⚠ MP-C	Gen500 (kW)	EnergyPod	Converter (kW)	Dispatch	COE ($)	NPC ($)	Operating cost ($)	Initial capital ($)	Ren Frac (%)	Hours	Production	Fuel (L)
2	500	1	100	CC	$0.462	$5.28M	$304,204	$1.35M	18	3,562	722,955	201,335
2	500	1	50.0	CC	$0.464	$5.30M	$344,256	$850,000	14	4,369	757,603	215,438
2	500	2	100	CC	$0.468	$5.35M	$302,062	$1.45M	18	3,521	720,608	200,475
2	500	2	50.0	CC	$0.471	$5.38M	$342,618	$950,000	15	4,343	755,093	214,644
2	500	1	150	CC	$0.472	$5.39M	$273,856	$1.85M	17	2,700	735,895	198,458
2	500	2	150	CC	$0.474	$5.42M	$268,468	$1.95M	17	2,583	732,618	196,840
2	500	3	100	CC	$0.477	$5.45M	$301,555	$1.55M	19	3,510	720,258	200,313
2	500	3	50.0	CC	$0.479	$5.48M	$342,583	$1.05M	15	4,343	754,949	214,609
2	500	3	150	CC	$0.480	$5.48M	$265,541	$2.05M	17	2,519	731,005	195,998
1	500	1	50.0	CC	$0.501	$5.72M	$376,907	$850,000	4.4	4,655	845,633	238,919

Figure 18.6 Overall optimization results table showing system configurations sorted by total net present cost.

Each row in the table represents a feasible system configuration. The first four columns contain icons indicating the presence of the different components, the next four columns indicate the number or size of each component, and the next five columns contain a few of the key simulation results, namely, the total capital cost of the system, the total NPC, the levelized cost of energy (cost per kilowatt-hour), the annual fuel consumption, and the number of hours the generator operates per year. The modeler can access the complete simulation results, including hourly data, for any particular system configuration; this table is a summary of the simulation results for many different configurations.

The first row in Figure 18.6 is the optimal system configuration, meaning the one with the lowest NPC. In this case, the optimal configuration contains one wind turbine, the 135-kW generator, 64 batteries, and a 30-kW converter. The second-ranked system is the same as the first except that it contains two wind turbines instead of one. The third-ranked system is the same as the first except that it contains fewer batteries. The eighth- and tenth-ranked systems contain no wind turbines.

HOMER can also show a subset of these overall optimization results by displaying only the least-cost configuration within each system category or type. In the overall list shown in Figure 18.6, the top-ranked system is the least-cost configuration within the wind–diesel–battery system category. Similarly, the eighth-ranked system is the least-cost configuration within the diesel–battery system category.

The categorized optimization results list, shown in Figure 18.7, makes it easier to see which is the least-cost configuration for each category by eliminating the need to scroll through the longer list of systems displayed in the overall list.

The results in Figure 18.7 show that under the assumptions of this analysis, adding wind turbines and a battery bank would indeed reduce the life-cycle cost of the system. The initial investment of $1.35 M for the optimal system

⚠ ↑ ⋯	MP-C	Gen500 (kW)	EnergyPod	Converter (kW)	Dispatch	COE ($)	NPC ($)	Operating cost ($)	Initial capital ($)	Ren Frac (%)	Hours	Production	Fuel (L)
↑ ⋯	2	500	1	100	CC	$0.462	$5.28M	$304,204	$1.35M	18	3,562	722,955	201,335
⋯		500	1	50.0	CC	$0.549	$6.28M	$419,756	$850,000	0.0	4,775	1,002,269	277,978

Figure 18.7 Categorized optimization results table.

leads to a savings in NPC of $1 M compared with the existing system, shown in the second row. The optimization results tables allow this kind of comparison because they display more than just the optimal system configuration. The overall optimization results table in particular tends to show many system configurations whose total NPC is only slightly higher than that of the optimal configuration. The modeler may decide that one of these suboptimal configurations is preferable in some way to the configuration that HOMER presents as optimal. For example, the simulation results may show that a system configuration with a slightly higher total NPC than the optimal configuration does a better job of avoiding deep and extended discharges of the battery bank, which can dramatically shorten battery life in real systems. This technical detail is beyond the scope of the model, but one that the modeler can take into consideration in making a final design decision.

18.4 Sensitivity Analysis

In Section 18.3 we described the optimization process, in which HOMER finds the system configuration that is optimal under a particular set of input assumptions. In this section, we describe the sensitivity analysis process, in which HOMER performs multiple optimizations, each using a different set of input assumptions. A sensitivity analysis reveals how sensitive the outputs are to changes in the inputs.

In a sensitivity analysis, the HOMER user enters a range of values for a single input variable. A variable for which the user has entered multiple values is called a sensitivity variable. Almost every numerical input variable in HOMER that is not a decision variable can be a sensitivity variable. Examples include the grid power price, the fuel price, the interest rate, or the lifetime of the PV array. As described in Section 18.4.2, the magnitude of an hourly data set, such as load and renewable resource data, can also be a sensitivity variable.

The HOMER user can perform a sensitivity analysis with any number of sensitivity variables. Each combination of sensitivity variable values defines a distinct sensitivity case, for example, if the user specifies six values for the grid power price and four values for the interest rate, which defines 24 distinct sensitivity cases. HOMER performs a separate optimization process for each sensitivity case and presents the results in various tabular and graphic formats.

One of the primary uses of sensitivity analysis is in dealing with uncertainty. If a system designer is unsure of the value of a particular variable, he or she can enter several values covering the likely range and see how the results vary across that range. However, sensitivity analysis has applications beyond coping with uncertainty. A system designer can use sensitivity analysis to evaluate trade-offs and answer such questions as: How much additional capital investment is required to achieve 50 or 100% renewable energy production? An energy planner can determine which technologies, or combinations of technologies, are optimal under different conditions. A market analyst can determine at what price, or under what conditions, a product (e.g., a fuel cell or a wind turbine) competes with the alternatives. A policy analyst can determine what level of incentive is needed to stimulate the market for a particular technology or what level of emissions penalty would tilt the economics toward cleaner technologies.

18.4.1 Dealing with Uncertainty

A challenge that often confronts the micropower system designer is uncertainty in key variables. Sensitivity analysis can help the designer understand the effects of uncertainty and make good design decisions despite uncertainty. For example, consider the wind–diesel system analysis in Section 18.3. In performing this analysis, the modeler assumed that the price of diesel fuel would be \$0.60/L over the 25-year project lifetime. There is obviously substantial uncertainty in this value, but many other inputs may be uncertain as well, such as the lifetime of the wind turbine, the maintenance cost of the diesel engine, the long-term average wind speed at the site, and even the average electric load. Sensitivity analysis can help the modeler to determine the effect that variations in these inputs have on the behavior, feasibility, and economics of a particular system configuration, the robustness of a particular system configuration (in other words, whether it is nearly optimal in all scenarios or far from optimal in certain scenarios), and how the optimal system configuration changes across the range of uncertainty.

The spider graph in Figure 18.8 shows the results of a sensitivity analysis on three variables. In this analysis, the modeler fixed the system configuration to the wind–diesel system that appears in the first row of the optimization table in Figure 18.6 but entered multiple values for three input variables: the diesel fuel price, the wind turbine lifetime, and the generator operating and maintenance (O&M) cost. For each variable, the modeler entered values ranging from 30% below to 30% above a best estimate. Figure 18.8 shows how sensitive the total NPC is to each of these three variables. The relative steepness of the three curves shows that the total NPC is more sensitive to the fuel price than to the other two variables. Such information can help a system designer to establish the bounds of a confidence interval or to prioritize efforts to reduce uncertainty.

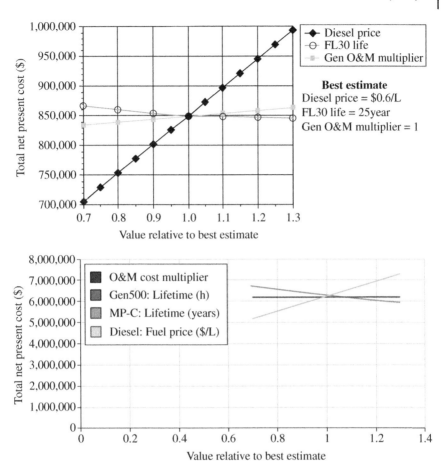

Figure 18.8 Spider graph showing the effect of changes in three sensitivity variables.

A sensitivity analysis can also incorporate optimization. Figure 18.9 shows the results of a second sensitivity analysis on the wind–diesel system from Section 18.3. The modeler performed this analysis to see whether the fuel price, which Figure 18.8 identified as the most important of the uncertain variables, affects the optimal system configuration. This time the modeler used the search space shown in Figure 18.5 and entered a range of fuel prices above and below the best estimate value of $0.60/L. The tabular results in Figure 18.9 show that as the fuel price increases, the optimal system configuration changes from a diesel–battery system to a wind–diesel–battery system comprising one wind turbine and, as the fuel price increases further, to a wind–diesel–battery system comprising two wind turbines.

Diesel ($/L)		FL30	Gen (kW)	Batt.	Conv. (kW)	Total NPC
0.420			135	48	30	$ 688,679
0.450			135	48	30	$ 721,987
0.480		1	135	64	30	$ 753,695
0.510		1	135	64	30	$ 777,748
0.540		1	135	64	30	$ 801,800
0.570		1	135	64	30	$ 825,852
→ 0.600		1	135	64	30	$ 849,905
0.630		2	135	64	30	$ 872,093
0.660		2	135	64	30	$ 889,525
0.690		2	135	64	30	$ 906,957
0.720		2	135	64	30	$ 924,389
0.750		2	135	64	30	$ 941,821
0.780		2	135	64	30	$ 959,253

Figure 18.9 Tabular sensitivity results showing optimal system configuration changing with fuel price.

	Architecture							Cost			
⚠ 🕊 📷 🔋 ☑	MP-C	Gen500 (kW)	EnergyPod	Converter (kW)	Dispatch	COE ($)	NPC ($)	Operating cost ($)		Initial capital ($)	
🕊 📷 🔋 ☑ 2		500	1	50.0	CC	$0.304	$3.48M	$203,265		$850,000	
📷 🔋 ☑		500	1	50.0	CC	$0.358	$4.10M	$251,090		$850,000	

Figure 18.10 Optimization results for the $0.42/L fuel price sensitivity case.

The arrow in Figure 18.9 highlights the best estimate scenario with a fuel price of $0.60/L. The optimal system configuration for this scenario (one wind turbine, a 135-kW generator, 64 batteries, and a 30-kW converter) is optimal for 5 of the 13 sensitivity cases. An investigation of the overall optimization results tables for the other sensitivity cases shows that this system configuration is nearly optimal for those sensitivity cases as well. For example, Figure 18.10 shows the optimization results table for the lowest fuel price scenario ($0.42/L). The system configuration that was optimal in the $0.60/L scenario ranks third in this scenario, with a total NPC of $705,590, which is only 2.4% higher than the total NPC of the optimal configuration. For the most expensive fuel scenario, it ranks fifth, with a total NPC only 3.6% higher than the optimal configuration. It is therefore a robust solution in that it performs well across the given range of fuel prices. By contrast, the diesel–battery system that is optimal in the lowest fuel price scenario ranks 36th in the highest fuel price scenario, with a total NPC 13.5% higher than the optimal configuration. The analysis shows that the diesel–battery system would have a higher risk than the wind–diesel–battery system because it is far from optimal under some fuel price scenarios.

With this kind of information, a modeler can make informed decisions despite uncertainty in important variables. As in the previous example, a sensitivity analysis can reveal how different system configurations perform over a wide range of possible scenarios, helping the designer assess the risks associated with each.

18.4.2 Sensitivity Analyses on Hourly Data Sets

One of HOMER's most powerful features is its ability to do sensitivity analyses on hourly data sets such as the primary electric load or the solar, wind, hydro, or biomass resource. HOMER's use of scaling variables enables such sensitivity analyses. Each hourly data set comprises 8760 values that have a certain average value. However, each hourly data set also has a corresponding scaling variable that the modeler can use to scale the entire data set up or down. For example, a user may specify hourly primary load data with an annual average of 120 kWh/day and then specify 100, 150, and 200 kWh/day for the primary load-scaling variable. In the course of the sensitivity analysis, HOMER will scale the load data so that it averages first 100 kWh/day, then 150 kWh/day, and finally 200 kWh/day. This scaling process changes the magnitude of the load data set without affecting the daily load shape, the seasonal pattern, or any other statistical properties. HOMER scales renewable resource data in the same manner.

Figure 18.11 shows the results of a sensitivity analysis over a range of load sizes and annual average wind speeds. The modeler specified eight values for the average size of the electric load and five values for the annual average wind speed. The axes of the graph correspond to these two sensitivity variables.

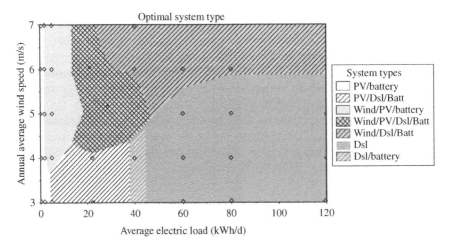

Figure 18.11 Optimal system graph.

At each of the 40 sensitivity cases, HOMER performed an optimization over a search space comprising more than 5000 system configurations. The diamonds in the graph indicate these sensitivity cases, and the color of each diamond indicates the optimal system type for that sensitivity case. At an average load of 22 kWh/day and an average wind speed of 4 m/s, for example, the optimal system type was PV–diesel–battery. At an average load of 40 kWh/day and the same wind speed, the optimal system type was diesel–battery. HOMER uses two-dimensional linear interpolation to determine the optimal system type at all points between the diamonds.

The graph in Figure 18.11 of optimal system type shows that for the assumptions used in this analysis, PV–battery systems are optimal for very small systems, regardless of the wind speed. At low wind speeds, as the load size increases, the optimal system type changes to PV–diesel–battery, diesel–battery, and then pure diesel. At high wind speeds, as the load size increases, the optimal system type changes to wind–PV–battery, wind–PV–diesel–battery, and finally, wind–diesel–battery. To an energy planner intending to provide electricity to many unelectrified communities in a developing country, such an analysis could help decide what type of micropower system to use for different communities based on the load size and average wind speed of each community.

18.5 Physical Modeling

In Section 18.2 we discussed the role of simulation and briefly described the process HOMER uses to simulate micropower systems. In this section, we provide greater detail on how HOMER models the physical operation of a system. In HOMER, a micropower system must comprise at least one source of electrical or thermal energy (such as a wind turbine, a diesel generator, a boiler, or the grid) and at least one destination for that energy (an electrical or thermal load or the ability to sell electricity to the grid). It may also comprise conversion devices such as an ac–dc converter or an electrolyzer and energy storage devices such as a battery bank or a hydrogen storage tank.

In the following subsections we describe how HOMER models the loads that the system must serve, the components of the system and their associated resources, and how that collection of components operates together to serve the loads.

18.5.1 Loads

In HOMER, the term load refers to a demand for electric or thermal energy. Serving loads is the reason for the existence of micropower systems, so the modeling of a micropower system begins with the modeling of the load or loads that the system must serve. HOMER models three types of loads. Primary load is electrical demand that must be served according to a particular schedule. Deferrable load is electrical demand that can be served at any time within a certain time span. Thermal load is demand for heat.

18.5.1.1 Primary Load

Primary load is electrical demand that the power system must meet at a specific time. Electrical demand associated with lights, radio, TV, household appliances, computers, and industrial processes are typically modeled as primary load. When a consumer switches on a light, the power system must supply electricity to that light immediately—the load cannot be deferred until later. If electrical demand exceeds supply, HOMER records a shortfall as unmet load.

The HOMER user specifies an amount of primary load in kilowatts for each hour of the year, either by importing a file containing hourly data or by allowing HOMER to synthesize hourly data from average daily load profiles. When synthesizing load data, HOMER creates hourly load values based on user-specified daily load profiles. The modeler can specify a single 24-hour profile that applies throughout the year or can specify different profiles for different months and different profiles for weekdays and weekends. HOMER adds a user-specified amount of randomness to synthesized load data so that every day's load pattern is unique. HOMER can model two separate primary loads, each of which can be ac or dc.

Among the three types of loads modeled in HOMER, primary load receives special treatment in that it requires a user-specified amount of operating reserve. Operating reserve is surplus electrical generating capacity that is operating and can respond instantly to a sudden increase in the electric load or a sudden decrease in the renewable power output. Although it has the same meaning as the more common term spinning reserve, we call it operating reserve because batteries, fuel cells, and the grid can provide it, but they do not spin. When simulating the operation of the system, HOMER attempts to ensure that the system's operating capacity is always sufficient to supply the primary load and the required operating reserve. Section 18.5.4 covers operating reserve in greater detail.

18.5.1.2 Deferrable Load

Deferrable load is an electrical demand that can be met anytime within a defined time interval. Water pumps, icemakers, and battery-charging stations are examples of deferrable loads because the storage inherent to each of those loads allows some flexibility as to when the system can serve them. The ability to defer serving a load is often advantageous for systems comprising intermittent renewable power sources, because it reduces the need for precise

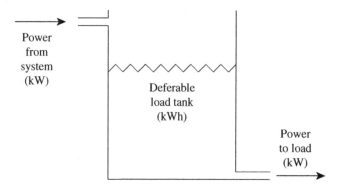

Figure 18.12 Deferrable load tank analogy.

control of the timing of power production. If the renewable power supply ever exceeds the primary load, the surplus can serve the deferrable load rather than going to waste.

Figure 18.12 shows a schematic representation of how HOMER models the deferrable load. The power system puts energy into a "tank" of finite capacity, and energy drains out of that tank to serve the deferrable load. For each month, the user specifies the average deferrable load, which is the rate at which energy drains out of the tank. The user also specifies the storage capacity in kilowatt-hours (the size of the tank) and the maximum and minimum rate at which the power system can put energy into the tank. Note that the energy tank model is simply an analogy; the actual deferrable load may or may not make use of a storage tank.

When simulating a system serving a deferrable load, HOMER tracks the level in the deferrable load tank. It will put any excess renewable power into the tank, but as long as the tank level remains above zero, HOMER will not use a dispatchable power source (a generator, the battery bank, or the grid) to put energy into the tank. If the level in the tank drops to zero, HOMER temporarily treats the deferrable load as a primary load, meaning that it will immediately use any available power source to put energy into the tank and avoid having the deferrable load go unmet.

18.5.1.3 Thermal Load

HOMER models thermal load in the same way that it models primary electric load, except that the concept of operating reserve does not apply to the thermal load. The user specifies the amount of thermal load for each hour of the year, either by importing a file containing hourly data or by allowing HOMER to synthesize hourly data from 24-hour load profiles. The system supplies the thermal load with either the boiler, waste heat recovered from a generator, or resistive heating using excess electricity.

18.5.2 Resources

The term resource applies to anything coming from outside the system that is used by the system to generate electric or thermal power. That includes the four renewable resources (solar, wind, hydro, and biomass) as well as any fuel used by the components of the system. Renewable resources vary enormously by location. The solar resource depends strongly on latitude and climate, the wind resource on large-scale atmospheric circulation patterns and geographic influences, the hydro resource on local rainfall patterns and topography, and the biomass resource on local biological productivity. Moreover, at any one location, a renewable resource may exhibit strong seasonal and hour-to-hour variability. The nature of the available renewable resources affects the behavior and economics of renewable power systems, since the resource determines the quantity and the timing of renewable power production. The careful modeling of the renewable resources is therefore an essential element of system modeling. In this section, we describe how HOMER models the four renewable resources and the fuel.

18.5.2.1 Solar Resource

In order to model a system containing a PV array, the HOMER user must provide solar resource data for the location of interest. Solar resource data indicate the amount of global solar radiation (beam radiation coming directly from the sun and diffuse radiation coming from all parts of the sky) that strikes the Earth's surface in a typical year. The data can be in one of three forms: hourly average global solar radiation on the horizontal surface (kW/m^2), monthly average global solar radiation on the horizontal surface (kWh/m^2day), or monthly average clearness index. The clearness index is the ratio of the solar radiation striking the Earth's surface to the solar radiation striking the top of the atmosphere. A number between 0 and 1, the clearness index is a measure of the clearness of the atmosphere.

If the user chooses to provide monthly solar resource data, HOMER generates synthetic hourly global solar radiation data using an algorithm developed by Graham and Hollands [5]. The inputs to this algorithm are the monthly average solar radiation values and the latitude. The output is an 8760-hour data set with statistical characteristics similar to those of real measured data sets. One of those statistical properties is autocorrelation, which is the tendency for one day to be similar to the preceding day and for one hour to be similar to the preceding hour.

18.5.2.2 Wind Resource

To model a system comprising one or more wind turbines, the HOMER user must provide wind resource data indicating the wind speeds the turbines would experience in a typical year. The user can provide measured hourly wind speed data if available. Otherwise, HOMER can generate synthetic hourly data from 12 monthly average wind speeds and four additional statistical parameters: the Weibull shape factor, the autocorrelation factor, the diurnal pattern strength, and the hour of peak wind speed. The Weibull shape factor is a measure of the distribution of wind speeds over the year. The autocorrelation factor is a measure of how strongly the wind speed in one hour tends to depend on the wind speed in the preceding hour. The diurnal pattern strength and the hour of peak wind speed indicate the magnitude and the phase, respectively, of the average daily pattern in the wind speed. HOMER provides default values for each of these parameters.

The user indicates the anemometer height, meaning the height above ground at which the wind speed data were measured or for which they were estimated. If the wind turbine hub height is different from the anemometer height, HOMER calculates the wind speed at the turbine hub height using either the logarithmic law, which assumes that the wind speed is proportional to the logarithm of the height above ground, or the power law, which assumes that the wind speed varies exponentially with height. To use the logarithmic law, the user enters the surface roughness length, which is a parameter characterizing the roughness of the surrounding terrain. To use the power law, the user enters the power law exponent.

The user also indicates the elevation of the site above sea level, which HOMER uses to calculate the air density according to the US Standard Atmosphere, described in Section 2.3 of White [6]. HOMER makes use of the air density when calculating the output of the wind turbine, as described in Section 18.5.3.

18.5.2.3 Hydro Resource

In order to model a system comprising a run-of-river hydro turbine, the HOMER user must provide stream flow data indicating the amount of water available to the turbine in a typical year. The user can provide measured hourly stream flow data if available. Otherwise, HOMER can use monthly averages under the assumption that the flow rate remains constant within each month. The user also specifies the residual flow, which is the minimum stream flow that must bypass the hydro turbine for ecological purposes. HOMER subtracts the residual flow from the stream flow data to determine the stream flow available to the turbine.

18.5.2.4 Biomass Resource

A biomass resource takes various forms (e.g., wood waste, agricultural residue, animal waste, energy crops) and may be used to produce heat or electricity. HOMER models biomass power systems that convert biomass into electricity. Two aspects of the biomass resource make it unique among the four renewable resources that HOMER models. First, the availability of the resource depends in part on human effort for harvesting, transportation, and storage. It is consequently not intermittent, although it may be seasonal. It is also often not free. Second, the biomass feedstock may be converted to a gaseous or liquid fuel to be consumed in an otherwise conventional generator. The modeling of the biomass resource is therefore similar in many ways to the modeling of any other fuel.

The HOMER user can model the biomass resource in two ways. The simplest way is to define a fuel with properties corresponding to the biomass feedstock and then specify the fuel consumption of the generator to show electricity produced versus biomass feedstock consumed. This approach implicitly, rather than explicitly, models the process of converting the feedstock into a fuel suitable for the generator, if such a process occurs. The second alternative is to use HOMER's biomass resource inputs, which allow the modeler to specify the availability of the feedstock throughout the year and to model explicitly the feedstock conversion process. In the remainder of this section, we focus on this second alternative. In the next section, we address the fuel inputs.

As with the other renewable resource data sets, the HOMER user can indicate the availability of biomass feedstock by importing an hourly data file or using monthly averages. If the user specifies monthly averages, HOMER assumes that the availability remains constant within each month.

The user must specify four additional parameters to define the biomass resource: price, carbon content, gasification ratio, and the energy content of the biomass fuel. For greenhouse gas analyses, the carbon content value should reflect the net amount of carbon released to the atmosphere by the harvesting, processing, and consumption of the biomass feedstock, considering the fact that the carbon in the feedstock was originally in the atmosphere. The gasification ratio, despite its name, applies equally well to liquid and gaseous fuels. It is the fuel conversion ratio indicating the ratio of the mass of generator-ready fuel emerging from the fuel conversion process to the mass of biomass feedstock entering the fuel conversion process. HOMER uses the energy content of the biomass fuel to calculate the thermodynamic efficiency of the generator that consumes the fuel.

Fuel HOMER provides a library of several predefined fuels, and users can add to the library if necessary. The physical properties of a fuel include its density, lower heating value, carbon content, and sulfur content. The user can also choose the most appropriate measurement units, either L, m^3, or kg. The two remaining properties of the fuel are the price and the annual consumption limit, if any.

18.5.3 Components

In HOMER, a component is any part of a micropower system that generates delivers, converts, or stores energy. HOMER models 10 types of components. Three generate electricity from intermittent renewable sources: PV modules, wind turbines, and hydro turbines. PV modules convert solar radiation into dc electricity. Wind turbines convert wind energy into ac or dc electricity. Hydro turbines convert the energy of flowing water into ac or dc electricity. HOMER can only model run-of-river hydro installations, meaning those that do not comprise a storage reservoir.

Another three types of components—generators, the grid, and boilers—are dispatchable energy sources, meaning that the system can control them as needed. Generators consume fuel to produce ac or dc electricity. A generator may also produce thermal power via waste heat recovery. The grid delivers ac electricity to a grid-connected system and may also accept surplus electricity from the system. Boilers consume fuel to produce thermal power.

Two types of components, converters and electrolyzers, convert electrical energy into another form. Converters convert electricity from ac to dc or from dc to ac. Electrolyzers convert surplus ac or dc electricity into hydrogen via the electrolysis of water. The system can store the hydrogen and use it as fuel for one or more generators. Finally, two types of components store energy: batteries and hydrogen storage tanks. Batteries store dc electricity. Hydrogen tanks store hydrogen from the electrolyzer to fuel one or more generators.

In this section, we explain how HOMER models each of these components and discuss the physical and economic properties that the user can use to describe each.

18.5.3.1 PV Array

HOMER models the PV array as a device that produces dc electricity in direct proportion to the global solar radiation incident upon it, independent of its temperature and the voltage to which it is exposed. HOMER calculates the power output of the PV array using the equation

$$P_{PV} = f_{PV} Y_{PV} \left(\frac{I_T}{I_s} \right) \tag{18.1}$$

where:

f_{PV} = The PV derating factor
Y_{PV} = The rated capacity of the PV array (kW)
I_T = The global solar radiation (beam plus diffuse) incident on the surface of the PV array (kW/m^2)
I_S = 1 kW/m^2, which is the standard amount of radiation used to rate the capacity of the PV array

In the following paragraphs, we describe these variables in more detail.

The rated capacity (sometimes called the peak capacity) of a PV array is the amount of power it would produce under standard test conditions of 1 kW/m^2 irradiance and a panel temperature of 25 °C. In HOMER, the size of a PV array is always specified in terms of rated capacity. The rated capacity accounts for both the area and the efficiency of the PV module, so neither of those parameters appears explicitly in HOMER. A 40-W module made of amorphous silicon (which has a relatively low efficiency) will be larger than a 40-W module made of polycrystalline silicon (which has a relatively high efficiency), but that size difference is of no consequence to HOMER.

Each hour of the year, HOMER calculates the global solar radiation incident on the PV array using the HDKR model, explained in Section 2.16 of Duffie and Beckman [7]. This model takes into account the current value of the solar resource (the global solar radiation incident on a horizontal surface), the orientation of the PV array, the location on the Earth's surface, the time of year, and the time of day. The orientation of the array may be fixed or may vary according to one of several tracking schemes.

The derating factor is a scaling factor meant to account for effects of dust on the panel, wire losses, elevated temperature, or anything else that would cause the output of the PV array to deviate from that expected under ideal conditions. HOMER does not account for the fact that the power output of a PV array decreases with increasing panel temperature, but the HOMER user can reduce the derating factor to (crudely) correct for this effect when modeling systems for hot climates.

In reality, the output of a PV array does depend strongly and nonlinearly on the voltage to which it is exposed. The maximum power point (the voltage at which the power output is maximized) depends on the solar radiation and the temperature. If the PV array is connected directly to a dc load or a battery bank, it will often be exposed to a voltage different from the maximum power point, and performance will suffer. A maximum power point tracker (MPPT) is a solid-state device placed between the PV array and the rest of the dc components of the system that decouples the array voltage from that of the rest of the system and ensures that the array voltage is always equal to the maximum power point. By ignoring the effect of the voltage to which the PV array is exposed, HOMER effectively assumes that an MPPT is present in the system.

To describe the cost of the PV array, the user specifies its initial capital cost in dollars, replacement cost in dollars, and O&M cost in dollars per year. The replacement cost is the cost of replacing the PV array at the end of its useful lifetime, which the user specifies in years. By default, the replacement cost is equal to the capital cost, but the two can differ for several reasons. For example, a donor organization may cover some or all of the initial capital cost but none of the replacement cost.

18.5.3.2 Wind Turbine

HOMER models a wind turbine as a device that converts the kinetic energy of the wind into ac or dc electricity according to a particular power curve, which is a graph of power output versus wind speed at hub height. Figure 18.13 shows

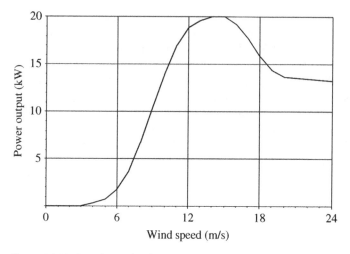

Figure 18.13 Sample wind turbine power curve.

an example power curve. HOMER assumes that the power curve applies at a standard air density of $1.225\,\text{kg/m}^3$, which corresponds to standard temperature and pressure conditions.

Each hour, HOMER calculates the power output of the wind turbine in a four-step process. First, it determines the average wind speed for the hour at the anemometer height by referring to the wind resource data. Second, it calculates the corresponding wind speed at the turbine's hub height using either the logarithmic law or the power law. Third, it refers to the turbine's power curve to calculate its power output at that wind speed assuming standard air density. Fourth, it multiplies that power output value by the air density ratio, which is the ratio of the actual air density to the standard air density. As mentioned in Section 18.5.2, HOMER calculates the air density ratio at the site elevation using the US Standard Atmosphere [6]. HOMER assumes that the air density ratio is constant throughout the year.

In addition to the turbine's power curve and hub height, the user specifies the expected lifetime of the turbine in years, its initial capital cost in dollars, its replacement cost in dollars, and its annual O&M cost in dollars per year.

18.5.3.3 Hydro Turbine

HOMER models the hydro turbine as a device that converts the power of falling water into ac or dc electricity at a constant efficiency, with no ability to store water or modulate the power output. The power in falling water is proportional to the product of the stream flow and the head, which is the vertical distance through which the water falls down. Information on the stream flow available to the hydro turbine each hour comes from the hydro resource data. The user also enters the available head and the head loss that occurs in the intake pipe due to friction. HOMER calculates the net head, or effective head, using the equation

$$h_{net} = h \cdot (1 - f_h)$$ (18.2)

where:

h = The available head
f_h = The pipe head loss

The user also enters the turbine's design flow rate and its acceptable range of flow rates. HOMER calculates the flow through the turbine by

$$\dot{Q}_{turbine} = \begin{cases} MIN\left(\dot{Q}_{stream} - \dot{Q}_{residual}, w_{max} \cdot \dot{Q}_{nom}\right), & \dot{Q}_{stream} - \dot{Q}_{residual} \geq w_{min} \cdot \dot{Q}_{nom} \\ 0, & \dot{Q}_{stream} - \dot{Q}_{residual} < w_{min} \cdot \dot{Q}_{nom} \end{cases}$$

(18.3)

where:

$\dot{Q}_{stream}$ = The stream flow
$\dot{Q}_{residual}$ = The residual flow
$\dot{Q}_{nom}$ = The turbine design flow rate
w_{min} and w_{max} = The turbine's minimum and maximum flow ratios

The turbine does not operate if the stream flow is below the minimum, and the flow rate through the turbine cannot exceed the maximum.

Each hour of the simulation, HOMER calculates the power output of the hydro turbine as

$$P_{hyd} = \eta_{hyd} \cdot \rho_{water} \cdot g \cdot h_{net} \cdot \dot{Q}_{turbine}$$ (18.4)

where:

η_{hyd} = The turbine efficiency
ρ_{water} = The density of water
g = The gravitational acceleration
h_{net} = The net head
$\dot{Q}_{turbine}$ = The flow rate through the turbine

The user specifies the expected lifetime of the hydro turbine in years, as well as its initial capital cost in dollars, replacement cost in dollars, and annual O&M cost in dollars per year.

18.5.3.4 Generators

A generator consumes fuel to produce electricity and possibly heat as a by-product. HOMER's generator module is flexible enough to model a wide variety of generators, including internal combustion engine generators, microturbines, fuel cells, Stirling engines, thermophotovoltaic generators, and thermoelectric generators. HOMER can model a power system comprising as many as three generators, each of which can be ac or dc and can consume a different fuel.

The principal physical properties of the generator are its maximum and minimum electrical power output, its expected lifetime in operating hours, the type of fuel it consumes, and its fuel curve, which relates the quantity of fuel consumed to the electrical power produced. In HOMER, a generator can consume any of the fuels listed in the fuel library (to which users can add their own fuels) or one of two special fuels: electrolyzed hydrogen from the hydrogen storage tank or biomass derived from the biomass resource. It is also possible to cofire a generator with a mixture of biomass and another fuel.

HOMER assumes the fuel curve is a straight line with a y-intercept and uses the following equation for the generator's fuel consumption:

$$F = F_0 \cdot Y_{gen} + F_1 \cdot P_{gen} \tag{18.5}$$

where:

F_0 = The fuel curve intercept coefficient
F_1 = The fuel curve slope
Y_{gen} = The rated capacity of the generator in kW
P_{gen} = The electrical output of the generator in kW

The units of F depend on the measurement units of the fuel. If the fuel is denominated in L, the units of F are L/h. If the fuel is denominated in m^3 or kg, the units of F are m^3/h or kg/h, respectively. In the same way, the units of F_0 and F_1 depend on the measurement units of the fuel. For fuels denominated in L, the units of F_0 and F_1 are L/h/kW.

For a generator that provides heat as well as electricity, the user also specifies the heat recovery ratio. HOMER assumes that the generator converts all the fuel energy into either electricity or waste heat. The heat recovery ratio is the fraction of that waste heat that can be captured to serve the thermal load. In addition to these properties, the modeler can specify the generator emissions coefficients, which specify the generator's emissions of six different pollutants in grams of pollutant emitted per quantity of fuel consumed.

The user can schedule the operation of the generator to force it on or off at certain times. During times that the generator is forced neither on nor off, HOMER decides whether it should operate based on the needs of the system and the relative costs of the other power sources. During times that the generator is forced on, HOMER decides at what power output level it operates, which may be anywhere between its minimum and maximum power output.

The user specifies the generator's initial capital cost in dollars, replacement cost in dollars, and annual O&M cost in dollars per operating hour. The generator O&M cost should account for oil changes and other maintenance costs, but not fuel cost because HOMER calculates fuel cost separately. As it does for all dispatchable power sources, HOMER calculates the generator's fixed and marginal cost of energy and uses that information when simulating the operation of the system. The fixed cost of energy is the cost per hour of simply running the generator, without producing any electricity. The marginal cost of energy is the additional cost per kilowatt-hour of producing electricity from that generator.

HOMER uses the following equation to calculate the generator's fixed cost of energy:

$$c_{gen,fixed} = c_{om,gen} + \frac{C_{rep,gen}}{R_{gen}} + F_0 \cdot Y_{gen} \cdot c_{fuel,eff} \qquad (18.6)$$

where:

$c_{om,gen}$ = The O&M cost in dollars per hour
$C_{rep,gen}$ = The replacement cost in dollars
R_{gen} = The generator lifetime in hours
F_0 = The fuel curve intercept coefficient in quantity of fuel per hour per kW
F_0 = The capacity of the generator in kW
$c_{fuel,eff}$ = The effective price of fuel in dollars per quantity of fuel

HOMER calculates the marginal cost of energy of the generator using the following equation:

$$c_{gen,mar} = F_1 \cdot c_{fuel,eff} \qquad (18.7)$$

where:

F_1 = The fuel curve slope in quantity of fuel per hour per kWh
$c_{fuel,eff}$ = The effective price of fuel (including the cost of any penalties on emissions) in dollars per quantity of fuel

18.5.3.5 Battery Bank

The battery bank is a collection of one or more individual batteries. HOMER models a single battery as a device capable of storing a certain amount of dc electricity at a fixed round-trip energy efficiency, with limits as to how quickly it can be charged or discharged, how deeply it can be discharged without causing damage, and how much energy can cycle through it before it needs replacement. HOMER assumes that the properties of the batteries remain constant throughout its lifetime and are not affected by external factors such as temperature.

In HOMER, the most important physical properties of a battery are its nominal voltage, capacity curve, lifetime curve, minimum state of charge, and round-trip efficiency. The capacity curve shows the discharge capacity of the battery in ampere-hours versus the discharge current in amperes. Manufacturers determine each point on this curve by measuring the ampere-hours that can be discharged at a constant current out of a fully charged battery. Capacity typically decreases with increasing discharge current. The lifetime curve shows the number of discharge–charge cycles the battery can withstand versus the cycle depth. The number of cycles to failure typically decreases with increasing cycle depth. The minimum state of charge is the state of charge below which the battery must not be discharged to avoid permanent damage. In the system simulation, HOMER does not allow the battery to be discharged any deeper than this limit. The round-trip efficiency indicates the percentage of the energy going into the battery that can be drawn back out.

In order to calculate the battery maximum allowable rate of charge or discharge, HOMER uses a kinetic battery model [8]. In this model the battery is considered as a two-tank system as illustrated in Figure 18.14. According to the kinetic battery model, part of the battery's energy storage capacity is immediately available for charging or discharging, but the rest is chemically bound. The rate of conversion between available energy and bound energy depends on the difference in "height" between the two tanks. Three parameters describe the battery. The maximum capacity of the battery is the combined size of the available and bound tanks. The capacity ratio is the ratio of the size of the available tank to the combined size of the two tanks. The rate constant is analogous to the size of the pipe between the tanks.

The kinetic battery model explains the shape of the typical battery capacity curve, such as the example shown in Figure 18.15. At high discharge rates, the available tank empties quickly, and very little of the bound energy can be converted to available energy before the available tank is empty, at which time the battery can no longer withstand the high discharge rate and appears fully discharged. At slower discharge rates, more bound energy can be converted to available energy before the available tank empties, so the apparent capacity

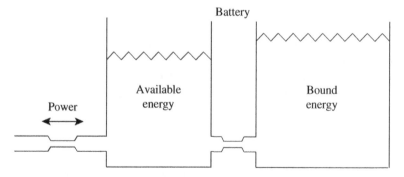

Figure 18.14 Kinetic battery model concept.

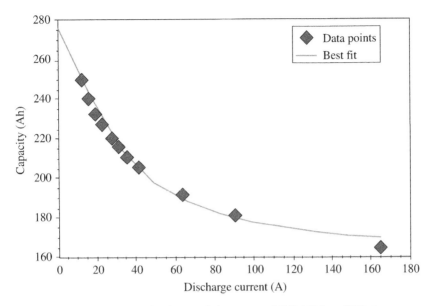

Figure 18.15 Capacity curve for deep-cycle battery model US-250 from US Battery Manufacturing Company (www.usbattery.com). *Source*: © US government.

increases. HOMER performs a curve fit on the battery's discharge curve to calculate the three parameters of the kinetic battery model. The line in Figure 18.15 corresponds to this curve fit.

Modeling the battery as a two-tank system rather than a single-tank system has two effects. First, it means the battery cannot be fully charged or discharged all at once; a complete charge requires an infinite amount of time at a charge current that asymptotically approaches zero. Second, it means that the battery's ability to charge and discharge depends not only on its current state of charge but also on its recent charge and discharge history. A battery rapidly charged to 80% state of charge will be capable of a higher discharge rate than the same battery rapidly discharged to 80%, since it will have a higher level in its available tank. HOMER tracks the levels in the two tanks each hour and models both these effects.

Figure 18.16 shows a lifetime curve typical of a deep-cycle lead–acid battery. The number of cycles to failure (shown in the graph as the lighter-colored points) drops sharply with increasing depth of discharge. For each point on this curve, one can calculate the lifetime throughput (the amount of energy that cycled through the battery before failure) by finding the product of the number of cycles, the depth of discharge, the nominal voltage of the battery, and the aforementioned maximum capacity of the battery. The lifetime throughput curve, shown in Figure 18.16 as black dots, typically shows a much weaker

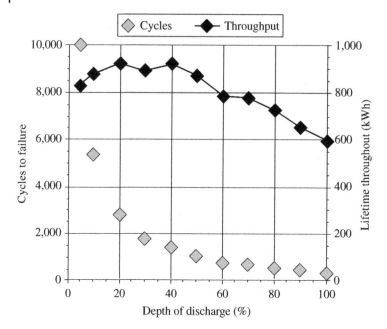

Figure 18.16 Lifetime curve for deep-cycle battery model US-250 from US Battery Manufacturing Company (www.usbattery.com). *Source:* © US government.

dependence on the cycle depth. HOMER makes the simplifying assumption that the lifetime throughput is independent of the depth of discharge. The value that HOMER suggests for this lifetime throughput is the average of the points from the lifetime curve above the minimum state of charge, but the user can modify this value to be more or less conservative.

The assumption that lifetime throughput is independent of cycle depth means that HOMER can estimate the life of the battery bank simply by monitoring the amount of energy cycling through it, without having to consider the depth of the various charge–discharge cycles. HOMER calculates the life of the battery bank in years as

$$R_{batt} = MIN\left(\frac{N_{batt} \cdot Q_{lifetime}}{Q_{thrpt}}, R_{batt,f}\right) \tag{18.8}$$

where:

N_{batt} = The number of batteries in the battery bank
$Q_{lifetime}$ = The lifetime throughput of a single battery
Q_{thrpt} = The annual throughput (the total amount of energy that cycles through the battery bank in one year)
$R_{batt,f}$ = The float life of the battery (the maximum life regardless of throughput)

The user specifies the battery bank's capital and replacement costs in dollars and the O&M cost in dollars per year. Since the battery bank is a dispatchable power source, HOMER calculates its fixed and marginal cost of energy for comparison with other dispatchable sources. Unlike the generator, there is no cost associated with "operating" the battery bank so that it is ready to produce energy; hence its fixed cost of energy is zero. For its marginal cost of energy, HOMER uses the sum of the battery wear cost (the cost per kilowatt-hour of cycling energy through the battery bank) and the battery energy cost (the average cost of the energy stored in the battery bank). HOMER calculates the battery wear cost as

$$c_{bw} = \frac{C_{rep,batt}}{N_{batt} \cdot Q_{lifetime} \cdot \sqrt{\eta_{rt}}} \tag{18.9}$$

where:

$C_{rep,batt}$ = The replacement cost of the battery bank in dollars
N_{batt} = The number of batteries in the battery bank
$Q_{lifetime}$ = The lifetime throughput of a single battery in kWh
η_{rt} = The round-trip efficiency

HOMER calculates the battery energy cost each hour of the simulation by dividing the total year-to-date cost of charging the battery bank by the total year-to-date amount of energy put into the battery bank. Under the load-following dispatch strategy, the battery bank is only ever charged by surplus electricity, so the cost associated with charging the battery bank is always zero. Under the cycle-charging strategy, however, a generator will produce extra electricity (and hence consume additional fuel) for the express purpose of charging the battery bank, so the cost associated with charging the battery bank is not zero. In Section 18.5.4 we discuss dispatch strategies in greater detail.

18.5.3.6 Grid
HOMER models the grid as a component from which the micropower system can purchase ac electricity and to which the system can sell ac electricity. The cost of purchasing power from the grid can comprise an energy charge based on the amount of energy purchased in a billing period and a demand charge based on the peak demand within the billing period. HOMER uses the term grid power price for the price (in dollars per kilowatt-hour) that the electric utility charges for energy purchased from the grid and the demand rate for the price (in dollars per kilowatt per month) the utility charges for the peak grid demand. A third term, the sellback rate, refers to the price (in dollars per kilowatt-hour) that the utility pays for power sold to the grid.

The HOMER user can define and schedule up to 16 different rates, each of which can have different values of grid power price, demand rate, and sellback rate. The schedule of the rates can vary according to month, time of day, and weekday/weekend. For example, HOMER could model a situation where an expensive rate applies during weekday afternoons in July and August, an intermediate rate applies during weekday afternoons in June and September and weekend afternoons from June to September, and an inexpensive rate applies at all other times.

HOMER can also model net metering, a billing arrangement whereby the utility charges the customer based on the net grid purchases (purchases minus sales) over the billing period. Under net metering, if purchases exceed sales over the billing period, the consumer pays the utility an amount equal to the net grid purchases times the grid power cost. If sales exceed purchases over the billing period, the utility pays the consumer an amount equal to the net grid sales (sales minus purchases) times the sellback rate, which is typically less than the grid power price and often zero. The billing period may be one month or one year. In the unusual situation where net metering applies to multiple rates, HOMER tracks the net grid purchases separately for each rate.

Two variables describe the grid's capacity to deliver and accept power. The maximum power sale is the maximum rate at which the power system can sell power to the grid. The user should set this value to zero if the utility does not allow sellback. The maximum grid demand is the maximum amount of power that can be drawn from the grid. It is a decision variable because of the effect of demand charges. HOMER does not explicitly consider the demand rate in its hour-by-hour decisions as to how to control the power system; it simply calculates the demand charge at the end of each simulation. As a result, when modeling a grid-connected generator, HOMER will not turn on the generator simply to save demand charges. However, it will turn on a generator whenever the load exceeds the maximum grid demand. The maximum grid demand therefore acts as a control parameter that affects the operation and economics of the system. Because it is a decision variable, the user can enter multiple values, and HOMER can find the optimal one.

The user also enters the grid emissions coefficients, which HOMER uses to calculate the emissions of six pollutants associated with buying power from the grid, as well as the grid emissions avoided, resulting from the sale of power to the grid. Each emissions coefficient has units of grams of pollutant emitted per kilowatt-hour consumed.

Because it is a dispatchable power source, HOMER calculates the grid's fixed and marginal cost of energy. The fixed cost is zero, and the marginal cost is equal to the current grid power price plus any cost resulting from emission penalties. Since the grid power price can change from hour to hour as the applicable rate changes, the grid's marginal cost of energy can also change from hour to hour. This can have important effects on HOMER's simulation of the system's behavior. For example, HOMER may choose to run a generator only during times of high grid power price when the cost of grid power exceeds the cost of generator power.

18.5.3.7 Boiler

HOMER models the boiler as an idealized component able to provide an unlimited amount of thermal energy on demand. When dispatching generators to serve the electric load, HOMER considers the value of any waste heat that can be recovered from a generator to serve the thermal load, but it will not dispatch a generator simply to serve the thermal load. It assumes that the system can always rely on the boiler to serve any thermal load that the generators do not. To avoid situations that violate this assumption, HOMER ensures that a boiler exists in any system serving a thermal load, it does not allow any consumption limit on the boiler fuel, and it does not allow the boiler to consume biomass or stored hydrogen (since either of those fuels could be unavailable at times).

The idealized nature of HOMER's boiler model means that the user must specify only a few physical properties of the boiler. The user selects the type of fuel the boiler consumes and enters the efficiency with which it converts that fuel into heat. The only other properties of the boiler are its emissions coefficients, which are in units of grams of pollutant emitted per quantity of fuel consumed.

As it does for all dispatchable energy sources, HOMER calculates the fixed and marginal cost of energy from the boiler. The fixed cost is zero. HOMER calculates the marginal cost using the equation

$$c_{boiler,mar} = \frac{3.6 \cdot c_{fuel,eff}}{\eta_{boiler} \cdot LHV_{fuel}} \tag{18.10}$$

where:

$c_{fuel,eff}$ = The effective price of the fuel (including the cost of any penalties on emissions) in dollars per kg
η_{boiler} = The boiler efficiency
LHV_{fuel} = The lower heating value of the fuel in MJ/kg

18.5.3.8 Converter

A converter is a device that converts electric power from dc to ac in a process called inversion and/or from ac to dc in a process called rectification. HOMER can model the two common types of converters: solid state and rotary. The converter size, which is a decision variable, refers to the inverter capacity, meaning the maximum amount of ac power that the device can produce by inverting dc power. The user specifies the rectifier capacity, which is the maximum amount of dc power that the device can produce by rectifying ac power, as a percentage of the inverter capacity. The rectifier capacity is therefore not a separate decision variable. HOMER assumes that the inverter and rectifier capacities are not surge capacities that the device can withstand for only short periods of time, but rather continuous capacities that the device can withstand for as long as necessary.

The HOMER user indicates whether the inverter can operate in parallel with another ac power source such as a generator or the grid. Doing so requires the inverter to synchronize to the ac frequency, an ability that some inverters do not have. The final physical properties of the converter are its inversion and rectification efficiencies that HOMER assumes to be constant. The economic properties of the converter are its capital and replacement cost in dollars, its annual O&M cost in dollars per year, and its expected lifetime in years.

18.5.3.9 Electrolyzer

An electrolyzer consumes electricity to generate hydrogen via the electrolysis of water. In HOMER, the user specifies the size of the electrolyzer, which is a decision variable, in terms of its maximum electrical input. The user also indicates whether the electrolyzer consumes ac or dc power and the efficiency with which it converts that power to hydrogen. HOMER defines the electrolyzer efficiency as the energy content (based on higher heating value) of the hydrogen produced divided by the amount of electricity consumed. The final physical property of the electrolyzer is its minimum load ratio, which is the minimum power input at which it can operate, expressed as a percentage of its maximum power input. The economic properties of the electrolyzer are its capital and replacement cost in dollars, its annual O&M cost in dollars per year, and its expected lifetime in years.

18.5.3.10 Hydrogen Tank

In HOMER, the hydrogen tank stores hydrogen produced by the electrolyzer for later use in a hydrogen-fueled generator. The user specifies the size of the hydrogen tank, which is a decision variable, in terms of the mass of hydrogen it can contain. HOMER assumes that the process of adding hydrogen to the tank requires no electricity and that the tank experiences no leakage.

The user can specify the initial amount of hydrogen in the tank either as a percentage of the tank size or as an absolute amount in kilograms. It is also possible to require that the year-end tank level must equal or exceed the initial tank level. If the user chooses to apply this constraint, HOMER will consider infeasible any system configuration whose hydrogen tank contains less hydrogen at the end of the simulation than it did at the beginning of the simulation. This ensures that the system is self-sufficient in terms of hydrogen. The economic properties of the hydrogen tank are its capital and replacement cost in dollars, its annual O&M cost in dollars per year, and its expected lifetime in years.

18.5.4 System Dispatch

In addition to modeling the behavior of each individual component, HOMER must simulate how those components work together as a system. That requires hour-by-hour decisions as to which generators should operate and at what power level, whether to charge or discharge the batteries, and whether to buy from or sell to the grid. In this section we describe briefly the logic HOMER uses to make such decisions. A discussion of operating reserve comes first because the concept of operating reserve significantly affects HOMER's dispatch decisions.

18.5.4.1 Operating Reserve

Operating reserve provides a safety margin that helps ensure reliable electricity supply despite variability in the electric load and the renewable power supply. Virtually every real micropower system must always provide some amount of operating reserve, because, otherwise, the electric load would sometimes fluctuate above the operating capacity of the system, and an outage would result.

At any given moment, the amount of operating reserve that a power system provides is equal to the operating capacity minus the electrical load. Consider, for example, a simple diesel system in which an 80-kW diesel generator supplies an electric load. In that system, if the load is 55 kW, the diesel will produce 55 kW of electricity and provide 25 kW of operating reserve. In other words, the system could supply the load even if the load suddenly increased by 25 kW. In HOMER, the modeler specifies the required amount of operating reserve, and HOMER simulates the system to provide at least that much operating reserve.

Each hour, HOMER calculates the required amount of operating reserve as a fraction of the primary load that hour, plus a fraction of the annual peak primary load, a fraction of the PV power output that hour, and a fraction of the wind power output that hour. The modeler specifies these fractions by considering how much the load or the renewable power output is likely to fluctuate in a short period and how conservatively he or she plans to operate the system. The more variable the load and renewable power output and the more conservatively the system must operate, the higher the fractions the modeler should specify. HOMER does not attempt to ascertain the amount of operating reserve required to achieve different levels of reliability; it simply uses the modeler's specifications to calculate the amount of operating reserve the system is obligated to provide each hour.

Once it calculates the required amount of operating reserve, HOMER attempts to operate the system to provide at least that much operating reserve. Doing so may require operating the system differently (at a higher cost) than would be necessary without consideration of operating reserve. Consider, for example, a wind–diesel system for which the user defines the required operating reserve as 10% of the hourly load plus 50% of the wind power output. HOMER will attempt to operate that system so that at any time, it can supply the load with the operating generators even if the load suddenly increased by 10% and the wind power output suddenly decreased by 50%. In an hour where the load is 140 kW and the wind power output is 80 kW, the required operating reserve would be 14 kW + 40 kW = 54 kW. The diesel generators must therefore provide 60 kW of electricity plus 54 kW of operating reserve, meaning that the capacity of the operating generators must be at least 114 kW. Without consideration of operating reserve, HOMER would assume that a 60-kW diesel would be sufficient.

HOMER assumes that both dispatchable and no dispatchable power sources provide operating capacity. A dispatchable power source provides operating capacity in an amount equal to the maximum amount of power it could produce at a moment's notice. For a generator, that is equal to its rated capacity if it is operating or zero if it is not operating. For the grid, that is equal to the maximum grid demand. For the battery, that is equal to its current maximum discharge power, which depends on state of charge and recent charge–discharge history, as described in Section 18.5.3. In contrast to the dispatchable power sources, the operating capacity of a no dispatchable power source (a PV array, wind turbine, or hydro turbine) is equal to the amount of power the source is currently producing, as opposed to the maximum amount of power it could produce.

For most grid-connected systems, the concept of operating reserve has virtually no effect on the operation of the system because the grid capacity is typically more than enough to cover the required operating reserve. Unlike a generator, which must be turned on, incurring fixed costs to provide operating capacity, the grid is always "operating" so that its capacity (which is usually very large compared to the load) is always available to the system. Similarly, operating reserve typically has little or no effect on autonomous systems with large battery banks, since the battery capacity is also always available to the system, at no fixed cost. Nevertheless, HOMER still calculates and tracks operating reserve for such systems.

If a system is unable to supply the required amount of load plus operating reserve, HOMER records the shortfall as capacity shortage. HOMER calculates the total amount of such shortages over the year and divides the total annual capacity shortage by the total annual electric load to find the capacity shortage fraction. The modeler specifies the maximum allowable capacity shortage fraction. HOMER discards as infeasible any system whose capacity shortage fraction exceeds this constraint.

18.5.4.2 Control of Dispatchable System Components

Each hour of the year, HOMER determines whether the (no dispatchable) renewable power sources by themselves are capable of supplying the electric load, the required operating reserve, and the thermal load. If not, it determines how best to dispatch the dispatchable system components (the generators, battery bank, grid, and boiler) to serve the loads and operating reserve. This determination of how to dispatch the system components each hour is the most complex part of HOMER's simulation logic. The no dispatchable renewable power sources, although they necessitate complex system modeling, are themselves simple to model because they require no control logic—they simply produce power in direct response to the renewable resource available. The dispatchable sources are more difficult to model because they must be controlled to match supply and demand properly and to compensate for the intermittency of the renewable power sources.

The fundamental principle that HOMER follows when dispatching the system is the minimization of cost. HOMER represents the economics of each dispatchable energy source by two values: a fixed cost in dollars per hour and a marginal cost of energy in dollars per kilowatt-hour. These values represent all costs associated with producing energy with that power source that hour. The previous sections on the generator, battery bank, grid, and boiler detail how HOMER calculates the fixed and marginal costs for each of these components. Using these cost values, HOMER searches for the combination of dispatchable sources that can serve the electrical load, thermal load, and the required operating reserve at the lowest cost. Satisfying the loads and operating reserve is paramount, meaning that HOMER will accept any cost to avoid capacity shortage. But among the combinations of dispatchable sources that can serve the loads equally well, HOMER chooses the one that does so at the lowest cost.

For example, consider the hydro–diesel–battery system shown in Figure 18.17. This system comprises two dispatchable power sources, the battery bank and the diesel generator. Whenever the net load is negative (meaning the power output of the hydro turbine is sufficient to serve the load), the excess power charges the battery bank. But whenever the net load is positive, the system must either operate the diesel or discharge the battery, or both, to serve the load. In choosing among these three alternatives, HOMER considers the ability of each source to supply the ac net load and the required operating reserve and the cost of doing so. If the diesel generator is scheduled off or has run out of fuel, it has

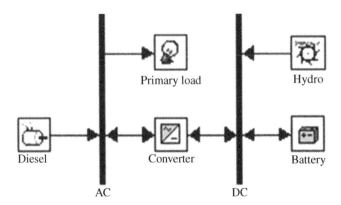

Figure 18.17 Hydro–diesel–battery system.

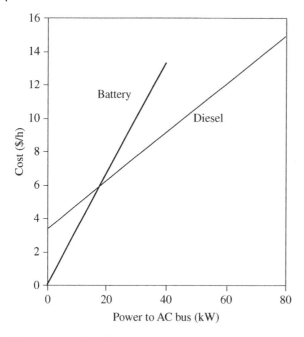

Figure 18.18 Cost of energy comparison.

no ability to supply power to the ac load. Otherwise, it can supply any amount of ac power up to its rated capacity. The battery's ability to supply power and operating reserve to the ac load is constrained by its current discharge capacity (which depends on its state of charge and recent charge–discharge history, as described in Section 18.5.3) and the capacity and efficiency of the ac–dc converter. If both the battery bank and the diesel generator are capable of supplying the net load and the operating reserve, HOMER decides which to use based on their fixed and marginal costs of energy.

Figure 18.18 shows one possible cost scenario, where the diesel capacity is 80 kW and the battery can supply up to 40 kW of power to the ac bus, after conversion losses. This scenario is typical in that the battery's marginal cost of energy exceeds that of the diesel. However, because of the diesel's fixed cost, the battery can supply small amounts of ac power more cheaply than the diesel. In this case the crossover point is around 20 kW. Therefore, if the net load is less than 20 kW, HOMER will serve the load by discharging the battery. If the net load is greater than 20 kW, HOMER will serve the load with the generator instead of the battery, even if the battery is capable of supplying the load.

HOMER uses the same cost-based dispatch logic regardless of the system configuration. When simulating a system comprising multiple generators, HOMER will choose the combination of generators that can most cheaply supply the load and the required operating reserve. When simulating a grid-connected microturbine supplying both heat and electricity, HOMER will operate the microturbine whenever doing so would save money compared with the alternative, which is to buy electricity from the grid and produce heat with a boiler.

HOMER's simulation is idealized in the sense that it assumes the system controller will operate the system to minimize total life-cycle cost, when in fact a real system controller may not. Nevertheless, HOMER's "economically optimal" scenario serves as a useful baseline with which to compare different system configurations.

18.5.4.3 Dispatch Strategy

The economic dispatch logic described in the preceding section governs the production of energy to serve loads and hence applies to all systems that HOMER models. However, for systems comprising both a battery bank and a generator, an additional aspect of system operation arises, which is whether (and how) the generator should charge the battery bank. One cannot base this battery-charging logic on simple economic principles, because there is no deterministic way to calculate the value of charging the battery bank. The value of charging the battery in one hour depends on what happens in future hours. In a wind–diesel–battery system, for example, charging the battery bank with diesel power in one hour would be of some value if doing so allowed the system to avoid operating the diesel in some subsequent hour. However, it would be of no value whatsoever if the system experienced more than enough excess wind power in subsequent hours to charge the battery bank fully. In that case, any diesel power put into the battery bank would be wasted because the wind power would have fully charged the battery bank anyway.

Rather than using complicated probabilistic logic to determine the optimal battery-charging strategy, HOMER provides two simple strategies and lets the user model them both to see which is better in any particular situation. These dispatch strategies are called load following and cycle charging. Under the load-following strategy, a generator produces only enough power to serve the load and does not charge the battery bank. Under the cycle-charging strategy, whenever a generator operates, it runs at its maximum rated capacity (or as close as possible without incurring excess electricity) and charges the battery bank with the excess. Barley and Winn [9] found that over a wide range of conditions, the better of these two simple strategies is virtually as cost-effective as the ideal predictive strategy. Because HOMER treats the dispatch strategy as a decision variable, the modeler can easily simulate both strategies to determine which is optimal in a given situation.

The dispatch strategy does not affect the decisions described in the preceding section as to which dispatchable power sources operate each hour. Only after these decisions are made does the dispatch strategy come into play. If the load-following strategy applies, whichever generators HOMER selects to operate in a given hour will produce only enough power to serve the load. If the cycle-charging strategy applies, those same generators will run at their rated output or as close as possible without causing excess energy.

An optional control parameter called the set-point state of charge can apply to the cycle-charging strategy. If the modeler chooses to apply this parameter, once the generator starts charging the battery bank, it must continue to do so until the battery bank reaches the set-point state of charge. Otherwise, HOMER may choose to discharge the battery as soon as it can supply the load. The set-point state of charge helps avoid situations where the battery experiences shallow charge–discharge cycles near its minimum state of charge. In real systems, such situations are harmful to battery life.

18.5.4.4 Load Priority

HOMER makes a separate set of decisions regarding how to allocate the electricity produced by the system. The presence of both an ac and a dc bus complicates these decisions somewhat. HOMER assumes that electricity produced on one bus will go first to serve primary load on the same bus and on the opposite bus and then deferrable load on the same bus and on the opposite bus, then to charge the battery bank, to serve the grid sales and electrolyzer, and to go to the dump load, which optionally serves the thermal load.

18.6 Economic Modeling

Economics play an integral role both in HOMER's simulation process, wherein it operates the system so as to minimize total NPC, and in its optimization process, wherein it searches for the system configuration with the lowest total NPC. This section describes why life-cycle cost is the appropriate metric with which to compare the economics of different system configurations, why HOMER uses the total NPC as the economic figure of merit, and how HOMER calculates total NPC.

Renewable and nonrenewable energy sources typically have dramatically different cost characteristics. Renewable sources tend to have high initial capital costs and low operating costs, whereas conventional nonrenewable sources tend to have low capital and high operating costs. In its optimization process, HOMER must often compare the economics of a wide range of system configurations comprising varying amounts of renewable and nonrenewable energy sources. To be equitable, such comparisons must account for both capital and operating costs. Life-cycle cost analysis does so by including all costs that occur within the life span of the system.

HOMER uses the total NPC to represent the life-cycle cost of a system. The total NPC condenses all the costs and revenues that occur within the project lifetime into one lump sum in today's dollars, with future cash flows discounted back to the present using the discount rate. The modeler specifies the discount rate and the project lifetime. The NPC includes the costs of initial construction, component replacements, maintenance, fuel, plus the cost of buying power from the grid, and miscellaneous costs such as penalties resulting from pollutant emissions. Revenues include income from selling power to the grid plus any salvage value that occurs at the end of the project lifetime. With the NPC, costs are positive and revenues are negative. This is the opposite of the net present value. As a result, the NPC is different from net present value only in sign.

HOMER assumes that all prices escalate at the same rate over the project lifetime. With that assumption, inflation can be factored out of the analysis simply by using the real (inflation-adjusted) interest rate rather than the nominal interest rate when discounting future cash flows to the present. The HOMER user therefore enters the real interest rate, which is roughly equal to the nominal interest rate minus the inflation rate. All costs in HOMER are real costs, meaning that they are defined in terms of constant dollars.

For each component of the system, the modeler specifies the initial capital cost, which occurs in year zero; the replacement cost, which occurs each time the component needs replacement at the end of its lifetime; and the O&M cost, which occurs each year of the project lifetime. The user specifies the lifetime of most components in years, but HOMER calculates the lifetime of the battery and generators as described in Section 18.5.3. A component's replacement cost may differ from its initial capital cost for several reasons. For example, a modeler might assume that a wind turbine nacelle will need replacement after 15 years, but the tower and foundation will last for the life of the project. In that case, the replacement cost would be *economic modeling* considerably less than the initial capital cost. Donor agencies or buy-down programs might cover some or all of the initial capital cost of a PV array but none of the replacement cost. In that case, the replacement cost may be greater than the initial capital cost. When analyzing a retrofit of an existing diesel system, the initial capital cost of the diesel engine would be zero, but the replacement cost would not.

To calculate the salvage value of each component at the end of the project lifetime, HOMER uses the equation

$$S = C_{rep} \cdot \frac{R_{rem}}{R_{comp}} \tag{18.11}$$

where:

S = The salvage value
C_{rep} = The replacement cost of the component
R_{rem} = The remaining life of the component
R_{comp} = The lifetime of the component

For example, if the project lifetime is 20 years and the PV array lifetime is also 20 years, the salvage value of the PV array at the end of the project lifetime will be zero because it has no remaining life. On the other hand, if the PV array lifetime is 30 years, at the end of the 20-year project lifetime, its salvage value will be one-third of its replacement cost.

For each component, HOMER combines the capital, replacement, maintenance, and fuel costs, along with the salvage value and any other costs or revenues, to find the component's annualized cost. This is the hypothetical annual cost that if it occurred each year of the project lifetime would yield an NPC equivalent to that of all the individual costs and revenues associated with that component over the project lifetime. HOMER sums the annualized costs of each component, along with any miscellaneous costs, such as penalties for pollutant emissions, to find the total annualized cost of the system. This value is an important one because HOMER uses it to calculate the two principal economic figures of merit for the system: the total NPC and the leveled cost of energy.

HOMER uses the following equation to calculate the total NPC:

$$C_{NPC} = \frac{C_{ann,tot}}{CRF(i, R_{proj})} \tag{18.12}$$

where:

$C_{ann,tot}$ = The total annualized cost
i = The annual real interest rate (the discount rate)
R_{proj} = The project lifetime
$CRF(i, R_{proj})$ = The capital recovery factor, given by the following equation:

$$CRF(i, N) = \frac{i(1+i)^N}{(1+i)^N - 1} \tag{18.13}$$

where:

i = The annual real interest rate
N = The number of years

HOMER uses the following equation to calculate the leveled cost of energy:

$$COE = \frac{C_{ann,tot}}{E_{prim} + E_{def} + E_{grid,sales}} \tag{18.14}$$

where:

$C_{ann,tot}$ = The total annualized cost
E_{prim} and E_{def} = The total amounts of primary and deferrable load, respectively, that the system serves per year
$E_{grid,sales}$ = The amount of energy sold to the grid per year

The denominator in equation (18.14) is an expression of the total amount of useful energy that the system produces per year. The leveled cost of energy is therefore the average cost per kilowatt-hour of useful electrical energy produced by the system.

Although the leveled cost of energy is often a convenient metric with which to compare the costs of different systems, HOMER uses the total NPC instead as its primary economic figure of merit. In its optimization process, for example, HOMER ranks the system configurations according to NPC rather than leveled cost of energy. This is because the definition of the leveled cost of energy is disputable in a way that the definition of the total NPC is not. In developing the formula that HOMER uses for the leveled cost of energy, we decided to divide by the amount of electrical load that the system actually serves rather than the total electrical demand, which may be different if the user allows some unmet load. We also decided to neglect thermal energy but to include grid sales as useful energy production. Each of these decisions is somewhat arbitrary, making the definition of the leveled cost of energy also somewhat arbitrary. Because the total NPC suffers from no such definitional ambiguity, it is preferable as the primary economic figure of merit.

References

1 J.F. Manwell and J.G. McGowan, A combined probabilistic/time series model for wind diesel systems simulation, Solar Energy, Vol. 53, pp. 481–490, 1994.
2 Maui Solar Energy Software Corporation, PV-DesignPro, http://www.mauisolarsoftware.com, accessed February 2, 2005.
3 PV*SOL, http://www.valentin.de, accessed February 2, 2005.
4 RETScreen International, http://www.retscreen.net, accessed February 2, 2005.
5 V.A. Graham and K.G.T. Hollands, A method to generate synthetic hourly solar radiation globally, Solar Energy, Vol. 44, No. 6, pp. 333–341, 1990.
6 F.M. White, Fluid Mechanics, 2nd ed., McGraw-Hill, New York, 1986.
7 J.A. Duffie and W.A. Beckman, Solar Engineering of Thermal Processes, 2nd ed., John Wiley & Sons, Inc., New York, 1991.
8 J.F. Manwell and J.G. McGowan, Lead acid battery storage model for hybrid energy systems, Solar Energy, Vol. 50, pp. 399–405, 1993.
9 C.D. Barley and C.B. Winn, Optimal dispatch strategy in remote hybrid power systems, Solar Energy, Vol. 58, pp. 165–179, 1996.

Appendix A

Diesel Power Plants

A.1 Introduction

By the end of the nineteenth century, there were mountains of useless coal dust piled up in the Ruhr valley in Germany. Therefore, Rudolf Diesel started to work on an engine that would burn coal dust. Eventually the attempts to design such engine failed, but in 1892 Rudolf Diesel was issued a patent for a proposed system that air would be compressed so much that the temperature would far exceed the ignition temperature of an oil fuel. Ever since, internal combustion engines have been providing shaft power from the beginning of the twentieth century.

There are basically two main types categorized by the type of fuel used: gasoline or diesel. The vast majority of those engines power automobiles, but they have been used for ships, boats, agricultural and agro processing machinery, and many other industrial applications. At the last quarter of the twentieth century, abundant fossil fuel production and distribution enabled the commercial application of diesel-powered electricity generation for several applications. In addition, hybrid schemes were deployed to integrate and complement intermittent distributed generation systems [1, 2].

Diesel-based small power is constituted basically by a diesel engine coupled to an electric power generator and to a field-exciting generator. The arrangement is very compact: it goes online in a very short lead time, requires only routine maintenance, and is easily available from practicing professionals in mechanical workshops and garages. Diesel generation has some disadvantages: it is noisy, pollutant, and economically driven by the fuel costs (suffering from worldwide politics), it needs fuel storage close to the power plant, and it requires logistics and infrastructure for transportation.

Figure A.1 depicts a typical arrangement of a small diesel engine-powered plant. Since a conventional synchronous generator is used, special attention is required with frequency and synchronization (see Chapter 10).

Integration of Renewable Sources of Energy, Second Edition. Felix A. Farret and M. Godoy Simões.

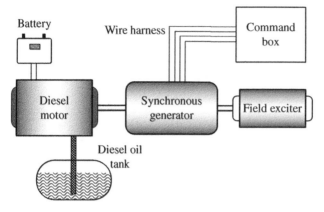

Figure A.1 Block diagram of a diesel oil power plant.

A.2 The Diesel Engine

A diesel engine converts the chemical energy of fuel into thermal energy that charges a cylinder in consequence of self-ignition and combustion of the fuel after compression of air. A slider–crank mechanism converts the thermal energy into mechanical work with a crankshaft. The combustion process has some significant features: in the gasoline engine the fuel and air mixture is drawn into the cylinder, compressed from 4:1 to 10:1, and ignited by a spark, while in the diesel engine air alone is drawn into the cylinder, compressed to a high ratio (14:1 to 25:1), causing a raise of the air temperature on the range of 700–900 °C, only then diesel fuel is injected and by a nozzle and ignites spontaneously. The diesel engine has some advantages over spark ignition motors (gasoline or alcohol), such as better fuel economy and longer engine life, that is, over the life of the engine, there is a trend to save money with diesel but initial high cost of the engine must be taken into account. Therefore, only long-time operation enables the fuel economy to overcome the increased price of the engine.

Because of the weight and compression ratio, diesel engines have lower maximum RPM ranges than gasoline engines, making them suitable for high torque rather than high acceleration. This is a good feature for generators that have to rather work in constant high speed. Diesel engines are also considered to have high efficiency when compared with spark ignition engines because for higher compression rate the calorific energy conversion into mechanical work has lower losses.

A.3 Main Components of a Diesel Engine

Basically, a diesel engine has the following parts: fixed parts, moving parts, and auxiliary systems.

A.3.1 Fixed Parts

Bedplate and Base: It is a foundation mounted on vibration absorbers providing support for the main bearings and engine crankcase. It is called a bedplate if an oil pan is bolted on it and a base if the oil pan is an integral part of the assembly.

Main Bearing Caps: They cover the crankshaft from the top as an upper part of the engine. They are the cylinder lid and must withstand the peak pressure of the piston in the combustion chamber below the piston.

Cylinder: Its cylindrical format requires die casting. It is manufactured in melted iron, and the base is of manganese or nickel–chrome–molybdenum to impress characteristics such as hardness and corrosion resistance.

Crankcase: It serves as a housing for the crankshaft and is located between the bedplate and the cylinder block. It usually incorporates the main bearing saddles and the reservoir for the lubricating oil. In some engines the crankcase is one piece of cast iron, while in others it is a welded steel.

A.3.2 Moving Parts

Piston: It receives direct impact of the combustion for transmission to the connecting rod. To protect the piston against seizure, the diameter must be reduced with an optimal gap. There are troughs in the head of the piston with camped segment rings. The top rings act as pressure seal, and the middle rings as wiper removal of oil film, and the bottom rings ensure even deposition of oil on the cylinder walls.

Connecting Rod: It is alternatively loaded in compression and tension owing to the cylinder firing pressure, consequently transmitting power from the pistons to the crankshaft.

Crankshaft: It changes the movement of the pistons and the connecting rod; there are eccentric offset rod bearings that convert the reciprocating motion into a rotating motion. It must be very strong and machined from forged alloy, carbon steel, or cast iron alloy.

Flywheel: It is connected on one end of the crankshaft and through its inertia reduces vibration and allows bolting the engine to an external load, and sometimes it has teeth that engages starting motors for starting up; increasing the number of cylinders increases the frequency of the firing strokes, and so it allows smaller flywheels.

Valves: They control the flow of the air–fuel mixture or the air inside and outside of the cylinder.

A.3.3 Auxiliary Systems

Air Intake System: Such system is responsible for providing a cool filtered air at the right fuel mixture to be fed to the cylinders. It is composed of the fuel tank, canalization, injector pump, filters, and injecting nozzles.

Cooling System: The majority of diesel engines have water cooling to transfer waste heat out of the block. It is very rare to have air-refrigerated diesel engines.

Lubrication System: Oil serves two purposes—to lubricate the bearing surfaces and to absorb friction-generated heat. A pressure relief valve maintains oil pressure in the galleries, returning through a filter to the oil pan.

A.4 Terminology of Diesel Engines

Lower Point (LDP): It is the lowest point the piston reaches in its descending course.

Upper Point (UDP): It is the highest point the piston reaches in its ascending course.

Cylinder Capacity: It is the volume capacity of the cylinder corresponding to the maximum acceptable volume of air in the cylinder. It is calculated by

$$V = \pi S r^2$$

where:

$r = $ The internal radius of the cylinder
$S = $ The piston course

Compression Rate (CR): It is the relationship between the total volume of the cylinder (v_a) and the volume of the compression chamber (v_e) given by

$$\rho_c = \frac{v_a + v_e}{v_e}$$

In order to increase the output power, it is possible to increase the compression rate, reducing v_e or increasing v_a.

A.4.1 The Diesel Cycle

There are two possible cycles: a four-stroke cycle, where the camshaft is geared to rotate at half of the speed of the crankshaft with one event per stroke, and a two-stroke cycle, in which more events have to be timed per stroke. Four-stroke diesel engines are used in diesel–electric facilities for small loads such as for powering installations with personal computer, mini power plants, and, more recently, small power plants. On the other hand, two-stroke diesel engines are used in large diesel–electric facilities. Since this book is more directed to small power plants, the four-stroke diesel engine will be discussed next.

The calorific energy transformed in mechanical work is generated by the combustion of a fuel (diesel) inside the cylinder. Four processes, illustrated in

(a)

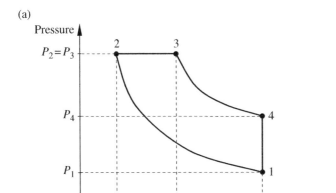

(b)

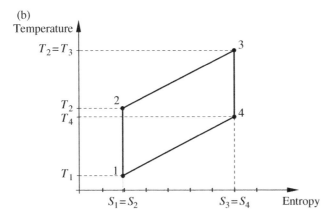

1, Isentropic pressure (process 1–2)
2, Admission (isobaric heat transfer) (process 2–3)
3, Isentropic expansion (process 3–4)
4, Discharge (process 4–1)

Figure A.2 Diesel cycle: (a) P–V diagram and (b) T–S diagram.

Figure A.2a and b, describe the operation. The thermal efficiency of the diesel engine is given by

$$\eta_{TD} = 1 - \frac{1}{r_c^{k-1}} \frac{r_c^k - 1}{k(r_c - 1)}$$

where:

r_c = Fuel cut rate
k = Boltzmann's constant

A.4.2   Combustion Process

Figure A.3 illustrates the operation of a four-stroke diesel engine. The first cycle begins with admission of air to completely fill the cylinder (which is at 180° turn in the crank). In the diesel engine, the fuel is previously injected in the chamber in which air is compressed up to 1/16 of its original volume and consequently heated to about 500–600 °C. During injection, the air temperature is high enough to provoke self-ignition in the dispersed droplets of fuel in the chamber. Some vaporization of at least part of the fuel is indeed necessary in order to establish zones of suitable air–fuel composition.

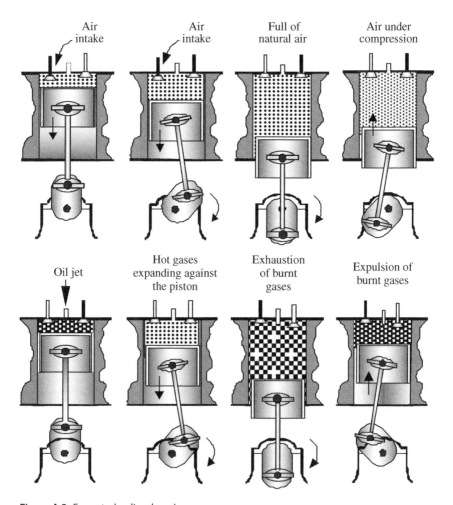

Figure A.3 Four-stroke diesel engine.

A.4.2.1 Four-Stroke Diesel Engine

Four-stroke models are built from small power sources up to 4000 HP. Since they are very efficient and have satisfactory durability, they are recommended for heavy-duty applications with a high daily rate use [3, 4].

High-speed models are predominant because they allow appreciable reduction in weight and in costs for a given power level, being very appropriate for electric power generation. For low-speed loads, the advantage of direct coupling (without reducers) justifies the construction of motors using lower rotating speed, especially when the application power is high. There are several adaptations of the fuel. Spark ignition engines can be easily adapted to run on ethanol or methanol, while vegetable oils can substitute diesel fuels and are easily integrated with biomass generating systems.

The necessary time to atomize, vaporize, and mix the fuel is called physical delay. Soon after the physical delay, a reaction chain begins for the combustion, constituting the so-called chemical delay. The total delay should be as small as possible, because fuel accumulates in the chamber as it is being injected.

At the onset of the combustion, the fuel burns and causes elevation of the pressure. In general, the injection continues beyond that point, and the remaining fuel burns as it is being injected. The combustion only finishes after having already traveled part of the expansion course. As a result, the combustion timing in diesel engine is defined in four phases:

1) Delay (physical and chemical)
2) Fast increase of pressure at the initial combustion
3) Controlled combustion to limit the maximum pressure
4) Burning during the expansion at a constant pressure

A.5 Cycle of the Diesel Engine

Diesel engines may operate at very fast combustion, approaching constant volume for most of the fuel. Such operation is obtained when the delay period is long enough so that most of the injected fuel is quite mixed and evaporated before the combustion. However, such operation is undesirable due to the high resulting maximum pressures and the high rates of pressure elevation.

In practical operation, the fuel injection system and the operating conditions are carefully selected to limit the rates of pressure elevation and maximum pressures below the attainable maximum. The equivalent cycle of fuel/air at constant volume represents the maximum power and efficiency supplied in the diesel engine and can be adequately used as a basis for comparison. However, with intentional limitation of maximum pressure, the equivalent cycle of fuel/air is chosen as the basis for the evaluation of diesel cycles.

The equivalent cycle of fuel/air at limited pressure has the following common characteristics with the current cycle:

Compression rate
Fuel/air rate
Maximum pressure
Point of mixture density
Composition of the mixture in all points

A.5.1 Relative Diesel Engine Cycle Losses

The losses relative to the cycle of a diesel engine are very variable because they are dependent on the operating conditions of the fuel, the design, the injection system, and the combustion chamber.

While in spark ignition motors the burn speed of the mixture at the beginning is relatively low and, by accelerating, reaches maximum, in the diesel engine the opposite occurs. Because the available supply of oxygen decreases with burning, the mixture process and burning process tend to reduce in the final instants.

Most of the combustion processes have a relatively constant temperature. The changes in the amount of fuel per cycle (fuel/air rate) are partially responsible for the constant temperature of the process. In addition, they seem to have minor effect on the maximum temperature and are not related to the crank angular position.

A.5.2 Classification of the Diesel Engine

Diesel engines can be used as motive machines for electrical generators. They are of various types and characteristics, and a systematic classification becomes very difficult. In this appendix they are classified in accordance with their construction, cycle, speed, and overall operation, since those features are more relevant for generating systems.

The diesel engine can be of vertical or horizontal construction according to disposition of the cylinders, whereas the diesel engine can be fast, medium, and slow according to its rated speed.

Fast motors have a relatively low weight/power ratio, high specific power, and reduced dimensions, with upper maximum rotation above 1200 rpm. They are more economical because of high speed, but technical difficulties limit the diesel engine construction for medium and small power range. They are used as prime movers in the auxiliary services of power plants and are mostly suitable for driving electrical generators.

The diesel engine of medium speed operates at a rotating range between 600 and 1000 rpm. They are used for quite high-power applications and relatively low ration weight/power ratio. In order to have good durability characteristics, these motors should operate at full power and at constant rotating speed for long periods of time.

The slow diesel engine is frequently used for auxiliary reserve power plant applications where the weight/power rate is not of fundamental importance. These motors are bulky, operating between 400 and 450 rpm, and are built for maximum power up to 10,000 HP. They do not reach the cylinder capacity-to-weight rate, but their operation in a low rpm translates in better mechanical reliability, a very important requirement for reserve prime movers in utility power plants. Those engines are usually adapted to burn heavy liquid fuels of several characteristics.

According to the operation mode of pistons, engines can be classified as:

- Diesel engine of single effect—The combustion pressure only acts on one face of the piston.
- Diesel engine of double effect—The combustion pressure acts alternately on each face of the piston.

As for the operation cycle, the diesel engine can be:

- Four-stroke diesel engine—The operating cycle follows four stages: aspiration, compression, combustion, and escape.
- Two-stroke diesel engine—The operating cycle is reduced to aspiration–compression and combustion–escape.

Automatic control techniques are applied for the regulation of speed of diesel engine for grid-connected applications. The regulator act directly, or through a servomotor, continually on the command of the valve that varies the flow of fuel.

A.6 Types of Fuel Injection Pumps

Injector pumps are built for dosing the amount of fuel and adjusting it to the load, in agreement with the control set point from the centrifugal regulator. According to the way the regulator acts on the pump, they are classified in three types:

Pump with regulation by free retreat In those pumps, the path traveled by the piston is variable. The injection of the fuel uses only part of the path traveled by the piston, and the fuel injected after that is returned to the aspiration chamber.

Pump with regulation by course control In this pump type, the regulation of the amount of fuel injected takes place by varying the course of the piston through a sloped output. In such system, the piston is impelled during its entire course. The pump with regulation by course control has the advantage of simplicity of its piston and sleeve, but it presents the inconvenience of fast wear-out of the contact surface due to the high specific pressure on the contact points. Besides, the exit shaft moves easily in a sense but needs strong effort to move backward.

Pumps with regulation by strangled retreat The pump operates in accordance with the principle of constant course of the piston, the same as the case of regulation by free retreat. The regulation of the amount of impelled fuel communicates the cylinder of the pump with the aspiration through an appropriate conduit. A small valve commanded by the regulator controls that conduit. If the communication orifice is completely opened by the valve, all fuel is injected up to the pump cylinder capacity. If it is necessary to reduce the amount of injected fuel, the valve controls that fuel, allowing the excess to return to the aspiration conduit. Therefore, varying the valve aperture also changed the relationship between aspirated and injected fuel, where the maximum aperture corresponds to a null impulse because all the impelled fuel returns to the aspiration conduit.

A.7   Electrical Conditions of Generators Driven by Diesel Engines

All reciprocating engines have flywheels to compensate for the oscillations and vibrations. For electric power generation, it is important to smooth out the mechanical shaft movement; otherwise voltage and frequency oscillations may affect the electric performance. Such situation is worse in similar-sized power plants coupled in parallel, because the voltage and power fluctuations may impair and disable the parallel connection [3–5].

Parallel operation is only satisfactory if the oscillation period of the generator is far apart from the fundamental oscillation of the force pulses of the prime mover. The minimum moments of inertia free from any resonance, necessary to damp the movement of 50/60-Hz parallel generators driven by four-stroke diesel engines, can be taken from Tables A.1 and A.2. The values in these tables depend on the short-circuit characteristics of the generators, and they vary according to their constructing features. To obtain values of the necessary inertia moment, one should multiply it by the value of nominal power in kVA. In Tables A.1 and A.2 the values are only valid for parallel operation with machines of the same driving class and speed. Operation of different operating classes such as coupling of a generator driven by a diesel engine with another one driven by a steam turbine should be carefully designed because the different speeds and reverse influences might cause mechanical resonance.

The flywheel is always coupled to the generator rotor. For slow generators, the rotor itself can act as the flywheel if its weight and diameter are well designed. In small groups of medium speed (500–1500 rpm), the flywheel should also be coupled to the prime mover, or additional flywheels must be installed between coupling brakes of the prime mover and the generator. The effect of this flywheel on the motor is to reduce vibration by smoothing out the power stroke as each cylinders fires.

TableA.1 Minimum moment of inertia to avoid resonance (50/60/Hz) in four-stroke diesel engines.

Speed (rpm)	Moment of inertia (kgm²/kVA)	
	(50 Hz)	(60 Hz)
150	350.00	243.06
167	228.00	158.33
188	144.00	100.00
214	84.50	58.68
250	45.50	31.60
300	21.90	15.21
375	9.00	6.25
428	5.30	3.68
500	2.90	2.01
600	1.40	0.97
750	0.57	0.40
1000	0.18	0.13

Table A.2 Minimum moment of inertia to avoid resonance (50/60/Hz) in two-stroke diesel engines.

Speed (rpm)	Moment of inertia (kgm²/kVA)	
	(50 Hz)	(60 Hz)
150	87.50	60.76
167	57.00	39.58
188	36.00	28.00
214	21.20	14.72
250	11.40	7.92
300	5.50	3.82
375	2.25	1.56
428	1.32	0.92
500	0.72	0.50
600	0.35	0.24
750	0.14	0.01
1000	0.06	—

Another aspect to be taken into account is the possibility of overloading the diesel engines. Therefore, it is recommended not to exceed 5–10% of its full load. It is possible to count on this overload capacity in frequent intervals or in operation periods with certain duration. Therefore, instead of using very heavy flywheels to regularize the variations of the load, the diesel engine should use special devices preferentially for regulation of the load. The required mechanical power of a diesel engine to drive an electrical generator can be analyzed as follows.

As electrical generators easily support overload of up to 5%, they can be adopted for a 5% lower power, resulting in

$$N_g = 0.95 N_{di} \eta_g \cdot 0.736$$

where:

N_g = Power of the generator in kW
N_{di} = Power of the diesel engine in C.V.
η_g = Efficiency of the generator

When a diesel engine-powered electric generator works in parallel with other generators driven by prime movers of different types, the operation is such that the diesel groups must operate at constant load even if the turbogenerators compensate the overloads. As an example, the overload values used by some manufacturers of diesel engine are :

6% of overload for 30 minutes
12.5% of overload for 5 minutes
Up to 20% for temporary load oscillations

References

1 R. Stone, Introduction to Internal Combustion Engines, The MacMillan Press Ltd, London, UK, 1992.
2 A.G. Domschke and O. Garcia, Internal combustion motors, Department of Mechanical Engineering, Polytechnic School of the Universidade de São Paulo, São Paulo, Brazil, 1968.
3 J.R. Vázquez, Maquinas motrices y generadores de energia elétrica, Publisher CEAC S.A., Barcelona, Spain, 1996.
4 E.J. Kates, Diesel and High-Compression Gas Engines Fundamentals, American Technical Society, Chicago, IL, 1965.
5 Engineers Edge, http://www.engineersedge.com/power_transmission/power_transmission_menu.shtml, accessed March 1, 2017.

Appendix B

The Stirling Engine

B.1 Introduction

The Stirling engine is an external heat engine that is vastly different from the internal combustion engines of ordinary cars. Invented by Reverend Robert Stirling in 1816, the Stirling engine has the potential to be much more efficient than a gasoline or diesel engine. It was invented to create a safer alternative to the steam engines of the time, whose boilers often exploded due to the high pressure of the steam and the primitive materials used to build them. Presently, to generate electricity for homes and businesses, Stirling generators fueled by either solar energy or natural gas have been tested in research. They run on solar power during sunny weather and automatically convert to clean burning natural gas at night or when the weather is cloudy. However, today's Stirling engines are used only in some very specialized applications, like in submarines, intelligent buildings, or auxiliary power generators for yachts, where quiet operation is important. As an example of industrial interest, between 1958 and 1970, General Motors began projects involving Stirling engines in automotive applications [1, 2].

In broad terms, Stirling engines can be useful only in places where silence is required. When there is time for the slow warming-up procedure, there are some plentiful heating and cooling sources. When there is a demand for low speed motor, constant output power, and no power surges, there are even some abundant sources of fuel that are already available or easily feasible. Such restrictions interest energy researchers because of their potential capabilities for clean, quiet, and diverse sources and efficient exchange of energy, although there has not been a successful mass-market application for them [1, 2].

Stirling engines are known to run on low temperature differences. In this case, they tend to be rather large for the amount of power they put out. However, this may not be a significant drawback since these engines can be largely manufactured from lightweight and cheap materials such as plastics.

Integration of Renewable Sources of Energy, Second Edition. Felix A. Farret and M. Godoy Simões.
© 2018 John Wiley & Sons, Inc. Published 2018 by John Wiley & Sons, Inc.

These engines can be used for applications such as irrigation and remote water pumping. For wider temperature differences, they can also be used in revertible applications, if mechanical work is input into the engine by connecting an electric motor to the power output shaft. As a result, one end will get hot and the other end will get cold. In a correctly designed Stirling cooler, the cold end will get extremely cold. Stirling coolers (built for research use) can cool below 10 K [3].

B.2 The Stirling Cycle

A Stirling engine uses the Stirling cycle, which is unlike the cycles used in internal combustion engines. Three major points should be observed:

1) The gasses used inside a Stirling engine never leave the engine. There are no exhaust valves that vent high-pressure gasses, as in gasoline or diesel engines, and there are no explosions taking place. Thus, Stirling engines are very quiet.
2) A flywheel or propeller with gentle spin is needed to start the engine.
3) The Stirling cycle uses an external heat source, which can be anything from sunshine, geothermal heat, gasoline, and solar energy to the heat produced by decaying plants. No combustion takes place inside the cylinders of the engine.

There are several ways to construct a Stirling engine, but this appendix shows two different configurations of this engine operation.

The key principle of a Stirling engine is that a fixed amount of a gas is sealed inside the engine. Stirling engines are external combustion engines, since the heat is supplied to the air inside the engine from a source outside the cylinder, instead of being supplied by a fuel burning inside the cylinder. But experiments with hydrogen and helium as working fluids led to vast improvements and modifications of Stirling's invention due to their lower specific heats. Using gasses with different properties led to reduction in size of the engine as well as to new applications for the machine [4–6].

The Stirling cycle involves a series of events that change the pressure of the gas inside the engine, causing it to do work. It utilizes the four basic thermodynamics processes of rotating energy conversion: compression, expansion, heat addition, and heat rejection. The Stirling cycle is an idealized cycle very similar to the Carnot cycle with some key process specifications. The first process is an isothermal compression, which occurs in the cold space, where heat is transferred from the working fluid to the surroundings to maintain the temperature during compression (see Figure B.1). The second process is a constant volume heat addition. This is followed by the isothermal expansion of the working fluid, where heat is added to the system during the expansion to

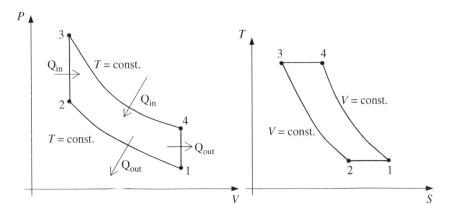

Figure B.1 P-V and T-S diagrams for the ideal Stirling cycle.

maintain temperature. The last process is heat rejected from the working fluid at constant volume. All four processes are illustrated in the P-V and T-S diagrams in Figure B.1.

There are several properties of gasses that are critical to the operation of Stirling engines:

- If there is a fixed amount of gas in a fixed volume of space and the temperature of that gas is raised, the pressure will increase according to the Gay-Lussac gas law:

$$\frac{p_1 V_1}{T_1} = \frac{p_2 V_2}{T_2} = nR \tag{B.1}$$

where:

p, V, and T = The pressure, volume, and temperature inside a cylinder, respectively
Subindexes 1 and 2 = Two different states of these variables
n = The number of molecules-gram
$R = 0.8207$ L-atm/mol·K = The universal constant of the gasses

- If a fixed amount of gas is compressed (decreasing the volume of its space), the temperature of that gas will increase, but equation (B.1) holds for every intermediary state.

To understand how the Stirling cycle works, it is better to examine a simplified model of this engine using two cylinders. One half cylinder is heated by an external heat source (such as fire), and the other half is cooled by an external cooling source (such as ice). The gas chambers of the two cylinders are

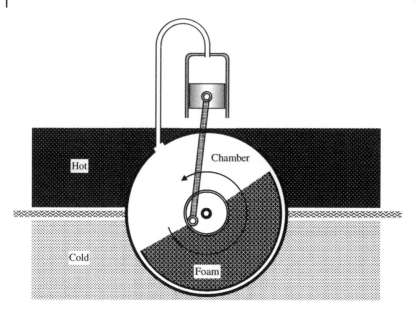

Figure B.2 Operating principles of the Stirling engine.

connected, and the piston and half piston are mechanically connected to each other with a linkage that determines how they will move in relation to one another [6–8].

There are four portions to the Stirling cycle with the two pistons, accomplishing each part of the cycle. As depicted in Figure B.2, heat is added to the gas in the left cylinder, causing pressure to build up. This forces the piston down, producing work in steps that follow [9–11]:

1) Heat is added to the gas inside the heated chamber (bottom), causing pressure to build up. The air is in the hot upper portion, expanding it. This forces the piston to move down. This is the part of the Stirling cycle that can be done via the crankshaft, with two connecting rods at 90° phase shift. This part of the cycle continues until most of the air is still in the hot upper portion, expanding and pushing the piston down.
2) The spinning foam half piston moves up, while the top piston—connected to the same crankshaft—moves down. Most of the air is in the cold lower portion, contracting and sucking the piston up. This pushes the hot gas into the cooled cylinder, which quickly cools the gas to the temperature of the cooling source, lowering its pressure. This makes it easier to compress the gas in the next part of the cycle.
3) The piston in the cooled cylinder is forced, via the crankshaft, to compress the gas. Heat generated by this compression is removed by the cooling source.

4) The half piston moves up, while the top piston moves down at some 90° phase lag. This forces the gas into the heated portion of the chamber, where it quickly heats up, and builds pressure. At this point the cycle repeats.

The Stirling engine only creates power during the first part of the cycle. The amount of work produced is represented by integration of the area between the higher and lower temperatures in the Stirling cycle diagram. Its movement is smoothed out by a flywheel connected to the crankshaft [10, 11].

There are two main ways to increase the power output of a Stirling cycle:

1) *Increase power output in stage 1*—In stage 1 of the cycle, the pressure of the heated gas pushing against the piston performs work. Increasing the pressure during this part of the cycle will increase the power output of the engine. One way to increase the pressure is by increasing the temperature of the gas, as predicted by equation (B.1). When we take a look at a two-piston Stirling engine later in this appendix, we will see how a device called regenerator can improve the power output of the engine by temporarily storing heat.
2) *Decrease pressure usage in stage 3*—In stage 3 of the cycle, the pistons perform work on the gas, using some of the power produced in stage 1. Lowering the pressure can decrease the power used during this stage of the cycle (effectively increasing the power output of the engine). One way to decrease the pressure is to cool the gas to a lower temperature.

This section describes the ideal Stirling cycle. Actual working engines vary the cycle slightly because of the physical limitations of their design, as dealt with in the following sections.

B.3 Displacer-Type Stirling Engine

Instead of having two pistons, a displacer-type engine has one piston and a displacer. The displacer serves to control when the gas chamber is heated and cooled [8–10]. This type of Stirling engine is sometimes used in classroom demonstrations and it can run using as little energy as human body heat. In order to run, the engine previously mentioned requires a temperature difference between the top and the bottom of the large cylinder. One part of the engine is kept hot, while another part is kept cold, separated by a displacer. This displacer is carefully mounted so it does not touch the walls of the cylinder. A mechanism then moves the air back and forth through and around the displacer, between the hot side and the cold side as in Figure B.3. When the air is moved to the hot side, it expands and pushes up on the piston, and when the air is moved back to the cold side, it contracts and pulls down on the piston. After the hot air expands and pushes the piston as far as the connecting rod

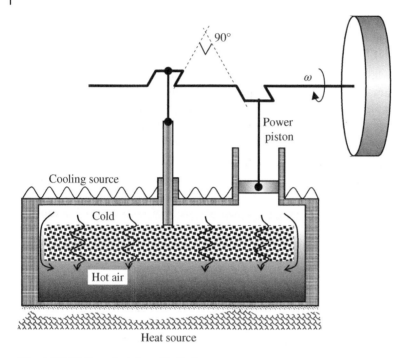

Figure B.3 Stirling principles with displacer.

allows, the air still has quite a bit of heat energy left in it. In this case, the difference between the temperature of one's hand and the air around it is enough to run the engine.

In Figure B.3, two pistons can be seen:

1) *The power piston*—This is the smaller piston at the top of the engine. It is a tightly sealed piston that moves up as the gas inside the engine expands.
2) *The displacer*—This is the large piston in the drawing. This piston is very loose in its cylinder so air can move easily between the heated and cooled sections of the engine as the piston moves up and down.

The displacer moves up and down to control whether the gas in the engine is being heated or cooled. There are two positions:

1) When the displacer is near the top of the large cylinder, most of the gas inside the engine is heated by the heat source and it expands. Pressure builds inside the engine, forcing the power piston up.
2) When the displacer is near the bottom of the large cylinder, most of the gas inside the engine cools and contracts. This causes the pressure to drop, making it easier for the power piston to move down and compress the gas.

The engine repeatedly heats and cools the gas, extracting energy from the gas's expansion and contraction with some waste heat. Some Stirling engines are made to store some of the waste heat by making the air flow through economizer tubes that absorb some of the heat from the air. This precooled air is then moved to the cold part of the engine where it cools very quickly, and as it cools, it contracts, pulling down on the piston. Next, the air is mechanically moved back through the preheating economizer tubes to the hot side of the engine where it is heated even further, expanding and pushing up on the piston. This type of heat storage is used in many industrial processes and is currently called "regeneration." Stirling engines do not require regenerators to work, but well-designed engines will run faster and put out more power if they have a regenerator [3, 12].

B.4 Two-Piston Stirling Engine

In this engine, the heated cylinder is heated by an external flame. The cooled cylinder is air-cooled and has fins on it to aid in the cooling process. A rod stemming from each piston is connected to a small disk that is, in turn, connected to a larger flywheel (see Figure B.4). This keeps the pistons from moving when no power is being generated by the engine. The flame continually heats the bottom cylinder [1, 2, 10]:

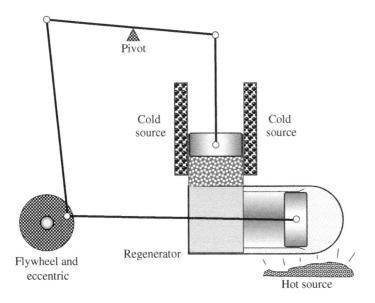

Figure B.4 The two-piston Stirling engine.

(a)

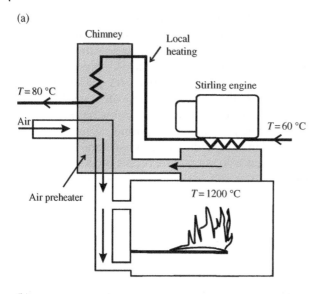

(b)

Figure B.5 CHP plant demonstrated with Stirling engine, (a) CHP plant demonstrated by H. Carlsten, (b) 35 kWe prototype for biomass combined plants. *Source*: BIOS Energy Systems & H. Carlsten. © Courtesy of Henrik Carlson, Denmark.

1) In the first part of the cycle, pressure builds up, forcing piston 1 to move to the left and do work. The cooled piston stays approximately stationary because it is at the point in its revolution where it changes direction.
2) In the next stage, both pistons move. The heated piston moves to the right and the cooled piston moves up. This moves most of the gas through the regenerator and into the cooled piston. The regenerator is a device that can temporarily store heat. It might be wire mesh that the heated gasses pass through. The large surface area of the wire mesh quickly absorbs most of the heat. This leaves less heat to be removed by the cooling fins.
3) Next, piston 2 in the cooled cylinder starts to compress the gas. Heat generated by this compression is removed by the cooling fins.
4) In the last phase of the cycle, both pistons move; the cooled piston moves down, while the heated piston moves to the left. This forces the gas across the regenerator (where it picks up the heat that was stored there during the previous cycle) and into the heated cylinder. At this point, the cycle begins again.

The possibilities for mass-market applications of Stirling engines are still limited. There are a couple of key characteristics that make Stirling engines impractical for use in many applications, including in most cars and trucks. Because the heat source is external, it takes a little while for the engine to respond to changes in the amount of heat being applied to the cylinder; it takes time for the heat to be conducted through the cylinder walls and into the gas inside the engine. This means:

• The engine requires some time to warm up before it can produce useful power.
• The engine cannot change its power output quickly.

These shortcomings prevent this engine from replacing automotive internal combustion engines. However, a Stirling-engine-powered hybrid car might be feasible as in the process diagram of a CHP plant demonstrated by H. Carlsten. A newly developed 35 kWe prototype for biomass combustion plants (BIOS Energy Systems & H. Carlsten) is shown in Figure B.5.

References

1 K. Nice, How Stirling Engines Work, American Stirling Co., http://auto.howstuffworks.com/stirling-engine1.htm, accessed March 29, 2017.
2 C.M. Hargreaves, The Phillips Stirling Engine, Elsevier Science Publishers B.V., Amsterdam, The Netherlands, 1991.
3 M.J. Collie (ed.), Stirling Engine: Design and Feasibility for Automotive Use, Noyes Data Corporation, Park Ridge, NJ, 1979.
4 G.T. Reader and C. Hooper, Stirling Engines, E. & F.N. Spon, New York, 1983.

5 Volunteers in Technical Assistance, http://www.appropedia.org/Stirling_
motor, accessed March 29, 2017.

6 J. Lewis, New simplified heat engine, http://www.emachineshop.com/engine/
animation.htm, accessed March 21, 2005.

7 Stirling Energy Systems: Leader in alternative or green energy,
http://www.stirlingenergy.com, accessed March 10, 2017.

8 Smart engine user manual, American Stirling Company, http://lib.store.
yahoo.net/lib/discoverthis/smart-stirling-engine-instructions.pdf, accessed
March 29, 2007.

9 C.D. West, Principles and Applications of Stirling Engines, Van Nostrand
Reinhold Company, New York, 1986.

10 J.R. Senft, An Introduction to Low Temperature Differential Stirling Engines,
Moriya Press, River Falls, WI, 1996.

11 G. Walker, Stirling Engines, Clarendon Press, Oxford, 1980.

12 A. Winkelmann and E.J. Barth, Design, modeling, and experimental validation
of a Stirling pressurizer with a controlled displacer piston, IEEE/ASME
Transactions on Mechatronics, Vol 21, No. 3, pp. 1754–1764, June 2016.

Index

Integration of Renewable Sources of Energy, Second Edition. Felix A. Farret and
M. Godoy Simões.
© 2018 John Wiley & Sons, Inc. Published 2018 by John Wiley & Sons, Inc.